QUALITATIVE THEORY OF DIFFERENTIAL EQUATIONS

QUALITATIVE THEORY OF DIFFERENTIAL EQUATIONS

BY

V. V. NEMYTSKII

AND

V. V. STEPANOV

PRINCETON, NEW JERSEY

PRINCETON UNIVERSITY PRESS

Published, 1960, by Princeton University Press

L. C. CARD 60-12240
ISBN 0-691-08020-8

Second Printing 1964
Third Printing 1967
Fourth Printing 1972

The translation and editing of this text was supported by the United States Air Force, Office of Scientific Research, of the Air Research and Development Command, under Contract AF 33 (038)-9993. Reproduction, translation, publication, use and disposal, in whole or in part, by or for The United States Government is permitted.

Printed in the United States of America

Preface to the English Language Edition

The English language edition of the Nemickii-Stepanov treatise has gone through many vicissitudes. Several years ago when only the first edition was available in this country a complete translation was made by Dr. Thomas Doyle, at the time a member of the faculty of Dartmouth College. This translation was edited by Donald Bushaw and John McCarthy, at the time graduate students at Princeton University. Hardly was this done when there appeared a much enlarged second edition of the book. Dr. Arnold Ross of the University of Notre Dame undertook to prepare an English translation of the first four chapters which he actually had to rewrite for the most part. Undoubtedly American mathematicians are greatly in debt to Dr. Ross for the enormous amount of work which he has done in this connection. The last two chapters, which did not differ too much in the two editions, were finally put in proper shape by Dr. Robert Bass, who utilized in the process translations of the few new sections by Dr. McCarthy and by Dr. Lawrence Markus. It seems fair to say that this edition contains all the material of the second Russian edition of the book.

A couple of years ago there appeared a brief summary written by Nemickii giving a resumé of the recent work done under his guidance by the very active Moscow school. The English language version of this resumé, prepared by Dr. McCarthy, is included at the end of Part One.

The book falls naturally into two parts: Part One on classical differential equations, and Part Two on topological dynamics and ergodic theory. The first part has its own bibliography and index, and the last two chapters, making up the second part, as well as the Appendix, have individual bibliographies.

Readers may be interested in the supplement to Chapter 5, written by Nemickii, which was published by the American Mathematical Society as Translation No. 103 (1954).

In conclusion we wish to say that the work was done under the auspices of the Air Research and Development Command under Contract AF 18(600)–332.

January 1, 1956

Princeton, N. J. S. LEFSCHETZ

TABLE OF CONTENTS

PART TWO

PART ONE

CHAPTER I

Existence and Continuity Theorems

1. Existence Theorems

In the qualitative theory of differential equations one considers systems of differential equations of the form

$$(1.01)\qquad \frac{dx_i}{dt} = f_i(x_1, x_2, \ldots, x_n), \qquad i = 1, 2, \ldots, n,$$

or

$$(1.02)\qquad \frac{dx_i}{dt} = f_i(x_1, \ldots, x_n, t), \qquad i = 1, 2, \ldots, n,$$

where the f_i are assumed to be continuous functions of their arguments in a certain domain G of the Euclidean space $R^n = \{(x_1, \ldots, x_n)\}$ the *phase space,* and in an interval $a < t < b$.

1.11. THEOREM. (*Existence of solutions* [45], [52], [54] [1]). *Consider a system of differential equations* (1.01) *where the functions $f_i(x_1, \ldots, x_n)$ are assumed to be continuous in a certain closed and bounded domain $\overline{G}$. Let $A_0(x_{10}, x_{20}, \ldots, x_{n0})$ be an arbitrary interior point of $\overline{G}$. Then there exists a solution of the system* (1.01), *which passes through A_0 at the time t_0 and which is defined in the interval*

$$t_0 - \frac{D}{M\sqrt{n}} \leqq t \leqq t_0 + \frac{D}{M\sqrt{n}},$$

where D is the distance of A_0 from the boundary of the domain $\overline{G}$ and M is an upper bound of $|f_i(x_1, \ldots, x_n)|$ in the domain $\overline{G}$.

The proof of this theorem follows.

1.12. ε-solutions. We call a system of n functions $\bar{x}_1(t), \ldots, \bar{x}_n(t)$ defined on $a \leqq t \leqq b$ a solution of the system (1.01) up to the error ε or simply an *ε-solution* of (1.01) if each of these functions is

[1]The numbers in square brackets refer to the bibliography at the end of each part of the book.

continuous, sectionally smooth[2], and satisfies the following system of integral equations

$$(1.121) \qquad \bar{x}_i(t) = \bar{x}_{i0} + \int_{t_0}^{t} f_i(\bar{x}_1, \bar{x}_2, \ldots, \bar{x}_n)\, dt + \int_{t_0}^{t} \theta_i(t)\, dt,$$

where $\theta_i(t)$ are piecewise continuous functions on $[a, b]$, less than ε in absolute value.

1.13. Euler polygons. Consider a point $A_0(x_{10}, \ldots, x_{n0})$ of $\bar{G}$ at distance $D > 0$ from the boundary. Let M be an upper bound of $|f_i(x_1, \ldots, x_n)|$ in the domain $\bar{G}$. In view of the uniform continuity of the functions $f_i(x_1, \ldots, x_n)$ in the domain $\bar{G}$, for every $\varepsilon > 0$ there exists a $\delta \geqq 0$ such that the inequality $|x_i' - x_i''| \leqq \delta$ implies that for all i

$$|f_i(x_1', x_2', \ldots, x_n') - f_i(x_1'', x_2'', \ldots, x_n'')| < \varepsilon \quad (i = 1, 2, \ldots, n).$$

We subdivide our domain $\bar{G}$ into cubes with sides of length δ.

Proceeding in the direction of increasing t we draw a segment along the straight line

$$x_i = x_{i0} + f_i(x_{10}, x_{20}, \ldots, x_{n0})(t - t_0)$$

from A_0 to the intersection $A_1(x_{11}, x_{21}, \ldots, x_{n1}) \neq A_0$, say at time t_1, of this line with one of the faces, say l, of a cube containing A_0. We write

$$x_{i1} = x_{i0} + f_i(x_{10}, x_{20}, \ldots, x_{n0})(t_1 - t_0), \quad t_0 < t_1.$$

Through the point A_1 we draw the line

$$x_i = x_{i1} + f_i(x_{11}, x_{21}, \ldots, x_{n1})(t - t_1),$$

and proceed in the direction of increasing t until we reach the point $A_2(x_{12}, x_{22}, \ldots, x_{n2}) \neq A_1$ of intersection of our line and a face different from l of a cube containing A_1.

This construction yields a polygon (an Euler polygon)

$$x_i = \bar{x}_i(t), \qquad t \geqq t_0$$

where, if $\bar{x}_i'$ is the right derivative of $\bar{x}_i$,

$$(1.131) \qquad \bar{x}_i'(t) = f_i(x_{1j}, x_{2j}, \ldots, x_{nj}) \quad \text{for} \quad t_j \leqq t < t_{j+1}.$$

A similar construction carried out in the direction of decreasing

[2]A function defined in an interval $[a, b]$ is sectionally smooth if it is continuous in this interval and is differentiable at every point of the interval except for at most a finite number of points where it has right and left derivatives. Moreover, we assume that the right and the left derivatives are bounded in the whole interval $[a, b]$.

t yields a polygon

$$x_i = \bar{x}_i(t), \qquad t \leqq t_0$$

with successive vertices $A_1(t_1'),\ A_2(t_2'), \ldots, t_0 > t_1' > t_2' \ldots$. In a manner similar to the above

$$\bar{x}_i'(t) = f_i(x_{1j}, x_{2j}, \ldots, x_{nj}) \quad \text{for} \quad t_j' > t \geqq t_{j+1}'. \tag{1.132}$$

Let us determine how far we may continue the above construction in either direction without leaving the domain $\bar{G}$. Our polygon remains in $\bar{G}$ as long as

$$\left| \int_{t_0}^{t} \sqrt{\Sigma_i(\bar{x}_i'(t))^2}\, dt \right| \leqq D.$$

But by (1.131) and (1.132), we have

$$\left| \int_{t_0}^{t} \sqrt{\Sigma(\bar{x}_i'(t))^2}\, dt \right| \leqq |t - t_0|\, M\sqrt{n}.$$

Thus our construction may be continued as long as

$$|t - t_0| M\sqrt{n} \leqq D,$$

that is as long as

$$t_0 - \frac{D}{M\sqrt{n}} \leqq t \leqq t_0 + \frac{D}{M\sqrt{n}}. \tag{1.133}$$

1.14. Let us show that our polygon is an "ε-solution". By construction each of the functions $\bar{x}_i(t)$ is continuous and sectionally smooth. It remains to verify that these functions satisfy equations (1.121).

The system of integral equations which the functions $\bar{x}_i(t)$ must satisfy is equivalent to the system

$$\bar{x}_i'(t) = f_i(\bar{x}_1(t), \bar{x}_2(t), \ldots, \bar{x}_n(t)) + \theta_i(t),$$

where $\bar{x}_i'(t)$ designates, say, the right-hand derivative of $\bar{x}_i(t)$.

Consider a fixed value of t and the corresponding point $B(t) = (\bar{x}_1(t), \ldots, \bar{x}_n(t))$ of our polygon. Then

$$\bar{x}_i(t) = \tilde{x}_i + f_i(\tilde{x}_1, \tilde{x}_2, \ldots, \tilde{x}_n)(t - \tilde{t}),$$

where $\tilde{x}_1, \ldots, \tilde{x}_n$ are the coordinates of the vertex immediately preceding B. Let us denote this vertex by $C(t)$. Thus, for the given value of t, $\bar{x}_i'(t) = f_i(\tilde{x}_1, \tilde{x}_2, \ldots, \tilde{x}_n)$.

If we define step functions $\tilde{x}_i(t)$ by the equalities $\tilde{x}_i(t) = \bar{x}_i(\bar{t}) = \tilde{x}_i$, then

$$\bar{x}_i'(t) = f_i(\tilde{x}_1(t), \tilde{x}_2(t), \ldots, \tilde{x}_n(t)).$$

If we let

$$\theta_i(t) = f_i(\tilde{x}_1(t), \ldots, \tilde{x}_n(t)) - f_i(\bar{x}_1(t), \bar{x}_2(t), \ldots, \bar{x}_n(t)),$$

then

$$\bar{x}_i'(t) = f_i(\bar{x}_1(t), \bar{x}_2(t), \ldots, \bar{x}_n(t)) + \theta_i(t).$$

Since the points $C(t)$ and $B(t)$ lie in the same cube of our partition, we have

$$|\theta_i(t)| = \left| f_i(\tilde{x}_1(t), \ldots, \tilde{x}_n(t)) - f_i(\bar{x}_1(t), \bar{x}_2(t), \ldots, \bar{x}_n(t)) \right| < \varepsilon.$$

Moreover, since

$$f_i(\bar{x}_1(t), \bar{x}_2(t), \ldots, \bar{x}_n(t))$$

are continuous and

$$f_i(\tilde{x}_1(t), \tilde{x}_2(t), \ldots, \tilde{x}_n(t))$$

assume only a finite number of values, the functions $\theta_i(t)$ are piecewise continuous. This completes the proof of our assertion.

The following observation will be useful in the sequel.

1.15. In constructing ε-solutions we may replace Euler polygons by what we shall call "universal polygons". Let the domain $\bar{G}$ be partitioned into cubes with sides of length $\delta/2$. We take a point in each one of these cubes, say the center, and determine the value of the functions $f_i(x_1, x_2, \ldots, x_n)$, $i = 1, 2, \ldots, n$, at each of these points. Beginning at a point A_0 we construct a polygon by a method similar to that used in the construction of Euler polygons. Here, however, the direction of each of the sides of our polygon is determined by the value of $f_i(x_1, x_2, \ldots, x_n)$ at the previously selected point of the corresponding cube.

1.16. Let us now take a sequence of positive numbers $\varepsilon_1, \varepsilon_2, \ldots, \varepsilon_n, \ldots$ tending to zero and proceeding as in 1.13, let us construct consecutively an ε_1-solution, through a point A interior to G, an ε_2-solution through A, and so on. Since the interval (1.131) for which our approximate solutions are defined does not depend on ε, all of these solutions can be constructed for one and the same interval, say,

$$t_0 - \frac{D}{M\sqrt{n}} \leq t \leq t_0 + \frac{D}{M\sqrt{n}}.$$

We denote an ε_k-solution by $\{x_i^{\varepsilon_k}(t)\}$. We shall prove that the family of solutions

$$\{x_i^{\varepsilon_k}(t)\}, \qquad k = 1, 2, \ldots,$$

forms an equicontinuous and uniformly bounded family of functions.

Since

$$x_i^{\varepsilon_k}(t) = x_{i0} + \int_{t_0}^{t} f_i(x_1^{\varepsilon_k}, x_2^{\varepsilon_k}, \ldots, x_n^{\varepsilon_k})dt + \int_{t_0}^{t} \theta_i^{\varepsilon_k}(t)\, dt,$$

we have

$$|x_i^{\varepsilon_k}(t)| \leqq L + M\frac{D}{M\sqrt{n}} + \varepsilon_k \frac{D}{M\sqrt{n}},$$

where L is an upper bound of the absolute values of the coordinates of points in $\overline{G}$. Furthermore,

$$x_i^{\varepsilon_k}(t+h) - x_i^{\varepsilon_k}(t) = \int_{t}^{t+h} f_i(x_1^{\varepsilon_k}, x_2^{\varepsilon_k}, \ldots, x_n^{\varepsilon_k})dt + \int_{t}^{t+h} \theta_i^{\varepsilon_k}(t)\, dt,$$

and therefore

$$|x_i^{\varepsilon_k}(t+h) - x_i^{\varepsilon_k}(t)| \leqq hM + h\varepsilon_k.$$

The last two inequalities establish our assertion.

1.17. In view of Arzela's theorem [3] there exists a sequence of indices $n_1, n_2, \ldots, n_k, \ldots$ such that the n sequences $x_i^{\varepsilon_{n_k}}(t)$, $i = 1, 2, \ldots, n$, converge in the interval

$$t_0 - \frac{D}{M\sqrt{n}} \leqq t \leqq t_0 + \frac{D}{M\sqrt{n}}$$

to continuous functions

$$x_1(t), \ldots, x_n(t).$$

Passing to the limit in the equalities

$$x_i^{\varepsilon_{n_k}}(t) = x_{i0} + \int_{t_0}^{t} f_i(x_1^{\varepsilon_{n_k}}, x_2^{\varepsilon_{n_k}}, \ldots, x_n^{\varepsilon_{n_k}})\, dt + \int_{t_0}^{t} \theta_i^{\varepsilon_{n_k}}(t)\, dt,$$

and observing that the $f_i(x_1, x_2, \ldots, x_n)$ are uniformly continuous in $\overline{G}$ and that

$$|\theta_i^{\varepsilon_{n_k}}(t)| < \varepsilon_{n_k}, \quad \text{for every } \varepsilon_{n_k},$$

[3]This theorem states that every infinite family of functions uniformly bounded and equicontinuous on a closed interval [a, b] contains a uniformly convergent sequence of functions. Cf. *Memorie Acad. Bologna* (5) vs. 5 (1895) and 8 (1899).

we obtain

$$x_i(t) = x_{i0} + \int_{t^0}^{t} f_i(x_1(t), \ldots, x_n(t))\, dt,$$

or

$$\frac{dx_i(t)}{dt} = f_i(x_1(t), \ldots, x_n(t)).$$

This completes the proof of Theorem 1.11.

1.2. We shall extend our existence theorem to systems of type (1.02).

Let f_i be defined and continuous for points $A(x_1, x_2, \ldots, x_n)$ of a closed and bounded domain $\overline{G}$ and for values of t in an interval $[t_0 - b, t_0 + b]$. We introduce a new independent variable τ such that $dt/d\tau = 1$. Then the given system (1.02) may be written in the form

$$\text{(1.201)} \qquad \begin{cases} \dfrac{dx_i}{d\tau} = f_i(x_1, x_2, \ldots, x_n, t), \\[2ex] \dfrac{dt}{d\tau} = 1. \end{cases}$$

Applying our existence theorem to the closed and bounded domain of the $(n + 1)$-dimensional space $(x_1, \ldots, x_n, t)$ determined by $\overline{G}$ and the interval $t_0 - b \leqq t \leqq t_0 + b$, we assure the existence of a solution of system (1.02) in the interval $(t_0 - h, t_0 + h)$, where

$$h = \frac{\min(D, b)}{1 + M\sqrt{n}}.$$

We shall speak of (1.201) as the *parametric system* corresponding to the system (1.02).

1.21 As a simple corollary of our existence theorem, we obtain the following result which is very important for the theory of dynamical systems.

1.21. THEOREM. *If as time increases, a given trajectory (an integral curve) remains in a closed bounded region Γ imbedded in an open domain G for which the conditions of our existence theorem are fulfilled, then the motion (the solution) may be continued for the whole infinite interval $[t_0, +\infty]$.*

Let $2D$ be the distance of the boundary of G from the boundary of Γ. Then successive applications of our existence theorem always lead to points whose distance from the boundary of G is not less than D. Consequently, at each step we can continue our solution for another interval of at least the length $D/M\sqrt{n}$.

1.3. Theorem 1.21 does not allow us to decide from the form of a given system of equations whether or not its solutions can be continued for the infinite interval $-\infty < t < +\infty$. We indicate several sufficient conditions for such continuation. [58], [59].

1.31. THEOREM. *If the functions*

$$f_1(x_1, x_2, \ldots, x_n), \ldots, f_n(x_1, x_2, \ldots, x_n)$$

are continuous for $-\infty < x_i < +\infty$, *and, moreover, if*

$$f_i(x_1, x_2, \ldots, x_n) = O(|x_1| + |x_2| + \ldots + |x_n|)$$

for $|x_1| + \ldots + |x_n| \to +\infty$, *then the solutions of the system*

$$\frac{dx_i}{dt} = f_i(x_1, x_2, \ldots, x_n)$$

are defined on the whole axis $-\infty < t < +\infty$.

1.32. It follows from the hypotheses of our theorem that

$$(1.321) \qquad |f_i(x_1, x_2, \ldots, x_n)| < A \max(|x_1|, \ldots, |x_n|, 1),$$

where A is some positive constant. For, if $|x_1| + \ldots + |x_n| > D > 0$, where D is some sufficiently large number, then the ratios

$$\frac{|f_i(x_1, x_2, \ldots, x_n)|}{\Sigma |x_j|}$$

remain bounded, whereas the functions $f_i(x_1, \ldots, x_n)$ themselves are bounded in the region $|x_1| + \ldots + |x_n| \leqq D$.

1.33. Let us consider first the cube $|x_i - x_{i0}| \leqq b$ $(i = 1, 2, \ldots, n)$, and let M be an upper bound of $|f_i(x_1, x_2, \ldots, x_n)|$ in this cube. According to the existence theorem, the solution passing through A_0 is defined in the whole interval $[t_0, t_0 + (b/M\sqrt{n})]$.

Set x_{i0}, t_0, and b equal to c_i, 0, and 1 respectively. Then it follows from the inequality (1.321) and the condition $|x_i(t) - c_i| \leqq 1$ that we may take $M = A(c+1) = A \max[c+1, 1]$ with $c = \max |c_i|$ $(i = 1, 2, \ldots, n)$. Write

$$t_1 = \frac{b}{M\sqrt{n}} = \frac{1}{M\sqrt{n}} = \frac{1}{A(c+1)\sqrt{n}}.$$

Then our solution is defined for $0 \leqq t \leqq t_1$, and in this interval

$$|x_i(t)| \leqq c + 1.$$

Next, let us take x_{i0}, t_0, b equal to $x_1(t_1)$, t_1, 1 respectively. Then we may take $M = A \max(c + 2, 1) = A(c + 2)$. We write

$$t_2 = \frac{b}{M\sqrt{n}} = \frac{1}{M\sqrt{n}} = \frac{1}{A(c+2)\sqrt{n}},$$

and observe that the solution is defined in $t_1 \leqq t \leqq t_1 + t_2 = \tau_2$.

Combining both of the above results we see that our solution is defined in the interval $[0, \tau_2]$. The inequality $|x_i(t) - x_i(t_1)| \leqq 1$ for $t_1 \leqq t \leqq \tau_2 = t_1 + t_2$, implies that $|x_i(\tau_2)| \leqq c + 2$. Continuing this process for m steps we obtain a number $t_m = 1/(c + m)A\sqrt{n}$ such that our solution is defined in the interval $[0, \tau_m]$ where $\tau_m = t_1 + t_2 + \ldots + t_m$ and $|x_i(\tau_m)| \leqq c + m$. The series

$$\frac{1}{A\sqrt{n}} \sum_{m=0}^{\infty} \frac{1}{c+m+1}$$

diverges. Therefore by means of a sufficiently large number of steps we can continue our solution for an interval of arbitrarily large length.[4]

1.34. COROLLARY. *If*

$$f_i(x_1, x_2, \ldots, x_n, t) = O(|x_1| + |x_2| + \ldots + |x_n|)$$

uniformly in t, then solutions of the system $dx_i/dt = f_i$ may be continued to the whole t-axis.

Indeed, let us consider the corresponding parametric system

$$\frac{dx_i}{d\tau} = f_i(x_1, x_2, \ldots, x_n, t),$$

$$\frac{dt}{d\tau} = 1.$$

[4]From the estimates given in the proof it follows that

$$|x_i(t)| = O(e^{ct})$$

where the constant c may be chosen independently of the initial conditions. In fact, after the mth step in the process of continuation we have

$$|x_i(t)| \leqq c + m,$$

for

$$t = t_1 + t_2 + \ldots + t_m = \frac{1}{A\sqrt{n}} \sum_{j=0}^{m-1} \frac{1}{c+j+1}.$$

Thus t is asymptotically equal to $A^{-1}n^{-1/2} \log m$ or m is asymptotically equal to $e^{An^{\frac{1}{2}}t}$. This proves our assertion since A is chosen independently of the initial conditions.

Since

$$|f_i(x_1, x_2, \ldots, x_n, t)| < A \max(|x_1|, \ldots, |x_n|, 1)$$

where A is independent of t, then obviously

$$|f_i(x_1, x_2, \ldots, x_n, t)| < A \max(|x_1|, \ldots, |x_n|, |t|, 1).$$

Consequently the conditions of Theorem 1.31 are fulfilled by the parametric system.

1.35. The last result may be somewhat generalized.

If functions $f_1, f_2, \ldots, f_n$ are continuous in an $(n+1)$-dimensional domain $0 < t < +\infty$ and $-\infty < x_i < +\infty$, if there exists a function $L(r)$ continuous for $0 < r < +\infty$ and such that $\int_0^\infty (1/L(r))\, dr = \infty$, and if $|f_i(x_1, \ldots, x_n, t)| < L(r)$, where $r^2 = x_1^2 + \ldots + x_n^2$, then all the solutions of the system $dx_i/dt = f_i$ may be continued over the entire t-axis.

We omit the proof of this theorem even though it is quite simple and refer the reader to the original work of Wintner [58].

2. Certain Uniqueness and Continuity Theorems

In what follows we shall consider systems of equations (1.01) in which the functions $f_i(x_1, \ldots, x_n)$ satisfy Lipschitz conditions in a bounded closed domain $\overline{G}$ called the Lipschitz domain. That is

$$|f_i(x_1', x_2', \ldots, x_n') - f_i(x_1'', x_2'', \ldots, x_n'')| < L \sum_{i=1}^{n} |x_i' - x_i''|.$$

The number L is called a Lipschitz constant. To indicate explicitly the connection between the domain $\overline{G}$ and the constant L we shall write $\overline{G}_L$ instead of $\overline{G}$.

We establish first the following simple lemma [5] which is quite essential for what follows.

2.11 LEMMA. *If a function $y(t)$ satisfies the inequality*

$$|y(t)| < M\left(1 + k\int_{t_0}^{t} |y(t)|\, |f(t)|\, dt\right) \tag{2.111}$$

where $f(t)$ is continuous, then we have the inequality

$$|y(t)| < M e^{kM\int_{t_0}^{t}|f(t)|dt} \quad (t > t_0). \tag{2.112}$$

Multiplying (2.111) by $|f(t)|$, we get

$$|y(t)|\, |f(t)| < M\, |f(t)|\left(1 + k\int_{t_0}^{t} |y(t)|\, |f(t)|\, dt\right). \tag{2.113}$$

Let $v(t) = \int_{t_0}^t |y(t)f(t)|\,dt$. Then the inequality (2.113) may be written in the form

$$v'(t) < M\,|f(t)|\,(1 + kv),$$

or

$$\frac{v'(t)}{1 + kv} < M\,|f(t)|.$$

Thus

$$\log\,(1 + kv(t)) < kM \int_{t_0}^t |f(t)|\,dt,$$

and hence

$$1 + k\int_{t_0}^t |f(t)y(t)|\,dt < e^{kM\int_{t_0}^t |f(t)|dt}.$$

By hypothesis

$$\frac{|y(t)|}{M} < 1 + k\int_{t_0}^t |f(t)y(t)|\,dt,$$

whence

$$|y(t)| < Me^{kM\int_{t_0}^t |f(t)|dt}.$$

2.12. We shall use our lemma to establish a fundamental inequality.

Consider two ε-solutions

$$\{x_i^{(1)}(t)\}, \qquad \{x_i^{(2)}(t)\}.$$

In view of (1.121),

$$x_i^{(1)}(t) - x_i^{(2)}(t)$$
$$= (x_{i0}^{(1)} - x_{i0}^{(2)}) + \int_{t_0}^t [f_i(x_1^{(1)}, x_2^{(1)}, \ldots, x_n^{(1)}) - f_i(x_1^{(2)}, x_2^{(2)}, \ldots, x_n^{(2)})]dt$$
$$+ \int_{t_0}^t [\theta_i^{(1)}(t) - \theta_i^{(2)}(t)]\,dt.$$

Making use of Lipschitz inequalities, we get

$$|x_i^{(1)}(t) - x_i^{(2)}(t)| < |x_{i0}^{(1)} - x_{i0}^{(2)}|$$
$$+ \int_{t_0}^t L \cdot \sum_{j=1}^n |x_j^{(1)} - x_j^{(2)}|\,dt + \int_{t_0}^t |\theta_i^{(1)}(t) - \theta_i^{(2)}(t)|\,dt$$

for $i = 1, 2, \ldots, n$, and $t > t_0$. Adding these inequalities and writing

$$\delta = \max |x_{i0}^{(1)} - x_{i0}^{(2)}| \qquad (i = 1, 2, \ldots, n),$$

and remembering that

$$|\theta_i^{(1)}(t)| \leqq \varepsilon \quad \text{and} \quad |\theta_i^{(2)}(t)| \leqq \varepsilon,$$

we obtain the inequality

$$\sum_{i=1}^{n} |x_i^{(1)}(t) - x_i^{(2)}(t)| < n\delta + n\int_{t_0}^{t} L \sum_{i=1}^{n} |x_i^{(1)}(t) - x_i^{(2)}(t)|\, dt$$

$$+ 2n\varepsilon(t - t_0) \leqq (2n\varepsilon(T - t_0) + n\delta)\left[1 + \frac{n}{2n\varepsilon(T-t_0) + n\delta}\right.$$

$$\left. \cdot \int_{t_0}^{t} L \sum_{i=1}^{n} |x_i^{(1)}(t) - x_i^{(2)}(t)|\, dt\right], \quad t_0 < t \leqq T.$$

Applying Lemma 2.11, we obtain

$$\sum_{i=1}^{n} |x_i^{(1)}(t) - x_i^{(2)}(t)| < [2n\varepsilon(T - t_0) + n\delta]\, e^{n\int_{t_0}^{t} L\, dt}$$

for $t_0 < t \leqq T$. If $t_0 > t$ we assume that $t_0 > t \geqq T$ and invert the order of integration throughout. Setting $t = T$ and simplifying, we obtain, in either case,

$$\sum_{i=1}^{n} |x_i^{(1)}(t) - x_i^{(2)}(t)| < 2n|t - t_0|\, \varepsilon\, e^{nL|t-t_0|} + n\delta\, e^{nL|t-t_0|}. \tag{2.124}$$

In what follows we shall refer to this estimate as *the fundamental inequality*.

2.2. We shall discuss next a number of immediate consequences of the fundamental inequality, all of which are of basic importance in the theory of differential equations.

2.21. THEOREM (Uniqueness). *If the right-hand members of system* (1.01) *satisfy Lipschitz conditions, then there exists a unique solution satisfying given initial conditions.*

Let $\{x_i^{(1)}(t)\}$, $\{x_i^{(2)}(t)\}$ be two solutions defined on a segment $[t_0, t_1]$ and satisfying the same initial conditions at t_0 (or at t_1). We may consider these solutions as ε-solutions for an arbitrarily small ε. Applying the fundamental inequality and observing that $\delta = 0$, we obtain

$$\sum_{i=1}^{n} |x_i^{(1)}(t) - x_i^{(2)}(t)| < 2n(t_1 - t_0)\, \varepsilon e^{nL|t_1 - t_0|}.$$

Since ε is arbitrarily small, we have

$$\sum_{i=1}^{n} |x_i^{(1)}(t) - x_i^{(2)}(t)| = 0 \quad \text{for} \quad t_0 \leqq t \leqq t_1,$$

which proves our assertion.

2.22. THEOREM (*continuity in the initial conditions*). *Let the right-hand members of* (1.01) *satisfy Lipschitz conditions in a domain* G_L. *If a solution* $\{x_i\} = \{x_i(t, t_0, x_{i0}, \ldots, x_{n0})\} = x(t)$ *is defined for* $t_0 \leqq t \leqq T$, *then for every* $\eta > 0$ *there is a* $\delta > 0$ *such that for* $|\bar{x}_{i0} - x_{i0}| < \delta$ $(i = 1, 2, \ldots, n)$ *the solution* $x_i = x_i(t, t_0, \bar{x}_{i0}, \ldots, \bar{x}_{n0}) = \bar{x}_i(t)$ *is also defined for* $t_0 \leqq t \leqq T$ *and for all values of t in this interval* $|\bar{x}_i(t) - x_i(t)| < \eta$.

The fundamental inequality (2.124) with $\varepsilon = 0$ yields

$$\sum_{i=1}^{n} |\bar{x}_i(t) - x_i(t)| < n\delta e^{nL(t-t_0)} \tag{2.221}$$

for every value of t in $t_0 \leqq t \leqq T$ for which $\bar{x}_i(t)$ is defined. For some $d > 0$, the d-neighborhood of the segment $C : x_i(t)$, $t_0 \leqq t \leqq T$, lies in the interior of G_L. If $\eta < d$, we let

$$\delta \leqq \frac{\eta}{ne^{nL(T-t_0)}}. \tag{2.222}$$

If we take the (closed) d-neighborhood of C as the domain $\overline{G}$ of Theorem 1.11, then $D \geqq d - \eta > 0$, we see at once that the solution $\bar{x}(t)$ through $(\bar{x}_{i0}, \ldots, \bar{x}_{n0})$ can be extended at least as far as T in view of the inequality (2.221) and the choice of δ. Also, throughout the interval $t_0 \leqq t \leqq T$, we have

$$|\bar{x}_i(t) - x_i(t)| < \eta.$$

2.3. One should note that the choice of δ depends not only upon the degree of the desired approximation, that is upon η, but also upon the length $(T - t_0)$ of our time interval. In many problems of mechanics, it is essential to seek solutions in which δ can be chosen independently of the length of the time interval. Such motions possess a certain degree of stability with respect to the change in the initial conditions. Detailed study of such motions and of the methods of their characterization was carried out by the inspired Russian scientist Liapounoff. We shall meet these ideas and methods in the subsequent chapters.

2.4. Stability of solutions with respect to changes in the right-hand members of our system. Let a system (1.01) be replaced by a system

$$\frac{d\bar{x}_i}{dt} = f_i(\bar{x}_1, \bar{x}_2, \ldots, \bar{x}_n) + \theta_i(\bar{x}_1, \bar{x}_2, \ldots, \bar{x}_n), \tag{2.411}$$

and let $|\theta_i| \leqq \varepsilon$ for all values of $\bar{x}_i$ in a closed domain $\bar{G}_L$. Then every solution $\bar{x}_i(t)$ of systems (2.411) is obviously an ε-solution of system (1.01). If $x_i(t)$ is a solution of system (1.01) satisfying the same initial condition as a solution $\bar{x}_i(t)$ of system (2.411), then, in view of the fundamental inequality, we obtain

$$(2.412) \qquad |\bar{x}_i(t) - x_i(t)| \leqq 2n\,|t - t_0|\,\varepsilon e^{nL|t-t_0|}.$$

It follows from this estimate that for a fixed interval of time we may make the difference of the above solutions arbitrarily small by choosing ε sufficiently small.

2.42. Frequent use is made of the process of linearization, i.e., of a replacement of a given nonlinear system by a linear system. In particular, such a method is considered permissible if the nonlinear terms have small parameters. The above inequality (2.412) makes it possible to obtain a numerical estimate of the error resulting from linearization.

2.5. A method of approximate integration [35]. In deriving the fundamental inequality we required that the functions $\theta_i(t)$ should be piecewise continuous.

Observing this, one may develop the following method of approximate integration of (1.01).

2.51. For a given $\varepsilon > 0$ we partition the domain G_L into cubes of side δ, where δ is so small that the inequalities $|x_i' - x_i''| \leqq \delta (i = 1, \ldots, n)$ imply

$$|f_i(x_1', x_2', \ldots, x_n') - f_i(x_1'', x_2'', \ldots, x_n'')| \leqq \frac{\varepsilon}{2}.$$

We construct new functions $\bar{f}_i(x_1, x_2, \ldots, x_n)$ which assume throughout each cube the values of the corresponding functions $f_i(x_1, \ldots, x_n)$ at the center of the cube. Obviously,

$$|f_i(x_1, x_2, \ldots, x_n) - \bar{f}_i(x_1, x_2, \ldots, x_n)| \leqq \varepsilon.$$

On the boundaries of the cubes we allow each $\bar{f}_i$ to be manyvalued.

Let us consider next the system of equations

$$(2.511) \qquad \frac{d\bar{x}_i}{dt} = \bar{f}_i(\bar{x}_1, \bar{x}_2, \ldots, \bar{x}_n).$$

Within each cube the solutions of (2.511) form a family of parallel straight line segments whose direction is determined by the values of $f_i(x_1, \ldots, x_n)$ at the center of this cube.

By a solution $\bar{x}_i(t)$ of the system (2.511) we shall mean a polygon constructed as follows: Given a point A_0 we choose one (there may be more than one) of the above segments A_0A_1, say, $A_{-1}A_0A_1$ passing through A_0. If A_1 is the initial point of a segment solution A_1A_2 of (2.511), we choose A_1A_2 as the second link, and so on until we exhaust that interval of time for which we seek an approximation to a solution of (1.01).

2.52. A solution of (2.511) is an ε-solution of the system (1.01). For, if $\bar{x}_j = \bar{x}_j(t)$ $(j = 1, \ldots, n)$ is a solution of (2.511), this system may be written

$$\frac{d\bar{x}_i}{dt} = f_i(\bar{x}_1, \bar{x}_2, \ldots, \bar{x}_n) + [\bar{f}_i(\bar{x}_1(t), \bar{x}_2(t), \ldots, \bar{x}_n(t)) - f_i(\bar{x}_1(t), \bar{x}_2(t), \ldots, \bar{x}_n(t))]$$

where the difference in the brackets is numerically smaller than ε in the domain $\bar{G}_L$ and is piecewise continuous in t. For $\bar{x}_j(t)$, as well as $f_i(x_1, x_2, \ldots, x_n)$ are continuous, and $\bar{f}_i(x_1, x_2, \ldots, x_n)$ assumes only a finite number of values.

2.53. We observe that in constructing an approximate solution we need not start our polygon at the given initial point of the desired solution.

Let us construct polygon solutions of (2.511) starting at the center of each cube of our partition. Let $\Lambda_1, \ldots, \Lambda_s$ be the family of all such solutions. Then, for every solution $\{x_i(t)\}$ of (1.01) defined for a time interval T, there exists a polygon $\Lambda_j = \{\bar{x}_i^{(j)}(t)\}$ such that

$$|x_i(t) - \bar{x}_i^{(j)}(t)| \leqq 2n\varepsilon T e^{nLT} + n\delta e^{nLT}. \tag{2.531}$$

Since we may assume that $\delta \leqq \varepsilon$, the inequality (2.531) yields

$$|x_i(t) - \bar{x}_i^{(j)}(t)| \leqq 2n\varepsilon(T + \tfrac{1}{2})e^{nLT}.$$

Thus, for a fixed T, the error may be made arbitrarily small by choosing ε sufficiently small.

2.6. Toroidal and cylindrical phase spaces. We shall conclude this section with a few remarks regarding the generality of the theorems considered above.

In all of our proofs we considered an n-tuple $(x_1, x_2, \ldots, x_n)$ as a point in an n-dimensional Euclidean space. This assumption was not necessary. We may assume that our solution space is a manifold every point of which has a neighborhood homeomorphic to an n-dimensional sphere of an n-dimensional Euclidean space R^n. In particular it can be an arbitrary domain in an n-dimensional Euclidean space. In case the space is only locally Euclidean, then the estimates of the interval of existence of solutions must obviously be changed.

A special role is played by systems of differential equations (1.01) in which the right-hand members are defined for all values of the variables $x_1, x_2, \ldots, x_n$ but in which certain of these variables are cyclic, i.e., they take values only in a finite interval of length γ_i. The domains of definition of these variables may be extended to the whole infinite line. Here we shall identify points whose ith coordinates differ by γ_i.

Consider for example a system of two equations

$$\frac{dx}{dt} = P(x, y), \qquad \frac{dy}{dt} = Q(x, y).$$

If (x, y) are plane coordinates then the solution space is a plane. If x varies from $-\infty$ to $+\infty$ but y is a cyclic coordinate, then the solution space is a cylinder. If both coordinates are cyclic, then the space is a torus. The theorem on unlimited continuation of solutions applies to the cylindrical as well as to the toroidal solution space.

3. Dynamical systems defined by a system of differential equations.

We shall give here only a few basic definitions and elementary results pertaining to dynamical systems.

3.1. First, we study an important property of systems of differential equations satisfying the uniqueness conditions of (2.21).

3.11. To indicate the dependence of solutions upon the initial conditions explicitly, we write

$$x_i = x_i(t, t_0, x_1^{(0)}, \ldots, x_n^{(0)}) \tag{3.111}$$

for that solution of (1.01) which passes through the point $x_i^{(0)}$ when $t = t_0$. If $t_0 = 0$, then we abbreviate (3.111) by writing

$$x_i = x_i(t, x_1^{(0)}, \ldots, x_n^{(0)}). \tag{3.112}$$

Next, we consider

$$x_i = x_i(t - t_0, x_1^{(0)}, \ldots, x_n^{(0)}). \tag{3.113}$$

Since the right-hand members of (1.01) do not contain t explicitly, (3.113) is a solution of (1.01). Moreover, we observe that it is the solution which passes through $x_i^{(0)}$ for $t = t_0$.

In particular, we have the important relation

$$x_i(t_2, x_1(t_1, x_1^{(0)}, \ldots, x_n^{(0)}), \ldots, x_n(t_1, x_1^{(0)}, \ldots, x_n^{(0)})) = x_i(t_1 + t_2, x_1^{(0)}, \ldots, x_n^{(0)}). \tag{3.114}$$

For both the right-hand member and the left-hand member of (3.114), considered as functions of t_2, represent solutions passing through the same point $x_i(t_1, x_1^{(0)}, \ldots, x_n^{(0)})$ for $t_2 = 0$.

3.12. Let us denote the solution (3.112) passing through the point $p(x_1^{(0)}, \ldots, x_n^{(0)})$ by the symbol $f(p, t)$. Thus for every t, $f(p, t) = q$ is a definite point on the trajectory through p and in particular $f(p, 0) = p$. Moreover, if for every p in $\overline{G}_L$ the function $f(p, t)$ is defined for $t \in T = (-\infty, +\infty)$ then

$$\begin{cases} f(p, t) \text{ is continuous in both of its arguments in} \\ \overline{G}_L \times T, \end{cases} \tag{3.121}$$

and in view of (3.114),

$$f(p, t_1 + t) = f(f(p, t_1), t). \tag{3.122}$$

Thus $f(p, t)$ defines a one-parameter group of transformations of the solution space $\overline{G}_L$ into itself. It is customary to speak of the set of all the transformations of this group as a *dynamical system* and of the totality of all the points $f(p, t)$ for a fixed p and $-\infty < t < +\infty$ as a *trajectory* of this dynamical system.

3.2. In general, even if the $f_i(x_1, \ldots, x_n)$ satisfy Lipschitz conditions or other conditions assuring uniqueness of solutions in a domain G of an n-dimensional Euclidean space or of a locally Euclidean manifold, the corresponding system (1.01) does not necessarily define a dynamical system, since it may have solutions which cannot be continued for all values of t. Some sufficient conditions for unlimited continuation were given in Sections **1.2** and **1.3.**

We shall show, however, that by merely changing the independent variable, i.e., by changing the parametrization of the integral curves of the given system, we can arrive at a system whose solutions do determine a dynamical system.

In other words, if we are interested only in the geometrical or, more precisely, topological properties of individual integral curves or of the whole family of integral curves, then we may limit ourselves to the study of differential equations which define dynamical systems.

3.21. DEFINITION. *Two systems* (1.01) *are called equivalent if their solutions (including the singular points) coincide geometrically. A system* (1.01) *will be called a D-system if its solutions define a dynamical system.*

A point $p(x_{10}, \ldots, x_{n0})$ is called a *singular point* of (1.01) if $f_i(x_{10}, \ldots, x_{n0}) = 0$ simultaneously for all the right-hand members of f_i of a system (1.01).

3.22. THEOREM (R. E. Vinograd) [55]. *Consider a system* (1.01) *satisfying Lipschitz conditions in an open domain $G_L \subset R^n$. There exists a D-system defined over the whole R^n and equivalent to* (1.01) *in G_L.*

3.23. Let us prove first that every system (1.01) may be replaced by an equivalent system with bounded right-hand members.

We define $\varphi_i(x)$ so that [5]

$$\varphi_i(x) = 1 \quad \text{if} \quad |f_i(x)| \leqq 1,$$

$$\varphi_i(x) = \frac{1}{f_i(x)} \quad \text{if} \quad f_i(x) > 1,$$

$$\varphi_i(x) = \frac{-1}{f_i(x)} \quad \text{if} \quad f_i(x) < -1,$$

and we write $\varphi(x) = \Pi_{i=1}^n \varphi_i(x)$. Obviously $0 < \varphi_i(x) \leqq 1$, $|f_i(x)\varphi_i(x)| \leqq 1$, and $\varphi_i(x)$ are continuous. Therefore $0 < \varphi(x) \leqq 1$, $|f_i(x)\varphi(x)| \leqq 1$ and $\varphi(x)$ is continuous.

The system

$$\frac{dx_i}{dt} = f_i(x)\varphi(x)$$

is equivalent to the given system in G_L and its right-hand members are bounded.

3.24. We may assume therefore that system (1.01) has bounded right-hand members.

[5]Here x is an abbreviation for $(x_1, x_2, \ldots, x_n)$.

We observe that in this case we have

$$\int_0^{t'} v(x)\,dt \leqq |t'|\,c,$$

where

$$v(x) = \sqrt{\sum_{i=1}^{n} f_i^2(x)}, \qquad c = M\sqrt{n}.$$

Thus,

3.241. *For a finite t' the length of the trajectory*

$$x(t, x^{(0)}), \qquad 0 \leqq t \leqq t',$$

is finite.

3.242. Now let $x(t, x^{(0)})$ be a solution of (1.01) which cannot be continued beyond $t = t_1$. The trajectory defined by this solution must have a limit point $x^{(1)}$ on the boundary B of G_L for otherwise, in view of Theorem 1.21, the solution could have been continued indefinitely.

By 3.241, the above limit point $x^{(1)}$ on the boundary is unique and it is approached along the trajectory as $t \to t_1$.

3.243. Write $F = R^n - G_L$ and let

$$\psi(x) = \frac{\varrho(x, F)}{\varrho(x, F) + \varrho(x, x^0) + 1}$$

where $\varrho(x, y)$ is the distance between x and y, $\varrho(x, F) = \min_{y \in F} \varrho(x, y)$, and $x^{(0)}$ is a fixed point of G_L.

The function $\psi(x)$ is continuous everywhere in R^n, $0 \leqq \psi(x) < 1$, and $\psi(x) = 0$ in F and nowhere else.

Let us consider the system

$$\frac{dx_i}{dt} = f_i(x)\psi(x). \tag{3.243}$$

It is equivalent to the given system and has bounded right-hand members in G_L. To prove that (3.243) determines a dynamical system, it suffices, in view of (3.242), to show that we can extend indefinitely solutions corresponding to half-trajectories of finite length s_0 terminating at a point $x^{(1)} \in B \subset F$.

We observe that

$$t = \int_0^s \frac{ds}{v(x)\psi(x)},$$

where we may for definiteness assume $s > 0$. Next, for every point $x = x(s)$ on our trajectory

$$\psi(x) = \frac{\varrho(x, F)}{\varrho(x, F) + \varrho(x, x^{(0)}) + 1} < \varrho(x, F) = \min_{y \in F} \varrho(x, y)$$
$$\leqq \varrho(x, x^{(1)}) \leqq s_0 - s.$$

Since $0 < v(x) \leqq c$, then

$$t \geqq \frac{1}{c}\int_0^s \frac{ds}{s_0 - s} = -\frac{1}{c}\log\frac{s_0 - s}{s_0}.$$

Thus $t \to \infty$ as $s \to s_0$.

3.244. We extend the domain of definition of the right-hand members of (3.243) by setting them equal to zero in F. Since $f_i(x)$ are bounded in G_L and $\psi(x)$ is continuous and vanishes on the boundary B, the extended system has continuous right-hand members. The new system is a D-system for which all points of F are singular points. This completes the proof of Theorem 3.22. The above reasoning also yields the following:

3.245. Theorem. *Given a system* (1.01) *and a closed set* $\Phi \subset G$, *there exists a D-system which is equivalent to the given system on* $G - \Phi$ *and has all the points of* Φ *as equilibrium points.*

3.25. We shall return now to the study of the properties of dynamical systems.

3.251. Definition. A point q is called an ω-limit point of a trajectory $f(p, t)$ if there exists a sequence $t_1, t_2, \ldots, t_n \to +\infty$ such that $\lim \varrho(f(p, t_n), q) = 0$. A point q is called an α-limit point of a trajectory $f(p, t)$ if there exists a sequence $t_1, t_2, \ldots, t_n, \ldots, \to -\infty$ such that $\lim \varrho(f(p, t_n), q) = 0$. The set of all ω-limit points of a given trajectory we shall call its ω-limit set, and we shall denote this set by Ω_p. Similarly, the α-limit set A_p of a given trajectory is the set of all its α-limit points. Both Ω_p and A_p are closed sets.

3.252. Theorem. *If q is either an ω- or an α-limit point of a trajectory $f(p, t)$, then all other points of the trajectory $f(q, t)$ are also ω- or α-limit points respectively of the given trajectory $f(p, t)$.*

Let $r = f(q, \bar{t})$ be a point on the trajectory $f(q, t)$. Since q is an ω-limit point, there exists a sequence $\{t_n\}$ with $t_n \to +\infty$ and such that $f(p, t_n) \to q$. Then by Theorem 2.22, $f(p, t_n + \bar{t}) \to f(q, \bar{t})$ and since $t_n + \bar{t} \to +\infty$, the point r is an ω-limit point.

This theorem may also be stated as follows:

3.253. *Both ω- and α-limit sets of a trajectory consist of whole trajectories.*

3.254. We now classify trajectories according to the properties of their α- and ω-sets.

3.2541. We say that a solution (or a trajectory) recedes in the positive direction if it has no ω-limit points.

3.2542. A solution (or a trajectory) $f(p, t)$ is called asymptotic in the positive direction if there exist ω-limit points, but they do not belong to this solution.

3.2543. A solution (or a trajectory) $f(p, t)$ is called stable in the positive direction in the sense of Poisson if it has ω-limit points which belong to this solution.

We introduce similar definitions describing behavior of solutions as $t \to -\infty$.

3.255. We now consider two important classes of solutions of (1.01) stable in the sense of Poisson. These are the singular points and the periodic solutions.

It is clear that if a point $p(x_{10}, \ldots, x_{n0})$ is a singular point, then the set of functions $x_i(t) = x_{i0}$ $(i = 1, \ldots, n)$ is a solution of (1.01). Thus a singular point $p(x_{10}, \ldots, x_{n0})$ is a trajectory and $f(p, t) = p$ for all t. Therefore every singular point is its own α- as well as ω-limit point, and hence is a trajectory stable in the sense of Poisson.

The set of all singular points is a closed set, and by 3.245, it can be an arbitrary closed set.

If a trajectory $f(p, t)$ has a unique limit point either for $t \to +\infty$ or for $t \to -\infty$ then, in view of 3.253, this limit point is a singular point.

3.256. THEOREM. *If every neighborhood, however small, of a point p contains a trajectory traversed over an arbitrarily long time span, then p is a singular point.*

If p is not a singular point, there exists a t_1 such that $p_1 = f(p, t_1) \neq p$. Then $p_{-1} = f(p, -t_1) \neq p$ as well. Let $d = \min\,[\varrho(p, p_1), \varrho(p, p_{-1})]$. By the continuity in the initial conditions we can find a $\delta > 0$ such that $\varrho(p, x) < \delta$ implies that $\varrho(f(p, t), f(x, t)) < d/3$ for $-t_1 \leqq t \leqq t_1$. We may assume that $\delta < d/3$. Then the trajectory $f(x, t)$ through any point x in the δ-neighborhood of p does not remain in this neighborhood for $|t| = |t_1|$.

3.257. Consider next the periodic solutions of (1.01), i.e., the solutions $x_i(t)$ in which all the functions $x_i(t)$ are periodic with a common period T. The trajectory of a periodic solution $f(p, t)$ is a closed curve in the phase space and $f(p, t+T) = f(p, t)$.

Thus every point of a periodic solution is an α- as well as an ω-limit point and therefore such a solution is stable in the sense of Poisson.

3.258. It is easy to find examples of systems of differential equations whose solutions are receding, periodic, or are singular points. The problem of constructing asymptotic solutions and solutions which are non-periodic and stable in the sense of Poisson is somewhat more difficult.

We note that a solution which is not a singular point and which has a single α- or ω-limit point is asymptotic. For, as was shown in 3.255, this limit point is a singular point and the solution cannot reach a singular point (which itself is a solution) in finite time in view of the uniqueness condition. It is clear therefore that any system whose singularities include a saddle point or a nodal point will have asymptotic solutions. Consider for example the system

$$\frac{dx}{dt} = x, \quad \frac{dy}{dt} = y, \quad x = C_1 e^t, \quad y = C_2 e^t.$$

The point $x = 0$, $y = 0$ is a singular point. All other solutions are asymptotic for $t \to -\infty$ and recede for $t \to +\infty$.

We shall consider next the more complicated examples of asymptotic solutions whose ω-limit sets contain more than one point.

3.26. Limit cycles. Consider a system

$$\frac{dx}{dt} = P(x, y), \qquad \frac{dy}{dt} = Q(x, y). \tag{3.2601}$$

A periodic solution of (3.2601) is called a *limit cycle* if it is either the α- or the ω-limit set of another solution of this system. Let C be the closed trajectory of a limit cycle. If C is the ω-limit set for solutions contained in its interior, as well as for solutions lying in its exterior, then the limit cycle is called *stable*. If C is the α-limit set for trajectories in the interior and for those in the exterior of C, then the limit cycle is called *unstable*. If, however, C is the α-limit set for the trajectories in the interior (exterior), but is the ω-limit set of the trajectories in the exterior (interior) of C, then the limit cycle is called *semi-stable*.

3.261. *Example.* Given the system

$$\frac{dx}{dt} = -y + \frac{x}{\sqrt{x^2 + y^2}} (1 - (x^2 + y^2)),$$

$$\frac{dy}{dt} = x + \frac{y}{\sqrt{x^2 + y^2}} (1 - (x^2 + y^2)).$$

Passing to polar coordinates, we let $x = r \cos \theta$ and $y = r \sin \theta$. Then

$$\frac{dx}{dt} = -y + \frac{x}{r} (1 - r^2); \quad \frac{dy}{dt} = x + \frac{y}{r} (1 - r^2).$$

Multiplying the first of these equations by x and the second by y and adding, we obtain

$$\frac{dr}{dt} = 1 - r^2 \qquad (r > 0).$$

Next, multiplying the first equation by y, the second one by x, subtracting, and making use of the identity

$$x \frac{dy}{dt} - y \frac{dx}{dt} = r^2 \frac{d\theta}{dt},$$

we get

$$\frac{d\theta}{dt} = 1.$$

Integrating the equation

$$\frac{dr}{1 - r^2} = dt$$

we get

$$\log \left| \frac{1+r}{1-r} \right| = 2t + \log A, \quad \text{where} \quad A = \left| \frac{1+r_0}{1-r_0} \right|.$$

Thus

$$r = \frac{Ae^{2t} - 1}{Ae^{2t} + 1} \quad \text{for} \quad 0 < r < 1, \quad \text{and } r = \frac{Ae^{2t} + 1}{Ae^{2t} - 1} \quad \text{for} \quad r > 1.$$

We observe that in both cases $r \to 1$ as $t \to +\infty$. Consequently all the solutions outside the circle $r = 1$, as well as all those inside, are spirals approaching this circle. Therefore the periodic solution $x = \cos(\theta_0 + t)$, $y = \sin(\theta_0 + t)$ is a stable limit cycle.

3.262. *Example.* Given the system

$$\frac{dx}{dt} = -y + x(x^2 + y^2 - 1), \quad \frac{dy}{dt} = x + y(x^2 + y^2 - 1),$$

which in polar coordinates has the form

$$\frac{dr}{dt} = r(r^2 - 1), \quad \frac{d\theta}{dt} = 1 \quad (r \geqq 0).$$

Integrating these equations we get $\theta = \theta_0 + t$ and

$$r = 0, \quad r = \frac{1}{\sqrt{1 + Ae^{2t}}} \quad \text{for} \quad 0 < r_0 < 1, \quad A = (1 - r_0^2)/r_0^2;$$

$$r = 1, \quad r = \frac{1}{\sqrt{1 - Ae^{2t}}} \quad \text{for} \quad r_0 > 1, \quad A = (r_0^2 - 1)/r_0^2.$$

The parameter A is always positive and we see at once that for the solutions outside the circle $r = 1$, as well as for the solutions inside this circle, we have $r \to 1$ as $t \to -\infty$. Thus the circle $r = 1$ is the α-limit set of the solutions originating outside the circle as well as for those originating inside. We note that these latter spiral toward the origin $r = 0$ as $t \to +\infty$. Thus the solution $x = \cos(\theta_0 + t)$, $y = \sin(\theta_0 + t)$ is an unstable limit cycle and the solution $x = y = 0$ is a position of stable equilibrium.

3.263. *Example.* Consider the system

$$\frac{dx}{dt} = x(x^2 + y^2 - 1)^2 - y,$$

$$\frac{dy}{dt} = y(x^2 + y^2 - 1)^2 + x.$$

In polar coordinates this becomes

$$\frac{dr}{dt} = r(r^2 - 1)^2, \qquad \frac{d\theta}{dt} = 1. \tag{3.2631}$$

If we let $r^2 = u$, we get

$$\frac{du}{dt} = 2u(u - 1)^2,$$

or

$$\frac{du}{u(u - 1)^2} = 2dt.$$

In view of the identity

$$\frac{1}{u(u-1)^2} = \frac{1}{u} - \frac{1}{u-1} + \frac{1}{(u-1)^2},$$

we get

$$\log\left|\frac{u}{u-1}\right| - \frac{1}{u-1} = \log C + 2t,$$

or

$$\left|\frac{u}{u-1}\right| e^{-\frac{1}{u-1}} = Ce^{2t}.$$

Finally, setting $u - 1 = v$, we get

$$\left(\frac{1}{v} + 1\right) e^{-1/v} = Ce^{2t} \quad \text{for} \quad v > 0 \ (r > 1), \tag{3.2632}$$

and

$$\left(-1 - \frac{1}{v}\right) e^{-1/v} = Ce^{2t} \quad \text{for} \quad v < 0 \ (r < 1). \tag{3.2633}$$

Let us consider the behavior of solutions $v = v(C, t)$ of (3.2632) and (3.2633) in the neighborhood of $v = 0$. For t positive and sufficiently large (3.2633) has a unique solution $v(C, t) < 0$. Moreover, as $t \to \infty$, $v(C, t) \to 0$ and hence $r = \sqrt{v+1} \to 1$. For t negative and sufficiently large numerically (3.2632) has a unique solution $v(C, t) > 0$. This solution $v(C, t) \to 0$ as $t \to -\infty$, whence $r \to 1$ in this case as well. Thus in this example the solution $x = \cos(\theta_0 + t)$, $y = \sin(\theta_0 + t)$ is a semi-stable limit cycle.

3.264. *Example.* Let us consider the system

$$\text{(3.2641)} \quad \begin{cases} \dfrac{dx}{dt} = -y + (x^2 + y^2 - 1)x \sin \dfrac{1}{x^2 + y^2 - 1} \\ \dfrac{dy}{dt} = x + (x^2 + y^2 - 1)y \sin \dfrac{1}{x^2 + y^2 - 1} \\ \text{for } x^2 + y^2 \neq 1, \text{ and} \\ \dfrac{dx}{dt} = -y, \quad \dfrac{dy}{dt} = x \quad \text{for} \quad x^2 + y^2 = 1. \end{cases}$$

In polar coordinates this system takes the form

(3.2642)
$$\frac{dr}{dt} = r(r^2 - 1) \sin \frac{1}{r^2 - 1} \quad \text{for} \quad r \neq 1,$$
$$\frac{dr}{dt} = 0 \quad \text{for} \quad r = 1,$$

and in both cases $d\theta/dt = 1$.

Thus in every neighborhood of the periodic solution

(3.2643)
$$x = \cos(\theta_0 + t), \quad y = \sin(\theta_0 + t)$$

of (3.2641) there are infinitely many periodic solutions

(3.2644)
$$x = r_k \cos(\theta_0 + t), \quad y = r_k \sin(\theta_0 + t)$$

where $r_k = \sqrt{1 + (1/k\pi)}$ satisfies the condition $\sin(r_k^2 - 1)^{-1} = 0$. In each ring-shaped region between two consecutive circles (3.2644), the trajectories are spirals approaching these two circles. Thus every solution (3.2644) is a limit cycle.

3.265. We should now give some examples of nonperiodic solutions stable in the sense of Poisson.

In the next chapter we shall see that there exist no such solutions either in the plane or on the surface of a two-dimensional cylinder. However, there do exist such solutions on a torus.

Let us introduce real Cartesian coordinates (φ, ϑ) in the plane and let us identify any two points (φ, ϑ) and $(\varphi + n, \vartheta + m)$ whose coordinates differ by integers n and m respectively.

On the resulting torus consider the system

(3.265)
$$\frac{d\varphi}{dt} = 1, \quad \frac{d\vartheta}{dt} = \alpha.$$

Whenever we are interested only in the geometrical arrangement of integral curves, we may consider the one equation $d\vartheta/d\varphi = \alpha$. There are two essentially different cases: one in which $\alpha = p/q$ is a rational number and the other in which α is irrational.

3.266. *Example.* Consider the integral curves of the equation

(3.2661)
$$\frac{d\vartheta}{d\varphi} = \frac{p}{q}$$

where q is a natural number, p is an integer, and the fraction p/q is irreducible. The solution corresponding to the initial conditions

$\varphi = 0$, $\vartheta = \vartheta_0$ has the form

$$\vartheta = \vartheta_0 + \frac{p}{q}\varphi. \tag{3.2662}$$

As φ takes on the value q, the coordinate ϑ in (3.2662) takes the value $\vartheta_0 + p$, the resulting point of our integral curve on the torus coincides with the initial point $(0, \vartheta_0)$, and the curve is closed. Thus the torus is covered by closed integral curves of (3.2661).

3.267. *Example.* We consider next the equation

$$\frac{d\vartheta}{d\varphi} = \alpha \tag{3.2671}$$

where α is an irrational number.

In this case there are no closed curves among the integral curves

$$\vartheta = \vartheta_0 + \alpha\varphi \tag{3.2672}$$

of (3.2671). For, suppose that a point (φ_1, ϑ_1) on the integral curve (3.2672) coincides with the initial point $(0, \vartheta_0)$. Then

$$\vartheta_1 = \vartheta_0 + \alpha\varphi_1 = \vartheta_0 + n\alpha = \vartheta_0 + m$$

(m, n integers), whence $n\alpha = m$, and $\alpha = m/n$ is a rational number.

Since all the trajectories can be obtained from the trajectory $\vartheta = \alpha\varphi$ by a translation along the ϑ axis, we need to consider only this trajectory in detail. Its intersections with the meridian $\varphi = 0$ are $\varphi = 0$, $\vartheta_n = n\alpha$, $n = 0, \pm 1, \pm 2, \ldots,$. *These points are everywhere dense in this meridian.*

Write $(\alpha) = \alpha - [\alpha]$, where $[\alpha]$ is the greatest integer in α. To prove our assertion we need only to show that the set $(n\alpha)$, $n = 0, 1, 2, \ldots,$ is everywhere dense in the interval $[0, 1]$. Since α is irrational, the $p + 1$ numbers

$$0, (\alpha), \ldots, (p\alpha) \tag{3.2673}$$

are all distinct and since they are all distributed among the p intervals

$$I_h : \frac{h}{p} \leqq \vartheta < \frac{h+1}{p} \qquad (h = 0, 1, \ldots, p-1), \tag{3.2674}$$

one of these intervals must contain at least two of the numbers (3.2673). Let $(k_1\alpha)$ and $(k_2\alpha)$ be two such numbers. They differ by less than $1/p$ since each of the intervals I_h is of length $1/p$.

If $k_2 > k_1$, we write $k = k_2 - k_1$. Then either

$$(k\alpha) \in I_0 \text{ or } (k\alpha) \in I_{p-1}.$$

In either case, the sequence

$$(k\alpha),\ (2k\alpha),\ (3k\alpha), \ldots,$$

continued as long as may be necessary, will partition the interval $[0, 1]$ into segments of length less than $1/p$.

To show that every ε-neighborhood of a point in $[0, 1]$, contains a point of the set $(n\alpha)$, it suffices to take $p > 1/\varepsilon$ in the above discussion.

Thus the set $(n\alpha)$ is everywhere dense in $[0, 1]$, and therefore every point of the meridian $\varphi = 0$ is a limit point for the set of points $\varphi = n$, $\vartheta = n\alpha$ of our trajectory. Similarly, every point $\varphi = \varphi_0$, $\vartheta = \vartheta_0$, is a limit point for the set of points

$$\varphi = n + \varphi_0, \quad \vartheta = \alpha(n + \varphi_0)$$

of the same trajectory.

It follows that the trajectory $\vartheta = \alpha\varphi$ and hence every trajectory of (3.2671) is everywhere dense on the torus. *In particular, every trajectory, even though it is not closed, contains some of its ω-limit points.*

3.268. *Example.* Consider the system

$$\frac{d\varphi}{dt} = (\varphi^2 + \vartheta^2), \qquad \frac{d\vartheta}{dt} = \alpha(\varphi^2 + \vartheta^2). \tag{3.268}$$

Trajectories of this system lie on the trajectories of the system (3.265). However, system (3.268) has a singular point at $\varphi = 0$, $\vartheta = 0$. This singular point splits the trajectory of (3.265) through the origin (stable in the sense of Poisson) into three trajectories of (3.268), viz., the singular point $(0, 0)$, and two other trajectories each of which is asymptotic in one direction and stable according to Poisson in the other.

3.27. The qualitative theory of differential equations whose right-hand members do not contain time explicitly, concerns itself with the solution of the following two problems.

3.271. The classification of solutions and the study of relationships between different classes of solutions. This problem is essentially solved and the results of such investigations will be presented in the following chapters.

3.272. The search for methods of determining the types of solutions admitted by a given system of differential equations on the basis of information supplied by the analytic properties of the right-hand members of this system. This problem is far from being completely solved. The reader will find the basic known results in the subsequent chapters of our book.

4.1. Families of Integral Curves

We consider now a family S of integral curves filling either a region G or a closed region $\overline{G}$ in R^n.

4.11. Definition. *A family S of trajectories filling a domain G* (not necessarily open) *in R^n, is called a regular family* (a notion due to Hassler Whitney [58]) *if there exists a homeomorphism (one to one and bicontinuous mapping) of the domain G onto a set $E \subset R^n$ or R^{n+1}, which maps trajectories into parallel straight lines so that the images of different integral curves lie on different straight lines.*

It is clear that a regular family of trajectories cannot contain trajectories which are either stable in the sense of Poisson or are asymptotic. On the other hand, there exist dynamical systems whose integral curves recede in both directions but whose families of trajectories are, nevertheless, not regular. Consider, for example, the system

$$\frac{dx}{dt} = \sin y, \quad \frac{dy}{dt} = \cos^2 y.$$

The integral curves of this system are the curves $x + c = (\cos y)^{-1}$ and the straight lines $y = k\pi + \pi/2$, $k = 0, \pm 1, \ldots$. We consider only the strip

$$\overline{G} : -\frac{\pi}{2} \leqq y \leqq \frac{\pi}{2} \cdot$$

Although all the integral curves situated within this strip recede in both directions (cf. Fig. 1), the family of integral curves filling this strip is not regular.

To prove this, draw a segment PQ with the endpoints P and Q on the lines $y = -\pi/2$ and $+\pi/2$ respectively, and consider a sequence of points P_n on this segment, converging to P. Write L_n for the trajectory passing through P_n, and L and L' respectively for the lower and the upper boundaries of the strip. Assume that our family of trajectories is regular, and let f be a homeomorphism of Definition 4.11. Then the sequence $f(P_n) \in f(L_n)$ converges to the point $f(P) \in f(L)$. Moreover, since $f(L_n)$ and $f(L)$ are parallel straight lines, any convergent sequence of points $y_n \in f(L_n)$ has its limit point of $f(L)$. To obtain a contradiction we observe that if $\{Q_n\}$ is a sequence such that $Q_n \in L_n$ and $Q_n \to Q \in L'$, then $y_n = f(Q_n) \in f(L_n)$ and $f(Q_n) \to f(Q) \in f(L')$. Thus $f(Q)$ must lie on $f(L)$ as well as on $f(L')$. This is a contradiction.

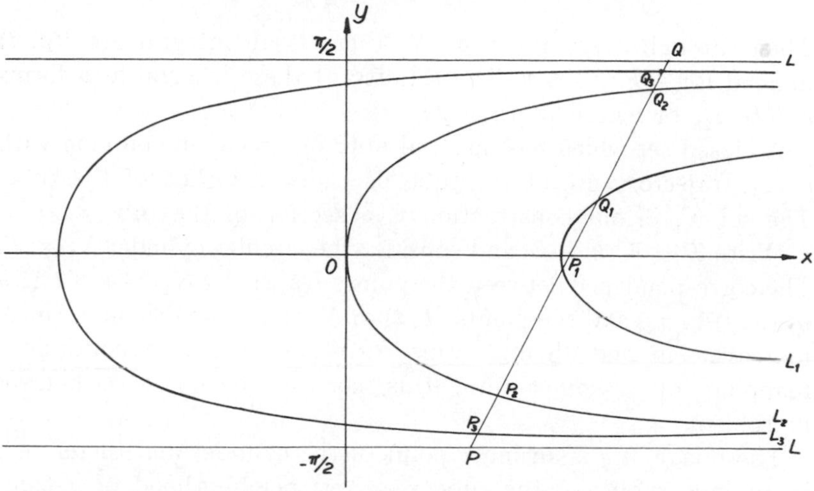

Fig. 1

The following theorem elucidates the part played by regular families in the theory of differential equations.

4.12. THEOREM. *Let G be a domain in which the system* (1.01) *satisfies both the uniqueness and the existence conditions* (*cf.* *Sections* 1 *and* 2) *and let q be a non-singular point of G. Then there exists a neighborhood of q such that the family of integral curves filling this neighborhood forms a regular family.*

Since q is not a singular point, then at q the integral curve L passing through q has a well-defined tangent and hence a well-defined normal hyperplane N as well.

By the continuity of the right-hand members of (1.01), there exists a closed spherical neighborhood $S_0(q, R) \subset G$ with center at q and of radius R, such that the directions of tangent vectors to integral curves at any point inside or on the boundary of the sphere S_0 deviate from the direction of the tangent vector at q by less than $\pi/4$.

Consider the closed $(n-1)$-sphere $N_1(q, R) = N \cap S_0$. Through every point $p \in N_1$ there passes a solution $f(p, t)$ defined for $|t| \leqq h_p$. Since N_1 is closed, then for sufficiently small R, $0 < h_0 = g.l.b\ h_p$ in view of Theorem 1.11. Also, the periods of the periodic solutions passing through N_1 (if there are such) have a lower bound t_0, provided R is chosen small enough. Write

$$h = \min\left(h_0, \frac{t_0}{2}\right).$$

Then through every point $p \in N_1$ there is an integral arc $f(p, t)$ defined for $-h \leqq t \leqq h$. The totality of these integral arcs forms a *tube* τ_{2h} of length of time $2h$.

A closed set which has one and only one point in common with every trajectory arc of the tube, is called a *section* of the tube. The set N_1 in our construction is a section of the tube τ_{2h}.

Write $T = [-h, +h]$ and consider the circular cylinder $N_1 \times T$. The correspondence between the points $f(p, t)$, $p \in N_1$, $-h \leqq t \leqq h$ of the tube τ_{2h} and the points (p, t) of $N_1 \times T$ is one-to-one. Since, moreover, in one direction this correspondence is a continuous mapping of a compactum, this correspondence is a homeomorphism.

The image of q is an inner point of the cylinder and therefore q is an inner point of the tube τ_{2h}. Any neighborhood of q completely contained in τ_{2h} will serve as the desired neighborhood.

4.2 We shall discuss now conditions under which a system of differential equations will define a regular dynamical system (i.e. with a regular family of trajectories).

4.21. THEOREM. (E. A. Barbashin [3]). *Consider a system of differential equations* (1.01) *which defines a dynamical system in a domain G. If there exists a single-valued function* $u(x_1, \ldots, x_n)$ *satisfying the condition*

$$\sum_{i=1}^{n} \frac{\partial u}{\partial x_i} f_i = 1 \tag{4.211}$$

in G, then our dynamical system is regular.

Let $u(p) = u(x_1, \ldots, x_n)$ be a single-valued function defined in a domain G, having in G derivatives of the first order, and satisfying the condition (4.211). If $f(p, t) = (x_1(t), \ldots, x_n(t))$ is a trajectory of a dynamical system defined by (1.01), then

$$\sum_{i=1}^{n} \frac{\partial u(x_1(t), \ldots, x_n(t))}{\partial x_i} f_i(x_1(t), \ldots, x_n(t)) = 1,$$

or

$$1 = \sum_{i=1}^{n} \frac{\partial u(x_1(t), \ldots, x_n(t))}{\partial x_i} \cdot \frac{dx_i}{dt} = \frac{du(t)}{dt}$$

where

$$u(t) = u(f(p, t)) = u(x_1(t), \ldots, x_n(t)).$$

Integrating, we get $u(t) = u(0) + t$, or

$$u(f(p, t)) = u(p) + t. \tag{4.212}$$

Let F be the set of all points q for which $u(q) = 0$. It follows from (4.212) that every trajectory has one and only one point in common with F.

Consider the topological product Z of the set F and the real axis T.[6]

If $p \in G$, then by (4.212), $f(p, -u(p)) = q \in F$. The mapping ψ defined by $\psi(p) = (q, t_p)$, where $t_p = -u(p)$ is a one-to-one mapping (since $u(p)$ is single-valued) of G onto Z, and maps trajectories into parallel straight lines in $Z \subset R^{n+1}$. We shall show next that both ψ and ψ^{-1} are continuous.

Let a sequence of points $p_1, p_2, \ldots, p_k, \ldots$ converge to a point p_0. From the continuity of $u(p)$ it follows that $u(p_k)$ tends to $u(p_0)$ and from the continuity of $f(p, t)$ it follows that

$$f(p_k, -u(p_k)) = q_k \to f(p_0, -u(p_0)) = q_0.$$

Thus,

$$\psi(p_k) \to \psi(p_0).$$

Conversely, if a sequence $(q_k, t^{(k)})$ converges, say, to $(q, \bar{t})$, then $q_k \to q$ and $t^{(k)} \to \bar{t}$. Thus the initial points q_k of the integral arcs

$$f(q_k, t), \qquad 0 \leqq t \leqq t^{(k)} \tag{4.213}$$

tend to the initial point q of the integral arc

$$f(q, t), \qquad 0 \leqq t \leqq \bar{t}, \tag{4.214}$$

and the time intervals $t^{(k)}$ tend to $\bar{t}$. By the continuity of $f(p, t)$,

[6]The set Z consists of the points of an $(n + 1)$-dimensional space, situated on parallel straight lines passing through the points of the set F.

the endpoints

$$p_k = \psi^{-1}((q_k, t^{(k)})) = f(q_k, t^{(k)}),$$

of the arcs (4.213) tend to the endpoint

$$p = \psi^{-1}(q, \bar{t}) = f(q, \bar{t})$$

of the arc (4.214).

Theorem 4.21 yields important corollaries.

4.22 *If solutions of* (1.02) *are defined for* $-\infty < t < +\infty$, *then the associated parametric system* (1.201) *is regular.*

The conditions (4.211) corresponding to the parametric system (1.201) will read

$$\sum_{i=1}^{n} \frac{\partial u}{\partial x_i} f_i + \frac{\partial u}{\partial t} = 1.$$

It will be satisfied by the single-valued and continuous function $u = t$.

4.23. *The conclusions of Theorem* 4.21 *still hold if we replace condition* (4.211) *by*

$$N = \sum_{i=1}^{n} \frac{\partial u}{\partial x_i} f_i \geqq K^2 > 0. \tag{4.231}$$

Let u satisfy the condition (4.231). Make the substitution

$$t' = \int_0^t N\, dt.$$

If $|t| \to \infty$, then $|t'| \to \infty$. The new system

$$\frac{dx_i}{dt'} = \frac{1}{N} f_i \tag{4.232}$$

is equivalent to the original system and also defines a dynamical system. The condition (4.211) for (4.232) becomes

$$\sum_{i=1}^{n} \frac{\partial u}{\partial x_i} \frac{1}{N} f_i = 1. \tag{4.233}$$

It is clear that u satisfies (4.233).

4.24. A system of the form

$$\frac{dx_i}{dt} = \frac{\partial F}{\partial x_i}. \tag{4.241}$$

is said to possess a velocity potential $F(x_1, \ldots, x_n)$. Corollary 4.23 yields

4.241. *If*

$$\sum_{i=1}^{n} \left(\frac{\partial F}{\partial x_i}\right)^2 \geqq K^2 > 0,$$

then the system (4.241) *is regular*

5.1. Fields of Linear Elements

Consider again a system of type (1.01). Such a system assigns a vector $(f_1, f_2, \ldots, f_n)$ to every point $p(x_1, \ldots, x_n)$ at which all the functions $f_i(x_1, \ldots, x_n)$ are defined and at which they do not all vanish. In a domain G in which f_i are all continuous, system (1.01) defines a vector field continuous except at the singular points. It may sometimes happen that we may augment the definition of our vector field so that it will become continuous everywhere. More precisely, we may sometimes find a function $\psi(x_1, \ldots, x_n)$ continuous everywhere except possibly at the singular points and such that the functions $f_i\psi$ are continuous everywhere and do not vanish simultaneously.

5.11. The theory of differential equations also studies systems in the so-called *symmetric* form:

$$\text{(5.11)} \qquad \frac{dx_1}{X_1(x_1, \ldots, x_n)} = \frac{dx_2}{X_2(x_1, \ldots, x_n)} = \ldots = \frac{dx_n}{X_n(x_1, \ldots, x_n)}.$$

We note that system (1.01) assigns to each point p a *vector* $(f_1, \ldots, f_n)$ whereas system (5.11) assigns to each point a *linear element* (line position) $dx_1 : dx_2 : \ldots : dx_n = X_1 : X_2 : \ldots : X_n$. This linear element is associated with two vectors, the vector $(X_1, \ldots, X_n)$ and the vector $(-X_1, \ldots, -X_n)$.

5.12. We now ask if there exists a system of type (1.01) whose trajectories are the integral curves of our initial system (5.11) and which has no singular points other than those of (5.11).

To state the problem geometrically, we ask if it is possible to choose a positive sense on each linear element and a suitable value for vector length so that the resulting vector field should be continuous everywhere except at the singular points.

Analytically, the problem consists in finding a function $\psi(x_1, \ldots, x_n)$

such that the products $X_1\psi, X_2\psi, \ldots, X_n\psi$ are continuous in D_1 and do not vanish simultaneously anywhere in D_1, if D_1 is a domain where (5.11) has no singular points.

5.13. We note that the problem of *orientation* of a field of linear elements defined in a domain D_1 is equivalent to the problem of establishing a positive direction along each integral curve of a system of type (5.11) in such a way that every two integral curves which are near each other must agree in direction.

5.14. *It is not always possible to orient a field of elements in the plane.* This can be seen from the following example.

Consider the field of linear elements in the plane defined by the differential equation

$$\frac{dy}{dx} = \cot \frac{\varphi}{2} \tag{5.141}$$

where φ is the polar angle. As is usual, in a neighborhood of a point near which the absolute value of the right-hand member is not bounded, we consider the equation

$$\frac{dx}{dy} = \tan \frac{\varphi}{2}.$$

It is clear that the field of linear elements is defined and is continuous everywhere except at the point (0, 0). Introducing polar coordinates, we obtain

$$\cos \frac{3\varphi}{2}\, dr = r \sin \frac{3\varphi}{2}\, d\varphi.$$

Solving this equation, we obtain three integral half-lines

$$\varphi = \frac{\pi}{3}, \quad \varphi = \pi, \quad \varphi = \frac{5\pi}{3}, \qquad r > 0,$$

and three families of similar curves (Fig. 2)

$$r = \frac{a}{\left(\cos\dfrac{3\varphi}{2}\right)^{2/3}}$$

with the parameter a and

$$\text{(I)}\quad -\frac{\pi}{3} < \varphi < \frac{\pi}{3}, \qquad \text{(II)}\quad \frac{\pi}{3} < \varphi < \pi, \qquad \text{(III)}\quad \pi < \varphi < \frac{5\pi}{3}.$$

The above field cannot be oriented. For, let us choose the direction away from the origin on the half-line $\varphi = \pi/3$, $r > 0$ as positive. Take a point p on this half-line and draw a circle through p with the center at $(0, 0)$. As we move along the circumference, say in the counterclockwise direction, considerations of continuity will

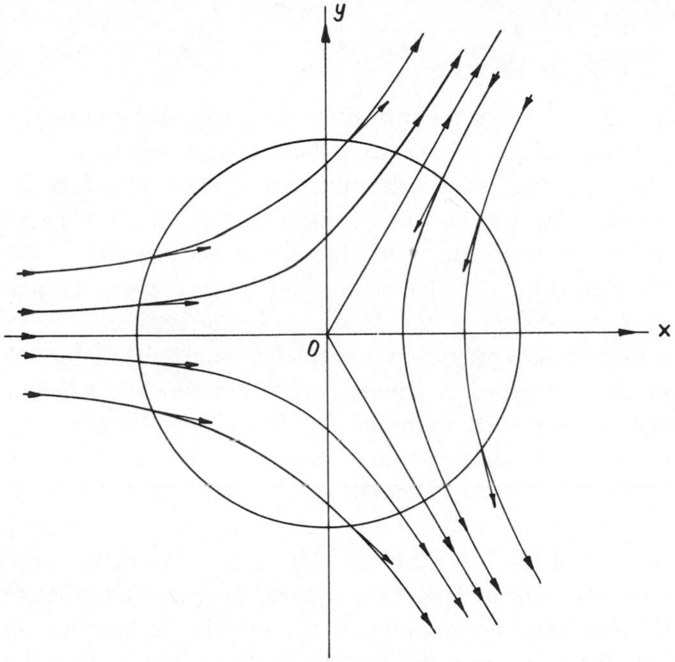

Fig. 2

assign as positive the direction toward the origin on the halfline $\varphi = \pi$, $r > 0$ and the direction away from the origin on the halfline $\varphi = 5\pi/3$, $r > 0$. We find that in view of the orientation induced in the family (III) by considerations of continuity as we move along the circumference, we shall arrive at the initial point with orientation opposite to that chosen originally.

5.141. The selected positive direction for the linear element through a point q on the circle in Fig. 2 may be indicated by a tangent unit vector $v(q)$. The angle ($2n\pi$ or $2n\pi + \pi$) through which v rotates as q spans the circle once in the positive sense, is a property of our field of directions. The rotation angle for the circle in our example is $-\pi$. In this discussion the circle could be replaced by any other simply-closed curve containing the origin.

5.142. We observe that if we write system (5.141) in the form

$$\frac{dy}{dx} = \frac{\sin \varphi}{1 - \cos \varphi} = \frac{r \sin \varphi}{r(1 - \cos \varphi)} = \frac{y}{\sqrt{x^2 + y^2} - x}$$

and then replace it by the system

$$\frac{dx}{dt} = \sqrt{x^2 + y^2} - x \qquad \frac{dy}{dt} = y$$

of type (1.01), then we introduce new singular points (points of equilibrium) filling the positive half of the x-axis.

5.143. In the above example the *domain* D (cf. 5.13) *is not simply-connected* (it does not contain the point $(0, 0)$). As will be seen from the next theorem, this is not accidental.

5.15. THEOREM. *A continuous field of linear elements in a simply-connected domain* D *of the plane can be oriented.*

The domain D can be approximated from within by a bounded domain D_1 composed of squares, and therefore it suffices to prove our theorem for such domains D_1. Let us choose the sides of our squares so small that within each one of them the oscillation of the direction of linear elements is less than $\pi/4$. If we choose a positive direction on one of the elements, then considerations of continuity will lead to a unique definition of positive direction for all other elements of the same square. To extend the definition to the element through a point p outside the square we again use considerations of continuity and proceed stepwise along a chain of adjacent squares until we reach p.

The positive direction for the linear element through p is defined uniquely, for if we proceed toward p along two different paths, which yield different orientations at p, then the closed path defines a reversal of the direction of the vector field. Indeed in this case our vector field turns through an angle of $\pi + 2k\pi$ (k is an integer) around some closed curve consisting of the edges of the squares; however the algebraic sum of the rotation angle around each of the squares in the interior of our path must equal the rotation around the outside path, and this is zero modulo 2π. This is a contradiction and the theorem is proved.

CHAPTER II

Integral Curves of a System of Two Differential Equations

1. General Properties of Integral Curves in the Plane.

1.1 Consider the system

$$\frac{dx}{dt} = P(x, y), \qquad \frac{dy}{dt} = Q(x, y). \tag{1.01}$$

Let the functions $P(x, y)$ and $Q(x, y)$ satisfy Lipschitz conditions in some domain G_L (a Lipschitz domain) of the plane. We shall study the behavior of the integral curves of this system. In view of I, 3.2 we may assume without loss of generality that the system (1.01) defines a dynamical system in G_L.

The basic results in this case were obtained by Bendixson [7] and Poincaré [47].

We make use of the continuity of the vector field $[P(x, y), Q(x, y)]$ defined by the right-hand members of (1.01), through the following basic lemmas.

1.11. Lemma. *If a point P_0 of G_L is not a singular point, then there exists an $\varepsilon > 0$ such that the circle $S(P_0, \varepsilon)$ with center at P_0 and of radius ε does not contain singular points either on its boundary or in its interior, and such that the angle between the vector of the field $[P(x, y), Q(x, y)]$ at P_0 and the vector of the field at an arbitrary point of the circle, $S(P_0, \varepsilon)$ is less than $\pi/4$.*

In what follows we shall speak of such a circle as a *small neighborhood of P_0*.

We write $f^+(Q_0)$ for the half-trajectory (semi-trajectory) $f(Q_0, t)$, $0 \leqq t < +\infty$ and $f^-(Q_0)$ for the half-trajectory $f(Q_0, t)$, $-\infty < t \leqq 0$.

1.12. Lemma. *Let $S(P_0, \varepsilon)$ be a small neighborhood of a point P_0. Let N and N' be the points of intersection of the circumference of $S(P_0, \varepsilon)$ with the normal at P_0 to the trajectory through P_0. Then* *there exists a positive* $\delta < \varepsilon$ *such that for every point* Q_0 *of* $S(P_0, \delta)$, *either the half-trajectory* $f^+(Q_0)$ *or the half-trajectory* $f^-(Q_0)$ *cuts across the segment* NN' *of the normal, before leaving* $S(P_0, \varepsilon)$.

This lemma is an immediate consequence of the theorem on local regularity of the family of integral curves in a neighborhood of a nonsingular point (cf. I.4.12).

Lemma 1.12 can be proved directly as well. Let $Q_0 \in S(P_0, \varepsilon)$. Then the trajectory through Q_0 lies within two vertical right angles whose common vertex is Q_0 and whose bisector is parallel to the

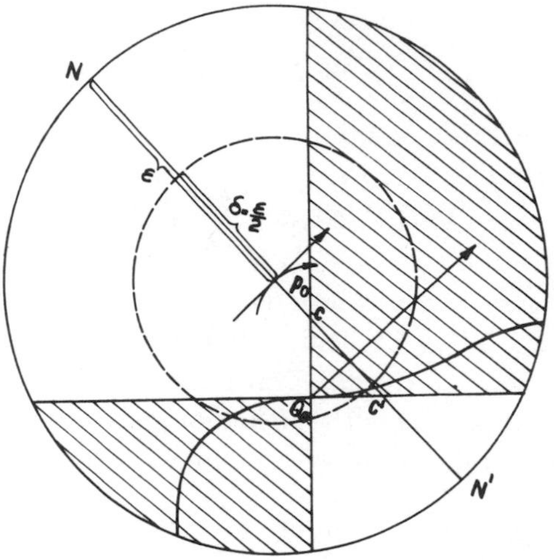

Fig. 3

tangent at P_0 of the trajectory through P_0 (cf. Fig. 3). Let $\delta = \varepsilon/2$ and let $Q_0 \in S(P_0, \delta)$. Then the points C, C' of the intersection of the sides of one of the above vertical angles and the normal, lie on the segment NN' of this normal. The half-trajectory which lies within this angle cuts across the segment NN' before leaving $S(P_0, \varepsilon)$.

We note that both half-trajectories $f^+(Q_0)$ and $f^-(Q_0)$ do leave the small neighborhood $S(P_0, \varepsilon)$. For suppose that $f^+(Q_0)$, say, remains in $S(P_0, \varepsilon)$. Then its ω-limit set contains exactly one point which must therefore be a singular point, which contradicts our hypothesis regarding $S(P_0, \varepsilon)$. The same conclusion may be arrived at by observing that since the velocity vector $V(p)$ defined by the right-hand members of (1.01), does not vanish in $\bar{S}(P_0, \varepsilon)$, then $||V(p)|| \geqq \mu > 0$ for $p \in S(P_0, \varepsilon)$. Moreover, the component $V_\tau(p)$

of $V(p)$ along the tangent at P_0, does not change its direction and $|V_\tau| \geqq \mu/2\sqrt{2}$. Therefore for $|t| > 2\varepsilon\sqrt{2}/\mu$, $f(P_0, t)$ lies outside of $S(P_0, \varepsilon)$.

1.13. The normal NP_0N' divides the small neighborhood $S(P_0, \varepsilon)$ of P_0 into two parts, D_1 and D_2. Suppose that the trajectory through P_0 cuts across the segment NN' from D_1 to D_2 with increasing t. We shall speak of D_1 and D_2 as the negative and the positive sides of the segment NN' respectively. *With increasing t all the trajectories cutting across NN' pass from the negative to the positive side of NN'.*

1.2. THEOREM. *Every trajectory of* (1.01) *possessing at least one-sided stability in the sense of Poisson, is either a singular point or a periodic solution.*

Let $f(A, t)$ be a trajectory, not a singular point, stable in the sense of Poisson for, say, $t \geqq 0$, and let P_0 be an ω-limit point of $f(A, t)$ lying on this trajectory. Since P_0 is not a singular point, it has a small (cf. 1.11) neighborhood $S(P_0, \varepsilon)$. Choose δ as in Lemma 1.12. Since P_0 is an ω-limit point of $f(A, t)$, every half-trajectory $f(P_0, t)$, $t \geqq t_0 > 0$ enters $S(P_0, \delta)$ (reenters—if t_0 is sufficiently large) and therefore, by the choice of δ, it cuts the segment NN'. Let P_1 be the first such intersection following P_0 on $f(P_0, t)$.

If $P_1 = P_0$ then the solution $f(A, t)$ is periodic.

Suppose $P_1 \neq P_0$. We shall show that this contradicts the hypothesis that P_0 is an ω-limit point.

If $P_1 \neq P_0$, only the two arrangements indicated in Fig. 4 are possible. Denote by $\overline{G}_1$ the closed domain bounded by the arc P_0P_1 of our trajectory and by the segment P_1P_0 of the normal. The arrangement in Fig. 4 (a) implies that the trajectory $f(P_1, t)$ remains in $\overline{G}_1$ for all $t > 0$. Moreover, if we take a small neighborhood $S(P_0, \varepsilon_1)$ not containing P_1, then $f^+(P_1)$ cannot enter the corresponding δ_1-neighborhood of the point P_0. For if $f^+(P_1)$ should enter the δ_1-neighborhood of P_0, then by Lemma 1.12 it would have to cut across the normal segment P_0P_1 from the positive to the negative side of NN' (cf. 1.13), which is impossible. We dispose of the alternative in Fig. 4(b) in a similar manner.

1.3. Assume that a Lipschitz domain $\overline{G}_L$ is bounded, closed, and *contains no singular points*. Then, in view of the uniform continuity of $P(x, y)$ and $Q(x, y)$, there exists $\varrho_0 > 0$ such that

$\varrho(P_1, P_2) \leqq \varrho_0$ for $P_1P_2 \epsilon \bar{G}_L$, implies that the angle formed by the vectors of the field at P_1 and P_2, is less than $\pi/4$.

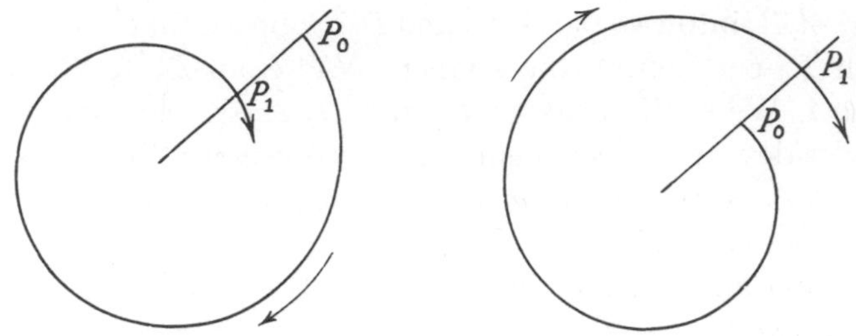

Fig. 4

Let L_{ab} be the arc $a \leqq t \leqq b$ of a (nonsingular) trajectory $f(P_0, t)$, and let $(L_{ab})_\varepsilon$ be an ε-neighborhood of L_{ab}. Since the set of singular points is closed, ε may be chosen so small that $(L_{ab})_\varepsilon$ contains no singular points *either* in its interior or on its boundary. Taking $(L_{ab})_\varepsilon$ as the domain $\bar{G}_L$ we choose $\varepsilon_1 \leqq \varrho_0$. If NP_1N' is a normal to L_{ab} at $P_1 = f(P_0, t_1)$, $a \leqq t_1 \leqq b$, then all the trajectories cutting across NP_1N' within $(L_{ab})_{\varepsilon_1}$ *agree in direction* (cf. 1.13) with L_{ab} at P_1.

We shall speak of such an ε_1-neighborhood of L_{ab} as a small neighborhood of L_{ab} (cf. 1.11).

1.41. THEOREM. *Let L_1 be a closed nonsingular trajectory of a dynamical system in the plane. Then for every $\varepsilon > 0$ there exists a $\delta > 0$ such that for every point P_1 in the δ-neighborhood of L_1, at least one of the half-trajectories $f^+(P_1)$ or $f^-(P_1)$ is contained in the ε-neighborhood of L_1.*

Let T be the period of L_1. Then $L_1 = L_{0T} = \{f(P_0, t)\}$, $0 \leqq t \leqq T$, $P_0 \epsilon L_1$, and the discussion in 1.3 assures the existence of a small neighborhood $(L_1)_{\varepsilon_2}$ of L_1. We may assume that $\varepsilon_2 \leqq \varepsilon$. Also, $S(P_0, \varepsilon_2)$ is a small neighborhood of P_0. Let $S(P_0, \delta_2)$ be a corresponding $S(P_0, \delta)$-neighborhood as in 1.12. Next, let $\gamma > 0$ be such that for every $P \epsilon S(P_0, \gamma)$ we have

$$\varrho(f(P_0, t), \ f(P, t)) < \delta_2 \quad \text{for} \quad 0 \leqq t \leqq T. \tag{1.4}$$

Take $P_1 \epsilon S(P_0, \gamma)$ on the normal NP_0N'. Since $f(P_0, T) = P_0$, then $f(P_1, T) \epsilon S(P_0, \delta_2)$, and the trajectory $f(f(P_1, T), t)$ cuts the normal segment NN' before leaving $S(P_0, \varepsilon_2)$. Let $P_2 = f(P_1, T_1)$, $T_1 \neq 0$, be the first such intersection following P_1.

Let C be the closed curve formed by the arc $f(P_1, t)$, $0 \leqq t \leqq T_1$ and by the segment P_2P_1 of the normal to L_1 at P_0. The curve C together with L_1 form a ring-shaped region Γ contained in the ε_2-neighborhood $(L_1)_{\varepsilon_2}$ of L_1.

If $P_2 = P_1$, both half-trajectories $f^+(Q_0)$ and $f^-(Q_0)$ lie in $(L_1)_{\varepsilon_2} \subset (L_1)_\varepsilon$ for every point Q_0 in the ring-shaped region Γ bounded by L_1 and $f(P_1, t)$.

Let $P_2 \neq P_1$. We distinguish two arrangements on the normal at P_0, viz., (1) $P_0P_2P_1$ and (2) $P_0P_1P_2$.

In the first case, for every $Q_0 \in \Gamma$, the positive half-trajectory $f^+(Q_0)$ cannot leave the region Γ, and in the second case, the negative half-trajectory $f^-(Q_0)$, $Q_0 \in \Gamma$, cannot leave Γ.

Let d be the distance between the arcs $f(P_0, t)$, $0 \leqq t \leqq T$, and $f(P_1, t)$, $0 \leqq t \leqq T_1$, in case $P_1 \neq P_0$ lies on NP_0 and let d' be the distance between these arcs for a choice of $P_1 \neq P_0$ on $N'P_0$. Let the ring-shaped regions corresponding to these choices of P_1, be Γ and Γ'. Then both d and d' are positive. Take $0 < \delta < \min(d, d')$. Then $(L_1)_\delta \subset \Gamma \cup \Gamma' \subset (L_1)_{\varepsilon_2} \subset (L_1)_\varepsilon$, and this δ fulfills the requirements of our theorem.

1.411. The above theorem implies that integral curves in the plane cannot approach a periodic solution arbitrarily close and then recede both for $t \to +\infty$ and for $t \to -\infty$. However, this may occur in R^3, for example.

1.42. Definition. *We shall say that a half-trajectory $f^+(Q_0)$ $[f^-(Q_0)]$ approaches a trajectory Λ spirally if for any point $P_0 \in \Lambda$ and an arbitrarily small segment $P_1P_0P_2$ of the normal to Λ at the point P_0, there exists t_0 such that the half-trajectory $f(Q_0, t)$, $t > t_0$ $[t < t_0]$ intersects $P_1P_0P_2$ infinitely many times in such a way that either all the points of intersection lie on P_1P_0 or all of them lie on P_0P_2.*

A trajectory L is said to approach a trajectory Λ spirally if a half-trajectory of L approaches Λ spirally.

1.43. Theorem. *If a closed integral curve L is contained in the ω-limit set Ω of some trajectory $f(P_0, t)$, then $\Omega = L$ and, if $f(P_0, t)$ is not closed, the half-trajectory $f^+(P_0)$ approaches L spirally.*

Take a point $Q_0 \in L$. Since $L \subset \Omega$, there exists a sequence $P_j = f(P_0, t_j)$, $j = 1, 2, \ldots$, such that $t_j \to +\infty$ and $P_j \to Q_0$. Consider a sequence $\varepsilon_j > 0$, such that $P_{j-1} \notin (L)_{\varepsilon_j}$. By Theorem

1.41, for each ε_j there exists a δ_j such that for $P_{m_j} \in (L)_{\delta_j}$ either $f^-(P_{m_j})$ or $f^+(P_{m_j})$ lies in $(L)_{\varepsilon_j}$. We may take $m_1 < m_2 < \ldots$. Since $P_{j-1} \notin (L)_{\varepsilon_j}$ we have $f^+(P_{m_j}) \subset (L)_{\varepsilon_j}$. We note that $\varepsilon_j \to 0$. Since, by the above, the half-trajectory $f^+(P_0)$ remains outside $(L)_{\varepsilon_j}$ for only a finite duration $0 \leqq t \leqq T_{\varepsilon_j} < t_{m_j}$, we have $\Omega \subset L$ and hence $\Omega = L$. Also, there exists a j_0 such that every neighborhood $(L)_{\varepsilon_j}$, $j \geqq j_0$ is a small neighborhood of L. Hence L is approached spirally by $f^+(P_0)$.

1.44. The mode of approach to a nonsingular ω-limit point. Let $P_0 \in \Omega(f^+(Q_0))$ be a nonsingular point. Let $S(P_0, \varepsilon)$ be a small neighborhood of P_0 (cf. 1.11 and 1.12). Since P_0 is an ω-limit point of $f^+(Q_0)$, there exists arbitrarily near P_0 a point $Q_1 \in S(P_0, \varepsilon)$ of intersection of this half-trajectory and the segment NN' (cf. 1.11) of the normal to $f(P_0, t)$ at P_0.

1.441. *If Q_1 lies on NP_0, then the successive intersections $Q_1Q_2Q_3, \ldots$ of $f^+(Q_0)$ with NN' all lie on NP_0, are arranged on NP_0 in that order, and tend to P_0.*

Write $Q_2 = f(Q_1, t_1)$. The arc $C_{12} = f(Q_1, t)$, $0 \leqq t \leqq t_1$ and the subsegment Q_1Q_2 of the segment NN' form a closed curve C. We shall see that

1.442. *If $P_0 \notin f(Q_0, t)$ then the closed curve C above separates the point P_0 and every half-trajectory $f(Q_1, t)$, $t \leqq -\delta < 0$. Moreover, no trajectory can pass from the domain containing the point P_0 into the domain containing the half-trajectory $f(Q_1, t)$, $t \leqq -\delta < 0$, with increasing t. The domains referred to above are the exterior and the interior of C.*

The set of points not on C is decomposed by C into two domains D^+ and D^-. Since no trajectory cuts across the arc C_{12}, a trajectory can pass from one of these domains into the other only by cutting across the segment Q_1Q_2 of the normal. Since, moreover, a trajectory cutting across Q_1Q_2 must agree in direction with $f(Q_1, t)$ at Q_1, the half-trajectories $f^-(Q_1)$ and $f^+(Q_2)$ except for the points Q_1 and Q_2, lie in the different domains, say D^- and D^+ respectively, and no trajectory can pass from D^+ into D^- with increasing t.

To prove 1.441, it suffices to show that $Q_2 \in Q_1P_0$. If $Q_2 \in P_0N'$, then by Lemma 1.12, $f^+(Q_2)$ in coming close to P_0 would have to cut across NN' in the wrong direction, i.e., from D^+ into D^- with increasing t. If $Q_2 \in NQ_1$, then $P_0 \in D^-$ and hence is bounded away from $f^+(Q_2)$.

To complete the proof of 1.442, we need only to observe that in view of 1.441, $P_0 \in D^+$.

1.45. The mode of approach to an arc of a trajectory contained in an ω-limit set. Let $P_0 \in \Omega(f^+(Q))$ be a nonsingular point not on $f^+(Q)$, and let $S(P_0, \varepsilon)$ be a small neighborhood of P_0. Use the notation of Lemma 1.12 and of Section 1.44.

We observe first that the whole half-trajectory $f^+(P_0)$ lies in D^+ and hence is bounded away from the (negative) half-trajectory $f(Q_1, t)$, $t \leqq -\delta < 0$.

Next, consider an arc $L_{01} = f(P_0, t)$, $0 \leqq t \leqq t_0$. For all Q_i sufficiently close to P_0, say for $i \geqq i_0$, the whole arc $L_{i1} = f(Q_i, t)$, $0 \leqq t \leqq t_0$ lies in a small neighborhood of L_{01}. Thus each of these arcs L_{i1} in cutting a normal to L_{01}, cuts it in the same direction as L_{01}, and L_{i1} tend to L_{01} as i tends to $+\infty$.

1.46. LEMMA. *Consider a trajectory $f(p, t)$. Let $f(Q_0, t) \subset \Omega(f(p, t))$. If $P_0 \in \Omega(f(Q_0, t))$, then either $P_0 \in f(Q_0, t)$ or P_0 is a singular point (or both).*

If $P_0 \in \Omega(f(Q_0, t))$, then $P_0 \in \Omega(f(p, t))$. If $P_0 \notin f(Q_0, t)$ and if P_0 is a nonsingular point, then in view of 1.442, $f^+(p)$ will enter the region D^+ containing the ω-limit point P_0 and will be bounded away from every point of the half-trajectory $f(Q_0, t)$, $t \leqq -\delta < 0$. Hence no point on this half-trajectory can be an ω-limit point of $f^+(p)$, which is a contradiction.

1.47. THEOREM. *If the ω-limit set of a trajectory $f(p, t)$ is bounded and contains no singular points, then it consists of exactly one closed trajectory L. If $f(p, t)$ is not a closed trajectory, then it spirals toward L as t tends to $+\infty$.*

Let $Q_0 \in \Omega(f(p, t))$ and let $P_0 \in \Omega(f(Q_0, t))$. (Here $\Omega(f(Q_0, t))$ is not empty in view of the boundedness of $\Omega(f(p, t))$). Since P_0 is nonsingular, then $P_0 \in f(Q_0, t)$ by Lemma 1.46, and hence $f(Q_0, t)$ is a closed curve by Theorem 1.2. The conclusions of our theorem follow at once from Theorem 1.43.

1.471. If a trajectory L is not closed, then it may have a limit cycle for $t \to +\infty$ as well as for $t \to -\infty$. In view of the discussion in 1.45, if both of the limit cycles exist, they are distinct.

1.48. THEOREM. *If L_1 consists of the ω-limit points of a trajectory L and is not a closed trajectory, then the set of the ω-limit points of L_1, if not empty, consists of singular points.*

This theorem follows at once from Theorem 1.2 and Lemma 1.46.

1.5. THEOREM. *Let a closed trajectory L be contained together with its interior in a Lipschitz domain G_L. Then there is at least one singular point in the interior of L.*

Let L_ν be either L or a closed trajectory in the interior G of L. We write G_ν for the interior of L_ν. The nonempty set $\{\overline{G}_\nu\}$ of closed domains is partially ordered by the inclusion relation. With Hausdorff[1] we assert that there exists a maximal chain $\{\overline{G}_\alpha\}$ of this partially ordered set. Let $\Gamma = \cap \overline{G}_\alpha$. Then Γ is not empty in view of the compactness of $\overline{G}$ and it contains singular points. For, let $P_0 \in \Gamma$. Then both $f^+(P_0)$ and $f^-(P_0)$ will be contained in each $\overline{G}_\alpha$ and hence in Γ. Since Γ is closed, both α- and ω-limit sets will be contained in Γ. If Γ contains no singular points, both of these limit sets will be closed trajectories. At least one of these closed trajectories is not in $\{L_\alpha\}$, and hence $\{\overline{G}_\alpha\}$ is not maximal. This contradiction proves our assertion.[2]

1.61. LEMMA. *If the ring-shaped region Γ contained between two closed trajectories L_1 and L_2, does not contain either singular points or closed trajectories, then one of the closed trajectories, say L_1 is the ω-limit set of the trajectory $f(P_0, t)$ through any point P_0 of Γ and the other closed trajectory L_2 is the α-limit set of every such trajectory.*

Let P_0 be an arbitrary point of Γ. Since the whole trajectory $f(P_0, t)$ lies in Γ and Γ is bounded, both the ω-limit set of $f(P_0, t)$ and its α-limit set are not empty. Moreover, both of these limit sets lie in the closed and bounded domain $\overline{\Gamma}$ which contains no singular points and no closed trajectories other than L_1 and L_2. However, in view of Theorem 1.47, the above limit sets consist of two distinct (cf. 1.471) closed trajectories. Therefore one of these trajectories is L_1 and the other one is L_2.

We prove next that if L_1 is the ω-limit set of $f(P_0, t)$, then L_1 is the ω-limit set of every trajectory $f(P_1, t)$, $P_1 \in \Gamma$. Choose a small neighborhood $(L_1)_\varepsilon$ of L_1 (cf. 1.3) so that $P_1 \notin (L_1)_\varepsilon$, and consider $(L_1)_\varepsilon \cap \overline{\Gamma}$. Choose δ as in Theorem 1.41, choose a point $Q_0 \in L_1$ and let $f_0 = f(P_0, t_0)$, $f_1 = f(P_0, t_1)$, $t_0 < t_1$, be two successive intersections of $f(P_0, t)$ with the normal to L_1 at Q_0. We may choose t_1 and t_0 so that both $f_0 = f(P_0, t_0)$ and $f_1 = f(P_0, t_1)$ lie in $S(Q_0, \delta)$. The ring-shaped region Γ_1 bounded by L_1, the closed curve formed by the arc $f(P_0, t)$, $t_0 \leqq t \leqq t_1$, and the segment $f_1 f_0$ of the normal

[1] Grundzüge der Mengenlehre, 1st. ed., p. 140.

[2] Bendixson [7] proves this theorem without using Hausdorff's result.

at Q_0, can be entered by the trajectory $f(g_1, t)$ only if this trajectory cuts across the segment f_1f_0 in the positive sense, i.e., with increasing t. Thus, $f^-(P_1) \subset \Gamma$ lies outside Γ_1 and hence L_1 is not an α-limit set of $f(P_1, t)$.

1.62. Lemma. *If the ω-limit set of a trajectory $f(P, t)$ is a closed trajectory L_1, then for any point $P_0 \in f(P, t)$, there exists a sufficiently small neighborhood $S(P_0, \delta)$ of P_0, such that $\Omega(f(Q, t)) = L_1$ for every point $Q \in S(P_0, \delta)$.*

It will suffice to prove that there exists a neighborhood $S(P_0, \delta)$ such that every trajectory $f(Q, t)$ through $Q \in S(P_0, \delta)$, enters a ring-shaped region Γ_1 constructed as in the proof of the preceding lemma. To this end consider a point $f_2 = f(P_0, t_2) \in \Gamma_1$, $t_2 > t_0$, and choose a neighborhood $S(f_2, \eta) \subset \Gamma_1$. By the continuity in the initial conditions (I, 2.22), we can choose $S(P_0, \delta)$ so that $f(Q, t_2) \in S(f_2, \eta)$ for every $Q \in S(P_0, \delta)$. Then $\Omega(f(Q, t)) \subset \bar{\Gamma}_1$ and by Theorem 1.47 we have $\Omega(f(Q, t)) = L_1$, L_1 being the only closed trajectory in $\bar{\Gamma}_1$. Here Γ_1, assumed to lie in a small (cf. 1.3) neighborhood of L_1, not only contains no singular points but it contains no closed trajectories as well. For, every closed trajectory would have to be contained in Γ_1, would cut every normal to L_1 in but one point, and hence would separate P_0 and L_1, which contradicts the assumption that $L_1 = \Omega(f(P_0, t))$.

1.63. A classification of closed trajectories. A closed trajectory L may possess a small neighborhood $(L)_\varepsilon$ containing no other closed trajectories either in the interior or on the boundary. In this case, if $(L)_\delta$ is the δ-neighborhood of L in Theorem 1.41, then for every point $P_1 \in (L)_\delta$, either $f^+(P_1)$ or $f^-(P_1)$ is contained in $(L)_\varepsilon$, and hence L as the only closed trajectory in $(L)_\delta$ is either the ω-limit set or the α-limit set of $f(P_1, t)$ (cf. Theorem 1.47). More precisely, in the light of the discussion in 1.61, if L is the ω-limit set (the α-limit set) of the trajectory $f(P_1, t)$ through one point P_1 in the interior of L and in the small neighborhood $(L)_\delta$ of L_1, then this is true for every interior point P in $(L)_\delta$ and interior to L.

Thus a closed trajectory L which has a small neighborhood $(L)_\varepsilon$ containing no other closed trajectories may be one of the following three kinds:

1.631. *A stable limit cycle*, that is a closed trajectory L which is the ω-limit set of every trajectory $f(P_1, t)$ for P_1 in a sufficiently small neighborhood $(L)_\delta$ of L.

1.632. *An unstable limit cycle,* that is a closed trajectory L which is the α-limit set of every trajectory $f(P_1, t)$ for P_1 in a sufficiently small neighborhood $(L)_\delta$ of L.

1.633. *A semi-stable limit cycle,* that is a closed trajectory L which is the α-limit set of every trajectory $f(P_1, t)$ for a point P_1 in the interior (exterior) of L and in a sufficiently small neighborhood $(L)_\delta$ of L, and which at the same time is the ω-limit set of every trajectory $f(P_2, t)$ for a point P_1 in the exterior (interior) of L and in $(L)_\delta$.

If every ε-neighborhood, however small, of a closed trajectory L, contains other closed trajectories, then this closed trajectory may be of the following two kinds:

1.634. *A periodic ring,* that is a closed trajectory with a small (ring-shaped) neighborhood consisting entirely of closed trajectories.

1.635. *A composite limit cycle,* which is a closed trajectory every small neighborhood of which contains closed trajectories as well as trajectories which spiral toward closed trajectories (their ω- and their α-limit sets).

1.7. To achieve a unified and a natural formulation of the properties of limit sets of dynamical systems it is convenient to adjoin to the (locally compact [3]) Euclidian plane E^2 a point P_∞ so that the resulting (compact [3]) space E_2^* is homeomorphic to the two dimensional sphere S^2.

A given dynamical system D (cf. I, 3) in E^2 defined by (1.01) may be extended to a dynamical system D^* in E_2^*. To accomplish this we let

$$f^*(P, t) = f(P, t) \quad \text{if} \quad P \neq P_\infty, \tag{1.701}$$

and

$$f^*(P_\infty, t) = P_\infty \quad \text{for} \quad -\infty < t < +\infty. \tag{1.702}$$

This one parameter group of transformations of E_2^* defines a dynamical system D^* in E_2^* with P_∞ as a singular point.

1.71. An alternative definition of the limit sets of a trajectory (Lefschetz [27]). It is easy to see that our definitions of

[3]Locally bi-compact, bi-compact respectively in the classical terminology.

the ω- and α-limit sets Ω and A may be stated in the following very useful form:

$$(1.711) \qquad \Omega(f(P_0, t)) = \bigcap_{t' \geqq 0} \overline{f^+(P(t'))}, \quad P(t') = f(P_0, t'),$$

and

$$(1.712) \qquad A(f(P_0, t)) = \bigcap_{t' \leqq 0} \overline{f^-(P(t'))}.$$

If we apply these definitions to a dynamical system D^* in E_2^* by taking the closure in E_2^* and replacing f by f^* in (1.711) and (1.712), then we see at once (Cantor) that each of the above limit sets is not empty, is closed, and is connected, since it is the intersection of a sequence of steadily decreasing closed connected sets of a compact Hausdorff space E_2^*. Thus, we have

1.72. THEOREM. *Each limit set of a trajectory $f^*(P_0, t)$ of a dynamical system D^* in E_2^*, is not empty, is closed, and is connected. Since P_∞ does not belong to the limit set of a bounded half-trajectory $f^+(P_0)$ (or $f^-(P_0)$) such a limit set has the above properties and is bounded as well.*

Point sets which are the limit sets of trajectories of dynamical systems in E_2^* may be characterized more precisely. Let Ω^* be the ω-limit set of a trajectory $f^*(P_0, t)$ of a dynamical system D^*. If $f^*(P_0, t)$ and Ω^* have points in common, Ω is a singular point or a closed trajectory by Theorem 1.2, and hence is the boundary of a simply-connected region in E_2^*.

Consider next the case when $f^*(P_0, t) \cap \Omega$ is empty. The complement $G = E_2^* - \Omega$ of Ω is an open set not in general connected, whose components $G_1, G_2, \ldots$ are open in E_2^*. The connected set $f(P_0, t)$ is contained in G and hence is entirely contained in one of the components, say G_1, of G. By (1.711), $\Omega \subset \overline{G}_1$ and since $\Omega \cap G_1 \subset \Omega \cap G = 0$, Ω is contained in the boundary $\dot{G}_1$ of G_1. On the other hand $\dot{G}_1 \subset \Omega$. For, if we write $H = \bigcup_{i \neq 1} G_i$ and denote by a primed letter the complement in E_2^*, we have $G_1 \subset H'$, H' is closed, $\overline{G}_1 \subset H'$, and thus

$$\dot{G}_1 = \overline{G}_1 - G_1 = \overline{G}_1 \cap G_1' \subset H' \cap G_1' = (H \cup G_1)' = G' = \Omega.$$

Thus Ω is the boundary of a connected region G_1, and since Ω is connected, it follows that G_1 is simply connected. This discussion yields the second part of the following.

1.73. THEOREM (P. E. Vinograd [55]). *A set Σ in E_2^* is a limit set of a trajectory $f^*(P_0, t)$ of a dynamical system D^* in E_2^* if and only if Σ is the boundary of a simply connected region G in E_2^*.*

The first part of our theorem obviously holds true for every Σ consisting of only one point.

Next, let Σ be the boundary of a simply-connected region G and let Σ contain at least two points.

Consider the system of equations

$$\frac{d\varrho}{dt} = \varrho(1 - \varrho), \qquad \frac{d\varphi}{dt} = 1 \tag{1.731}$$

where ϱ and φ are the absolute value and the argument of a complex variable $z = \varrho e^{i\varphi} = x + iy$. This system has a singular point at $\varrho = 0$. All other trajectories L of (1.731) in the interior C of the unit circle $|z| < 1$, spiral toward the circumference S: $|z| = 1$ as $t \to +\infty$.

Let $w = \Phi(z) = u(x, y) + iv(x, y)$ be a one-to-one conformal mapping of C onto the simply-connected region G (the Riemann mapping theorem). This mapping transforms our system (1.731) into a system

$$\frac{du}{dt} = P(u, v), \qquad \frac{dv}{dt} = Q(u, v) \tag{1.732}$$

whose trajectories are the images $\Phi(L)$ in G of the trajectories L of (1.731) in C. We construct next (cf. I, 3.2) a dynamical system D^* defined over the whole E_2^*, equivalent to (1.732) in G, and such that every point outside of G is an equilibrium (singular) point. *We assert that the boundary Σ of G is the ω-limit set of every trajectory L^* of D^* which is the image $\Phi(L)$ of a spiral trajectory L of* (1.731) *in C.*

We are required to show that if $w_0 \in \Sigma$, then every ε-neighborhood of w_0 contains a point of L^*. If w_0 is not arc-wise accessible from G, we may find a point w' of Σ which is arc-wise accessible from G and which is arbitrarily close to w_0, and say, lies within the $\varepsilon/2$ neighborhood of w_0.[4] Let $J = J(w', w'')$ be a Jordan arc with w' as one of its endpoints and such that $J - w' \subset G$. Consider the half-closed arc $j = \Phi^{-1}(J - w')$. If a sequence $w_1, w_2, \ldots$ of points on $J - w'$, tends to w', then the image sequence $z_1 = \Phi^{-1}(w_1)$,

[4]The set of all points in Σ which are arc-wise accessible from G is dense in Σ.

$z_2 = \Phi^{-1}(w_2), \ldots$ on j has a limit point on S. It follows from this that if we choose J so that not all the points s of S are limit points of $\Phi^{-1}(J)$ then the points of intersection of our spiral trajectory $L \subset C$ form an infinite set $z = \{z'_\mu\}$ with limit points on S. The image set $W = \Phi(z) = \{\Phi(z'_\mu) = w'_\mu\}$ on $J - w'$ consisting of points of the trajectory $L^* = \Phi(L)$ must then have limit points in Σ. If w' were not one of the limit points of W, then a suitably chosen δ-neighborhood $S(w', \delta)$ of w' would contain no points of W, all points of W would lie on the closed sub-arc $J(w''', w'')$ of J, where we may take $w''' \notin (J - w') \cap S(w', \delta)$, and no point of Σ would be a limit point of W. Thus we must accept the contrary, and there exists a point $w_\nu \in W$ for which $|w_\nu - w'| < \varepsilon/2$, whence $|w_0 - w_\nu| < \varepsilon$. Hence there are points of L^* in every ε-neighborhood of w_0, and hence $\Sigma \subset \Omega(L^*)$. On the other hand, since $\Omega(L)$ has no points in C, $\Omega(L^*) = \Omega(\Phi(L))$ has no points in G and therefore $\Omega(L) \subset \Sigma$, which completes the proof of our theorem.

1.74. Definitions. As the next step in our study of the limit sets of dynamical systems we shall study the structure of such sets in terms of the behavior of their constituent nonsingular trajectories and of their singular points.

1.741. Let $L = f(P, t)$ be a trajectory of a dynamical system D. We write $\Omega(L) = \Omega_s(L) \cup \Omega_n(L)$ where $\Omega_s(L)$ consists of all the singular points in $\Omega(L)$. We shall denote the components of Ω_s by C_s.

We note that if D^* is the associated dynamical system in E_2^* and L is a trajectory of D and hence of D^*, then the ω-limit set $\Omega^*(L)$ in E_2^* contains $\Omega(L)$. More precisely, $\Omega_n^* = \Omega_n$ and $\Omega_s^* \supset \Omega_s$. The equality $\Omega^*(L) = \Omega(L)$ (or $\Omega_s^* = \Omega_s$) means that $\Omega(L)$ is bounded. The set Ω_s^* is closed.

1.75. Theorem (Solučev [51] -Vinograd [55]). *Consider a dynamical system D defined by* (1.01) *in E^2. Let D^* be the associated dynamical system in the compact space $E_2^* = E^2 \cup P_\infty$ and let $L = f(P, t)$ be a trajectory of D (and hence of D^*). Then,*

(1.751) *$\Omega_n^*(L) = \Omega^*(L)$ if and only if $\Omega_n(L)$ is a closed trajectory,*

(1.752) *if $\Omega_n(L) \neq \Omega^*(L)$, then each trajectory in Ω_n adheres at both ends to components of Ω_s^*,*

(1.753) *the trajectory L approaches spirally each of the trajectories in $\Omega_n = \Omega_n^*$, and*

(1.754) $\Omega_n(L) = \Omega_n^*(L)$ *consists of at most a denumerable number of trajectories.*

1.751. If $\Omega_n = \Omega_n^* = \Omega^*$, then $\Omega^* = \Omega$ and Ω^* consists of a single closed trajectory of Theorem 1.47. If Ω contains a closed trajectory, then by Theorem 1.43, Ω consists precisely of the points of this closed trajectory. Moreover, Ω is bounded and hence $\Omega = \Omega^*$.

1.752. Next let $\Omega_n^* \neq \Omega^*$, and let $L_1 \subset \Omega_n^* = \Omega_n$. Since Ω^* does not contain closed trajectories in this case, it follows from Lemma 1.46 that $\Omega(L_1)$ and hence $\Omega^*(L_1)$ contains only singular points. Since $\Omega^*(L_1) \subset \Omega^*(L)$, we have $\Omega^*(L_1) \subset \Omega_s^*(L)$. By Theorem 1.72, the set $\Omega^*(L_1)$ is connected and hence is contained in a component C_s of $\Omega_s^*(L)$. A similar argument applies to the α-limit set $A(L_1)$ of L_1.

1.753-4. Let $L_{01} = \{f(P_0, t), 0 \leqq t \leqq t_0\}$ be an arc of a trajectory L_μ in $\Omega_n(L)$. It was shown in Section 1.45 that L approaches L_{01} spirally. More precisely, there exists a small neighborhood $(L_{01})_\varepsilon$ of L_μ (cf. 1.3 and 1.45) such that the arcs $L \cap (L_{01})_\varepsilon$ lie on one side of L_μ, agree in direction with L_{01}, and cut a normal NP_0 to L_{01} at P_0 in a sequence $Q_0, Q_\nu = f(Q_0, t_\nu), t_{\nu+1} > t_\nu, \nu = 1, 2, \ldots,$ which tends steadily toward P_0 with increasing ν.

Thus $(L_{01})_\varepsilon$ contains no points of Ω_n other than those of L_μ. Let $R_\mu \in L_{01}$ be an interior point of $(L_{01})_\varepsilon$. Then the distance ϱ_μ from R_μ to $\Omega_n - L_\mu$ is positive. Let $\delta_\mu < \min(\varepsilon, \frac{1}{2}\varrho_\mu)$. Then $S_\mu = S(R_\mu, \delta_\mu)$ contains no points of $\Omega_n - L_\mu$ and no two circles S_μ and $S_{\mu'}$ corresponding to two different trajectories L_μ, $L_{\mu'}$ in Ω_n, have points in common. Since the number of such nonoverlapping circles is at most denumerable, the same is true of the trajectories L_μ.

Theorem 1.75 implies the first part of

1.76. THEOREM. *Consider a trajectory L of a dynamical system D^* in E_2^*. Let $\Omega_s^*(L)$ consist of a single point P_s. Then,* (1) *$\Omega_n^*(L)$ consists of at most a denumerable number of trajectories L_μ all of which adhere to P_s at both ends, and* (2) *the circumference C of every circle $S(P_s, r)$ with the center at P_s, intersects at most a finite number of trajectories $L_\mu \subset \Omega_n$.*

To prove the second part of this theorem, we select one point P_μ of intersection of L_μ and C for every L_μ intersecting C. If C is cut by an infinite number of distinct trajectories L_μ, then the number of distinct points P_μ is infinite. Let P_0 be a limit point of the set $\{P_\mu\}$. Since $\Omega(L)$ is closed and $\{P_\mu\} \subset \Omega(L)$, we see that $P_0 \in \Omega(L)$. Moreover, since $P_0 \neq P_s$, $P_0 \in \Omega_n(L)$ and $\Omega_n(L) \supset f(P_0, t) = L_{\mu_0}$. Let $S(P_0, \varepsilon)$ be a small neighborhood of P_0. Using the notation of Lemma 1.12, we observe that there are infinitely many points P_μ in $S(P_0, \delta)$. Among these, select three points P_{μ_1}, P_{μ_2}, P_{μ_3} arranged in that order on C (one of P_{μ_j} may be taken as P_0). Let Q_1, Q_2, Q_3 be the points of intersection of L_{μ_1}, L_{μ_2}, L_{μ_3} respectively with the normal NP_0N' to L_{μ_0} at P_0. We note that the trajectories L_{μ_1}, L_{μ_2}, L_{μ_3} all agree in direction. The closed and connected set $\overline{f^+(Q_1)} \cup \overline{f^+(Q_3)} \cup \overline{Q_1Q_3}$ is the common boundary of the domain D_2 containing the half-trajectory $f(Q_2, t)$, $t > 0$, and of the domain D_2' containing the half-trajectory $f(Q_2, t)$, $t < 0$. Since our trajectory L can pass from one of these domains into the other only by cutting across the arc $\overline{Q_1Q_3}$ and that only in one direction only, viz., in the common direction of the trajectories L_{μ_1}, L_{μ_2}, and L_{μ_3} (cf. 1.13), we see that once L enters D_2, it remains in D_2 and cannot come arbitrarily close to the ω-limit points on the half-trajectory $f(Q_2, t)$, $t \leqq \eta < 0$. This contradiction shows that the number of points P_μ is finite.

1.77. Corollary. *If a half-trajectory $f^+(P_0)$ of a dynamical system D, is not bounded and if its ω-limit set has no singular points in the finite plane, then $\Omega(f^+(P_0))$ consists of at most a denumerable number of trajectories each of which recedes to infinity in both directions. Moreover, each bounded domain of the plane is cut by at most a finite number of these trajectories.*

Here $\Omega^* = \Omega(f^+(P_0))$ contains but one singular point P_∞.

Corollary 1.77 extends the results of Theorem 1.47 and completes the discussion of the case of trajectories whose limit sets have no singular points in E^2.

1.78. The mode of approach of a trajectory to its ω-limit set.[5] It follows from the Heine-Borel lemma that for every $\varepsilon > 0$ there is a $T = T(\varepsilon) > 0$ such that the half-trajectory $f(p, t)$, $t > T$ is contained in the ε-neighborhood of its (bounded) ω-limit set Ω. We observe that Ω is the boundary of a simply-connected domain Γ. A conformal mapping of Γ onto the interior of the unit circle K defines a one-to-one bi-continuous mapping φ of $\Omega_n \subset \Omega$ into the circumference C of K since every point P of

[5]For a complete analysis see Solučev and Vinograd, loc. cit.

Ω_n is accessible along the normal to the trajectory $f(P, t)$ of Ω_n and every component $f(p, t)$ of Ω_n is free (Caratheodory). Thus (Caratheodory), every trajectory $f(P, t)$ in Ω will map onto one and only one open arc $\varphi(f(P, t))$ of the circumference of K. This correspondence induces a cyclic order among the trajectories in Ω_n. An ε-neighborhood of $\Omega f((p, t))$ in Γ is carried over into a narrow strip adjacent to the circumference of K. This strip contains the image of the half-trajectory $f(p, t)$, $t > T$.

Choose a positive sense on the circumference C. Consider the direction along the image arc of $f(P, t)$, which corresponds to the direction along $f(P, t)$ given by increasing t. *We assert that for every two trajectories $f(P_1, t)$, $f(P_2, t)$ in Ω_n the induced directions on the image arcs have the same sign.* For, suppose the contrary. Choose points $P'_i = f(P_i, t')$ $P''_i = f(P_i, t'')$, $t' < t''$, and consider small neighborhoods Γ_i, $i = 1, 2$, of the trajectory arcs $f(P_i, t)$, $t' \leqq t \leqq t''$. Assume that the order of the pairs of limit points $Q'_1Q''_1$ and $Q'_2Q''_2$ on C do not agree. A simple argument shows that the behavior of the image L of $f(p, t)$ contradicts the assumption that Q'_1 is in the image of Ω_n, since the images l'_i, l''_i in G_i of the normals at P'_i, P''_i in Γ_i are cut by $f(p, t)$ in one direction only (Γ_i were assumed to be small neighborhoods (cf. 1.3)).

Let θ_1 and θ_2, $\theta_1 < \theta_2$ be the polar angles of points Q_1 and Q_2 respectively on the circumference C of K. Let U_i be nonintersecting small neighborhoods of $\varphi^{-1}(\theta_i) = \varphi^{-1}(Q_i)$. Then there exists a $T > 0$ such that for $t > T$, our trajectory $f(p, t)$ does not reenter U_1 until it passes through U_2. This follows at once from the truth of this statement for the trajectory image $\varphi(f(p, t))$ in K and the neighborhoods $\varphi(U_1)$ and $\varphi(U_2)$ of Q_1 and Q_2.

1.79. In Section 1 we have studied the behavior of trajectories of a dynamical system in a plane. We should note that many of our results hold true for trajectories on a cylinder even though the proofs must be modified somewhat.

Closed contours on a cylinder may be of two types. The first type encompasses a bounded domain homeomorphic to a domain in the plane and the second type wraps around the cylinder and does not delimit a bounded domain. However, in both cases a closed contour C splits the cylinder into two domains which have C as the common boundary.

The normal at a point P to a trajectory in the plane must be replaced by the curve of intersection of the cylinder with the plane through P and normal to the trajectory.

We should note that every ring-shaped region associated with a closed contour of either type, is homeomorphic to a ring in the plane.

Consider a ring-shaped region R on a cylinder. Should this region contain no singular points, then if it contains a closed trajectory, this trajectory splits the cylinder into two domains one of which contains one of the boundary components of R and the other one the second component. Making use of these observations one can establish Theorems 1.2–1.47 and 1.72 for the cylinder. The characterization of the neighborhoods of the periodic solutions can be carried out as before, and Theorem 1.5 holds true for closed trajectories which encompass a domain homeomorphic to a circle. It seems that other results of this section should also hold for the cylinder — this has not yet been verified.

2. Trajectories on a Torus

Consider a parametric representation of a torus,

$$(2.01)\quad \begin{cases} x = (R + r \cos 2\pi\vartheta) \cos 2\pi\varphi, \quad y = (R + r \cos 2\pi\vartheta) \sin 2\pi\varphi, \\ z = r \sin 2\pi\vartheta, \quad 0 \leqq \varphi < 1, \quad 0 \leqq \vartheta < 1, \quad 0 < r < R. \end{cases}$$

We shall refer to the lines $\varphi =$ constant (the generating circles) as the meridians, the lines $\vartheta =$ constant as the parallels, and the coordinates φ and ϑ as the latitude and the longitude respectively.

2.11. Consider dynamical systems defined on the torus by a system of differential equations

$$(2.11)\qquad \frac{d\varphi}{dt} = \Phi(\varphi, \vartheta), \quad \frac{d\vartheta}{dt} = \Theta(\varphi, \vartheta).$$

The functions Φ, Θ are assumed to satisfy conditions which assure the uniqueness of solutions as well as the continuity with respect to the initial values. As we learned in Chapter I, Lipschitz conditions suffice for this purpose.

2.12. The qualitative theory of differential equations on the torus is due to Poincaré [47]. Following him, we assume that the domain of definition of the function Φ and Θ is the whole (φ, ϑ)-

plane and that these functions are periodic of period unity in each of the arguments φ and ϑ.

The trajectories in the (φ, ϑ)-plane of the resulting differential equation (2.11) yield the desired trajectories on the torus if we identify all points (φ, ϑ) with those points (φ', ϑ') of the square $0 \leqq \varphi' \leqq 1$, $0 \leqq \vartheta' \leqq 1$, for which the differences $\varphi - \varphi'$ and $\vartheta - \vartheta'$ are integers.

Here the square

$$(2.12) \qquad \Sigma\colon\ 0 \leqq \varphi \leqq 1, \quad 0 \leqq \vartheta \leqq 1$$

in which we identify the pairs of opposite sides $\varphi = 0$, $\varphi = 1$ and $\vartheta = 0$, $\vartheta = 1$ (all vertices are identified with the point $(0, 0)$), serves as a convenient representation of the torus.

2.13. In what follows we consider only those dynamical systems on the torus for which $\Phi(\varphi, \vartheta)$ is different from zero everywhere. Such systems (2.11) have no singular points. In studying the trajectories of a system (2.11) with a nonvanishing $\Phi(\varphi, \vartheta)$ one may replace this system by the equation

$$(2.13) \qquad \frac{d\vartheta}{d\varphi} = A(\varphi, \vartheta), \quad A(\varphi, \vartheta) = \frac{\Theta(\varphi, \vartheta)}{\Phi(\varphi, \vartheta)}.$$

When considered in the whole (φ, ϑ)-plane, $A(\varphi, \vartheta)$ is a continuous periodic function subject to additional restrictions, say the Lipschitz conditions, to assure uniqueness of solutions and their continuity with respect to the initial values.

2.14. Since $A(\varphi, \vartheta)$ is bounded, the solution through a point (φ_0, ϑ_0) can be extended to the whole range $-\infty < \varphi < +\infty$. Let us denote this solution by

$$(2.141) \qquad \vartheta = u(\varphi, \varphi_0, \vartheta_0).$$

By the uniqueness of solutions and the periodicity of $A(\varphi, \vartheta)$,

$$(2.142) \quad u_{m,n}(\varphi, \varphi_0, \vartheta_0) = u(\varphi, \varphi_0 + m, \vartheta_0 + n) = u(\varphi - m, \varphi_0, \vartheta_0) + n$$

for every pair of integers m and n. Thus the translation of an integral curve L in (2.141), by (m, n) is again an integral curve which we shall denote by $L(m, n)$.

Since every solution passes through a point $(0, \vartheta_0)$ on the axis $\varphi = 0$ (or on the meridian $\varphi = 0$ of the torus), we obtain all the solutions by varying the parameter ϑ_0 in

$$(2.143) \qquad \vartheta = u(\varphi, 0, \vartheta_0) - u(\varphi, \vartheta_0).$$

We shall denote the integral curve corresponding to (2.143) by $L(\vartheta_0)$.

2.15. For a fixed integer n the function $u(n, \vartheta)$ defines a mapping of the line $\varphi = 0$ onto the line $\varphi = n$. This mapping is one-to-one and order preserving (i.e., $u(n, \vartheta)$ is steadily increasing) in view of the uniqueness of solutions. It is continuous in view of the continuity in the initial values. Moreover

$$(2.151)\quad u(n, 0) \leqq u(n, \vartheta) \leqq u(n, 1) = u(n, 0) + 1 \quad \text{for} \quad 0 \leqq \vartheta \leqq 1,$$

by (2.142). This means that after the identification of points (cf. 2.12), $u(n, \vartheta)$ induces a one-to-one mapping

$$(2.152)\qquad (0, \vartheta) \to (0, (u(n, \vartheta))), \qquad 0 \leqq \vartheta < 1$$

of the (oriented) generating circle

$$C\colon \varphi = 0, \qquad 0 \leqq \vartheta < 1$$

of the torus onto itself. This mapping is continuous and order preserving. Here

$$(2.153)\qquad (u(n, \vartheta)) = u(n, \vartheta) - [u(n, \vartheta)],$$

where $[u(n, \vartheta)]$ is the greatest integer in $u(n, \vartheta)$.

The mapping (2.152) may be conveniently represented by a continuous steadily increasing function

$$(2.154)\qquad \bar{u}(n, \vartheta) = u(n, \vartheta) - [u(n, 0)] \quad \text{for} \quad 0 \leqq \vartheta \leqq 1.$$

We observe that

$$(2.155)\qquad u(m, u(n, \vartheta)) = u(m + n, \vartheta),$$

in view of (2.143) and since

$$\begin{aligned} u(m + n, 0, \vartheta) &= u(m, -n, \vartheta) = u(m, 0, u(0, -n, \vartheta)) \\ &= u(m, 0, u(n, 0, \vartheta)), \end{aligned}$$

by (2.142).

Let

$$(2.156)\qquad u = u(\vartheta) = (u(1, \vartheta)), \qquad 0 \leqq \vartheta < 1$$

define the transformation in (2.152) with $n = 1$. Then

$$(2.157)\qquad u^n(\vartheta) = (u(n, \vartheta)).$$

To prove (2.157) we observe that if it holds true for $n = k - 1 > 0$, then it holds true for $n = k$ by

$$u^k = u^{k-1}u = (u(k-1, (u(1, \vartheta)))) = (u(k-1, u(1, \vartheta) - [u(1, \vartheta)]))$$
$$= (u(k-1, u(1, \vartheta))) = (u(k, \vartheta)).$$

We note that for $n = 0$, (2.157) yields the identity mapping, since

$$u^0(\vartheta_0) = u(0, \vartheta_0) = \vartheta_0.$$

Also $u^{-n}(\vartheta) = (u(-n, \vartheta)) = (u^n)^{-1}(\vartheta)$ by (2.155).

2.16. A trajectory (2.143) on the torus, is closed if and only if

$$(2.161) \qquad u(n, \vartheta_0) = \vartheta_0 + m, \qquad 0 \leqq \vartheta_0 < 1,$$

for some positive integers m and n, that is if and only if

$$(2.162) \qquad u^n(\vartheta_0) = \vartheta_0,$$

and the nth power of the transformation (2.156) has a fixed point ϑ_0. For, (2.162) as well as (2.161) implies that the trajectory (2.143) returns to the initial point $(0, \vartheta_0)$ and conversely.

2.17. The points $(0, \vartheta_k)$ on the torus, determined by the successive iterations

$$(2.17) \qquad \vartheta_k = u^k(\vartheta_0), \quad 0 \leqq \vartheta_k < 1, \quad k = 0, \pm 1, \pm 2, \ldots,$$

of ϑ_0 by the transformation (2.156) are precisely the points of intersection of the trajectory (2.143) and the meridian $\varphi = 0$.

2.171. If ϑ_0 lies on a closed trajectory, then, by 2.16, the set (2.17) contains only a finite number of distinct elements, and conversely if only a finite number of values in (2.17) are distinct, then (2.143) is closed.

2.172. Assume next that some trajectories of the equation (2.13) are closed on the torus. Then if (2.143) is not closed, it approaches spirally a closed trajectory. (Proof to follow). Let

$$(2.1721) \qquad U(\vartheta) = u^m(\vartheta)$$

be the least positive power of u which has a fixed point ϑ'. There are fixed points of this mapping since there are closed trajectories. Write

$$(2.1722) \qquad \Theta_j = U^j(\vartheta_0), \qquad j = 0, 1, 2, \ldots,$$

and consider the arc $\alpha_0 = (\Theta_0, \Theta_1, \vartheta')$ of the meridian C. Since U is order preserving, the image

$$\alpha_1 = U(\alpha_0) = (U(\Theta_0), U(\Theta_1), U(\vartheta')) = (\Theta_1, \Theta_2, \vartheta')$$

is a subarc of α_0. Therefore

$$\alpha_{j+1} = U(\alpha_j) \subset \alpha_j \quad \text{for} \quad j = 0, 1, \ldots,$$

and the sequence (2.1722) of the left endpoints of α_j is monotone and bounded by the common right endpoint ϑ' of these arcs. Therefore the sequence (2.1722) has a unique limit point ϑ''. Since both $U(\vartheta)$ and $U^{-1}(\vartheta)$ are continuous, the point $U(\vartheta'')$ is the unique limit point of the image

$$U\{\Theta_j\} = \{U(\Theta_j)\} = \{\Theta_1, \Theta_2, \ldots\} \tag{2.1723}$$

of (2.1722), and hence

$$U(\vartheta'') = \vartheta''.$$

Thus, the trajectory $\vartheta = u(\varphi, \vartheta'')$ is closed on the torus and is approached by (2.143) spirally as φ tends to $+\infty$.

Considering the sequence (2.1722) for $j = 0, -1, -2, \ldots$ we prove the existence of a limit cycle approached spirally by (2.143) as φ tends to $-\infty$.

2.18 DEFINITION. *Two distinct points ϑ_i and ϑ_j of the set* (2.17) *are called neighbors if there are no points of* (2.17) *on one of the two open arcs into which ϑ_i and ϑ_j separate the meridian C.*

2.19. LEMMA. *If for some ϑ_0, the set* (2.17) *has a pair of neighbors, then some power of u has a fixed point ϑ'' and hence the trajectory defined by $\vartheta = u(\varphi, \vartheta'')$ is closed.*

Let ϑ_g and ϑ_h be a pair of neighbors in the set (2.17). Let $h > g$ and set $h - g = m$. Then

$$\vartheta_h = u^m(\vartheta_g) = U(\vartheta_g).$$

Let $\alpha_0 = (\vartheta_g, \vartheta_h)$ denote the closed arc (cf. 2.18) of the meridian C, containing no points (2.17) other than ϑ_g and ϑ_h. Consider the sequence of arcs

$$\alpha_j = U(\alpha_{j-1}) = U^j(\alpha_0) = (\vartheta_{g+mj}, \vartheta_{g+m(j+1)}). \tag{2.191}$$

All of these arcs are similarly directed and every pair of successive arcs $\alpha_{j-1}\alpha_j$ abut.

If the number of distinct arcs (2.191) is finite, say is equal to n, then $\vartheta'' = \vartheta_g$ is a fixed point of u^{mn} and the trajectory $\vartheta = u(\varphi, \vartheta_g)$ is closed.

If (2.191) contains infinitely many distinct elements, the sequence

$$\vartheta_h = \vartheta_{g+m}, \vartheta_{g+2m}, \ldots$$

is monotone, bounded by ϑ_g and therefore converges to a unique limit point ϑ'' which is thus a fixed point of $U = u^m$ (cf. 2.17). Hence $\vartheta = u(\varphi, \vartheta'')$ is a closed trajectory.

2.2 We consider now the case of differential equations (2.13) without closed trajectories on the torus. In this case no power of the associated transformation (2.156) has fixed points and the elements of the sequence

$$(2.21) \quad \mathscr{D}(\vartheta_0) = \{\ldots, \vartheta_{-1} = u^{-1}(\vartheta_0), \vartheta_0, \vartheta_1 = u(\vartheta_0), \vartheta_2 = u^2(\vartheta_0), \ldots\}$$

are all distinct. Denote by $\mathscr{D}'(\vartheta_0)$ the set of all limit points of the set $\mathscr{D}(\vartheta_0)$. The set $\mathscr{D}'(\vartheta_0)$ is not empty and is closed. Moreover, every point of $\mathscr{D}'(\vartheta_0)$ is its limit point. For, if $\vartheta' \in \mathscr{D}'(\vartheta_0)$, then every neighborhood of ϑ' contains infinitely many elements of $\mathscr{D}(\vartheta_0)$ and hence it contains an arc $(\vartheta_i, \vartheta_k)$ with $\vartheta_i, \vartheta_k \in \mathscr{D}(\vartheta_0)$ and not containing ϑ'. In view of Lemma 2.19, the arc $(\vartheta_i, \vartheta_k)$ contains infinitely many elements of $\mathscr{D}(\vartheta_0)$ and hence also a limit point $\vartheta'' \neq \vartheta'$. Thus

$$(2.22) \qquad \textit{every } \mathscr{D}'(\vartheta_0) \textit{ is a perfect set.}$$

21.2. LEMMA. *For every $\bar{\vartheta}_0$ and ϑ_0 in C, we have $\mathscr{D}'(\bar{\vartheta}_0) = \mathscr{D}'(\vartheta_0)$.*

If $\bar{\vartheta}_0 \in \mathscr{D}'(\vartheta_0)$, then $\bar{\vartheta}_0$ is a limit point of $\mathscr{D}(\vartheta_0)$, whence by the continuity of u^k, $u^k(\bar{\vartheta}_0)$ is a limit point of

$$(2.211) \qquad u^k(\mathscr{D}(\vartheta_0)) = \mathscr{D}(u^k(\vartheta_0)) = \mathscr{D}(\vartheta_k) = \mathscr{D}(\vartheta_0).$$

Thus

$$(2.212) \qquad u^k(\bar{\vartheta}_0) \in \mathscr{D}'(\vartheta_0), \quad k = 0, \pm 1, \pm 2, \ldots,$$

that is $\mathscr{D}(\bar{\vartheta}_0) \subset \mathscr{D}'(\vartheta_0)$. Since $\mathscr{D}'(\vartheta_0)$ is closed, $\mathscr{D}'(\bar{\vartheta}_0) \subset \mathscr{D}'(\vartheta_0)$.

We note that (2.212) implies that

$$(2.213) \qquad u^k(\mathscr{D}'(\vartheta_0)) \subset \mathscr{D}'(\vartheta_0), \quad k = 0, \pm 1, \pm 2, \ldots,$$

and hence $\mathscr{D}'(\vartheta_0) \subset u^{-k}(\mathscr{D}'(\vartheta_0))$. Therefore

$$(2.214) \qquad u^j(\mathscr{D}'(\vartheta_0)) = \mathscr{D}'(\vartheta_0), \quad j = 0, \pm 1, \pm 2, \ldots.$$

Next, let $\bar{\vartheta}_0 \in O = C - \mathscr{D}'(\vartheta_0)$. The set O is open and hence is the union of at most a denumerable number of distinct and non-overlapping arcs. Let A be the set of all such arcs. By (2.214) and the continuity of u^j, $u^j(\alpha) \in A$ for every $\alpha \in A$. Let $\bar{\vartheta}_0$ lie in $\alpha_0 \in A$. Then $\bar{\vartheta}_j = u^j(\bar{\vartheta}_0) \in u^j(\alpha_0) = \alpha_j$. Since no two arcs α_j overlap, no arc in A contains more than one point of $\mathscr{D}(\bar{\vartheta}_0)$. Hence no point of $\mathscr{D}'(\bar{\vartheta}_0)$ lies in O.

Thus $\mathscr{D}'(\bar{\vartheta}_0) \subset \mathscr{D}'(\vartheta_0)$ for every $\vartheta_0 \in C$.

Interchanging $\bar{\vartheta}_0$ and ϑ_0 in the above discussion we get $\mathscr{D}'(\vartheta_0) \subset \mathscr{D}'(\bar{\vartheta}_0)$, and (2.25) follows.

2.22. LEMMA. *If $\mathscr{D}'(\vartheta_0)$ contains an arc, then $\mathscr{D}'(\vartheta_0) = C$.*

Let $\mathscr{D}'(\vartheta_0)$ contain an arc β. Then β contains a subarc $\beta_0 = (\vartheta_g, \vartheta_h)$ with the endpoints ϑ_g and ϑ_h in $\mathscr{D}(\vartheta_0)$. Let $h - g = m$. Then

$$\vartheta_h = u^m(\vartheta_g) = U(\vartheta_g).$$

Consider the sequence of arcs

(2.221) $$\beta_j = U^j(\beta_0) = (\vartheta_{g+mj}, \vartheta_{g+m(j+1)}), \quad j = 1, 2, \ldots,.$$

All of these arcs are similarly directed and every pair of successive arcs β_{j-1} and β_j abut. Thus the sequence

(2.222) $$\Sigma = U^j(\vartheta_g)$$

is monotone. By (2.214), $\beta_j \subset \mathscr{D}'(\vartheta_0)$ for every j and the same holds for the union

$$B_i = \bigcup_{j=0}^{i-1} \beta_j$$

of the first i intervals in (2.221). The arc B_i, for i sufficiently large, covers all of C. Otherwise the monotone sequence (2.222) would be bounded, say by ϑ_g and its unique limit point ϑ'' would be a fixed point of $U = u^m$, contrary to our main hypothesis in 2.2.

The result of Lemmas 2.21 and 2.22 may be stated in the following form

2.23. THEOREM. *For a given transformation u, the set $\mathscr{D}'(\vartheta_0) = F$ is the same for all $\vartheta_0 \in C$. Moreover, either*

(2.231) $F = C$ *(the transitive or ergodic case)*

or

(2.232) *F is a nowhere dense perfect set (the intransitive case).*

2.24. If $\mathscr{D}'(\vartheta_0) = C$, $\mathscr{D}(\vartheta_0)$ is everywhere dense in C and the corresponding trajectory (2.143) is everywhere dense on the torus and conversely. It follows from Theorem 2.23, therefore, that if one trajectory of (2.13) is everywhere dense on the torus then the same is true for all the trajectories of this equation.

If, however, $\mathscr{D}'(\vartheta_0) = F \neq C$, then there are essentially two kinds of trajectories, those for which ϑ_0 in (2.143) is in F and those for which ϑ_0 is in $O = C - F$. Every trajectory of the first kind returns arbitrarily close to any point through which it passes. This is not true of the trajectories of the second kind. In fact, every point on one of these latter has a neighborhood not reentered by the trajectory. We note that every trajectory of either kind comes arbitrarily close to any point on a trajectory of the first kind.

2.25. *Example.* Consider the family of trajectories on the torus defined by the equation

$$\frac{d\vartheta}{d\varphi} = \mu.$$

Here,

$$u(\varphi, \vartheta_0) = \vartheta_0 + \mu\varphi \tag{2.2501}$$

and hence the integral curves in the (φ, ϑ)-plane are straight lines with the slope μ and passing through the point $(0, \vartheta_0)$.

Thus the induced transformation (cf. 2.15) of the meridian circle C of unit circumference is given by the rotation

$$\bar{\vartheta} = U(\vartheta) = U(1, \vartheta) = \vartheta + \mu$$

through the angle $2\pi\mu$.

2.251. Let $\mu = n/m$ where m and n are integers. Here u^m is the identity transformation. Since every point of our circle is fixed by u^m, every trajectory is closed in this case.

2.252. Let μ be irrational. Then no power of u has fixed points and therefore there are no closed trajectories. In this case, for every ϑ_0, the set

$$\vartheta_k = u^k(\vartheta_0) = \vartheta_0 + k\mu$$

is everywhere dense on our meridian circle (cf. I, 11.67) and hence, the integral curve defined by (2.2501) is everywhere dense on the torus.

Thus, if μ is irrational there are no closed trajectories and every integral curve is everywhere dense on the torus.

2.26. Some sufficient conditions for ergodicity (Denjoy [13] Siegel [49]). In what follows we shall write $\widehat{pq}$ for that arc on the oriented circumference C which is generated by a point moving from p to q in the positive sense along C.

Let $p_0 = p$, $q_0 = q$, $u^k(p_0) = p_k$, $u^k(q_0) = q_k$. In what follows one assumes that the arcs $\widehat{p_i q_i}$ and $\widehat{p_j q_j}$, $i \neq j$, are disjoint.

2.261. LEMMA. *If (2.13) has no closed trajectories, then for every integer $N > 0$, there exists an integer $n > N$ such that either the n arcs $\widehat{p_{-k}q_{n-k}}$ or the n arcs $\widehat{p_{n-k}q_{-k}}$ for $k = 1, \ldots, n$, are disjoint.*

We observe that the set of arcs $\widehat{p_{n-k}q_{-k}} (k = 1, 2, \ldots, n)$ coincides with the set of arcs $\widehat{p_k q_{k-n}}$ $(k = 0, 1, 2, \ldots, n-1)$.

Consider the $2N$ arcs $\widehat{p_0p_j}$ $(j = \pm 1, \ldots, \pm N)$ and suppose that $\widehat{p_0p_m}$ is the smallest among these. Thus for $j \neq m$, p_m lies inside the arc $\widehat{p_0p_j}$. Since (2.13) has no closed trajectories, by Lemma 2.19 there exists a point $p_h = u^h(p_0)$ inside the arc $\widehat{p_0p_m}$ with $|h| > N$. We may take the smallest such $|h|$ as n. For suppose that neither the n arcs $\widehat{p_{-k}q_{n-k}}$ $(k = 1, \ldots, n)$ nor the n arcs $\widehat{p_kq_{k-n}}$ $(k = 0, 1, 2, \ldots, n-1)$ are disjoint. Then if $h > 0$, we can find integers r, l among the integers $1, 2, \ldots, n$ and if $h < 0$ we can find r, l among the integers $0, -1, -2, \ldots, -n$, such that p_{-l} lies inside the arc $\widehat{p_{-r}q_{h-r}}$, whence $p_{r-l} = u^r(p_{-l})$ lies inside the arc $\widehat{p_0q_h}$ since $\widehat{p_hq_h}$ and $\widehat{p_mq_m}$ are disjoint by hypothesis and since $\widehat{p_0p_h} \subset \widehat{p_0p_m}$ it follows that $\widehat{p_0q_h} \subset \widehat{p_0q_m}$, and hence $p_{r-l} \in \widehat{p_0q_m}$. This contradicts the choice of m and h, since $|r - l| < |h| = n$.

2.262. If we write $u_k = u^k(\vartheta)$ and $u'(\vartheta) = du/d\vartheta$ (cf. (2.156)) we see at once that

$$(2.2621)\quad \left.\frac{du_n}{d\vartheta}\right|_{\vartheta=\vartheta_0} = \prod_{k=1}^{n} u'(\vartheta_{n-k}), \quad \left.\frac{d\vartheta}{du_{-n}}\right|_{\vartheta=\vartheta_0} = \prod_{k=1}^{n} u'(\vartheta_{-k}) \quad (n = 1, 2, \ldots).$$

Choose an arc $\widehat{pq}$ and let δ_k be the length of the arc $\widehat{p_kq_k}$, then applying the mean value theorem to the two ratios δ_n/δ_0, δ_0/δ_{-n} and using (2.2621) we get

$$(2.2622)\quad \begin{aligned} \log \frac{\delta_0^2}{\delta_n\delta_{-n}} &= \sum_{k=1}^{n} (\log u'(\vartheta_{-k}) - \log u'(\bar{\vartheta}_{n-k})) \\ &\leqq \sum_{k=1}^{n} |\log u'(\vartheta_{-k}) - \log u'(\bar{\vartheta}_{n-k})| = \sum\nolimits_n . \end{aligned}$$

If (2.13) has no periodic solutions and is not ergodic, then the set O in (2.215) is not empty. Choose an open arc $\alpha_0 = O$. Let δ_k be the length of $\alpha_k = u^k(\alpha_0)$. Since no two open arcs α_k overlap, then $\sum_{k=-\infty}^{+\infty} \delta_k \leqq 1$, whence $\lim_{n\to\infty} \delta_n\delta_{-n} = 0$. Thus $\delta_0^2/\delta_n\delta_{-n}$ and hence $\log \delta_0^2/\delta_n\delta_{-n}$ tends to $+\infty$ as $n \to +\infty$.

In case the right-hand member Σ_n of (2.2622) is bounded we arrive *at a contradiction which implies that O is in fact empty and* (2.13) *is ergodic.*

Since the open arcs α_k do not overlap, 2.261 applies to their end-points. Hence the intervals $\vartheta_{-k}\vartheta_{n-k}$ are disjoint. Hence $\Sigma_n \leqq$ total variation of $\log u'$. Hence if $\log u'$ is of bounded variation Σ_n is bounded. Finally, $\log u'$ will be of bounded variation if u' is both positive and of bounded variation. Thus:

2.263. THEOREM. *If* (2.13) *has no closed trajectories, then it is ergodic provided that* $u'(\vartheta)$ *is of bounded variation and is different from zero everywhere.*

2.264. The boundedness of Σ_n is also assured by the requirement that $\partial A/\partial\vartheta$ be continuous and of bounded variation in ϑ uniformly in φ for $0 \leqq \varphi \leqq 1$.

For,

$$\log \frac{\partial u(\varphi, \vartheta)}{\partial \vartheta} = \int^{\varphi} \frac{\partial A(\varphi, u)}{\partial u} d\varphi, \quad u = u(\varphi, \vartheta),$$

whence, if we write $\partial A(\varphi, u)/\partial u = A_u(\varphi, u(\varphi, \vartheta))$,

$$\log u'(\vartheta) = \int_{\varphi_0}^{\varphi_0+1} Au(\varphi, u(\varphi, \vartheta))\, d\varphi$$

and therefore

$$(2.2641) \qquad \log u'(\vartheta_{-k}) - \log u'(\bar{\vartheta}_{n-k})$$
$$= \int_{\varphi_0}^{\varphi_0+1} [Au(\varphi, u(\varphi, \vartheta_{-k})) - Au(\varphi, u(\varphi, \vartheta_{n-k}))]\, d\varphi.$$

The images

$$u(\varphi, \vartheta_{-k}), \quad u(\varphi, \vartheta_{n-k})$$

on the meridian φ of, say, the family $\{\vartheta_{-k}, \vartheta_{n-k}\}$ of disjoint arcs, are also disjoint and if $A_n(\varphi, \vartheta)$ is of bounded variation in ϑ uniformly in φ, then (2.2641) implies the boundedness of Σ_n.

If we replace the conditions in the beginning of this section by somewhat stronger conditions, we obtain the well-known

2.265. THEOREM. *If* (2.13) *has no closed trajectories and if* (*Denjoy*) $\partial^2 A/\partial\vartheta^2$ *is continuous* (*or even bounded*) *on the torus then* (2.13) *is ergodic. In particular, if* (*Poincaré*) $A(\varphi, \vartheta)$ *is analytic then* (2.13) *is either ergodic or it possesses closed trajectories.*

2.3. It is of interest to note that in case (2.13) is ergodic, then its family of trajectories is topologically equivalent to the family of the straight line trajectories in (2.252) for a suitable (unique) choice of μ. This number μ introduced by Poincaré [47] is called the *rotation number* of the original system (2.13). In what follows we shall treat this useful concept in a geometrical and intuitively satisfying manner [56] and make use of it to provide an alternative way of establishing some of the results in the preceding sections.

Let L be a fixed trajectory (2.143) of (2.13). Consider the family

(2.301) $$\{L(m, n)\} \qquad (m, n \text{ integers})$$

of all the curves $L(m, n)$ in the (φ, ϑ)-plane, obtained from L by translations (m, n) with integral m and n. We saw in 2.14 that every $L(m, n)$ is again a trajectory and that it passes through $(m, \vartheta + n)$.

If we take the intersection of (2.301) with the square Σ in (2.12) we obtain the whole trajectory through $(0, \vartheta_0)$ on our torus. In particular, each point $(0, \vartheta_k)$ in 2.17 is the intersection of

(2.302) $$L(-k, -[u(k, \vartheta_0)])$$

and the line $\varphi = 0$, in view of (2.157).

2.31. We shall say that a point (φ, ϑ_1) lies above a trajectory

(2.311) $$L\colon \vartheta = u(\varphi, \vartheta_0)$$

if $\vartheta_1 > u(\varphi, \vartheta_0)$, and that it lies below this trajectory if $\vartheta_1 < u(\varphi, \vartheta_0)$. Since only one trajectory passes through each point, it follows that if (φ_1, ϑ_1) is above (below) L, then the whole trajectory

$$\vartheta = u(\varphi, \varphi_1, \vartheta_1)$$

lies above (below) L.

It is clear that $L(0, 1)$ is above L and $L(0, -1)$ is below L.

If $L(m, n)$ is above L, then applying the translation (m, n) to this pair of curves, we see that $L(2m, 2n)$ is above $L(m, n)$ and for any positive integer k, $L(km, kn)$ is above $L((k-1)m, (k-1)n)$ and hence above L. Applying the translation $(-m, -n)$ to the same pair of curves we see that $L(-m, -n)$ and in general $L(-km, -kn)$ is below $L = L(0, 0)$.

The slope μ of a straight line (2.2501) with the ϑ intercept ϑ_0 is completely determined by the manner in which this line partitions the lattice $(m, \vartheta_0 + n)$, i.e., the integral lattice (m, n)

translated by $(0, \vartheta_0)$. We shall see that the manner in which every trajectory (2.143) partitions the lattice $(m, \vartheta_0 + n)$, again defines a real number μ, the *rotation number*. This rotation number has the same value for all the trajectories (2.143) of the equation (2.13) and takes the place of the slope of Example 2.25 in the description of the basic properties of the family of trajectories of this equation.

As in the case of the straight line, we put a rational number n/m (n and m are integers and $m > 0$) into class R_0 or into class R_1 depending on whether $L(m, n)$ is below L or is above L. From the discussion in the beginning of this section it follows that if $n/m = n'/m'$, $m > 0$, $m' > 0$, then $L(m, n)$ and $L(m', n')$ are both below L, both above L, or both coincide with L, so that there is no ambiguity in the classification. The two classes exist, because if n is positive and very large $L(1, n)$ is above L and $L(1, -n)$ is below L. If n/m is not in class R_0 and if $s/r > n/m$ then s/r is in R_1. The trajectory $L(rm, rn)$ is either above L or is coincident with L. Next $L(rm, sm)$ is above $L(rm, rn)$ since it is obtained from it by the translation $(0, sm - rn)$, and $sm - rn > 0$. Hence $L(rm, sm)$ is above L, and $sm/rm = s/r$ is in R_1. Similarly, if n/m is not in R_1 and if $s/r < n/m$, then s/r is in R_0. Thus all rational numbers with possibly one exception, are included either in R_0 or in R_1 and R_0 and R_1 define a real number μ_0, rational or irrational.

2.32. If n/m is neither in R_0 nor in R_1, then $\mu_0 = n/m$, $L(m, n)$ coincides with L, and L defines a closed trajectory on the torus. Here (cf. 2.143)

$$u(\varphi, \vartheta_0) - \frac{n}{m}\varphi$$

is periodic in φ.

2.33. Let p be given, and let q be the largest integer such that q/p is in R_0. Then $q \leqq \mu_0 p \leqq q + 1$. Therefore every point $(\varphi + p, u(\varphi, \vartheta_0) + q)$ of $L(p, q)$ is below $L = L(\vartheta_0)$ and every point $(\varphi + p, u(\varphi, \vartheta_0) + q + 1)$ is not below L. Thus

$$u(\varphi, \vartheta_0) + q < u(\varphi + p, \vartheta_0) \leqq u(\varphi, \vartheta_0) + q + 1. \tag{2.331}$$

If

$$\alpha \leqq u(\varphi, \vartheta_0) \leqq \beta \quad \text{for} \quad 0 \leqq \varphi \leqq 1, \tag{2.332}$$

then

$$(2.333)\quad \alpha + \mu_0(p+\varphi) - \mu_0 - 1 < u(\varphi + p, \vartheta_0) \leqq \beta + \mu_0(p+\varphi) + 1 \quad \text{for} \quad 0 \leqq \varphi \leqq 1.$$

Let $\bar{\varphi}$ be arbitrarily large and write $[\bar{\varphi}] = p$ and $(\bar{\varphi}) = \varphi$. Then (2.333) yields

$$(2.334)\qquad \mu_0\bar{\varphi} + \alpha - \mu_0 - 1 < u(\bar{\varphi}, \vartheta_0) < \mu_0\bar{\varphi} + \beta + 1.$$

The last inequality implies that

$$(2.335)\qquad \lim_{\varphi\to\infty} \frac{u(\varphi, \vartheta_0)}{\varphi} = \mu_0.$$

Consider two trajectories

$$L\colon \vartheta = u(\varphi, \vartheta_i), \quad i = 0, 1; \quad \vartheta_0 < \vartheta_1,$$

and let μ_i be their rotation numbers. Write $r = [\vartheta_1 - \vartheta_0]$. Then $u(\varphi, \vartheta_0) < u(\varphi, \vartheta_1) < u(\varphi, \vartheta_0 + r + 1) = u(\varphi, \vartheta_0) + r + 1$ by (2.142),

$$|u(\varphi, \vartheta_1) - u(\varphi, \vartheta_0)| < r + 1,$$

and hence

$$\lim_{\varphi\to\infty} \frac{u(\varphi, \vartheta_1) - u(\varphi, \vartheta_0)}{\varphi} = 0.$$

Thus $\mu_0 = \mu_1$. It follows that all the trajectories (2.143) of (2.13) have the same rotation number.

2.34. *If μ is rational then at least one of the trajectories of the equation* (2.13) *is closed on the torus.*

Let L be a fixed trajectory (2.143) of (2.13) and let $\mu = n/m$, $(n, m) = 1$, $m > 0$. If n/m is neither in R_0 nor in R_1, then (cf. 2.32) L defines a closed trajectory on the torus. Suppose next that n/m is in R_0. Then $L(m, n)$ lies below L. Moreover, no curve $L(p, q)$ lies between $L(m, n)$ and L. For, $L(rm, rn)$ is below $L((r-1)m, (r-1)n)$ for every positive integer r, $L(p, q)$, $p > 0$, is below $L(m, n)$ if $q/p < m/n$ (cf. 2.31) and $L(p, q)$ is above L if $q/p > m/n$.

We observe that the intercepts ϑ_0 and ϑ_{-m} of L and $L(m, n)$ respectively are neighbors and hence there exists a closed trajectory by Lemma 2.19.

2.35. In what follows we suppose that μ is irrational. We consider two distinct trajectories $L(m_1, n_1)$ and $L(m_2, n_2)$ of the family (2.301). Without the loss of generality we may assume that $m_2 \geqq m_1$. The second one of these trajectories is obtained from the first one by the translation $(m_2 - m_1, n_2 - n_1)$. Therefore the trajectory $L(m_2, n_2)$ lies above the trajectory $L(m_1, n_1)$ if

(2.351) $$m_2 = m_1 \quad \text{and} \quad n_2 > n_1$$

or if

(2.352) $$m_2 \neq m_1 \quad \text{and} \quad \frac{n_2 - n_1}{m_2 - m_1} > \mu.$$

Conditions (2.351) and (2.352) together are equivalent to the condition

(2.353) $$n_2 - m_2\mu > n_1 - m_1\mu.$$

We observe that

(3.354) $$\tau(m, n) = n - m\mu$$

is the ϑ intercept of the straight line Λ (m, n) with the slope μ and passing through the lattice point (m, n).

If μ is irrational, the correspondence

(2.355) $$\Lambda(m, n) \longleftrightarrow L(m, n)$$

is one-to-one. Moreover since (2.353) implies that $\Lambda(m_2, n_2)$ lies above $\Lambda(m_1, n_1)$, this correspondence is monotone. Correspondence (2.355) induces a monotone correspondence between the ϑ intercepts $\tau(m, n)$ of $\Lambda(m, n)$ and the ϑ intercepts $u(-m, \vartheta_0) + n$ (cf. (2.142)) of $L(m, n)$. In particular the kth iterate $\vartheta_k = (k\mu) \in \mathscr{D}(0)$ in 2.252, for the special case in our Example 2.25, is the ϑ intercept $\tau(-k, -[k\mu])$ of $\Lambda(-k, -[k\mu])$. Since, moreover,

$$\frac{[u(k, 0)]}{k} < \mu < \frac{[u(k, 0)] + 1}{k}$$

by the definition of μ, we have

(2.356) $$[u(k, 0)] = [k\mu]$$

and by 2.302), the kth iterate $\vartheta_k \in \mathscr{D}(0)$ in 2.17 is the ϑ intercept of $L(-k, -[k\mu])$. By the monotoneity of (2.355) the correspondence

$$\vartheta_k = (u(k, 0)) \longleftrightarrow \bar{\vartheta}_k = (k\mu)$$

is monotone in view of (2.302) and (2.356). That is if $u^{k_2}(0) > u^{k_1}(0) > 0$ then $(k_2\mu) > (k_1\mu)$ and conversely. If for some $0 < \vartheta_0 < 1$ we have $u^{k_2}(\vartheta_0) < u^{k_1}(\vartheta_0)$, then the difference

$$u^{k_2}(\vartheta) - u^{k_1}(\vartheta)$$

vanishes for some ϑ_1 in the interval $0 < \vartheta < \vartheta_0$. This means that $u^{k_2-k_1}(\vartheta_1) = \vartheta_1$, or that $u^{k_2-k_1}$ has a fixed point ϑ_1. This cannot happen if μ is irrational. Thus, as was first observed by Poincaré, *the circular order of the points in the set* (2.21), *coincides with the circular order of the points* $(k\mu)$ *and hence depends only upon* μ *in case* μ *is irrational.* That is, the mapping

$$\vartheta_k = (u(k, \vartheta_0)) \longleftrightarrow \bar{\vartheta}_k = (k\mu) \tag{2.357}$$

is monotone for every ϑ_0.

2.36. The mapping f of the set $\mathscr{D}(\vartheta_0)$ onto the set $\mathscr{D}(0)$ defined by (2.357) may be extended to the interval $0 \leqq \vartheta < 1$ (the meridian $\varphi = 0$ of the torus) by letting

$$f(\vartheta) = \underset{\vartheta_{mj} \leqq \vartheta}{\text{l.u.b.}}\, f(\vartheta_{mj}). \tag{2.361}$$

Then $f(\vartheta)$ is a single valued, monotone, and continuous function mapping the meridian $\varphi = 0$ onto itself.

The behavior of f may be used to differentiate between the transitive and the intransitive cases (cf. 2.23 2.24). In the transitive case the set $\mathscr{D}(\vartheta_0)$ is everywhere dense on the meridian and f is strictly monotone. In the intransitive case there are whole intervals not containing points of $\mathscr{D}(\vartheta_0)$. Throughout each such interval the function f is stationary.

If f is strictly monotone, then the mapping it induces is one-to-one and bicontinuous.

2.37. Next,

$$\mathrm{f}(\mathscr{D}(\vartheta)) = \{(\mathrm{f}(\vartheta) + \mathrm{k}\mu)\}, \mathrm{k} = 0, \pm 1, \pm 2, \ldots, \tag{2.371}$$

for every ϑ. For, since $u(\vartheta)$ in (2.156) is monotone we get, in view of (2.361)

$$\begin{aligned} f(u^k(\vartheta)) &= \underset{u^k(\vartheta_{mj}) \leqq u^k(\vartheta)}{\text{l.u.b.}}\, f(u^k(\vartheta_{mj})) = \underset{\vartheta_{mj} \leqq \vartheta}{\text{l.u.b.}}\, f(u^k(\vartheta_{mj})) \\ &= \underset{\vartheta_{mj} \leqq \vartheta}{\text{l.u.b.}}\, f(\vartheta_{mj+k}) = \underset{\vartheta_{mj} \leqq \vartheta}{\text{l.u.b.}}\, (m\mu + j + k\mu) \\ &= \underset{\vartheta_{mj} \leqq \vartheta}{\text{l.u.b.}}\, (m\mu + j) + (k\mu) = f(\vartheta) + (k\mu). \end{aligned}$$

2.38. Represent the torus as in (2.12) and consider the family of trajectories of a system (2.13) with a trajectory L everywhere dense on the torus. Let μ be the rotation number of L and consider the family of trajectories in the Example 2.252.

Consider the mapping

$$T : \varphi' = \varphi, \quad \vartheta = f(u(0, \varphi, \vartheta)) + \mu\varphi \tag{2.381}$$

of the torus Σ onto itself. This mapping is a homeomorphism. For it is one-to-one and it itself as well as its inverse are continuous by the continuity in the initial conditions.

One observes that in order to obtain the image $T(\varphi, \vartheta) = (\varphi', \vartheta)$ of a point (φ, ϑ) on Σ, one constructs the trajectory through (φ, ϑ), determines its ϑ intercept, takes the image of this intercept under the mapping f and draws the straight line of slope μ through this image. The point on this line with the abscissa equal to φ is the desired image.

Thus in the case of a transitive system the torus may be homeomorphically mapped onto itself in such a way that the images of the trajectories of our system are the trajectories of Example 2.252. It follows therefore that the behavior of trajectories of a transitive system is essentially that of the trajectories of the special case 2.252. In particular, this observation yields a proof independent of that of Theorem 2.23, that in the transitive case every trajectory is everywhere dense on the torus.

2.4. Kneser [25] studied systems (2.13) for which either $\Phi(\varphi, \vartheta)$ or $\Theta(\varphi, \vartheta)$ but not both were allowed to vanish. Maier [33] made a qualitative study of curve families on surfaces of genus $n \geqq 2$. The torus is a surface of genus 1 and is the only two dimensional manifold admitting vector fields without singular points.

3. Geometrical Classification of Singular Points

Let O be in isolated singular point.

3.1. We say that O is *stable* if in every circular neighborhood of O there is a closed trajectory containing O in its interior, otherwise we say that O is *unstable.*

3.2. THEOREM. *In a sufficiently small neighborhood $S(O, r)$ of an isolated singular point O, every closed trajectory contained in $S(O, r)$ contains O in its interior.*

The proof follows directly from Theorem 1.5.

3.3. Stable isolated singular points. Consider a small circle $S(O, r)$ with the center at O and containing no singular points other than O, either in its interior or on its boundary. Suppose that O is stable.

Take a fixed radius OP of $S(O, r)$ and consider on OP the set

Σ of all the points through which pass closed trajectories entirely contained in $S(O, r)$. These trajectories obviously encompass O. By the continuity in the initial conditions, the set $\Sigma \cup O$ is closed. Let M be the least upper bound (in the sense of distance from O) of Σ. Then $M \in \Sigma$, $\Sigma \cup O \subset OM$, and the complement Δ of $\Sigma \cup O$ in OM consists of at most a denumerable number of open intervals $I_j = (a_j, b_j)$.

If this complement Δ is not empty and if $p \in I_j$, then (cf. Lemma 1.61) the trajectory through p spirals toward the two closed trajectories $f(a_j, t)$ and $f(b_j, t)$. Such a stable singular point is called a *center-focus*.

If Δ is empty for a sufficiently small $S(O, r)$, then (after Poincaré) O is called a *center*.

3.4. Unstable isolated singular points. If O is unstable, we may assume, in view of Theorem 3.2, that $S(O, r)$ is small enough so that there are no other singular points and no periodic trajectories contained in $\bar{S}(O, r)$.

If a half-trajectory has O as its only ω- or α-limit point in $S(O, r)$ we say that it *adheres* to O.

Again we choose a fixed radius OP of $S(O, r)$ and consider a trajectory $f(q, t)$ through a point q on OP.

If either one of the half-trajectories $f^+(q)$ or $f^-(q)$ are contained in $S(O, r)$ then either the ω- or the α-limit set is contained in $S(O, r)$. Then O is in the ω-limit set (α-limit set), and in fact is its only singular point, and by Theorem 1.76, either O is the only point of the limit set or there exists a trajectory adhering to O at both ends.

Suppose next that both $f^+(q)$ and $f^-(q)$ leave $S(O, r)$ in finite time. Let $p = f(q, t(q))$ be the first intersection of $f^+(q)$ and the circumference of S. As q tends toward O, p moves monotonically along the circumference. Let $m = \lim_{q \to 0} p$. We assert that $f^-(m) \subset S(O, r)$.

Consider a sequence $q_1, q_2, \ldots, q_l, \ldots$ on OP tending toward O and write $p_j = f(q_j, t(q_j))$. Then $p_j \to m$. Since O is an equilibrium point, by the continuity in the initial conditions we have $t_j = t(q_j) \to + \infty$ as $q_j \to O$. Thus for every positive T, however large, we can find a point p_i sufficiently close to m so that $t(q_i) > T$. Should $f^-(m)$ leave $S(O, r)$ in finite time T, then for points p_i sufficiently close to m, $f^-(p_i)$ would leave S after the lapse of $T + \varepsilon_i$, where

$\varepsilon_i > 0$ and may be made arbitrarily small — a contradiction. This completes the proof of

3.42. THEOREM. *If an isolated singular point O is not stable then there exists a half-trajectory which has O as its only ω- or α-limit point.*

Moreover

3.42. THEOREM. *There exists a small enough neighborhood $S(O, r)$ of an unstable isolated singular point O, so that every half-trajectory $f^+(q)$, $q \in S(O, r)$ either has O as its only ω-* or *α-limit point or it leaves S in finite time.*

Let $f^+(q)$ or $f^-(q)$ be a half-trajectory which remains in $S(O, r)$. Then either O is the only limit point of our half-trajectory or there is a trajectory $f(q_1, t)$ which adheres to O at both ends. Let d be the diameter of $f(q_1, t)$. It is easily seen that $S(O, d/2)$ is the neighborhood called for in our theorem. For, let Op_1 be an arc of $f(q_1, t)$ joining O with a point p_1 on the circumference of $S(O, d/2)$. Let $p \in S(O, d/2)$. If one of the half-trajectories through p remains in $S(O, d/2)$, then either O is the only limit point of this half-trajectory or our half-trajectory spirals around (a curve adhering to) O. This last is impossible in view of the presence of the trajectory arc Op_1.

Thus in a sufficiently small neighborhood $S(O, r)$ of an unstable singular point O the trajectories $f(p, t)$, $p \in S(O, r)$, fall into the following three classes [11].

3.43. Elliptic trajectories, i.e., those which are contained in in $S(O, r)$ and which adhere to the origin at both ends.

3.44. Parabolic trajectories, whose one end, say $f^+(p)$, is contained in $S(O, r)$ and adheres to O and whose other half-trajectory $f^-(p)$ leaves the neighborhood in finite time.

3.45. Hyperbolic trajectories, for which both half-trajectories $f^+(p)$ and $f^-(p)$ leave $S(O, r)$ in finite time.

If all the points in a given set lie on elliptic trajectories we shall say that the point set is elliptic. Similarly this holds for the other types of trajectories.

3.46. THEOREM. *Consider a small neighborhood $S(O, r)$ of O called for in Theorem 3.42. The set of points in $S(O, r)$ lying on elliptic (hyperbolic) trajectories, if there exist such trajectories in $S(O, r)$ at all, has inner points and the set of all points lying on the hyperbolic trajectories form a domain.*

Consider an elliptic trajectory. We assert that all the trajectories $f(p, t)$ in its interior are elliptic. If the limit set of a half-trajectory of $f(p, t)$, say $f^+(p)$, contains no singular points, then by Theorem 1.47 it is a closed trajectory which (Theorem 1.5) must contain in its interior singular points different from O — a contradiction. Thus the limit set of $f^+(p)$, contains singular points and hence consists of a unique singular point, namely O. If O were not the only limit point of $f^+(p)$, then $f^+(p)$ would spiral around a curve adhering to O at both ends (1.76 and 1.78), which is impossible if $f(p, t)$ is contained in the interior of an elliptic trajectory. Thus every point in the *interior* of an elliptic trajectory is elliptic.

If p is a point on a hyperbolic trajectory $f(p, t)$, then both half-trajectories $f^+(p)$ and $f^-(p)$ leave $S(O, r)$ in finite time. In view of the continuity in the initial conditions the same is true for every point q in a sufficiently small neighborhood U_p of p.

3.47. Lemma. *Consider a small neighborhood $S(O, r)$ of an isolated unstable singular point O, chosen as in Theorem 3.42. Let $C(O, r)$ be the circumference of $S(O, r)$. Suppose that for two points p and q on $C(O, r)$ both half-trajectories $f^+(p)$ and $f^+(q)$ are contained in $S(O, r)$ and both adhere to O. If for no other point m on an arc pq of C the half-trajectory $f^+(m)$ adheres to the origin, then there is a point n on this arc such that $f^-(n)$ is contained in S and adheres to the origin.*

If $f^-(p)$ or $f^-(q)$ is contained in the "triangular" sector pOq of S bounded by our arc pq and the two half-trajectories $f^+(p)$ and $f^+(q)$, then we may take n to be p or q respectively. Suppose next that this is not true. Let p_j be a sequence of points on our arc such that $p_j \to p$ as $j \to \infty$. The half-trajectories $f^+(p_j)$ enter S for all p_j sufficiently close to p, but cannot remain in S without adhering to O. Let $\xi_j = f(p_j, t(p_j))$ be the point of intersection of $f^+(p_j)$ and C through which $f^+(p_j)$ first leaves S. As j increases, the points ξ_j move monotonically away from p and toward q. Let $\lim_{j \to \infty} \xi_j = q'$. We assert that $f^-(q')$ remains in S and adheres to O. Since $f(p, t)$ remains in S for all $t > 0$, by the continuity in the initial conditions $t(p_j) \to \infty$ and this in turn implies, again by the continuity in the initial conditions, that $f(q', t)$ remains in S for all finite negative values of t. It follows from this by the now familiar argument that O is the only α-limit point of $f(q', t)$.

3.471. We shall speak of two half-trajectories through points

p and q on C and adhering to O as *adjacent attached half-trajectories* when there are no halftrajectories through points on an arc pq, adhering to O, and contained in the corresponding "triangular" (cf. 3.47) sector pOq, bounded by our arc and the two given half-trajectories. We have the following

3.472. COROLLARY. *Two adjacent attached halftrajectories are of opposite sign. That is if one of them is $f^+(p)$ then the other is $f^-(q)$ and vice versa.*

3.48. LEMMA. *If a small neighborhood $S(O, r)$ of an isolated unstable singular point O is chosen as in Theorem 3.42 then the number of pairs of adjacent attached trajectories is finite.*

Suppose that there is an infinite number of pairs $(f^{\pm}(p_j), f^{\mp}(q_j))$ of adjacent attached half-trajectories (cf. 3,471). Let q be a limit point of q_j. Choose among the q_j a subsequence q_i which tends to q. Then the corresponding sequence p_i also tends to q. Since q is not a singular point, the arc $A : f(q, t)$, $-t_0 \leqq t \leqq t_0$ has, for a suitable choice of t_0 and an $\varepsilon > 0$, an ε-neighborhood (A_ε) (cf. 1.3) in which all the trajectories cut normals to A in one direction only. This is inconsistent with the fact that for sufficiently large i both p_i and q_i fall into $(A)_\varepsilon$.

3.49. We are in a position now to analyze the behavior of trajectories in a sufficiently small (cf. Theorem 3.42) neighborhood $S(O, r)$ of an isolated unstable singular point O.

First we observe that the set F of all points on the circumference $C(O, r)$ of $S(O, r)$ through which pass half-trajectories contained in S and adhering to the origin, is closed. Then the complement of F in C is open and hence, if not empty, consists of at most a denumerable number of non overlapping open arcs $(p_j q_j)$. For each j, the adjacent adhering half-trajectories through p_j and q_j are of opposite signs (3.472), whence the number of such pairs and hence of the above open arcs is finite. All the trajectories through points of each of the (at most) finite number of open arcs are of hyperbolic type (3.45). Each of the "triangular" (3.47) sectors p_jOq_j is called *a hyperbolic sector*. The number of points m in F for which both half-trajectories $f^+(m)$ and $f^-(m)$ are contained in S and adhere to O is finite (this can be shown by an argument similar to that in 3.48). Each such point m of F is associated with an *elliptic sector*, i.e., a closed region in $\bar{S}$ filled entirely with elliptic trajectories (3.43). Through all the remaining points of F

pass parabolic trajectories. Since F, as the complement of (at most) a finite number of open arcs, consists of (at most) a finite number of isolated points and of a finite number of closed arcs, the set of all parabolic points consists of (at most) a finite number of isolated half-trajectories adhering to O and of a finite number of *parabolic sectors* corresponding to subarcs of the closed arc-components of F.

We should note that S may contain a denumerable number of elliptic (closed) domains which do not reach $C(O, r)$.

4. Analytic criterica for various types of singular points.

We shall consider first two suggestive elementary examples.

4.1. Linear systems with constant coefficients. The origin is an isolated singular point (in fact the only singular point) of the system

$$\frac{dx}{dt} = Ax, \quad A = \begin{bmatrix} a_{11} & a_{12} \\ a_{21} & a_{22} \end{bmatrix}, \quad x = \begin{bmatrix} x_1 \\ x_2 \end{bmatrix}, \tag{4.101}$$

$$|A| = a_{11}a_{22} - a_{21}a_{12} \neq 0.$$

Following Poincaré we associate the basic forms of the behavior of the trajectories of (4.101) near the origin with a classification of the *characteristic roots* of A, i.e., the roots of the *characteristic equation*

$$|A - \lambda I| = \begin{vmatrix} a_{11}-\lambda & a_{12} \\ a_{21} & a_{22}-\lambda \end{vmatrix} = \lambda^2 - T\lambda + |A| = 0, \tag{4.102}$$

$$T = a_{11} + a_{22},$$

of A.

It is convenient to employ a nonsingular linear transformation

$$x = Ky, \quad K = \begin{bmatrix} k_{11} & k_{12} \\ k_{21} & k_{22} \end{bmatrix}, \quad y = \begin{bmatrix} y_1 \\ y_2 \end{bmatrix}, \quad |K| \neq 0,$$

which does not change the nature of the behavior of solutions near the origin, to reduce (4.101) to the Jordan cannonical form (cf. III. 1.46)

$$\frac{dy}{dt} = Jy, \quad J = K^{-1}AK. \tag{4.103}$$

If the characteristic roots λ_1 and λ_2 of A are distinct then

$$J = \begin{bmatrix} \lambda_1 & 0 \\ 0 & \lambda_2 \end{bmatrix}. \tag{4.104}$$

We note that the condition $|A| \neq 0$ in (4.101) implies, in view of (4.102), that neither one of the characteristic roots of the system (4.101) (i.e., those of its matrix A) can vanish.

If $\lambda_1 = \lambda_2$ is a multiple root of (4.102), then J is either given by (4.104) or

$$J = \begin{bmatrix} \lambda_1 & 0 \\ 1 & \lambda_1 \end{bmatrix}. \tag{4.105}$$

4.11. If the characteristic roots λ_1 and λ_2 in (4.104) are real and distinct, then all the real solutions of(4.103) are given by

$$y_1 = c_1 e^{\lambda_1 t}, \quad y_2 = c_2 e^{\lambda_2 t}. \tag{4.111}$$

Thus

$$\frac{dy_2}{dy_1} = \frac{\lambda_2\, y_2}{\lambda_1\, y_1} = \lambda c e^{\lambda_1(\lambda-1)t}, \quad \lambda = \frac{\lambda_2}{\lambda_1}, \quad c = \frac{c_2}{c_1}. \tag{4.112}$$

One can see from (4.111) that if λ_1 and λ_2 are of the same sign, then each trajectory (4.111) is of *parabolic type* and adheres (cf. Section 3) to the origin at one end. In this case we speak of the origin as a node.

If both λ_1 and λ_2 are negative, then each trajectory tends to the origin as $t \to +\infty$ and we say that the origin is a *stable node*.[6] Moreover, in view of (4.112), each trajectory is tangent at the origin[7] to the axis $y_2 = 0$ in case $\lambda > 1$ and is tangent to the axis $y_1 = 0$ in case $\lambda < 1$. These two axes contain the straight-line trajectories obtained from (4.111) by setting in turn $c_1 = 0$ and $c_2 = 0$. At infinity, i.e., as $t \to -\infty$, the trajectories tend to become parallel to the axis $y_1 = 0$ when $\lambda > 1$ and to the axis $y_2 = 0$ when $\lambda < 1$. (See Fig. 5).

If both λ_1 and λ_2 are positive, then each trajectory tends to the origin as $t \to -\infty$ and we say that the origin is an *unstable node*. Further analysis proceeds as above.

If $\lambda < 0$, i.e., if λ_1 and λ_2 are of opposite signs, say $\lambda_1 > 0$, then each trajectory (4.111) with $c_1 c_2 \neq 0$, is of *hyperbolic type* with the axes as asymptotes. The two trajectories on the axis $y_1 = 0$ tend to the origin as $t \to +\infty$ and the two trajectories on the axis

[6] One should guard against a misunderstanding due to a slight inconsistency in the accepted terminology. For, a *stable nodal point* is an *unstable singular point*.

[7] More precisely: the slope of the tangent line to the trajectory at $y(t)$, tends to zero as $t \to +\infty$ in case $\lambda > 1$.

$y_2 = 0$ tend to the origin as $t \to -\infty$. Every other trajectory tends to the axis $y_2 = 0$ as the asymptote as $t \to +\infty$. With increasing t every point first moves closer to the origin (if the motion starts far enough on the negative side of the trajectory) and after reaching

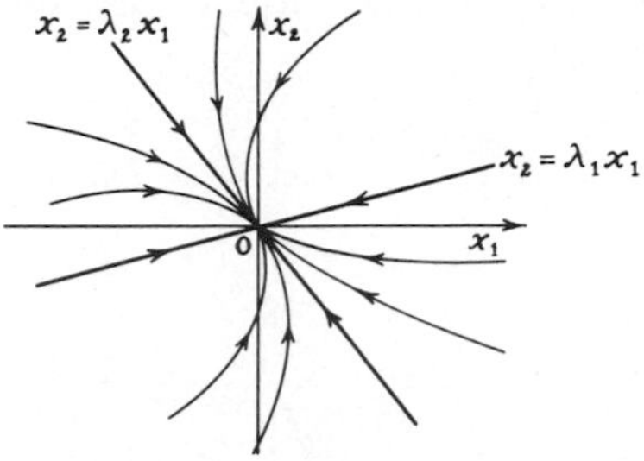

Fig. 5

a certain minimal distance the point moves away from the origin with its trajectory asymptotic to the axis $y_2 = 0$. In this case we say that the origin is a *saddle point* of the system (4.103) and hence of our original system (4.101). If $\lambda_1 < 0$, $\lambda_2 > 0$, the roles of the two axes are reversed. A saddle point is an unstable equilibrium (singular) point.

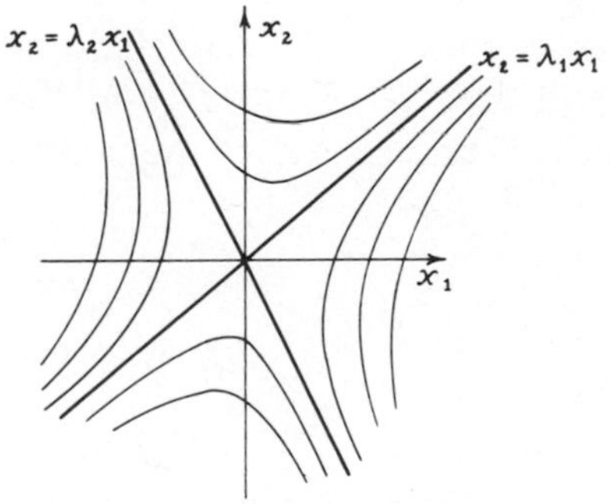

Fig. 6

4.12. If the characteristic roots λ_1, λ_2 of A are equal (and hence real) then all the real solutions of (4.103) *with J as in* (4.105) are given by

$$y_1 = c_1 e^{\lambda_1 t} + c_2 t e^{\lambda_1 t}, \quad y_2 = c_2 e^{\lambda_1 t}. \tag{4.121}$$

Thus,

$$y_1 = (c + t)y_2, \quad \frac{dy_1}{dy_2} = c' + t. \tag{4.122}$$

If $\lambda_1 < 0$, then the trajectories (4.121) tend toward the origin as $t \to +\infty$ and at the origin they are tangent to the axis $y_2 = 0$ by (4.122). As $t \to -\infty$ the trajectories tend to become parallel to this axis.

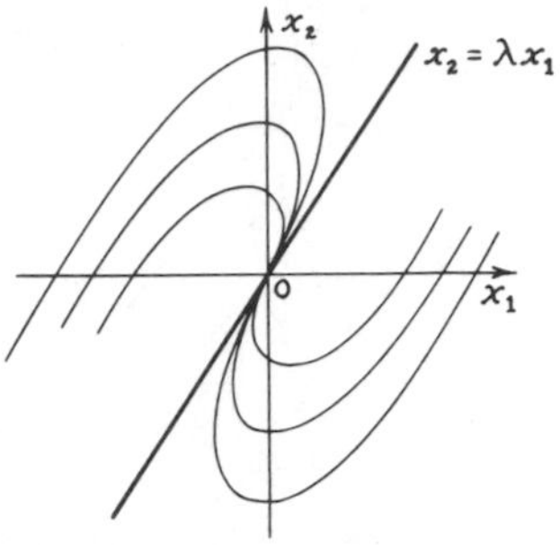

Fig. 7

If $\lambda_1 > 0$, then the trajectories (4.121) tend toward the origin as $t \to -\infty$, are tangent to the line $y_2 = 0$ at the origin, and tend to become parallel to this line as $t \to +\infty$.

The straight-line trajectories lie on the line $y_2 = 0$.

In both cases we speak of the origin as a *degenerate node* (Fig. 7).

If $\lambda_1 = \lambda_2$ and *J is given by* (4.104), then all the trajectories are given by (4.111) and consist of the familiy of all the rays which

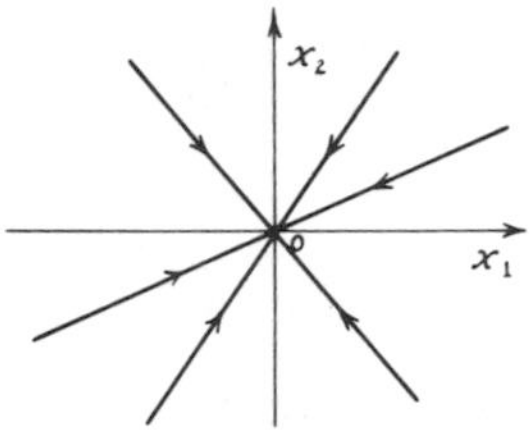

Fig. 8

terminate at the origin or the family of all those rays which point outward from the origin. In this case we say that the origin is a *singular node*.

4.13. If the characteristic roots of A are complex conjugates

$$(4.131) \qquad \lambda_1 = \alpha + \beta i = \varrho e^{i\vartheta}, \qquad \lambda_2 = \alpha - \beta i = \varrho e^{-i\vartheta},$$

then a suitable (cf. III, 2) linear transformation yields the very convenient real canonical system

$$(4.132) \qquad \frac{dw}{dt} = Rw, \quad R = \begin{bmatrix} \alpha & -\beta \\ \beta & \alpha \end{bmatrix}, \quad w = \begin{bmatrix} w_1 \\ w_2 \end{bmatrix},$$

all the solutions of which are given by

$$(4.133) \qquad \begin{aligned} w_1 &= e^{\alpha t}(c_1 \cos \beta t - c_2 \sin \beta t) & w_2 &= e^{\alpha t}(c_1 \sin \beta t + c_2 \cos \beta t) \\ &= e^{\alpha t} r \cos(\beta t + t_0), & &= e^{\alpha t} r \sin(\beta t + t_0). \end{aligned}$$

In case $\alpha = 0$, the trajectories (4.133) form a family of concentric circles (each of radius $r = \sqrt{c_1^2 + c_2^2}$) with centers at the origin. In this case the origin (our isolated singular point) is a *center* (cf. 3.3).

In case $\alpha < 0$, each of the trajectories (4.133) is a spiral which approaches the origin as $t \to +\infty$ winding around the origin either in the clockwise direction ($\beta > 0$) or in the counterclockwise direction ($\beta < 0$). The origin is called a *stable* focus [8]. If $\alpha > 0$, each spiral (4.133) winds around the origin and approaches it as $t \to -\infty$, thus yielding an *unstable focus*.

4.14. The behavior of the solutions of the original system (4.101) near the origin is essentially the same as the behavior of the solutions of its canonical form (4.103). However, the straight-line trajectories of the original system need not lie on the coordinate axes but are determined by the vectors k_i satisfying the conditions

$$(4.141) \qquad S(\lambda_i) : (A - \lambda_i \mathrm{I})k_i = 0, \quad k_i = \begin{bmatrix} k_{1i} \\ k_{2i} \end{bmatrix}.$$

If the system (4.141) is of rank 1 its solutions are proportional to a_{21}, $a_{22} - \lambda_i$. This is true except in the case of a singular node.

The discriminant Δ of (4.102) in terms of the elements a_{ij} of A, is given by

$$(4.142) \qquad \Delta = (a_{11} - a_{22})^2 + 4\, a_{12}\, a_{21}.$$

[8]Again, we should note that a *stable* focus is an *unstable singular point.*

The nature of the roots $\lambda_1, \lambda_2 = \frac{1}{2}[-T \pm \sqrt{\Delta}]$ can be easily discerned from the values of the elements of A. Thus if the trace T is positive and the discriminant Δ is negative, then the origin is a stable focus.

4.15. We may summarize the above discussion as follows:

4.151. If the characteristic roots λ_1 and λ_2 of A are real and $\lambda_1 \lambda_2 > 0$, then:

(a) In case $\lambda_1 \neq \lambda_2$, the straight line trajectories lie on two distinct straight lines determined by the two systems $S(\lambda_1)$ and $S(\lambda_2)$, each of rank one. All other trajectories are tangent to one of these straight lines at the origin and tend to become parallel to the other one at infinity (cf. Fig. 6).

(b) In case $\lambda_1 = \lambda_2$ and the system $S(\lambda_1)$ in (4.141) is of rank one (i.e., in case our matrix has one elementary divisor of degree two), then the origin is a degenerate node and the straight line trajectories lie on the straight line determined by the system $S(\lambda_1)$. All the trajectories are tangent to this line at the origin (cf. Fig. 8).

(c) In case $\lambda_1 = \lambda_2$ and the system $S(\lambda_1)$ in (4.141) is of rank zero, then every straight line through the origin contains two trajectories and the origin is a singular node (cf. Fig. 8).

4.152. If λ_1 and λ_2 are real and $\lambda_1 \lambda_2 < 0$, then the origin is a saddle point and the trajectories are asymptotic to two lines determined by the systems $S(\lambda_1)$ and $S(\lambda_2)$ respectively. The four straight-line trajectories lie on these two straight lines (cf. Fig. 6).

4.153. Let λ_1 and λ_2 be complex conjugates. If $\lambda_1 + \lambda_2 = 0$,

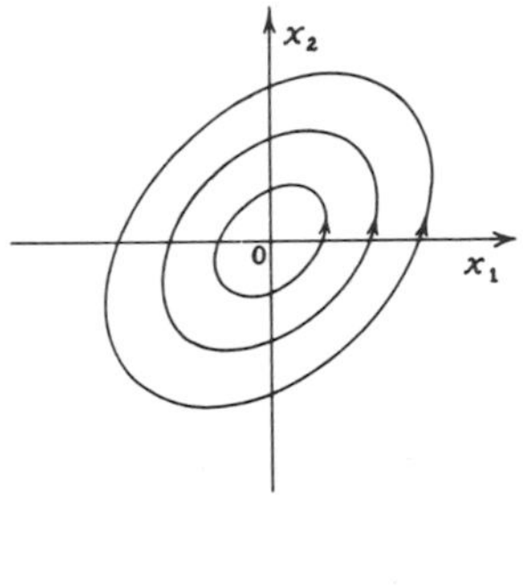

Fig. 9

the origin is a center (cf. Fig. 9) and if $\lambda_1 + \lambda_2 \neq 0$, the origin is a (stable or unstable) focus (cf. Fig. 10).

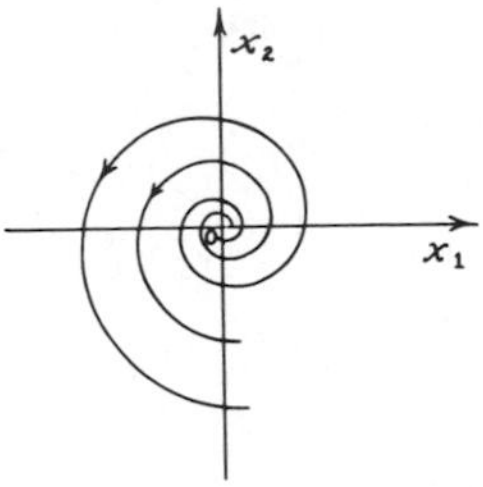

Fig. 10

4.154. Suppose next that $|A| = 0$, i.e., that the system (4.101) is singular. If the rank of A is zero, then every point is a singular point of the system. If A is of rank one, then all the singular points, i.e., all the solutions of $Ax = 0$ lie on one straight line and all the nonsingular trajectories lie on a family of parallel straight lines each of which contains one (and only one) singular point. This is easily seen if one considers the canonical form with, say, $\lambda_1 \neq 0$. The equations are $dy_1/dt = \lambda_1 y_1$, $dy_2/dt = 0$. The line $y_1 = 0$ is the line of singular points and all the trajectories are given by $y_1 = c_1 e^{\lambda_1 t}$, $y_2 = c_2$ and lie on lines $y_2 = c_2$. For $c_1 > 0$ the corresponding trajectory lies on the positive ray of the line $y_2 = c_2$ and tends to the singular point $y_1 = 0$, $y_2 = c_2$ as $t \to +\infty$ if $\lambda_1 < 0$ or as $t \to -\infty$ if $\lambda_1 > 0$.

4.2. *We consider next, following Forster [19], a homogeneous system of the form*

$$\frac{dy}{dx} = \frac{a_0 y^m + a_1 y^{m-1} x + \ldots + a_m x^m}{b_0 y^m + b_1 y^{m-1} x + \ldots + b_m x^m} = \frac{A_m(x, y)}{B_m(x, y)}, \tag{4.201}$$

and state without proof the results obtained by Forster for the case when $A_m(x, y)$ and $B_m(x, y)$ have no real linear factors in common and when the equality

$$\frac{A_m(x, y)}{B_m(x, y)} = \frac{y}{x} \tag{4.202}$$

does not hold identically in x and y.

Introducing polar coordinates, we write

$$x = r\cos\varphi, \quad y = r\sin\varphi, \quad \tau = \tan\varphi,$$
$$A_m(\varphi) = a_0 \sin^m\varphi + a_1 \sin^{m-1}\varphi\cos\varphi + \ldots + a_m \cos^m\varphi.$$
$$B_m(\varphi) = b_0 \sin^m\varphi + b_1 \sin^{m-1}\varphi\cos\varphi + \ldots + b_m \cos^m\varphi,$$

and

$$Z(\varphi) = A_m(\varphi)\sin\varphi + B_m(\varphi)\cos\varphi,$$
$$N(\varphi) = B_m(\varphi)\sin\varphi - A_m(\varphi)\cos\varphi = N_1(\tau)\cos^{m+1}\varphi.$$

Then, writing (4.201) in polar form, we have

(4.203) $$\frac{dr}{d\varphi} = -r\frac{Z(\varphi)}{N(\varphi)},$$

whence

(4.204) $$r(\varphi) = r_0\, e^{-\int_{\varphi_0}^{\varphi} \frac{Z(\varphi)}{N(\varphi)}\,d\varphi}.$$

To describe the behavior of trajectories near the origin we determine first the straight-line trajectories which adhere to the origin. In view of (4.203), these trajectories lie on the rays (integral rays) determined by the real roots of the equation

(4.205) $$N(\varphi) = 0.$$

If $\varphi = \varphi_1$ is a *real* root of (4.205), then $\varphi = \varphi_1 + \pi$ is also a real root and the resulting pair of corresponding rays lies on the same straight line. Since $N(\varphi)$ is a homogeneous polynomial in $\sin\varphi$, $\cos\varphi$ of degree $m + 1$ and is not identically zero by hypothesis (cf. (4.202)), *there are at most $m + 1$ lines through the origin determined by the above pairs of straight-line trajectories.*

4.21. THEOREM. When $N(\varphi) = 0$ *has no real roots, we have* (*cf.* (4.204)):

4.211. If $\int_0^{2\pi} \frac{Z(\varphi)}{N(\varphi)}\,d\varphi = 0$, *then all the trajectories are closed and the origin is a center.*

4.212. *If* $\int_0^{2\pi} \frac{Z(\varphi)}{N(\varphi)}\,d\varphi > 0$, *then* $\lim_{\varphi\to+\infty} r(\varphi) = 0$, *the trajectories are spirals, and the origin is a stable focus.*

4.213. *If* $\int_0^{2\pi} \frac{Z(\varphi)}{N(\varphi)}\,d\varphi < 0$, *then* $\lim_{\varphi\to-\infty} r(\varphi) = 0$ *and the origin is an unstable focus.*

We consider next the case when (4.205) has real roots. Let $\varphi = \varphi_1$ and $\varphi = \varphi_2$, $\varphi_1 < \varphi_2$, be two consecutive straight-line trajectories (integral rays). The trajectory $r = r(\varphi)$ through a point of the sector $r > 0$, $\varphi_1 < \varphi < \varphi_2$ is well defined for $\varphi_1 < \varphi < \varphi_2$ and as φ tends to one of φ_1 or φ_2, $r(\varphi)$ tends either to 0 or to $+\infty$. The behavior of trajectories within the angle $\varphi_1 < \varphi < \varphi$ is partially determined by the number of zeros of $Z(\varphi)$ in the angle, each being counted with its multiplicity).

Let φ_j be a real root of (4.205) of multiplicity ν_j, i.e., let $\tau_j = \tan\varphi_j$ be such a root of $N_1(\tau)$. Since $N(\varphi)$ is homogeneous in $\sin\varphi$ and $\cos\varphi$, we may write

(4.214) $N(\varphi) = \sin \nu_j\ (\varphi - \varphi_j) \cdot Q_j(\varphi),\ Q_j(\varphi_j) \neq 0,$

where $Q_j(\varphi)$ is a homogeneous polynomial of degree $m - \nu_j$ in $\sin\varphi$, $\cos\varphi$ and $Q_j(\varphi_j) \neq 0$. Also

$$\frac{Z(\varphi_j)}{Q_j(\varphi_j)} \neq 0. \tag{4.215}$$

The behavior of trajectories in the neighborhood of an integral ray $\varphi = \varphi_j$ is described by

4.22. Theorem. *Consider an integral ray corresponding to a real root φ_j of $N(\varphi) = 0$ of multiplicity ν_j.*

4.221. If ν_j is odd and $Z(\varphi_j)/Q_j(\varphi_j) < 0$, then $r(\varphi)$ tends toward zero as $\varphi \to \varphi_j + 0$ or $\varphi_j - 0$. Thus on both sides of our ray the trajectories tend toward the origin as $\varphi \to \varphi_j$. Moreover, these trajectories are tangent to the ray at the origin. Such a ray Forster calls a *nodal ray.*

4.222. If ν_j is odd and (cf. 4.214–5) $Z(\varphi_j)/Q_j(\varphi_j) > 0$, then on both sides of our ray the trajectories recede from the origin as $\varphi \to \varphi_j$. Forster speaks of such a ray as an *isolated ray.*

4.223. If ν_j is even then the trajectories do not behave alike on both sides of the ray given by $\varphi = \varphi_j$.

In case $Z(\varphi_j)/Q_j(\varphi_j) < 0$, we have: $r(\varphi) \to 0$ as $\varphi \to \varphi_j + 0$ and $r(\varphi) \to +\infty$ as $\varphi \to \varphi_j - 0$.

In case $Z(\varphi_j)/Q_j(\varphi_j) > 0$, we have: $r(\varphi) \to +\infty$ as $\varphi \to \varphi_j + 0$ and $r(\varphi) \to 0$ as $\varphi \to \varphi_j - 0$.

In all cases the trajectories which tend to the origin are tangent to the given ray at the origin, i.e., not only does $\varphi(t)$ tend toward

φ_j but the slope of the tangent to $(r(t), \varphi(t))$ at t tends to the slope of the polar ray φ_j.

4.3. *The non-integrable case. Comparison with the truncated equation* (4.201). We consider a differential equation

$$\frac{dy}{dx} = \frac{a_0 y^m + a_1 y^{m-1}x + \ldots + a_m x^m + d(x, y)}{b_0 y^m + b_1 y^{m-1}x + \ldots + b_m x^m + e(x, y)} = \frac{A_m(x, y) + d(x, y)}{B_m(x, y) + e(x, y)} \tag{4.301}$$

and seek conditions on $d(x, y)$ and $e(x, y)$ which would insure that the behavior of trajectories of (4.301) near the origin should be the same as the behavior of trajectories near the origin of the truncated equation (4.201).

In particular, if the truncated equation is linear, i.e., if it is of the form

$$\frac{dy}{dx} = \frac{a_1 x + a_0 y}{b_1 x + b_0 y}, \tag{4.302}$$

then the conditions which we seek tell when the study of trajectories of a nonlinear system near a singular point (the origin) may be reduced to the study of the behavior near the origin of the trajectories of a linear system with constant coefficients.

4.31. Examples and heuristic remarks. We should note that the assumption that near the origin $d(x, y)$ and $e(x, y)$ are infinitesimals of higher order than $A_m(x, y)$ and $B_m(x, y)$ does not suffice to guatantee that the behavior of trajectories near the origin of the original and of the truncated equations essentially agree.

4.311. Example. Consider the system [11]

$$\frac{dx}{dt} = -x + \frac{2y}{\log(x^2 + y^2)}, \quad \frac{dy}{dt} = -y - \frac{2x}{\log(x^2 + y^2)}.$$

The origin is a singular node (cf. (4.12)) of the corresponding truncated system

$$\frac{dy}{dx} = \frac{y}{x}.$$

Returning to the original system and passing to polar coordinates we obtain

$$\frac{dr}{dt} = -r, \quad \frac{d\varphi}{dt} = -\frac{1}{\log r},$$

whence

$$r = r_0 e^{-t}, \quad \varphi = \varphi_0 + \log(t - \log r_0), \quad t > \log r_0.$$

Thus, $r \to 0$ and $\varphi \to +\infty$ as $t \to +\infty$, and the origin is a stable focus of the given system.

We see that by adding to linear terms infinitesimals of higher order we may change the nature of the behavior of solutions near the origin. We shall see next that the addition of infinitesimals of higher order may lead to a type of behavior near the origin which does not occur either in linear systems or homogeneous systems (4.201).

4.312. *Example.* The characteristic roots of the system

$$\frac{dy}{dx} = -\frac{y}{x}$$

are $\lambda_1 = +1$ and $\lambda_2 = -1$. Hence the origin is a saddle point of this system. The origin is also a saddle point of the system

$$\frac{dy}{dx} = -\frac{y + \alpha y}{x - \beta x},$$

say, as long as α and β are numerically less than unity.

4.32. Some general results (Forster) [19]. Changing over to polar coordinates we may write (4.301) in the form

$$\frac{dr}{d\varphi} = -r \frac{Z(\varphi) + \Delta(r, \varphi)}{N(\varphi) + E(r, \varphi)}. \tag{4.321}$$

We shall make the following assumptions regarding the "perturbation terms" $\Delta(r, \varphi)$ and $E(r, \varphi)$.

4.321. Both $\Delta(r, \varphi)$ and $E(r, \varphi)$ are continuous in r and φ, periodic in φ with the period 2π and

$$\lim_{r \to 0} \Delta(r, \varphi) = 0, \quad \lim_{r \to 0} E(r, \varphi) = 0$$

uniformly in φ.

4.332. If the equation $N(\varphi) = 0$ has no real roots, then we suppose that there exists a ϱ_1 such that for every point (r_0, φ_0) in $0 < r \leqq \varrho_1$, $0 \leqq \varphi \leqq 2\pi$, there is a rectangle containing (r_0, φ_0) in which

$$|\Delta(r_1, \varphi) - \Delta(r_2, \varphi)| + |E(r_1, \varphi) - E(r_2, \varphi)| \leqq H(r_0, \varphi_0)|r_1 - r_2|.$$

In particular, this condition is satisfied if both $\Delta(r, \varphi)$ and $E(r, \varphi)$ have bounded partial derivatives with respect to r.

4.323. If $N(\varphi) = 0$ has real roots, we shall enumerate the n roots in the interval $0 \leqq \varphi < 2\pi$ in the order of increasing magnitude. Thus we write

$$\varphi_1(\geqq 0),\ \varphi_2, \ldots \varphi_n,\ \varphi_{n+1} = \varphi_1 + 2\pi(\geqq 2\pi).$$

We assume that there exist ϱ_1, and $\bar{\varphi}_j$, $\bar{\bar{\varphi}}_j$ for each j having the following properties:

(a) $\bar{\varphi}_j < \varphi_j < \bar{\bar{\varphi}}_j$.

(b) The intervals $\bar{\varphi}_j \leqq \varphi \leqq \bar{\bar{\varphi}}_j$ contain no roots of the equation $Z(\varphi) = 0$.

(c) For every point (r_0, φ_0) in

$$0 < r \leqq \varrho_1,\ \bar{\varphi}_j \leqq \varphi \leqq \bar{\bar{\varphi}}_j\ \ (j = 1, 2, \ldots, n),$$

there exists a rectangle containing it in which

$$|\Delta(r, \varphi') - \Delta(r, \varphi'')| + |E(r, \varphi') - E(r, \varphi'')| \leqq H(r_0, \varphi_0)|\varphi' - \varphi''|.$$

(d) For every point (r_0, φ_0) of the residual domains

$$0 < r \leqq \varrho_1,\ \bar{\bar{\varphi}}_j \leqq \varphi \leqq \varphi_{j+1}\ \ (j = 1, \ldots, n),$$

wherever these occur, there exists a rectangle containing (r_0, φ_0) and in which

$$|\Delta(r_1, \varphi) - \Delta(r_2, \varphi)| + |E(r_1, \varphi) - E(r_2, \varphi)| \leqq H(r_0, \varphi_0)|r_1 - r_2|.$$

4.324. If the differential equation (4.201) has an isolated ray $\varphi = \varphi_j$ of multiplicity ν_j, then we assume that there exist positive ω, ϱ_2, and μ, with $\mu < 1$, having the following properties:

(a) The partial derivatives $\partial\Delta(r, \varphi)/\partial\varphi$ and $\partial E(r, \varphi)/\partial\varphi$ are bounded and continuous in

$$\varphi_j - \omega \leqq \varphi \leqq \varphi_j + \omega, \quad 0 \leqq r \leqq \varrho_2, \quad j = 1, 2, \ldots, n.$$

(b) For the functions

$$g_j(r) = \max_{\varphi_j - \omega \leqq \varphi \leqq \varphi_j + \omega} |E(r, \varphi)|, \quad j = 1, 2, \ldots, n,$$

the integrals

$$\int_0^{\varrho_2} \frac{g_j(r)}{r}\, dr = E,$$

are finite.

(c) For the functions

$$h_j(r) = \min_{\varphi_j-\omega\leqq\varphi\leqq\varphi_j+\omega} \{\min[Z(\varphi_j)(\mu\nu_j Q_j(\varphi_j)(\varphi-\varphi_j)^{\nu_j-1} + \frac{\partial E(r,\varphi)}{\partial\varphi}),0]\},$$

the integrals

$$\int_0^{\varrho_2} \frac{h_j(r)}{r}\, dr$$

are finite. Since the integrands do not take on positive values, we can therefore find a $g > 0$ such that

$$0 \geqq \int_0^{\varrho_2} \frac{h_j(r)}{r}\, dr > -g, \quad j = 1, 2, \ldots, n.$$

4.325. If a ray $\varphi = \varphi_j$ is of even multiplicity ν_j (cf. 4.223), then we shall assume that there exist positive γ_1 and R' such that the integrals

$$\int_0^{R'} \frac{g_j(r)|\log r|}{r}\, dr, \quad j = 1, 2, \ldots, n$$

are finite. Here

$$g_j(r) = \max_{\varphi_j-\gamma_1\leqq\varphi\leqq\varphi_j+\gamma_1} |E(r, \varphi)|.$$

We note that all of the above conditions are fulfilled if there exists an $\alpha > 0$ such that

$$\frac{E(r, \varphi)}{r^\alpha} \to 0, \quad \frac{\Delta(r, \varphi)}{r^\alpha} \to 0$$

uniformly in φ, as $r \to 0$.

4.326. THEOREM (Forster) [19]). *If the equation* (4.321) *fulfills the conditions* 4.321–4.325, *then near the origin, the solutions of* (4.301) *behave essentially as do those of the corresponding truncated system* (4.201).

More precisely:

(a) If the origin is a focus of the truncated system (4.201), then it is also a focus of the original system (4.301).

(b) If the origin is a center (cf. 4.211) of the truncated system (4.201), then every trajectory of the original system (4.301) near the origin, is either closed or is a spiral.

(c) If $\lim\,[x(t)^2 + y(t)^2] = 0$ as $t \to +\infty$ (or $-\infty$) for a trajectory $x = x(t)$, $y = y(t)$ of (4.301), then either this trajectory is a spiral or it is tangent at the origin to a ray determined by a real root of the equation $N(\varphi) = 0$.

(d) If $\varphi = \varphi_j$ is a nodal ray of the truncated system (4.201), then a trajectory of (4.301) tends toward the origin and is tangent to the ray $\varphi = \varphi_j$ at the origin, as long as this trajectory passes through a point close enough to the origin and contained in a sufficiently narrow angular neighborhood $\varphi_j - \gamma < \varphi < \varphi_j - \gamma$ of our ray.

(e) If $\varphi = \varphi_j$ is an isolated integral ray of the truncated system (4.201), then the original differential equation (4.301) has only one trajectory $r = r(t)$, $\varphi = \varphi(t)$ which tends to the origin and for which

$$\lim_{t \to +\infty} r(t) = 0, \quad \lim_{t \to +\infty} \varphi(t) = \varphi_j.$$

(f) If $\varphi = \varphi_j$ is an integral ray of even multiplicity (cf. 4.223) of the truncated system (4.201), then there exists a trajectory $r = r(t)$, $\varphi = \varphi(t)$ of the original system (4.301) such that

$$\lim_{t \to +\infty} r(t) = 0, \qquad \lim_{t \to +\infty} \varphi(t) = \varphi_j,$$

and on one side of our ray every trajectory passing through a point adjacent to the ray and sufficiently close to the origin tends to the origin in the direction of the ray, i.e., is tangent to the ray at the origin. This does hold for the trajectories through points on the other side of the ray.

4.327. One should remark that in the case of a truncated (homogeneous) system we know the behavior of trajectories in the neighborhood of the origin as soon as we know their behavior near the integral rays. This is no longer true of the more general system (4.301).

4.4 A geometrical analysis of the behavior of trajectories near an isolated singular point (Frommer [20]). Considerable insight into the behavior of trajectories of a dynamical system near an isolated singular point can be gained through a preliminary geometrical analysis. In this section our dynamical system will be represented by its vector field $F(r, \varphi)$ and the basic restrictive hypothesis will be delineated in the process of our discussion.

4.41. Critical directions. In the case of a homogencous

system of type (4.201), each straight-line trajectory adhering to the origin for $t \to +\infty$ or for $t \to -\infty$ singles out a polar direction, i.e., the direction of the ray on which it lies. In this case the behavior of trajectories in each narrow sector containing a straight-line trajectory determines the over-all behavior of integral curves.

Although in the more general case (4.301) there may exist no straight-line trajectories and the over-all picture is not as simple as that for the homogeneous case, still the behavior of trajectories near the origin (an isolated singular point of the system) is closely associated with certain *critical* directions of the polar ray.

4.411. DEFINITION. *We consider a ray $0 < r$, $\varphi = \varphi_0$, and we say that its direction (determined by) φ_0 is critical if there exists a sequence of points $A_j = (r_j, \varphi_j)$ $(j = 1, 2, \ldots)$ with $r_j \to 0$, $\varphi_j \to \varphi_0$ and such that $\alpha_j = \alpha(r_j, \varphi_j) \to 0$, where $\alpha(r, \varphi) = F_\varphi/F_r$* (see 4.412), *is the tangent of the angle between the direction φ and the direction of the field vector $F(r, \varphi)$ at $A = (r, \varphi)$.*

If, as $t \to +\infty$ or as $t \to -\infty$, a given trajectory $r = r(t)$, $\varphi = \varphi(t)$ tends to the origin in the direction φ_0 (i.e., if it adheres to the origin, and if $\varphi(t) \to \varphi_0$, and $\alpha(t) = \alpha(r(t), \varphi(t)) \to 0$), then it is clear that φ_0 is a critical direction. Thus, among the critical directions are contained all those directions in which trajectories can tend to the origin. In particular, all the directions φ satisfying (4.205) are critical directions for the truncated system (4.201) and under certain conditions are critical directions for the system (4.301) with perturbation terms, as well.

4.412. We shall accept the usual conventions of sign. Namely, we shall assume that the positive direction along a polar ray $\varphi = \bar{\varphi}$, points away from the origin and that the normal $n(\bar{r}, \bar{\varphi})$ to the ray $\varphi = \bar{\varphi}$ at $r = \bar{r}$, (i.e., the tangent to the circle $r = \bar{r}$ at $\varphi = \bar{\varphi}$) is pointed in the direction of increasing φ. As is customary, by the φ-component of the vector field at $(\bar{r}, \bar{\varphi})$ we shall mean the projection F_φ of the vector $F(\bar{r}, \bar{\varphi}) = (\dot{x}(\bar{r}, \bar{\varphi}), \dot{y}(\bar{r}, \bar{\varphi}))$ of our field at $(\bar{r}, \bar{\varphi})$ on to the normal $n(\bar{r}, \bar{\varphi})$, and by the r-component the projection F_r of this vector onto the (oriented) polar ray $\varphi = \bar{\varphi}$.

4.413. If $\varphi = \bar{\varphi}$ is not a critical direction, then there exists a value $\bar{r}$ of r, such that the φ component F_φ of the field is either positive at all the points $0 < r \leqq \bar{r}$, $\bar{\varphi} = \varphi$, or F_φ is negative at all these points. Indeed, suppose that there exists no such $\bar{r}$, then there exists a sequence $r_1 > r_2 > \ldots$ which tends to zero and such that F_φ is positive at each point $(r_{2k}, \bar{\varphi})$ and is negative or zero at each point $(r_{2k+1}, \bar{\varphi})$. In view of the continuity of the field there exists a sequence of points $(\bar{r}_k, \bar{\varphi})$, $r_{2k} > \bar{r}_k > r_{2k+1}$, at which the φ-component F_φ is zero. This last, in turn, implies that φ is a critical direction (see Fig. 11).

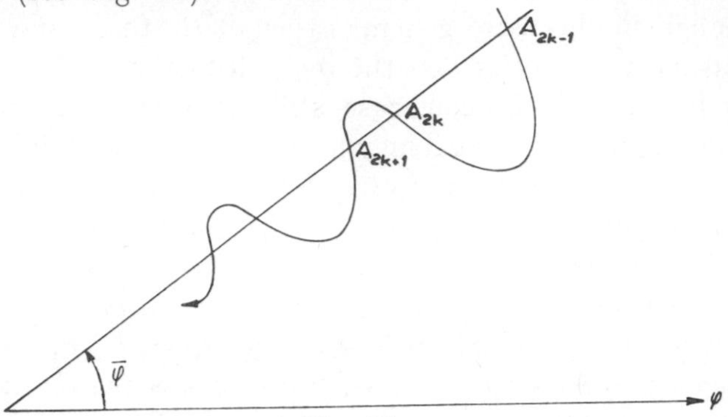

Fig. 11

In particular for large t a trajectory $r = r(t)$, $\varphi = \varphi(t)$ which tends to the origin as $t \to +\infty$ $(t \to -\infty)$, may cut a polar ray $\varphi = \bar{\varphi}$ whose direction is not critical in only one direction (say, that corresponding to a positive φ-component).

4.414. Theorem. *If there exists no interval $\bar{\varphi} < \varphi < \bar{\bar{\varphi}}$ consisting entirely of critical directions φ, then every half-trajectory $r(t)$, $\varphi(t)$, $t > 0$ $(t < 0)$, which comes arbitrarily near the origin (isolated singular point) either* (1) *tends to the origin in such a way that $\varphi \to \varphi_0$ where φ_0 is a critical direction or* (2) *is a spiral which tends toward the origin (Bendixson).*

Under suitable restrictions (cf. 4.326(c)) on the vector field, we have a stronger alternative to (2) in 4.414, viz., a trajectory which does not spiral toward the origin must actually be tangent to [9] a critical direction φ_0 (cf. 4.2 and 4.3) at the origin.

Here we say that a trajectory $\varrho = \varrho(t)$, $\varphi = \varphi(t)$ is a (generalized) spiral if $\overline{\lim}\, |\varphi(t)| = +\infty$ as $t \to +\infty$ or as $t \to -\infty$.

Suppose that $\overline{\lim}\, |\varphi(t)| < +\infty$ and $\varrho(t) \to 0$ as $t \to +\infty$. If $\overline{\lim}_{t\to+\infty}\, \varphi(t) = \varphi_s > \underline{\lim}_{t\to+\infty}\, \varphi(t) = \varphi_i$, then for every direction φ_1, $\varphi_i < \varphi_1 < \varphi_s$ our trajectory cuts across the polar ray $\varphi = \varphi_1$ infinitely many times in the positive as well as in the negative direction. This, however, would contradict (cf. 4.412) the hypothesis that there is

[9]We shall always use this term in the strong sense of $\varphi(t) \to \varphi_0$ and $\alpha(t) \to 0$ (cf. 4.411).

no interval consisting entirely of critical directions. Thus, either

(a) $\overline{\lim_{t\to+\infty}} |\varphi(t)| = +\infty.$

or

(b) $\overline{\lim_{t\to+\infty}} \varphi(t) = \underline{\lim_{t\to+\infty}} \varphi(t) = \lim_{t\to+\infty} \varphi(t) = \varphi_0.$

We note that φ_0 is a critical direction. For, if (b) holds, then for $t > t_0$ the trajectory is contained in an arbitrarily narrow sector $\varphi_0 - \varepsilon \leqq \varphi \leqq \varphi_0 + \varepsilon$, $0 < r \leqq \delta$. This would be impossible if the field vector $(\dot{x}(r, \varphi), \dot{y}(r, \varphi))$ should differ in direction from φ_0 by more than $\varepsilon_1 > 0$ throughout each sector defined by small ε and δ.

4.415. THEOREM. *Let $\varphi_1 \leqq \varphi \leqq \varphi_2$ be an interval (a sector) which does not contain a critical direction. Then there exists a (sufficiently small) r_0 such that for every trajectory $\varrho = \varrho(t)$, $\varphi = \varphi(t)$ through a point of the sector $\varphi_1 \leqq \varphi \leqq \varphi_2$, $0 < r \leqq r_0$, the φ component varies monotonically with t in such a way that our trajectory leaves this sector.*

Suppose first that for every $\bar{r} > 0$, there is an arc of trajectory L_0 in the sector $\varphi_1 \leqq \varphi \leqq \varphi_2$, $0 < r < \bar{r}$, along which $\varphi(t)$ is not monotone. Consider a sequence $r_1, r_2, \ldots$ of such values, tending to zero. Then in each sector $\mathfrak{S}_j = \mathfrak{S}(r_j;\ \varphi_1,\ \varphi_2) : 0 < r \leqq r_j$, $\varphi_1 \leqq \varphi \leqq \varphi_2$, there exists an arc of trajectory L_j along which $\varphi(t)$ is not monotone, and hence in every sector $\mathfrak{S}_j$ there is a point $(\bar{r}_j, \bar{\varphi}_j)$ where the direction of the field element coincides with $\bar{\varphi}_j$. Since every limit point φ_0 of the infinite sequence $\bar{\bar{\varphi}}_j$ bounded by φ_1 and φ_2, is a critical direction and $\varphi_1 \leqq \varphi_0 \leqq \varphi_2$, we arrive at a contradiction.

Thus $\varphi(t)$ is monotone along every trajectory in $\mathfrak{S}_0 = \mathfrak{S}(r_0, \varphi_1, \varphi_2)$ as long as r_0 is sufficiently small.

If we assume that r_0 is small enough so that there are no other singular points in $\mathfrak{S}_0$, then either our trajectory tends to the origin as $\varphi(t)$ tends to some critical direction φ_0, $\varphi_1 \leqq \varphi_0 \leqq \varphi_2$ (again a contradiction) or (the only remaining alternative) our trajectory leaves the sector $\mathfrak{S}_0$.

4.5. Normal domains. We study next the behavior of trajectories in a neighborhood of an isolated critical direction.

Consider a sector $\mathfrak{S}(r_0, \varphi_1, \varphi_2) : \varphi_1 \leqq \varphi \leqq \varphi_2$, $0 < r \leqq r_0$. We speak of the straight-line segments L_1, L_2 determined by $\varphi = \varphi_1$, $\varphi = \varphi_2$, $0 < r \leqq r_0$ as *the boundary segments* and of the arc $r = r_0$, $\varphi_1 \leqq \varphi \leqq \varphi_2$ as *the boundary arc* of this sector.

4.51. Definition. *A sector* $\mathfrak{S}(r_0, \varphi_1, \varphi_2)$ *is called a normal domain whenever*

4.511. *it contains only one critical direction* φ_0, $\varphi_1 < \varphi_0 < \varphi_2$, *with the directions* φ_1, φ_2 *of the boundary segments non critical, and*

4.512. *the direction of the field element at every point P either on the boundary or in the interior of the sector* $\mathfrak{S}$, *is not orthogonal to the direction of the radius vector OP, that is,* $r(t)$ *is strictly monotone in* $\mathfrak{S}$.

It follows from 4.512 that the r-component F_r of the vector field is either positive at every point (r, φ) of a normal domain or it is negative at every such point.

4.513. If we choose r_0 small enough we may assume that in a normal domain (cf. 4.413, 4.511) F_φ as well as F_r does not change sign and does not vanish on each of the boundary segments.

In case F_r is negative throughout $\mathfrak{S}$, we need to consider only the three essentially different types of trajectory behavior which are distinguished by the behavior of trajectories as they cross the boundary segments with increasing t (see Fig. 12, 13, 14).

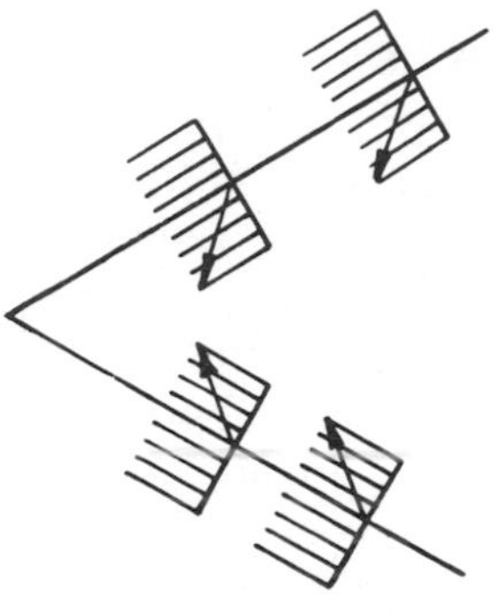

Fig. 12

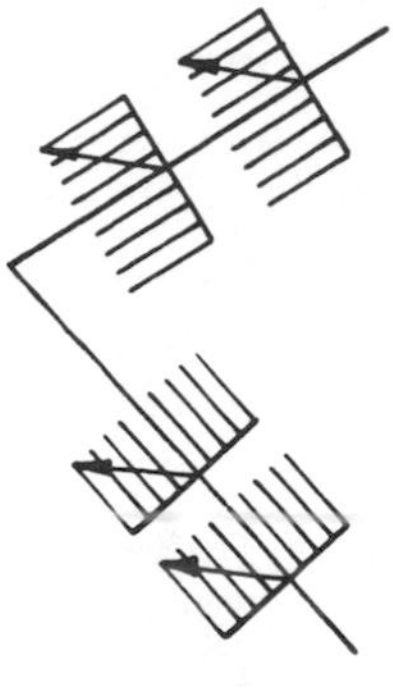

Fig. 13

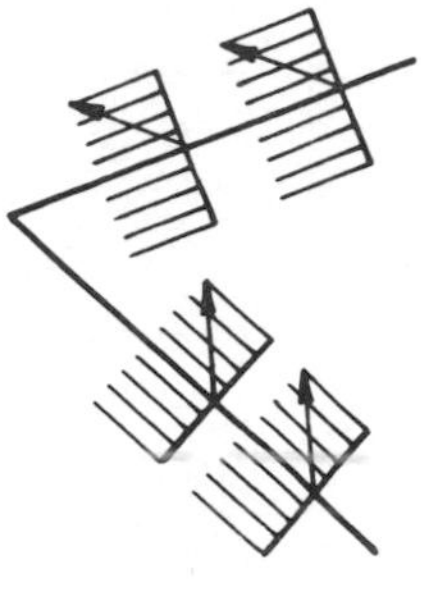

Fig. 14

If F_r is positive in $\mathfrak{S}$, then the vectors in Fig. 12 describe the behavior of trajectories as they cut the boundary segments with decreasing t.

4.514. Normal domains are defined thus: In a normal domain $\mathfrak{S}$ of Type I all the trajectories cutting across the boundary segments enter $\mathfrak{S}$ (with increasing t if $F_r < 0$ in $\mathfrak{S}$ and with decreasing t if $F_r > 0$ in $\mathfrak{S}$) (Fig. 12). In a normal domain of Type II all such trajectories leave $\mathfrak{S}$ with increasing $t(F_r < 0)$ or decreasing $t(F_r > 0)$ (Fig. 13). Finally in a normal domain of Type III (Fig. 14), all the trajectories cutting across one of the boundary segments enter $\mathfrak{S}$ and those cutting across the other one leave $\mathfrak{S}$.

4.515. THEOREM. *If as t increases (decreases) a trajectory enters a normal domain $\mathfrak{S}$ of the first type, then it remains in $\mathfrak{S}$ and tends to the origin* [Fig. 12].

If a trajectory $f(p, t)$ enters a normal domain $\mathfrak{S}$ of the first type, it can leave this domain only if the F_r component of the field changes its sign at some point P on the trajectory and inside $\mathfrak{S}$. This is impossible in a normal domain. Since $r(t)$ is monotone, the limit set of $f(p, t)$ must lie on an arc $f = \bar{r}$, $\varphi_1 \leqq \varphi \leqq \varphi_2$ $(0 \leqq \bar{r} < r_0)$; it cannot be a closed trajectory and hence it must contain singular points (cf. 1.47). Since $\mathfrak{S}$ contains no singular points, the origin is in the limit set of $f(p, t)$, whence $\bar{r} = 0$, and $f(p, t)$ tends tot he origin.

4.516. THEOREM. *If a trajectory enters a normal domain $\mathfrak{S}$ of the second type by cutting across a boundary segment then it ultimately leaves $\mathfrak{S}$. On the boundary arc, however, there exist* (1) *a point or* (2) *a whole (closed) subarc $\bar{\varphi}_1 \leqq \varphi \leqq \bar{\varphi}_2$ through which pass trajectories adhering to the origin as $t \to +\infty$ or as $t \to -\infty$.*

A trajectory $f(p, t)$ which enters (as $t \to +\infty$ if $F_r > 0$ or as $t \to -\infty$ as $F_r < 0$) a normal domain $\mathfrak{S}$ of the second type by cutting across a boundary segment, cannot leave by cutting across either one of the two boundary segments. The argument in the proof of the preceding theorem shows that $f(p, t)$ cannot remain in $\mathfrak{S}$. Thus $f(p, t)$ must leave $\mathfrak{S}$ across the boundary arc.

Every trajectory through a point $s = (r, \varphi_j)$, $0 < r < r_0$ of a boundary segment φ_j leaves $\mathfrak{S}$ at a point $a = a(s) = (r_0, \varphi)$, $\varphi_1 < \varphi < \varphi_2$, of the boundary arc. The continuous mapping a: $s \to a(s)$, maps each of the two open segments $\hat{\varphi}_j$ (the boundary segments φ_j without their end points): $0 < r < r_0$, $\varphi = \varphi_j (j=1, 2)$ into the open arc A: $\varphi_1 < \varphi < \varphi_2$, $r = r_0$. The images $a(\hat{\varphi}_1)$, $a(\hat{\varphi}_2)$ are two non-overlapping open arcs $\hat{A}_1$: $\varphi_1 < \varphi < \bar{\varphi}_1$ and $\hat{A}_2$: $\bar{\varphi}_2 < \varphi < \varphi_2$, $\bar{\varphi}_1 \leqq \bar{\varphi}_2$. Every trajectory $f(p, t)$ entering $\mathfrak{S}$ with increasing t (decreasing t) through a point of the nonempty closed set $\bar{A}_{12}$: $\bar{\varphi}_1 \leqq \varphi \leqq \bar{\varphi}_2$, $r = r_0$ (a point or a closed arc) complementary to $A_1 \cup A_2$ in $\bar{A}$, tends to the origin as $t \to +\infty$ $(t \to -\infty)$. For, $f(p, t)$ cannot leave a normal domain across the boundary segments or a boundary arc in view of the definition of $\hat{A}_{12}$; thus if it remains in $\mathfrak{S}$ it must go into the origin as is shown by the argument in the proof of Theorem 4.513.

4.517. THEOREM. *Consider a normal domain $\mathfrak{S}$. Let F_φ/F_r be positive (negative) on both of the boundary segments φ_1 and φ_2 of $\mathfrak{S}$* (Type III). (*Here F_r is the r-component and F_φ is the φ-component of our vector field $F(r, \varphi)$*). *There exists a point $\bar{p}_2 = (\bar{r}, \varphi_2)$ $0 \leqq \bar{r} \leqq r_0$ on $\varphi_2 \cup 0$ ($\bar{p}_1 = (\bar{r}, \varphi_1)$ on $\varphi_1 \cup 0$) such that every trajectory through a point $p = (r, \varphi_2)$, $0 \leqq r \leqq \bar{r}$, tends the origin and every trajectory which enters through a point $p = (r, \varphi_2)$, $\bar{r} < r$ leaves $\mathfrak{S}$.*

Every trajectory $f(l_2, t)$ which enters $\mathfrak{S}$ through a point $p_2 = (r, \varphi_2)$ $0 \leqq r \leqq r_0$, either leaves $\mathfrak{S}$ across φ_1 or tends to the origin. If $f((r', \varphi_2), t)$, $0 < r' \leqq r_0$, upon entering $\mathfrak{S}$ tends to the origin, then every $f((r, \varphi_2))$, $0 < r \leqq r'$ also tends to the origin. On the other hand, if $f((r'', \varphi_2), t)$ leaves $\mathfrak{S}$ so does $f((r, \varphi_2), t)$ for $r'' \leqq r \leqq r_0$. Moreover, by the continuity in the initial conditions $f((r'' - \varepsilon, \varphi_2), t)$ leaves across φ_1 or the boundary arc for all sufficiently small ε. Suppose that there are points $p' = (r', \varphi_2)$ on φ_2 for which $f(p', t)$, upon entering $\mathfrak{S}$, tends to the origin. Let $\bar{r}$ be the least upper bound of the set R' of all such points r'. Then $0 < \bar{r}$, and, upon entering $\mathfrak{S}$, $f(p(r, \varphi_2), t)$ tends to the origin for $0 < r \leqq \bar{r}$, and leaves $\mathfrak{S}$ for $\bar{r} < r \leqq r_0$. If the set R' is empty, we take $\bar{r} = 0$.

4.518. If $\bar{r} = r_0$, then an argument similar to the above shows that there exists a point $(r_0, \bar{\varphi})$ on the boundary arc such that $f((r_0, \varphi), t)$, upon entering $\mathfrak{S}$, leaves it across φ_1 for $\varphi_1 \leqq \varphi < \bar{\varphi}$ and tends to the origin if $\bar{\varphi} \leqq \varphi \leqq \varphi_2$.

4.6. Analytical criteria. In what follows we consider a system

$$\frac{dx}{dt} = P(x, y), \qquad \frac{dy}{dt} = Q(x, y), \tag{4.601}$$

where $P(x, y)$ and $Q(x, y)$ have continuous derivatives up to and including $(m + 1)$-st and $(n + 1)$-st orders respectively. We assume that all the partial derivatives of $P(x, y)$ of order less than m vanish at the origin, that the same is true of all the partial derivatives of $Q(x, y)$ of order less than n, and that there are partial derivatives of $P(x, y)$ of order m and of $Q(x, y)$ of order n which do not vanish at the origin. Such a system we shall call *algebroid*.

Let us write (cf. 4.3)

$$P(x, y) = A_m(x, y) + d(x, y), \quad Q(x, y) = B_n(x, y) + e(x, y) \tag{4.602}$$

where $A_m(x, y)$ and $B_n(x, y)$ are homogeneous polynomials of degrees m and n respectively, and where

$$\lim_{r\to 0} \frac{d(x, y)}{r^m} = \lim_{r\to 0} \frac{e(x, y)}{r^n} = 0, \quad r = \sqrt{x^2 + y^2}. \tag{4.603}$$

4.61. The Plan of Investigation. *To utilize effectively the results of our geometric study in Sections* 4.4–5 *we proceed as follows:*

4.611. We investigate the existence of critical directions.

4.612. We determine the critical directions if such exist.

4.613. We establish which of the critical directions are associated with a normal domain (cf. 4.511).

4.614. We determine the type of the normal domain associated with a given critical direction (cf. 4.513).

4.615. We establish which of the alternative behaviors allowed by Theorems 4.514–4.515 actually does occur in the case of normal domains of the second and the third types.

4.616. If there exist no critical directions, then, in general, the origin (our singular point) is either a focus (cf. 4.13) or a stable singular point (cf. 3.3). As we have seen, a stable singular point may be either a center or a focus. We shall seek analytical criteria differentiating between different kinds of stable points and, among other things, shall show that in the analytic case the composite singular point, center-focus, cannot occur.

We observe that the existence of a regular critical direction of the first and second types (cf. 4.6231–2) indicates that our singular point is neither a stable singular point nor a focus. That one cannot, in general, draw such conclusions from the existence of a critical direction, appears from the following:

4.617.. *Example.* The differential equation

$$\frac{dy}{dx} = - \frac{2x^3}{y}$$

has two critical directions $\varphi = 0$ and $\varphi = \pi$. On the other hand the integral curves of our equation are given by

$$x^4 + y^2 = c.$$

It should be noted that neither of the above critical directions is regular, for both are roots of multiplicity two of the characteristic equation (4.632).

No analytical criteria have been developed to date, for recognizing stability in cases such as the one in the above example.

4.618. Suppose that there are a finite number of critical directions. Then a parabolic domain (Fig. 15) (cf. 3.49) may arise in a normal domain associated with a regular critical direction of any of the three main types (cf. 4.513), a hyperbolic domain (Fig. 16) may be associated with a pair of successive regular critical directions of the second (or the third) type (Fig. 16), and an elliptic domain (Fig. 17) may arise in connection with a pair of successive regular critical directions of the first type.

If the number of critical directions is infinite then all the directions with but a finite number of exceptions, are regular critical directions. There is a trajectory tending to the origin along every polar ray with at most a finite number of exceptions.

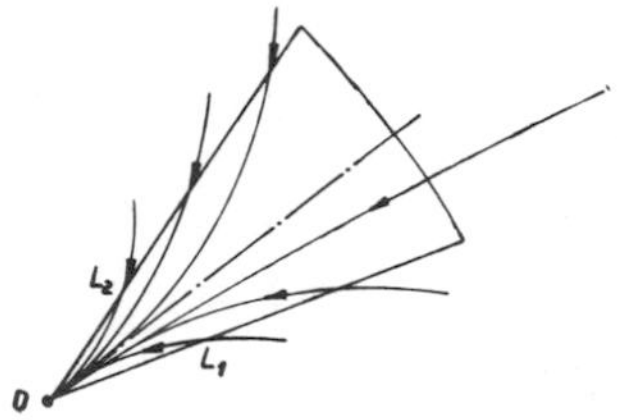

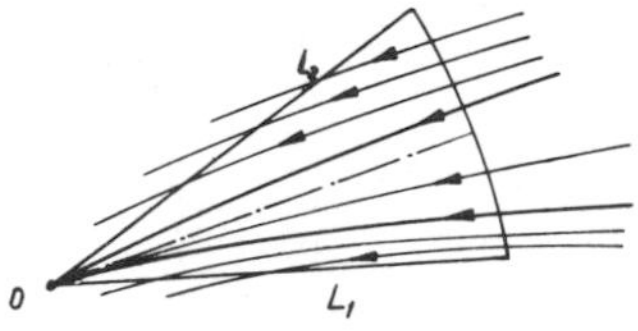

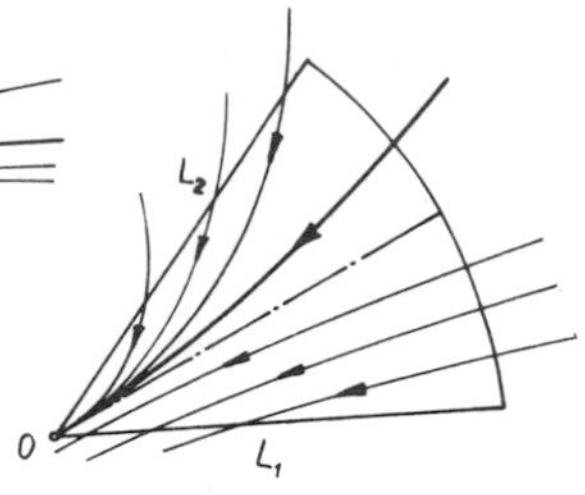

Fig. 15a Fig. 15b Fig. 15c

4.62. The case when $m = n$ and the number of critical directions is finite. Changing to polar coordinates we see that

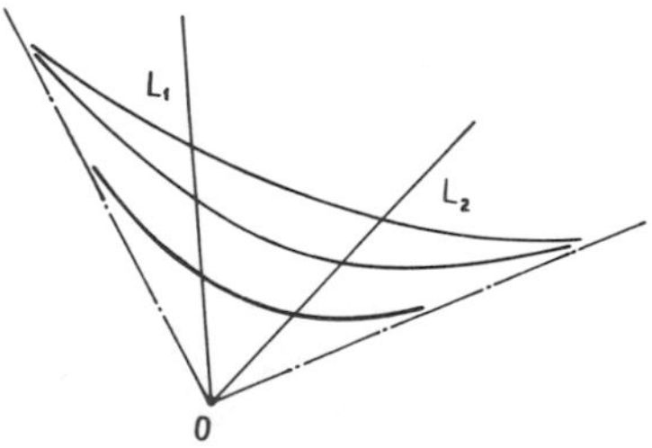

Fig. 16

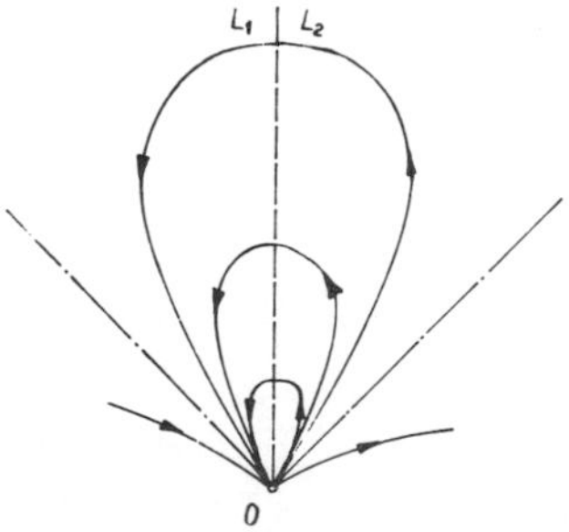

Fig. 17

$$F_r = \frac{dr}{dt} = (B_m(x, y)\cos\varphi + A_m(x, y)\sin\varphi) + (e(x, y)\cos\varphi + d(x, y)\sin\varphi),$$

$$F_\varphi = r\frac{d\varphi}{dt} = (A_m(x, y)\cos\varphi - B_m(x, y)\sin\varphi) + (d(x, y)\cos\varphi - e(x, y)\sin\varphi).$$

Here again F_r is the r-component and F_φ is the φ-component of our vector field $F(r, \varphi)$ at the point (r, φ). If we let

$$\xi(r, \varphi) = d\sin\varphi + e\cos\varphi,$$
$$\eta(r, \varphi) = d\cos\varphi - e\sin\varphi,$$

and use the notation of Section 4.2, we can write

$$F_\varphi = r^m\left(-N(\varphi) + \frac{\eta(r, \varphi)}{r^m}\right) \tag{4.6201}$$

$$F_r = r^m\left(Z(\varphi) + \frac{\xi(r, \varphi)}{r^m}\right) \tag{4.6202}$$

$$\overline{\Psi}(r, \varphi) = \frac{F_\varphi}{F_r} = r\frac{d\varphi}{dr} = \frac{-N(\varphi) + \dfrac{\eta(r, \varphi)}{r^m}}{Z(\varphi) + \dfrac{\xi(r, \varphi)}{r^m}} \tag{4.6203}$$

The ratio F_φ/F_r of (4.6203) is the tangent of the angle between the radius vector and the field vector $F(r, \varphi)$ at the point (r, φ) measured in the counterclockwise direction from the radius vector to the field vector. Therefore, if φ_0 is a critical direction (Definition 4.411) then $N(\varphi_0) = 0$. Conversely, if φ_0 is a root of the *characteristic equation*

$$N(\varphi) = 0 \tag{4.6204}$$

and if

$$Z(\varphi_0) \neq 0 \tag{4.6205}$$

then φ_0 is a critical direction.

Critical directions φ_0 which satisfy (4.6205) we shall call regular.[10]

A root φ_0 of (4.6204) may still be a critical direction even if it is also a root of $Z(\varphi)$, as long as the quotient $\overline{\Psi}(r, \varphi)$ in (4.6203) tends to zero as $r \to 0$ and $\varphi \to \varphi_0$.

[10]Frommer in his more general study [20] uses the term *regular* in a somewhat more general sense.

4.621. Normal domains associated with a regular critical direction. Consider a regular critical direction φ_0. Since $Z(\varphi_0) \neq 0$ and $\zeta(r, \varphi)/r^m$ is arbitrarily small for small r, the denominator of the right hand member of (4.6203) and hence F_r itself does not vanish (and hence does not change its sign) in the sector $\mathfrak{S}(r_0, \varphi_0, \delta)$: $0 < r \leqq r_0$, $\varphi_0 - \delta \leqq \varphi \leqq \varphi_0 + \delta$ as long as δ and $r_0 = r_0(\delta)$ are sufficiently small. This means that at no point P of $\mathfrak{S}(r_0, \varphi_0; \delta)$ is the field vector orthogonal to the radius vector OP (cf. 4.512).

If $N(\varphi) \not\equiv 0$, then the number of critical directions is finite (cf. 4.411 and 4.62). In this case for every critical direction φ_0 we can choose δ small enough so that $\mathfrak{S}(r_0, \varphi_0, \delta)$ contains no critical direction other than φ_0 (cf. 4.511). If, moreover, φ_0 is regular, then in view of the above discussion $\mathfrak{S}(r_0, \varphi_0; \delta)$ is normal provided that δ and r_0 are small enough.

If $N(\varphi) \equiv 0$, then every direction is critical. For, since $N^2 + Z^2 = (A_m^2 + B_m^2)/r^{2m} \not\equiv 0$ the equation $Z(\varphi) = 0$ has at most a finite number of roots. Hence all except at most a finite number of directions φ are critical. But the set of all critical directions is closed, whence follows our assertion.

4.622. We consider next an isolated regular critical direction φ_0 and an associated normal domain $\mathfrak{S}(r_0, \varphi_0; \delta)$. Let φ_0 be a root of multiplicity ν of (4.6204). Then

$$N(\varphi) = c(\varphi - \varphi_0)^\nu + o(|\varphi - \varphi_0|^\nu). \tag{4.6221}$$

$$Z(\varphi_0) = Z_0 \neq 0. \tag{4.6222}$$

In view of the discussion in 4.513 we may assume that r_0 is so small that the sign of F_φ and hence that of F_φ/F_r does not change along the boundary segments of $\mathfrak{S}$. If δ is small enough the sign of $N(\varphi_0 \pm \delta)$ agrees with that of its term $c(\pm\delta)^\nu$ of smallest order. If, in addition, $r_0 = r_0(\delta)$ is chosen small enough, then by (4.6201) the sign of F_φ on the boundary segments φ_j ($\varphi = \varphi_0 \pm \delta$) agrees with that of $-c(-1)^{\nu j}$. For δ and r_0 small enough, F_r retains the same sign, viz., that of Z_0, throughout $\mathfrak{S}$ and hence on φ_j as well and thus for a suitable choice of δ and of $r_0(\delta)$ the sign of F_φ/F_r along the boundary segment φ_j of the normal domain $\mathfrak{S}(r_0, \varphi_0, \delta)$ agrees with that of $-c\, Z_0(-1)^{\nu j}$.

These observations together with Theorems 4.515-6-7 yield

4.623. THEOREM. *Let φ_0 be a root of multiplicity ν of the charac-*

teristic equation (4.6204) *and such that* $Z(\varphi_0) = Z_0 \neq 0$.

4.6231. *If* ν *is odd and if* $c\,Z_0 < 0$, *then for a suitable choice of* δ *and* $r_0(\delta)$, *the ratio* F_φ/F_r *is positive on the boundary segment* L_2 *and is negative on the boundary segment* L_1 *of the associated normal domain* $\mathfrak{S}(r_0, \varphi_0, \delta)$. *Thus* $\mathfrak{S}$ *is a normal domain of the first type and all the trajectories entering* $\mathfrak{S}$ *tend to the origin. Here the sector* $\mathfrak{S}$ *is parabolic.*

4.6232. *If* ν *is odd and if* $cZ_0 > 0$, *then* F_φ/F_r *is negative on* L_2 *and is positive on* L_1. *Thus the normal domain* $\mathfrak{S}(r_0, \varphi_0, \delta)$ *associated with an appropriate choice of* δ *and of* $r_0 = r_0(\delta)$, *is a normal domain of the second type. Hence,* (1) *either there exists a unique trajectory which tends to the origin upon entering* $\mathfrak{S}$, *or* (2) *there are an infinite number of such trajectories, all of which cut across a subarc* $\bar{\varphi} \leqq \varphi \leqq \bar{\bar{\varphi}}$ *of the boundary arc.*

4.6233. *If* ν *is even, then the sign of* F_φ/F_r *agrees with that of* $-c\,Z_0$ *on both* L_1 *and* L_2 *and our normal domain* $\mathfrak{S}$ *is of the third kind. In this case* (1) *either there are infinitely many trajectories which tend to the origin upon entering* $\mathfrak{S}$ (*cf. Theorem* 4.517) *or* (2) *there exist no such trajectories.*

The distinction between the allowable alternatives in 4.6232 and in 4.6233 is dealt with in the next section.

4.63. The case $m \neq n$. Proceeding as in 4.62 we get

$$F_r = \frac{dr}{dt} = r^n B_n(\cos\varphi, \sin\varphi)\cos\varphi + r^m A_m(\cos\varphi, \sin\varphi)\sin\varphi + \xi(r,\varphi),$$

$$F_\varphi = r\frac{d\varphi}{dt} = r^m A_m(\cos\varphi, \sin\varphi)\cos\varphi - r^n B_n(\cos\varphi, \sin\varphi)\sin\varphi + \eta(r,\varphi).$$

If $m > n$, we write

$$F_r = \frac{dr}{dt} = r^n\left(B_n(\cos\varphi, \sin\varphi)\cos\varphi + r^{m-n}A_m(\cos\varphi, \sin\varphi)\sin\varphi + \frac{\xi(r,\varphi)}{r^n}\right),$$

$$F_\varphi = r\frac{d\varphi}{dt} = r^n\left(r^{m-n}A_m(\cos\varphi, \sin\varphi)\cos\varphi - B_n(\cos\varphi, \sin\varphi)\sin\varphi + \frac{\eta(r,\varphi)}{r^n}\right).$$

Thus, in view of (4.602) (4.603),

$$F_r = r^n(B_n(\cos\varphi,\ \sin\varphi)\cos\varphi + o(1)),$$

$$F_\varphi = r^n(-B_n(\cos\varphi,\ \sin\varphi)\sin\varphi + o(1)),$$

$$\frac{F_\varphi}{F_r} = -\frac{B_n(\cos\varphi,\ \sin\varphi)\sin\varphi + o(1)}{B_n(\cos\varphi,\ \sin\varphi)\cos\varphi + o(1)}. \tag{4.631}$$

By arguments similar to those used in 4.62, we show that every critical direction φ_0 satisfies the new *characteristic equation*

(4.632) $$B_n(\cos\varphi,\ \sin\varphi)\sin\varphi = 0.$$

A critical direction φ_0 is regular if and only if

$$\sin\varphi_0 = 0, \quad B_n(\cos\varphi_0,\ \sin\varphi_0) \neq 0.$$

Therefore $\varphi_0 = 0$ is a regular critical direction if $B_n(1,\ 0) \neq 0$, that is if 0 is a simple root of the characteristic equation. Similarly $\varphi_0 = \pi$ is a regular critical direction if $B_n(-1,\ 0) \neq 0$.

Suppose that $\varphi_0 = 0$ is a regular critical direction. Let its associated normal domain be $\mathfrak{S}(r_0,\ 0,\ \delta)$. If $r_0 = r_0(\delta)$ is small enough, then, by (4.631),

$$\frac{F_\varphi}{F_r} \approx -\tan(\pm\delta)$$

on the segments L_1 and L_2. Therefore, in this case $\mathfrak{S}$ is a normal domain of the second type, and consequently either there exists a unique trajectory which tends to the origin upon entering $\mathfrak{S}$ or there are infinitely many such trajectories (cf. 4.516). We shall see that under certain conditions the second alternative cannot occur.

If $m < n$,

$$\frac{F_\varphi}{F_r} = \frac{A_m(\cos\varphi,\ \sin\varphi)\cos\varphi + o(1)}{A_m(\cos\varphi,\ \sin\varphi)\sin\varphi + o(1)}.$$

Here every critical direction must be a root of the characteristic equation

(4.633) $$A_m(\cos\varphi,\ \sin\varphi)\cos\varphi = 0,$$

and the only possible regular critical directions are $\varphi_0 = \pi/2$ and $\varphi_0 = 3\pi/2$. The conditions that these be regular are $A_m(0,\ 1) \neq 0$ and $A_m(0,\ -1) \neq 0$ respectively. If $\varphi_0 = \pi/2$ is a regular critical direction, then

$$\frac{F_\varphi}{F_r} = \cot\left(\frac{\pi}{2} \pm \vartheta\right)$$

along the segments L_1, L_2 respectively of the associated normal domain $\mathfrak{S}(r_0,\ \pi/2,\ \delta)$, and here again $\mathfrak{S}$ is a normal domain of the second type. It is easily seen that $\mathfrak{S}(r,\ 3\pi/2,\ \delta)$ is also of the second type.

We note that since neither $A_m(x, y)$ nor $B_n(x, y)$ is identically zero, the number of critical directions in the case $m \neq n$, is finite.

4.64. (The case $m = n$ continued). The number of critical directions is infinite. Here $m = n$ and $N(\varphi) \equiv 0$. Therefore, as was shown in 4.621, every direction is a critical direction.

In this case we employ the (Briot-Bouquet) transformation $y = ux$. Then

$$\frac{du}{dx} x + u = \frac{P(x, ux)}{Q(x, ux)} = \frac{A_m(x, ux) + d(x, ux)}{B_m(x, ux) + e(x, ux)},$$

or

$$\frac{du}{dx} = \frac{A_m - uB_m + d - ue}{x(B_m + e).}.$$

Since, by hypothesis,

$$A_m - uB_m \equiv 0, \tag{4.6401}$$

we have

$$\frac{du}{dx} = \frac{d - ue}{x(B_m + e)} = \frac{\dfrac{d(x, ux)}{x^m} - u\dfrac{e(x, ux)}{x^m}}{x\left(\dfrac{B_m(x, ux)}{x^m} + \dfrac{e(x, ux)}{x^m}\right)}. \tag{4.6402}$$

Write

$$D(x, u) = \frac{d(x, ux)}{x^m}, \quad E(x, u) = \frac{e(x, ux)}{x^m}. \tag{4.6403}$$

We suppose that both $D(x, u)$ and $E(x, u)$ are continuous for $x = 0$. No general conclusions can be drawn regarding the form of the trajectories if we know merely that both D and E vanish at $x = 0$. We shall make a stronger assumption [11] that

$$D(x, u) = xF(x, u), \quad E(x, u) = x\,G(x, u), \tag{4.6404}$$

where $F(0, u)$ and $G(0, u)$ are continuous and bounded functions of u.

In view of (4.6401), $b_0 = 0$ (cf. 4.2) and

$$\frac{B_m(x, ux)}{x^m} = b_1u^{m-1} + b_2u^{m-2} + \dots + b_m = \Pi(u)$$

[11] Frommer [20] studied this problem under less restrictive conditions than those we impose here. He merely asks that $D(x, u)/x^\varepsilon = O(1)$, $E(x, y)/x^\varepsilon = O(1)$, $\varepsilon > 0$.

is a polynomial in u of degree not exceeding $m - 1$. The equation (4.6402) may therefore be written in the form

(4.6405) $$\frac{du}{dx} = \frac{F(x, u) - u\, G(x, u)}{\Pi(u) + x\, G(x, u)}.$$

4.641. Nonsingular directions. A point $(0, u_0)$ on the u-axis is a nonsingular point of (4.6405) if u_0 is not a root of the equation

(4.6411) $$\Pi(u) = 0.$$

Thus, there are at most $m - 1$ singular points of the equation (4.6405) on the u-axis.

Through every nonsingular point $(0, u_0)$ there goes one and only one trajectory of (4.6405) in the (x, u)-plane.

If u_0 is not a root of (4.6411), then our trajectory cuts across the u-axis at $(0, u_0)$, and for every polar direction φ_0 in the (x, y)-plane such that $\tan \varphi_0 = u_0$, there is a unique trajectory $r = r(t)$, $\varphi = \varphi(t)$ of the original system (4.601) which tends to the origin along φ_0, i.e., which tends to the origin so that $r \to 0$ and $\varphi \to \varphi_0$.

If u_0 is a root of (4.6411) but the numerator of the right-hand member of (4.6405) does not vanish at $(0, u_0)$, then $(0, u_0)$ is again a nonsingular point. The unique trajectory through $(0, u_0)$ is tangent to the u-axis. If our trajectory cuts across the u-axis at $(0, u_0)$, then there is a unique trajectory of (4.601) tending to the origin along φ_0 as well as a unique trajectory tending to the origin along $\varphi_0 + \pi$. Here again $\tan \varphi_0 = u_0$. If, however, our trajectory remains in the same halfplane near $(0, u_0)$, say in the half-plane $x \geqq 0$, then if $\tan \varphi_0 = u_0$, $-\pi/2 < \varphi_0 < \pi/2$, there are exactly two trajectories tending to the origin along the polar ray $\varphi = \varphi_0$, one on each side of the ray, and there is no trajectory of (4.601) tending to the origin along the ray $\varphi = \varphi_0 + \pi$.

4.642. Singular directions. If u_0 is a root of (4.6411) and the numerator of the right-hand member of (4.6405) does vanish at $(0, u_0)$, then $(0, u_0)$ is a singular point of (4.6405). The behavior of solutions near such a singular point may belong to any of the types discussed in Part 4 of this chapter. In case the characteristic equation is again identically zero, we employ, as before, the Briot-Bouquet transformation. Since

$$\Pi(u) = (u - u_0)^k \Pi_0(u), \quad \Pi_0(u_0) \neq 0,$$

the order of the denominator in (4.6405) does not exceed $k \leqq m-1 < m$ where m is the order of the denominator Q of (4.601). Thus, successive applications of the Briot-Bouquet transformation will lead either to the case where either there are no singular points on the y-axis or else the characteristic equation associated with every singular point on this axis, does not vanish identically.

4.643. To study the behavior of trajectories near the critical directions $+\pi/2$, $-\pi/2$, we employ the transformation $x = wy$ and proceed as above.

4.65. The refinement of Theorems 4.6232–3. In what follows we consider again systems of the form (4.301) in which $A_m(x, y)$ and $B_m(x, y)$ do not have real linear factors in common, in which both perturbation terms $d(x, y)$ and $e(x, y)$ are continuous and $o(r^m)$. We suppose again that the origin is an isolated singular point. Since.

$$A_m^2(\cos \varphi, \sin \varphi) + B_m^2(\cos \varphi, \sin \varphi) = N^2(\varphi) \neq Z^2(\varphi),$$

every root φ_0 common to both $N(\varphi)$ and $Z(\varphi)$, yields a common real linear factor of A_m and B_m. Since A_m and B_m have no such factors in common, every critical direction φ_0 is regular (cf. 4.62) and the discussion in 4.62 and 4.64 applies.

Thus every critical direction falls into one of the three types discussed in Theorems 4.6231–3.

If there exists a critical direction φ_0 of the first type (Theorem 4.6231) then every trajectory $r = r(t)$, $\varphi > \varphi(t)$ which enters a normal domain $\mathfrak{S}(r_0, \varphi_0, \delta)$ associated with φ_0, tends to the origin in the direction of φ_0, i.e., not only does $\varphi \to \varphi_0$ but the direction of the tangent line to our trajectory at (r, φ) also tends to the direction of the polar ray $\varphi = \varphi_0$.

If there exists a critical direction φ_0 of the second type then there is at least one trajectory which tends to the origin in the direction of φ_0. Under certain conditions, in the linear case, for example, there is only one such trajectory for each critical direction of the second kind. Theorem 4.6232 tells us that in general there are two alternatives: either there is a unique trajectory which tends to the origin in the direction of φ_0, or there are infinitely many such trajectories, viz., those which enter a normal domain $\mathfrak{S}(r_0, \varphi_0, \delta)$ by cutting across a certain subarc of the boundary arc. In what follows we shall give analytic criteria which will allow us to distinguish between these alternatives under rather general conditions.

If there exists a critical direction φ_0 of the third type then there may exist infinitely many trajectories tending to the origin in the direction of φ_0 or there may exist no such trajectories. Here again we shall seek analytical criteria which will allow us to distinguish between these alternatives.

If there exists a critical direction φ along which there is a trajectory tending to the origin, then we know that the origin can be neither a stable singular point nor a focal point.

If, however, there exist only those critical directions φ_0 for which the second one of the two alternatives in 4.6233 holds true (cf. 4.617) or if there exists no critical direction at all ($N(\varphi) = 0$ has no real roots) then the origin may be either a stable singular point or a focal point. We shall study analytical criteria for distinguishing which of the last two alternatives does occur in a given case.

4.651. A comparison of (4.301) **with the truncated equation** (4.201). We shall relate the question of the refinement of Theorem 4.6252–3 to the problem of the comparison of the behavior of trajectories of (4.301) near the origin with that of the trajectories of the truncated system (4.201).

For the truncated system (4.201), every critical direction φ_0 yields a straight-line trajectory lying on the ray $\varphi = \varphi_0$. Thus, it is obvious that only the first alternative in 4.6233 may hold here. It follows from this that in the case of the truncated system (4.201) the origin is a stable singular point or a focus only if the characteristic equation has no real roots, in which case the discussion in 4.21 shows that we do have either a focus or a center. A precise description of the trajectory behavior near the critical directions (integral rays) of the second and the third types is given in (4.222–3) where, if we set $j = 0$, $Q_0(\varphi_0) = c$ and hence $cZ_0 = Q_0(\varphi_0)Z(\varphi_0)$ is of the same sign as $Z(\varphi_0)/Q_0(\varphi_0)$. It is quite clear (cf. 4.222) that in 4.6232 also the first alternative alone may hold for (4.201).

In view of the above, we see that in the neighborhood of the origin (an isolated singular point of (4.301)) the trajectories of (4.301) behave essentially as do the trajectories of (4.201) in the following cases:

4.6511. near an isolated regular critical direction of the first type, always;

4.6512. near an isolated regular critical direction of the second type, if and only if the first alternative in 4.6232 holds:

4.6513. near an isolated regular critical direction of the third type, if and only if the first alternative in 4.6233 holds;

4.6514. in case there are no critical directions, solely if the origin is a center or a focus of (4.301).

4.6515. in case every direction is a critical direction and if the perturbation terms tend toward zero rapidly enough, say if $d(x, y)/x^{m+\varepsilon}$ and $e(x, y)/x^{m+\varepsilon}$ are equal to $O(1)$ for $\varepsilon > 0$. (Cf. 4.64 for $\varepsilon = 1$). Here the behavior is essentially the same except possibly for a finite number of polar directions.

The origin may be a focus or a stable point (even a center) of (4.301) if there exist only isolated regular critical directions of the third type for which the second alternative in 4.6233 holds, or if every direction is critical and the perturbation terms $d(x, y)$, $e(x, y)$ do not tend toward zero rapidly enough with r (cf. 4.514), (cf. Examples 4.311 where $e(x, y)/x^{1+\varepsilon} = 2/\{x^\varepsilon \log x^2(1+u^2)\} \to -\infty$) or if we weaken our restrictions on (4.301) and allow isolated critical directions which are not regular (Example 4.617). Here the addition of perturbation terms may induce a radical change in the behavior of trajectories.

4.652. Critical directions of the second type. In what follows we use the notations of Section 4.62. We write $d(x, y) = \eta_1(x, y)$ and $e(x, y) = \eta_2(x, y)$.

4.6521. Lemma (Lonn) [30]. *If there exists a continuous function $D(r)$ such that*

$$\int_0^{r_0} \frac{D(r)}{r}\, dr < +\infty, \qquad D(r) \geqq 0,$$

and (cf. 4.6203))

$$\frac{\overline{\Psi}(r, \varphi_1) - \overline{\Psi}(r, \varphi_2)}{\varphi_1 - \varphi_2} \leqq D(r) \quad \text{for} \quad r < r_0,$$

then the first alternative in 4.6232 (Fig. 18) *holds.*

Indeed, assume that there are two distinct trajectories $\varphi = \varphi_1(r)$ and $\varphi = \varphi_2(r)$ which go into the origin in the direction φ_0. Since these trajectories do not intersect we may assume that $\varphi_2(r) > \varphi_1(r)$ for $r > 0$. Then

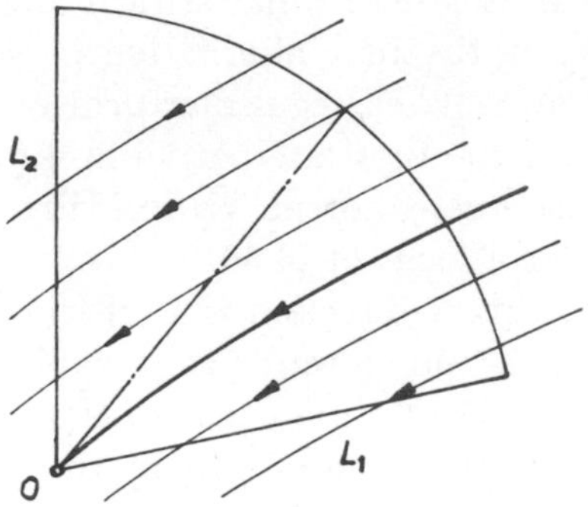

Fig. 18

$$\frac{d(\varphi_2 - \varphi_1)}{dr} = \frac{1}{r}\{\overline{\Psi}(r, \varphi_2) - \overline{\Psi}(r, \varphi_1)\} \leqq \frac{D(r)}{r}(\varphi_2 - \varphi_1).$$

Hence, if $0 < r' < r'' < r_0$, then

$$\log\,[\varphi_2(r) - \varphi_1(r)]_{r'}^{r''} \leqq \int_0^{r_0} \frac{D(r)}{r}\,dr,$$

whence

$\log\,[\varphi_2(r') - \varphi_1(r')] \geqq k$ (k a constant independent of r'), and therefore

$$\varphi_2(r') - \varphi_1(r') \geqq e^k > 0.$$

The last inequality holds for arbitrarily small r', which contradicts the hypothesis that both curves are tangent at the origin to the same ray $\varphi = \varphi_0$.

4.6522. Theorem. *If φ_0 is a simple root of the equation $N(\varphi) = 0$, and if $dN(\varphi_0)/d\varphi Z(\varphi_0) > 0$, then the first alternative in* 4.6232 *holds in case the perturbation terms η/r^m and ξ/r^m satisfy the Lipschitz conditions*

$$\frac{1}{r^m}\frac{\eta(r, \varphi_2) - \eta(r, \varphi_1)}{\varphi_2 - \varphi_1} \leqq C_\eta,$$

$$\frac{1}{r^m}\frac{\xi(r, \varphi_2) - \xi(r, \varphi_1)}{\varphi_2 - \varphi_1} \leqq C_\xi,$$

where the Lipschitz constant C_η of η/r^m is not too large, more precisely, where

$$C_\eta < \frac{dN(\varphi_0)}{d\varphi}.$$

In view of (4.6221), $dN(\varphi_0)/d\varphi = c$ and our hypothesis imply that every normal domain associated with φ_0, is of Type II (cf. 4.6232).

It is easily seen that

$$\begin{aligned}\Psi(r, \varphi_2) - \Psi(r, \varphi_1) = \left(Z_2 + \frac{\xi_2}{r^m}\right)^{-1}\left(Z_1 + \frac{\xi_1}{r^m}\right)^{-1} & \Big\{(N_1 - N_2)Z_1 + N_1(Z_2 - Z_1) \\ & + \frac{1}{r^m}[(N_1 - N_2)\xi_1 + N_1(\xi_2 - \xi_1)] \\ & + \frac{1}{r^m}[Z_1(\eta_2 - \eta_1) + (Z_1 - Z_2)\eta_1] \\ & + \frac{1}{r^{2m}}[(\eta_2 - \eta_1)\xi_1 + \eta_1(\xi_1 - \xi_2)]\Big\},\end{aligned}$$

where the subscript $i\,(i = 1, 2)$ indicates that we consider the value of the function at φ_i. By (4.6221), we choose δ small enough so that in $\mathfrak{S}(r_0, \varphi_0, \delta)$ the expression $(N_1 - N_2)Z_1$ is approximated with any desired accuracy by

$$Zc(\varphi_1 - \varphi_2) = -Z(\varphi_0)N'(\varphi_0)(\varphi_2 - \varphi_1).$$

Since η as well as $\xi = o(r^m)$, it follows from the hypotheses of our theorem that for r_0 and δ sufficiently small the difference $\Psi(r, \varphi_2) - \Psi(r, \varphi_1)$ has the sign of $(N_1 - N_2)Z_1$ that is the sign of $-Z(\varphi_0)N'(\varphi_0) < 0$ throughout $\mathfrak{S}(r_0, \varphi, \delta)$. Thus we may choose $D(r) \equiv 0$ in Lemma 4.6521. This completes the proof of our theorem.

One may wonder whether the first alternative for normal domains of Type II must necessarily hold if the right-hand members of our differential equations are analytic or, perhaps merely polynomials. The answer to this, however, is in the negative as can be seen from the following:

4.6523. *Example.* It was shown by Frommer [20] that the trajectories of the equation

$$\frac{dy}{dx} = \frac{y^3 - x^4 y}{-x^6}$$

behave as is indicated in Fig. 19. The critical directions $\varphi_0 = \pi/2$, $3\pi/2$ are regular. The critical directions $\varphi_0 = 0$, π are regular in the more general sense of Frommer and lie in normal domains.

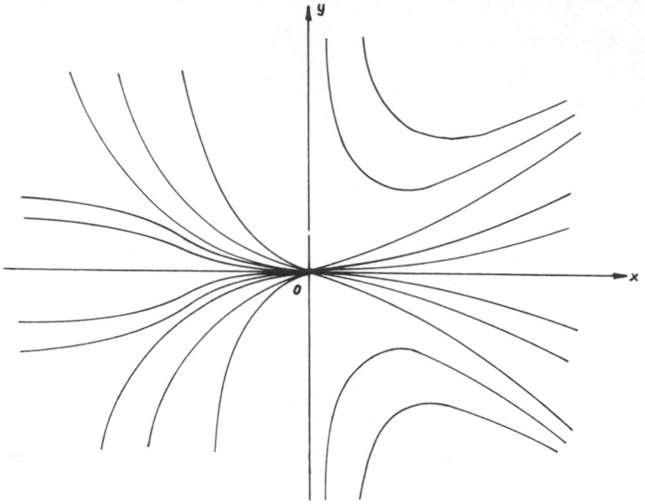

Fig. 19

The normal domain associated with the critical direction $\varphi_0 = 0$ *is of type II in which the second alternative in* 4.6232 *holds.*

4.653. Critical directions of the third type. Let φ_0 be a regular critical direction whose multiplicity ν as a root of the characteristic equation $N(\varphi) = 0$ (cf. 4.62) is even. Then, by Theorem 4.6233, every normal domain associated with φ_0 is of Type III. In what follows we shall give analytical criteria which distinguish between the two kinds of behavior of trajectories in a normal domain of Type III.

4.6531. Some preliminary considerations. We consider the following auxiliary differential equation

$$(4.6532) \qquad r\,\frac{d\bar{\varphi}}{dr} = a(\bar{\varphi} - \varphi_0)^\nu + b\,\frac{A(r)}{r^m} \qquad (\nu \text{ even}).$$

Here $a \geqq 0$, $b \geqq 0$, and $A(r)$ is some function of r such that $A(r) = o(r^m)$.

For $b = 0$, (4.6532) is satisfied by

$$\bar{\varphi} - \varphi_0 = \left[(\nu - 1)a \log \frac{c}{r}\right]^{-\frac{1}{\nu-1}}.$$

Substituting

$$\bar{\varphi} - \varphi_0 = z\left(\log \frac{1}{r}\right)^{-\frac{1}{\nu-1}} \tag{4.6533}$$

into (4.6532) we get

$$\begin{aligned} & r\frac{dz}{dr}\left(\log \frac{1}{r}\right)^{-\frac{1}{\nu-1}} + \frac{z}{\nu - 1}\left(\log \frac{1}{r}\right)^{-\frac{\nu}{\nu-1}} \\ & \quad = az^{\nu}\left(\log \frac{1}{r}\right)^{-\frac{\nu}{\nu-1}} + b\,\frac{A(r)}{r^m}. \end{aligned} \tag{4.6534}$$

We consider the domain $r < 1$, and we let

$$A(r) = r^m\left(\log \frac{1}{r}\right)^{-\frac{\nu}{\nu-1}} \qquad (= o\,(r^m)). \tag{4.6535}$$

For this choice of $A(r)$ our differential equation (4.6534) becomes

$$r\,\frac{dz}{dr}\left(\log \frac{1}{r}\right) = az^{\nu} - \frac{z}{\nu - 1} + b = M(z),$$

and we have

$$\int \frac{dz}{M(z)} = -\log\log \frac{1}{r} + c \quad (c = \text{a constant}).$$

The denominator $M(z)$ has either no real roots or exactly two real roots. For, since the second derivative $M''(z) = \nu(\nu - 1)az\ \nu^2$ is positive for $z \neq 0$, the graph of the polynomial $M(z)$ does not have points of inflection, and for all values of b which exceed a certain value b_0 the polynomial $M(z)$ has no real roots, for all values of $b < b_0$ this polynomial $M(z)$ has exactly two distinct roots, ξ_1 and ξ_2, and if $b = b_0$ then $M(z)$ has a double root ξ_0. Since $M(\xi_0) = a\xi_0^{\nu} - [\xi_0/(\nu - 1)] + b_0 = 0$ and $M'(\xi_0) = \nu a\xi_0^{\nu-1} - [1/(\nu - 1)] = 0$ we have

$$b_0 = \nu^{-\frac{\nu}{\nu-1}}\,[a(\nu - 1)]^{-\frac{1}{\nu-1}}.$$

In case $b < b_0$, the function

$$\Phi_0\,(z) = \int_{z_0}^{z} \frac{dz}{M(z)}$$

is monotonically decreasing in the interval $\xi_1 < z < \xi_2$ and becomes infinite both at ξ_1 and at ξ_2.

Let us look upon r and $\bar{\varphi}$ in (4.6533) as polar coordinates and let us consider the behavior of the curves L_k defined by (4.6533) and

$$\Phi_0(z) = -\log\log\frac{1}{r} + k,$$

for various values of the constant of integration k. Each such curve is given by these relations with z as a parameter varying between ξ_1 and ξ_2. As z varies continuously from ξ_1 to ξ_2, $\Phi(z)$ varies continuously from $+\infty$ to $-\infty$, while r varies continuously from 1 to 0. Moreover, as z tends to ξ_2, r tends to zero, and (cf. 4.6533) $\bar{\varphi}$ tends to φ_0. Thus, by (4.6532), the tangent to L_k at a point P, tends to the line $\bar{\varphi} = \varphi_0$ as P tends to the origin along L_k.

When $b > b_0$, the polynomial $M(z)$ has no real root and (since $\nu \geqq 2$) $\Phi_0(z)$ is a continuous, monotonically increasing, bounded function for $-\infty < z < +\infty$. Adding a suitable constant to $\Phi_0(z)$ by changing the limits of integration, we replace it by

$$\Phi_\infty(z) = \int_{+\infty}^{z} \frac{dz}{M(z)} \, (< 0)$$

and consider curves L_k defined by (4.6533) and

$$\Phi_\infty(z) = -\frac{1}{r} + k.$$

On L_k, $\log\log(1/r)$ is bounded, and r does not come arbitrarily near to either 0 or 1. Moreover, along such a curve the coordinate φ is a monotonically increasing function of r by (4.6532), and as z tends to either $+\infty$ or $-\infty$, so does $\bar{\varphi} - \varphi_0$ by (4.6533) since $\log 1/r$ remains between positive bounds. We shall need to consider the curves L_k only for $|\bar{\varphi} - \varphi_0| \leqq \delta$. In each such sector L_k goes from one boundary ray to the other and does not adhere to the origin. Also, there exist arcs of L_k arbitrarily close to the origin, since for $k = \log\log(1/r_1)$, we have $r < r_1$ along L_k.

The case $b = b_0$ will not be needed in what follows. We shall now prove

4.6536. Theorem (Lonn) [30]. *Let* φ_0 *be a root of even multiplicity* ν *of the characteristic equation* (4.6204), *and* (*cf.* 4.6221) *let* $\bar{c} = -c > 0$ *and* $Z_0 = Z(\varphi_0) > 0$. *Let* $A(r)$ *be given by* (4.6535), *and write*

$$D = \left(\frac{Z_0}{\nu}\right)^{\frac{\nu}{\nu-1}} [\bar{c}(\nu - 1)]^{-\frac{1}{\nu-1}}.$$

(i) *If in some sector* $|\varphi - \varphi_0| \leqq \delta$, *the perturbation function* $\eta(r, \varphi)$ *in* 4.62 *has an estimate*

$$\eta(r, \varphi) \leqq c_1 A(r) \quad \textit{for} \quad 0 < c_1 < D$$

for small r, *then the first alternative in* 4.6232 *holds and near the origin in our sector the behavior of the trajectories of the original system* (4.301) *agrees with that of the trajectories of the truncated system* (4.201) (Fig. 18).

ii) *If, however, in a sector* $|\varphi - \varphi_0| \leqq \delta$,

$$\eta(r, \varphi) \geqq c_2 A(r) \quad \textit{for} \quad c_2 > D,$$

then the second alternative in 4.6232 *holds true and near the origin in our sector the behavior of trajectories of* (4.301) *does not agree with that of the trajectories of the truncated system* (4.201) (Fig. 20).

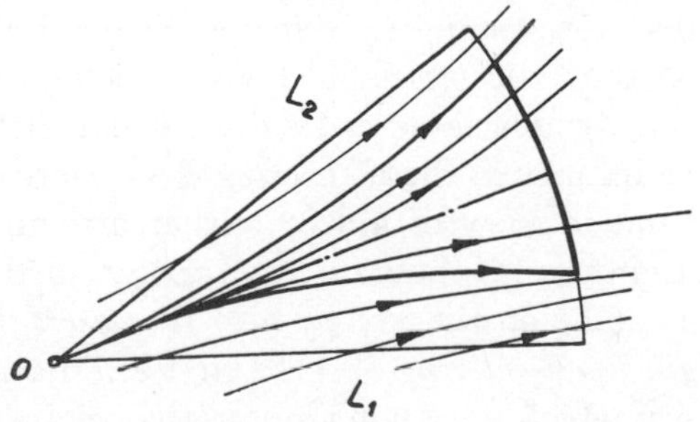

Fig. 20

Let $\mathfrak{S}(r_0, \varphi_0, \delta)$ be a normal domain so small that in it the estimate in (i) holds true for $\eta(r, \varphi)$. Then

$$(4.6537) \quad r\frac{d\varphi}{dr} = \frac{-N(\varphi) + \dfrac{\eta(r, \varphi)}{r^m}}{Z(\varphi) + \dfrac{\xi(r, \varphi)}{r^m}} < \frac{\bar{c}(\varphi - \varphi_0)^\nu + c_1 \dfrac{A(r)}{r^m}}{Z_0}(1+\vartheta) \equiv \Psi_1(r, \varphi),$$

where ϑ is positive and can be chosen arbitrarily small by taking r_0 and δ small enough.

The differential equation

$$r \frac{d\bar{\varphi}}{dr} = \Psi_1(r, \bar{\varphi}) \tag{4.6538}$$

is of the form (4.6532) with

$$a = \frac{\bar{c}}{Z_0}(1 + \vartheta), \qquad b = \frac{c_1}{Z_0} \quad (1 + \vartheta).$$

By hypothesis,

$$\frac{c_1}{Z_0} < \nu^{-\frac{\nu}{\nu-1}} \left[\frac{\bar{c}}{Z_0}(\nu - 1)\right]^{-\frac{1}{\nu-1}},$$

and hence

$$b = \frac{c_1}{Z_0}(1 + \vartheta) < \nu^{-\frac{\nu}{\nu-1}} \left[\frac{\bar{c}}{Z_0}(1 + \vartheta)(\nu - 1)\right]^{-\frac{1}{\nu-1}} = b_0$$

for sufficientely small ϑ. In view of the discussion in 4.6531, there exists a trajectory L_k: $\bar{\varphi} = \bar{\varphi}(r)$ of our auxiliary differential equation (4.6538), which is tangent to the ray $\varphi = \varphi_0$ at the origin. We may choose r_0 so that for $r < r_0$, L_k remains inside $\mathfrak{S}(r_0, \varphi_0, \delta)$.

Consider next the trajectories of (4.601) which enter our normal domain $\mathfrak{S}(r_0, \varphi_0, \delta)$ associated with the regular critical direction φ_0. There are an infinity of these entering $\mathfrak{S}$ say across the segment L_2 of $\varphi_0 - \delta$, and in view of 4.6233, either an infinity of these tend to the origin or else they all leave $\mathfrak{S}$ by cutting across the boundary segment L_k on the ray $\varphi_0 - \delta$. *We assert that all of them tend to the origin upon entering* $\mathfrak{S}$. (The first alternative in 4.6232). For, a trajectory which does not tend to the origin upon entering $\mathfrak{S}$ by way or L_2 must leave $\mathfrak{S}$ by way of L_1 on the boundary ray $\varphi_0 - \delta$, and in so doing must cut across the trajectory L_k of our auxiliary differential equation (4.6538). This is impossible since at the point of intersection

$$r \frac{d\varphi}{dr} \geqq r \frac{d\bar{\varphi}}{dr} = \Psi_1(r, \bar{\varphi}) = \Psi_1(r, \varphi),$$

which contradicts (4.6537).

Suppose next that in some normal domain $\mathfrak{S}(r_0, \varphi_0, \delta)$ the estimate in (ii) holds true for $\eta(r, \varphi)$. Then, when $\vartheta > 0$ can be made small enough by choosing Γ_0 and δ sufficiently small

$$r\frac{d\varphi}{dr}=\frac{-N(\varphi)+\dfrac{\eta(r,\varphi)}{r^m}}{Z(\varphi)+\dfrac{\xi(r,\varphi)}{r^m}}>\frac{\bar{c}(\varphi-\varphi_0)^\nu+c_2\dfrac{A(r)}{r^m}}{Z_0}(1-\vartheta)=\Psi_2(r,\varphi).$$

The differential equation

$$r\frac{d\bar{\varphi}}{dr}=\Psi_2(r,\bar{\varphi}) \tag{4.6539}$$

is of the form (4.6532) with

$$a=\frac{\bar{c}}{Z_0}(1-\vartheta),\qquad b=\frac{c_2}{Z_0}(1-\vartheta).$$

By hypothesis

$$\frac{c_2}{Z_0}>\nu^{-\frac{\nu}{\nu-1}}\left[\frac{\bar{c}}{Z_0}(\nu-1)\right]^{-\frac{1}{\nu-1}}$$

and hence for sufficiently small ϑ

$$b=\frac{c_2}{Z_0}(1-\vartheta)>\nu^{-\frac{\nu}{\nu-1}}\left[\frac{\bar{c}}{Z_0}(1-\vartheta)(\nu-1)\right]^{-\frac{1}{\nu-1}}=b_0.$$

Let $f(P,t)$ be a trajectory of (4.601) through a point $P=(r_1,\varphi_1)$ $\mathfrak{S}(r_0,\varphi_0,\delta)$. Since $b>b_0$, it follows from 4.6531 that for $\bar{r}<r_1$ the trajectory L_k of the auxiliary equation (4.6539), is contained in $\mathfrak{S}(\bar{r},\varphi_0,\delta)$ and goes from one boundary segment to the other. Thus in order that $f(P,t)$ should adhere to the origin it must cut L_k at some point at which

$$r\frac{d\varphi}{dr}\leqq r\frac{d\bar{\varphi}}{dr}=\Psi_2(r,\bar{\varphi})=\Psi_2(r,\varphi),$$

a contradiction.

4.6541. *Example.* Consider the differential equation

$$\frac{dy}{dx}=\frac{x+y+\eta_1}{x+\eta_2}.$$

Here, $N(\varphi)=-\cos^2\varphi$, with double roots $\varphi_0=\pi/2$ and $-\pi/2$. In view of the above discussion, the functions η_1 and η_2 may be chosen so that the alternative (ii) holds true near both $+\pi/2$ and $-\pi/2$, whence the trajectories of our equation will not approach the origin in a fixed direction and the origin will be a focus.

We note that the origin is a node of the corresponding truncated system.

4.6542. The analytic case. In this case $\eta(r, \varphi)/r^{n+1}$ remains bounded as $r \to 0$, whence

$$\eta(r, \varphi) \leqq Cr^{n+1} < c_1 r^n \left(\log \frac{1}{r}\right)^{-\frac{\nu}{\nu-1}}$$

for an arbitrary $c_1 > 0$ and for a sufficiently small r.

4.6543. Although Lonn's theorem cannot be applied directly to the case $m \neq n$ in 4.63, one can employ his Lemma 4.6521 as well as the method of comparison with the auxiliary equation described in 4.6531. In particular, for the case

$$r \frac{d\varphi}{dr} = -\tan \varphi + o(1)$$

discussed in (4.63), the first one of the two possible alternatives holds true (cf. 4.6232).

4.655. Analytical criteria for differentiating between a center and a focus. If the characteristic equation (4.6204) has no real roots then the origin is either a stable singular point (cf. 3.3) or a focus.

We first seek criteria which give assurance that our singular point is a center. We shall find that in general such criteria involve infinitely many operations [48], [32], [20].

Consider the equation

$$\text{(A)} \qquad y' = \frac{P_n(x, y) + P_{n+1}(x, y) + P_{n+2}(x, y) + \dots}{Q_n(x, y) + Q_{n+1}(x, y) + Q_{n+2}(x, y) + \dots}$$

where $P_i(x, y)$ and $Q_i(x, y)$ are homogeneous polynomials of degree i, and where both the numerator and the denominator of the right-hand member are analytic in x and y in a neighborhood of the origin. Changing over to polar coordinates r and φ we obtain

$$\text{(B)} \qquad \frac{dr}{d\varphi} = \frac{rp_{n+1} + r^2 p_{n+2} + r^3 p_{n+3} + \dots}{q_{n+1} + rq_{n+2} + r^2 q_{n+3} + \dots} \equiv \frac{L}{Z}$$

where

$$\begin{aligned} p_{n+i+1} &= P_{n+i}(\cos\varphi, \sin\varphi)\sin\varphi + Q_{n+i}(\cos\varphi, \sin\varphi)\cos\varphi \\ q_{n+i+1} &= P_{n+i}(\cos\varphi, \sin\varphi)\cos\varphi - Q_{n+i}(\cos\varphi, \sin\varphi)\sin\varphi. \end{aligned}$$

Since we have assumed that there are no critical directions, q_{n+1} has no real roots.

We consider next a family of closed curves

$$f(r, \varphi) = rf_0(\varphi) + r^2 f_1(\varphi) + r^3 f_2(\varphi) + \ldots = \text{const.},$$

and we seek periodic functions $f_0(\varphi)$, $f_1(\varphi)$, ... such that

$$f(r, \varphi) = \text{const.}$$

is a formal integral of our differential equation.

The differential equation for the above family has the form

$$\frac{dr}{d\varphi} = -\frac{rf_0' + r^2 f_1' + r^3 f_2' + \ldots}{f_0 + 2rf_1 + 3r^2 f_2 + \ldots} \equiv \frac{L_1}{Z_1}.$$

In order to obtain the desired result we must have

$$\frac{rp_{n+1} + r^2 p_{n+2} + r^3 p_{n+3} + \ldots}{q_{n+1} + rq_{n+2} + r^2 q_{n+3} + \ldots} = -\frac{rf_0' + r^2 f_1' + r^3 f_2' + \ldots}{f_0 + 2rf_1 + 3r^2 f_2 + \ldots}$$

or

$$r(q_{n+1}f_0' + p_{n+1}f_0) + r^2(q_{n+1}f_1' + 2p_{n+1}f_1 + f_0' q_{n+2} + f_0 p_{n+2}) + \ldots = 0.$$

In order that the left-hand member should vanish, it is necessary and sufficient that a denumerable number of differential equations should have periodic solutions. These differential equations may be written in the form

$$(4.6551) \quad q_{n+1}f_i' + (i+1)p_{n+1}f_i + R_i = 0 \qquad (i = 0, 1, 2, \ldots),$$

where R_i depend only upon f_k, f_k' with $k < i$.

4.6552. THEOREM. *The origin is a center of our differential equation if and only if the infinite system of differential equations* (4.6551) *has periodic solutions of period* 2π.

Indeed, for $i = 0$ we have

$$(4.6553) \qquad q_{n+1}f_0' + p_{n+1}f_0 = 0.$$

The solution

$$f_0 = Ce^{-\int_0^{\varphi}\frac{p_{n+1}}{q_{n+1}}d\varphi}$$

is periodic with period 2π for every C (we may take $C = 1$) if

$$\int_0^{2\pi} \frac{p_{n+1}}{q_{n+1}} d\varphi = 0.$$

This is the first necessary condition. If it is not fulfilled, then the equation

(4.6554) $$q_{n+1} F_0' + p_{n+1} F_0 = D_0$$

has a periodic solution for some D_0. To determine such a D_0 we examine the solution

$$F_0(\varphi) = e^{-\int_0^\varphi \frac{p_{n+1}}{q_{n+1}} d\varphi} \left[C + D_0 \int_0^\varphi \frac{1}{q_{n+1}} e^{\int_0^\varphi \frac{p_{n+1}}{q_{n+1}} d\varphi} d\varphi \right]$$

The condition of periodicity yields the relation

$$0 = F_0(2\pi) - F_0(0) =$$

$$(e^{-\int_0^{2\pi} \frac{p_{n+1}}{q_{n+1}} d\varphi} - 1)C + D_0 e^{-\int_0^{2\pi} \frac{p_{n+1}}{q_{n+1}} d\varphi} \int_0^{2\pi} \frac{1}{q_{n+1}} e^{\int_0^\varphi \frac{p_{n+1}}{q_{n+1}} d\varphi} d\varphi.$$

This equation can always be solved for D_0 whose coefficient is well defined and is different from zero since q_{n+1}, and hence the whole expression under the outer integral sign does not change its algebraic sign. It can be easily seen that F_0 does not vanish.

Next, let f_k be the first nonperiodic function in the sequence $f_1, f_2, \ldots, f_n, \ldots$. Then there exists a constant $D_k \neq 0$ such that the equation

(4.6555) $$q_{n+1} F_k' + (k+1) p_{n+1} F_k + R_k = D_k$$

has a periodic solution F_k. Its solution is given by

$$F_i = e^{-(i+1) \int_0^\varphi \frac{p_{n+1}}{q_{n+1}} d\varphi} \left[C - \int_0^\varphi \frac{R_i - D_i}{q_{n+1}} e^{(i+1) \int_0^\varphi \frac{p_{n+1}}{q_{n+1}} d\varphi} d\varphi \right]$$

for $i = k$. For $D_i = 0$, we have $F_i = f_i$. The condition for periodicity yields

$$0 = F_i(2\pi) - F_i(0)$$

(4.6556) $$= -\int_0^{2\pi} \frac{R_i}{q_{n+1}} e^{(i+1) \int_0^\varphi \frac{p_{n+1}}{q_{n+1}} d\varphi} dt + D_i \int_0^{2\pi} \frac{1}{q_{n+1}} \cdot e^{(i+1) \int_0^\varphi \frac{p_{n+1}}{q_{n+1}} d\varphi} d\varphi.$$

If

$$\int_0^{2\pi} \frac{R_i}{q_{n+1}} e^{(i+1) \int_0^\varphi \frac{p_{n+1}}{q_{n+1}} d\varphi} d\varphi = 0,$$

then there exists a periodic solution f_i of (4.6551) for this i. If this last integral is not zero, then since the coefficient of D_i in (4.6556) is not zero, this equation defines uniquely a $D_i \neq 0$.

Consider the family of periodic curves

$$rf_0 + r^2 f_1 + \ldots + r^k f_{k-1} + r^{k+1} F_k = \text{const.},$$

and compare the direction of the field determined by these curves with the direction of the field as determined by the given equation, by calculating the value of the tangent of the angle between the directions of these two fields. Our construction implies that in the numerator of this expression all the terms containing r^i with $i < k$, and (4.655) the coefficient of r^{k+1} will disappear. Hence we may choose a neighborhood of the origin in which the numerator $\neq 0$ and is say positive. It follows from this that the direction of the vector field determined by our equation does not coincide anywhere with the direction of the field associated with the above family of closed curves, and therefore the trajectories of our equation either tend to the origin in a fixed direction or spiral toward it. The first alternative is excluded by the hypothesis that there are no critical directions. Thus all the trajectories are spirals and we have a focal point.

Next, we prove the sufficiency of our criteria. Since $q_{n+1} = 0$ is our characteristic equation, the denominator Z of the right-hand member of our equation

$$\frac{dr}{d\varphi} = \frac{L}{Z}$$

with L, Z given by (B) in 4.655, does not vanish for small values of r. Determine a comparison equation (C), i.e.,

$$\frac{d\bar{r}}{d\varphi} = \frac{L_1}{Z_1}$$

so that in the numerator of the fraction giving the difference $dr/d\varphi - d\bar{r}/d\varphi$, all the terms of order not exceeding $2n + 1$ in r cancel out. Since the denominator does not vanish for small values of r, then, for small r,

$$\left| \frac{dr}{d\varphi} - \frac{d\bar{r}}{d\varphi} \right| < r^{2n+1} M',$$

where M' is constant.

Let $r < k$ be a domain in which the denominator Z of $dr/d\varphi$ does not vanish and hence does not change its sign. Then there exists a constant M such that $|dr/d\varphi| < M$ in this domain.

The constant M may be chosen arbitrarily small as long as k

is sufficiently small. Therefore if we choose r_0 small enough, then as φ varies from 0 to 2π our trajectory will remain in the domain $r < k$, i.e., in the domain in which the denominator does not change its sign.

Let us seek the trajectory through the point $(r_0, 0)$ by the method of successive approximations. It is clear that all approximations will remain in the domain in which the denominator does not change its sign. Their infinite series expansions will have as their initial terms the successive partial sums of the series

$$r = r_0 w_1 + r_0^2 w_2 + r_0^3 w_3 + \ldots,$$

where w_i are continuous functions depending only upon $\sin \varphi$ and $\cos \varphi$.

Since for $0 \leqq \varphi \leqq 2\pi$ all the approximating curves are contained in the domain $r < k$ and the approximations converge on $[0, 2\pi]$, then the series above converges at 2π.

Let

$$\varrho = r_0 w_1(2\pi) + r_0^2 w_2(2\pi) + \ldots,$$

and consider the function

$$\Psi(r_0) = r - \varrho = C_0 r_0 + C_1 r_0^2 + \ldots.$$

We shall show that all C_i are equal to zero.

Indeed, suppose n is an arbitrarily large positive integer and consider the comparison equation

$$\frac{d\bar{r}}{d\varphi} = \frac{L_1}{Z_1}$$

for which the numerator of the difference $dr/d\varphi - d\bar{r}/d\varphi$ does not contain terms of order $\leqq (2n + 1)$ in r. If we express a solution of this equation as an infinite series, i.e., if we set

$$\bar{r} = r_0 \bar{w}_1(\varphi) + r_0^2 \bar{w}_2(\varphi) + \ldots + r_0^n \bar{w}_n(\varphi) + \ldots$$

and let

$$\bar{\varrho} = \bar{r}(2\pi) = r_0 \bar{w}_1(2\pi) + r_0^2 \bar{w}_2(2\pi) + \ldots,$$

then the function

$$\overline{\Psi}(r_0) = \bar{r} - \bar{\varrho} = \sum_{i=0}^{\infty} \bar{C}_i r_0$$

is identically zero, i.e., $\bar{C}_i = 0$ for $i = 0, 1, \ldots.$

Since we assumed that the comparison equation was chosen so that

$$w_i'(\varphi) = \bar{w}_i'(\varphi) \qquad (i = 0, 1, \ldots, 2n + 1)$$

we have $C_i = \bar{C}_i = 0$ for $i \leqq 2n + 1$. Since n may be taken arbitrarily, large, $C_i = 0$ for all i. However, $C_i = w_i(2\pi) - w_i(0)$; therefore $w_i(2\pi) = w_i(0)$, and the sufficiency of our conditions is established.

4.656. The analytic case. The procedure for differentiating between a center and a focal point which we described above, involves transcendental considerations and requires the solving of differential equations. If, however, our equation contains terms of the first order, then in the analytic case our process may be replaced by an algorithm introduced by Poincaré and requiring only the solution of algebraic equations. We shall give a brief account of this method.

In the presence of terms of the first order the origin is a center only if we can reduce our system by a nonsingular transformation to the form

$$\text{(4.6561)} \qquad \frac{dx}{dt} = y + q(x, y), \quad \frac{dy}{dt} = -x - p(x, y)$$

where $p(x, y)$ and $q(x, y)$ are analytic functions beginning with terms of order greater than the first. Thus, let

$$q(x, y) = q_2(x, y) + q_3(x, y) + \ldots + q_i(x, y) + \ldots,$$
$$p(x, y) = p_2(x, y) + p_3(x, y) + \ldots + p_i(x, y) + \ldots,$$

where $p_i(x, y)$ and $q_i(x, y)$ are homogeneous polynomials of order i. As our comparison equation we take

$$\frac{dy_1}{dx} = -\frac{f_x'(x, y)}{f_y'(x, y)} = -\frac{2x + f_{3x}'(x, y) + f_{4x}'(x, y) + \ldots}{2y + f_{3y}'(x, y) + f_{4y}'(x, y) + \ldots}$$

where

$$f(x, y) = x^2 + y^2 + f_3 + f_4 + \ldots + f_k + \ldots$$

characterizes a family of closed curves. The homogeneous polynomials f_k of order k have indeterminate coefficients. Let

$$f_k(x, y) = \sum_{n=0}^{k} A_{n,\, k-n}\, x^n y^{k-n},$$

and form the difference

$$y' - y_1' = [-(x + p_2 + p_3 + \dots)(2y + f'_{3y} + f'_{4y} + \dots)$$
$$+ (y + q_2 + q_3 + \dots)(2x + f'_{3x} + f'_{4x} + \dots)][(y + q(x, y))f'_y(x, y)]^{-1}.$$

One sees at once that the terms of order 2 in the numerator disappear.

The terms of order 3 are given by

$$-xf'_{3y} - 2yp_2 + yf'_{3x} + 2xq_3$$

which may be written

$$(-xf'_{3y} + yf'_{3x}) + (2xq_2 - 2yp_2).$$

The polynomial in the second parentheses has coefficients which can be expressed linearly in terms of the coefficients $p_2(x, y)$ and $q_2(x, y)$. Denote this polynomial by

$$B_{30}x^3 + B_{21}x^2y + B_{12}xy^2 + B_{03}y^3.$$

The polynomial $-xf_{3y} + yf_{3x}$ in the first parentheses is of the form

$$\begin{aligned}
&-xf'_{3y} + yf'_{3x} \\
&\quad = -x(A_{21}x^2 + 2A_{12}xy + 3A_{03}y^2) + y(3A_{30}x^2 + 2A_{21}xy + A_{12}y^2) \\
&\quad = -A_{21}x^3 + (-2A_{12} + 3A_{30})x^2y + (-3A_{03} + 2A_{21})xy^2 + A_{12}y^3.
\end{aligned}$$

In order that the terms of order 3 disappear, the coefficients A_{03}, A_{12}, A_{21}, and A_{30} must satisfy the equations

$$\begin{aligned}
-A_{21} + B_{30} &= 0, \\
-2A_{12} + 3A_{30} + B_{21} &= 0, \\
-3A_{03} + 2A_{21} + B_{12} &= 0, \\
A_{12} + B_{03} &= 0.
\end{aligned}$$

Consider next the terms of the fourth order. These are given by

$$-xf'_{4y} - p_2f'_{3y} - 2yp_3 + yf'_{4x} + q_2f'_{3x} + 2xq_3,$$

or

$$(-xf'_{4y} + yf'_{4x}) + (-p_2f'_{3y} - 2yp_3 + q_2f'_{3x} + 2xq_3).$$

The second grouping of terms in the parentheses again has previously determined coefficients. We write them in the form

$$B_{40}x^4 + B_{31}x^3y + B_{22}x^2y^2 + B_{13}xy^3 + B_{04}y^4.$$

We shall write the first group of terms in the form

$$-x(A_{31}x^3 + 2A_{22}x^2y + 3A_{13}xy^2 + 4A_{04}y^3) + \\ + y(4A_{40}x^3 + 3A_{31}x^2y + 2A_{22}xy^2 + A_{13}y^3).$$

The system of linear equations which must be satisfied by the indeterminate coefficients above does not have a solution in general. However, we can always choose these coefficients so that all the terms of the fourth order reduce to the expression $D_1(x^4 + y^4)$.

Indeed, equating the coefficients of x^3y, x^2y^2, xy^3 to zero and equating the coefficients of x^4 and y^4, we obtain the system

$$\begin{aligned} -A_{31} + B_{40} &= A_{13} + B_{04} \\ -2A_{22} + 4A_{40} + B_{31} &= 0 \\ -3A_{13} + 3A_{31} + B_{22} &= 0 \\ -4A_{04} + 2A_{22} + B_{13} &= 0. \end{aligned}$$

These conditions uniquely determine A_{31} and A_{13}, whereas A_{40}, A_{22}, A_{04} will depend upon one arbitrary parameter.

The coefficient A_{31} is given by

$$A_{31} = \frac{1}{6}(3B_{40} - B_{22} - 3B_{04}),$$

whence

$$D_1 = -A_{31} + B_{40} = \frac{1}{6}(3B_{40} + B_{22} + 3B_{04}).$$

The common coefficient D_1 of x^4 and y^4 may or may not be equal to zero.

If $D_1 \neq 0$, then the origin has a circular neighborhood in which the direction of the field determined by the given equation differs throughout from the direction of the field determined by the auxiliary equation used for comparison. Hence all trajectories tend to the origin and since no critical directions exist, these trajectories spiral toward the origin either for $t \to +\infty$ ot $t \to -\infty$ depending on the sign of D_1. Thus the origin is a focus.

Let $D_1 = 0$. Then we consider the terms of fifth order and find that these may be made to vanish by a suitable choice of the indeterminate coefficients. Turning our attention to the terms of sixth order, we find that these can be reduced to the expression $D_2(x_6 + y_6)$. If $D_2 \neq 0$, then arguing as above we find that the origin is a focus. If $D_2 = 0$ we proceed to terms of higher order.

Thus, infinitely many conditions are needed to ascertain the fact that the origin is a center.

One can establish the sufficiency of the above conditions. As was shown by Lyapounoff, for equations with terms of the first order these conditions are algebraically equivalent to the conditions obtained by the first method described in this section. Moreover, Lyapounoff showed [31] that if all the $D_i = 0$, then the series expansion of $f(x, y)$ converges for sufficiently small $|x|$, $|y|$, i.e., that for systems of the form (4.6561) for which the origin is a center, there exists an analytic integral

$$f(x, y) = x^2 + y^2 + f_3(x, y) + f_4 + \dots = \text{const.}$$

Conversely, if the characteristic equation does not have real roots, the existence of such an analytic integral shows that the origin is a center.

If our equation does not contain terms of the first order, then there may exist no analytic integral even in the case when the origin is a center. For example, the equation

$$y' = -\frac{x[(2x^2 + y^2) + 2(x^2 + y^2)^2]}{y[(2x^2 + y^2) + (x^2 + y^2)^2]}$$

has

$$(2x^2 + y^2)e^{\frac{1}{x^2+y^2}} = \text{const.}$$

as its integral. This integral determines a family of closed curves each of which surrounds the origin. Thus the origin is a center. However, in the last equality neither the expression on the left nor any analytic function of this expression is analytic at the origin.

4.657. Some special criteria. We shall discuss one condition due to Poincaré, which assures us that the origin is a center.

4.6571. THEOREM. *If the vector field defined by the system* (4.6561) *is symmetric with respect to the x-axis, then the origin is a center.*

In this case, every trajectory cutting upward across the x-axis at a point $A(x_0, 0)$, $x_0 > 0$, sufficiently close to the origin, cuts across the x-axis again at a point $(B(-x_1, 0)$, $x_1 > 0$, and has vertical tangents at both points A and B. Because of the symmetry of our field, continuing our trajectory *beyond* B yields an arc which is a mirror image of the trajectory arc AB. Hence the trajectory will close at A.

Analytically, the above conditions mean that system (4.6561) does not change if we replace y by $-y$ and t by $-t$, i.e., that

$$p(x, -y) = p(x, y), \quad q(x, -y) = -q(x, y).$$

4.6572. If the right-hand members of our equations are given by power series, then, in case the origin is a center, we need to fulfill infinitely many conditions upon the infinitely many coefficients of this series in order to establish this analytically. If the right-hand members are polynomials, then the required infinitely many conditions relate but a finite number of indeterminate coefficients. We saw that these conditions were reduced to infinitely many equations of the form $D_i = 0$, where D_i are polynomials in the given coefficients. The totality of polynomials whose vanishing is both a necessary and a sufficient condition for the existence of a center, is an ideal in a ring of polynomials in a finite number of variables. Such an ideal has a finite basis and hence we have here only a finite number of algebraically independent conditions.

In order to make an effective use of these conclusions we must answer the following question: Given that the right-hand members of our equation are polynomials of degree n, to determine $N(n)$ such that all the equalities $D_i = 0$ for $i > N(n)$ are consequences of such equalities for $i \leqq N(n)$. The problem of the characterization of $N(n)$ is still unsolved.

4.6572. The ideas discussed above may be used to characterize all those equations in a family of equations of a given type, for which the origin is a center.

This problem has been solved completely for polynomials of second degree by Frommer [20], Bautin [14] and Sacharnikov [49]. Here the results are as follows: If we write

$$y' = -\frac{x + ax^2 + (2b + \alpha)xy + cy^2}{y + bx^2 + (2c + \alpha)xy + dy^2}$$

then the origin is a center if

(1) $a + c = 0, \quad b + d = 0$

(2) $\dfrac{\alpha}{\beta} = \dfrac{b + d}{a + c} = k, \quad ak^3 - (3b + \alpha)k^2 + (3c + \beta)k - d = 0$

(3) $\alpha = 0, \quad \beta = 0$

(4) $a = c = \beta = 0$

(5) $b + d = \alpha = \beta + 5a + 5c = ac + 2a^2 + d^2 = 0$ $\qquad (a + c \neq 0)$

(6) $a_1 + c_1 = \beta_1 = \alpha_1 + 5b_1 + 5d_1 = b_1d_1 + 2d_1^2 + a_1^2 = 0$ $\qquad (b + d \neq 0)$,

where $a_1, b_1, \ldots$ are coefficients of the equation obtained from the given one by the rotation

$$x = x_1 \cos \varphi - y_1 \sin \varphi, \quad y = x_1 \sin \varphi + y_1 \cos \varphi$$

in which

$$\tan \varphi = -\frac{a + c}{b + d}.$$

4.6573. For equations of the form

$$\frac{dy}{dx} = \frac{-x + F(x, y)}{y}$$

Kukles [26] established certain general criteria for the existence of a center. He applied his criteria to the cases when $F(x, y)$ is a polynomial of the third and of the fifth degrees. When $F(x, y)$ is of degree three the results are as follows: If we write

$$-x + F(x, y) = a_2^{(0)}y^2 + a_3^{(0)}y^3 + (a_0^{(1)} + a_1^{(1)}y + a_2^{(1)}y^2)x + (a_0^{(2)} + a_1^{(2)}y)x^2 + a_0^{(3)}x^3$$

then the origin is a center if and only if one of the following conditions holds:

(1) $\alpha = a_0^{(3)}(a_1^{(1)})^2 + a_1^{(2)}(a_2^{(0)}a_1^{(1)} + 3a_3^{(0)}) = 0,$

$\delta = [3a_3^{(0)}(a_2^{(0)}a_1^{(1)} + 3a_3^{(0)}) + (a_2^{(0)}a_1^{(1)} + 3a_3^{(0)})^2 + a_2^{(1)}(a_1^{(1)})^2]a_1^{(2)} - 3a_3^{(0)}(a_2^{(0)}a_1^{(1)} + 3a_3^{(0)})^3 - a_2^{(1)}(a_1^{(1)})^2(a_2^{(0)}a_1^{(1)} + 3a_3^{(0)}) = 0,$

$\chi = (a_2^{(0)} + a_0^{(2)})a_1^{(1)} + a_1^{(2)} + 3a_3^{(0)} = 0,$

$\tau = 9a_2^{(1)}(a_1^{(1)})^2 + 2(a_1^{(1)})^4 + 9(a_2^{(0)}a_1^{(1)} + 3a_3^{(0)})^2 + 27a_3^{(0)}(a_2^{(0)}a_1^{(1)} + 3a_3^{(0)}) = 0.$

(2) $a_3^{(0)} = \alpha = \delta = \chi = 0,$

(3) $a_3^{(0)} = a_1^{(1)} = a_1^{(2)} = 0,$

(4) $a_3^{(0)} = a_2^{(0)} = a_0^{(2)} = a_1^{(2)} = 0$

4.6574. Almuhamedov [1] studied differential equations of the form

$$\frac{dy}{dx} = \frac{X_h + X}{Y_h + Y},$$

where X_h and Y_h are homogeneous polynomials of degree h and the characteristic equation $xY_h - yX_h = 0$ has no real roots. He showed that the origin is a center in case even powers are absent in the expansion of

$$X = \sum_{m+k>h} b_{mk} x^m y^k$$

and if odd powers are absent in the expansion

$$Y = \sum_{m+k>h} c_{mk} x^m y^k.$$

4.66. The index of Poincaré [47]. It is sometimes convenient to consider a crude but easily calculable characteristic of a singular point, viz., its Poincaré index.

We denote by $\vec{V}(M)$ the vector whose initial point is $M(x, y)$ and whose components are $P(x, y)$ and $Q(x, y)$. The totality of such vectors is the vector field defined by the system

$$\text{(4.6601)} \qquad \frac{dx}{dt} = P(x, y), \qquad \frac{dy}{dt} = Q(x, y),$$

and $M(x, y)$ is a singular point if and only if $\vec{V}(M) = 0$.

Consider next a closed curve k not passing through a singular point. Take a point M on k. As M describes k once, say in the positive sense, the vector $\vec{V}(M)$ may go through a number of complete revolutions some in the positive and some in the negative sense. In the process the angle which $\vec{V}(M)$ makes with a fixed vector changes by an integral multiple $2\pi J_k$ of 2π. *The integer (positive, zero, or negative) J_k is called the index of k relative to the vector field $V = \{\vec{V}(M)\}$.*

The index J_k has the following three properties which are basic for what follows.

4.661. *The index of J_k does not change as k varies continuously without crossing singular points.*

Since the index J_k must vary continuously with a continuous deformation of k, and J_k is an integer, it follows that it does not vary at all.

4.662. *Consider two vector fields* $V = \{\vec{V}(M)\}$ *and* $V' = \{\vec{V}'(M)\}$. *If for every point M of a closed Jordan curve k, the vectors* $\vec{V}(M)$ *and* $\vec{V}'(M)$ *are never in opposition, then the index J_k relative to V is equal to the index J'_k relative to V'.*

For each λ in the interval $0 \leqq \lambda \leqq 1$ we define a vector field V_λ by setting

$$\vec{V}_\lambda(M) = (1 - \lambda)\vec{V}(M) + \lambda\vec{V}'(M).$$

Since $\vec{V}_\lambda(M) \neq 0$ for any M on k, the index of k relative to the vector field V_λ is well defined for every λ. Moreover, this index varies continuously with λ, and being an integer it must remain constant. Hence its values J_k (for $\lambda = 0$) and J'_k (for $\lambda = 1$) are equal.

4.6622. *If the angle θ between* $\vec{V}(M)$ *and* $\vec{V}'(M)$ *is constant for every M on k, then $J_k = J'_k$.*

In particular this is true if $\theta = \pi$, i.e., the vectors $\vec{V}(M)$ and $\vec{V}'(M)$ are always in opposition along k.

4.663. *Let $k_1 = (M_1ABM_1)$ and $k_2 = (M_2BAM_2)$ be two closed Jordan curves with a common arc AB and let the positive sense along k_1 and k_2 assign opposite orientations along the common arc AB. Let k_3 be the closed Jordan curve $(M_1AM_2BM_1)$ obtained by deleting the common arc AB. Then*

$$J_{k_1} + J_{k_2} = J_{k_3}.$$

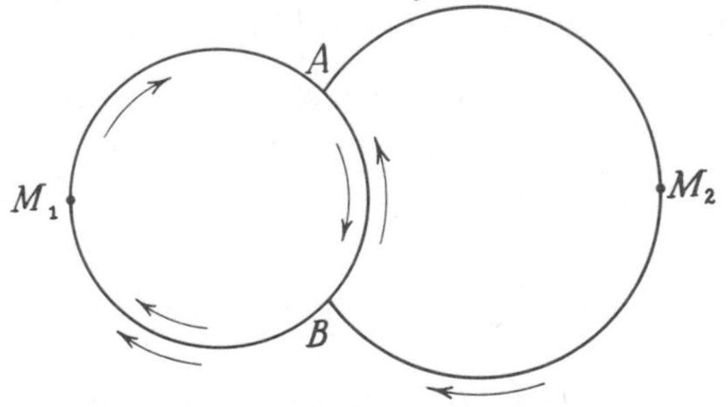

Fig. 21

To see this we consider a Jordan curve k (not necessarily closed) joining two points M_0 and M'. As M varies along k from M_0 to M_1, the angle $2\pi w(M_0M)$ between the variable vector $\vec{V}(M)$ and the fixed vector $\vec{V}(M_0)$ varies continuously from zero to a fixed value $2\pi w(M_0M_1)$. We call $w(M_0M')$ the *rotation of V along the arc* M_0M'.

We see that $w(M_0M') = J_k$ if $M_0 = M'$.

Also, if A, B, C are three points on a Jordan curve, then $w(AB) + w(BC) = w(AC)$. Furthermore with opposite descriptions of the same arc AB, we have $w(A, B) = - w(B, A)$.

Since

$$J_{k_2} = w(M_1A) + w(AB) + w(BM_1),$$
$$J_{k_1} = w(M_2B) + w(BA) + w(AM_2),$$

we have

$$J_{k_1} + J_{k_2} = w(M_1A) + w(AM_2) + w(M_2B) + w(BM_1) = J_{k_3}.$$

4.664. Since every affine transformation is a product of a rotation and two mutually perpendicular contractions, and since under all of these constituent transformations, the index J_k of k remains invariant, this index remains invariant under every affine transformation of k.

4.665. Index of a point. Consider a point M and choose a circular neighborhood $S(M, \varepsilon)$ containing no singular points with the possible exception of M itself. Let k be the circumference of S. *Then* J_k *is called the index of* M. In this definition the circumference k may be replaced by any other simply-closed Jordan curve k_1 containing in its interior our point M but no singular points with the possible exception of M. For, by 4.661, $J_k = J_{k_1}$. We write J_M for the index of the point M.

If M is a nonsingular point, then $J_M = 0$. To see this we need only to choose a small neighborhood $S(M, \varepsilon)$ (cf. 1.11). Then obviously $J_M = J_k = 0$ (see Fig. 22). The converse of this is not true. There exist singular points whose index is equal to zero.

As a simple illustration of this we may mention the "degenerate saddle point" of the system

$$\frac{dx}{dt} = Ax, \qquad \frac{dy}{dt} = Bx.$$

There are less trivial examples of singular points whose index is equal zero. We shall merely indicate one such through the sketch in Fig. 23. Here all the trajectories which tend to the origin do so in the direction of the y-axis.

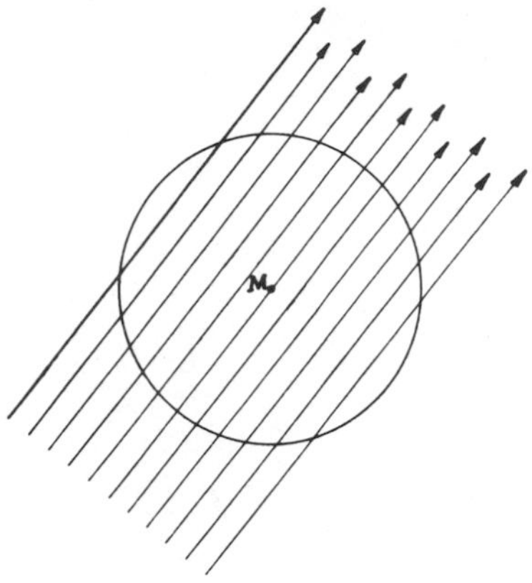

Fig. 22

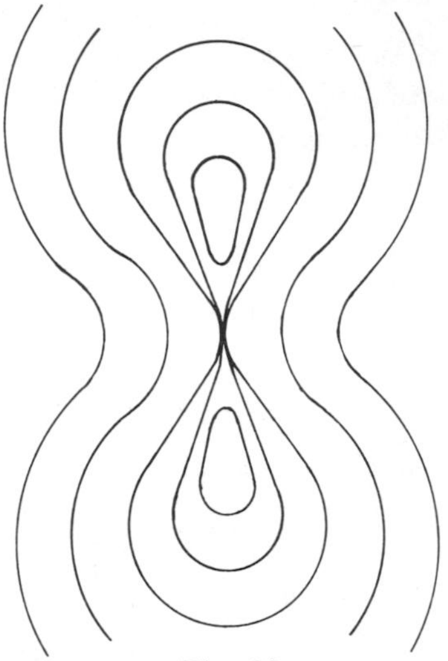

Fig. 23

4.6651. THEOREM. *Let k be a simply-closed Jordan curve containing in its interior (exactly) n singular points $M_1, \ldots, M_n$. Then*

$$J_k = \sum_{i=1}^{n} J_{M_i}.$$

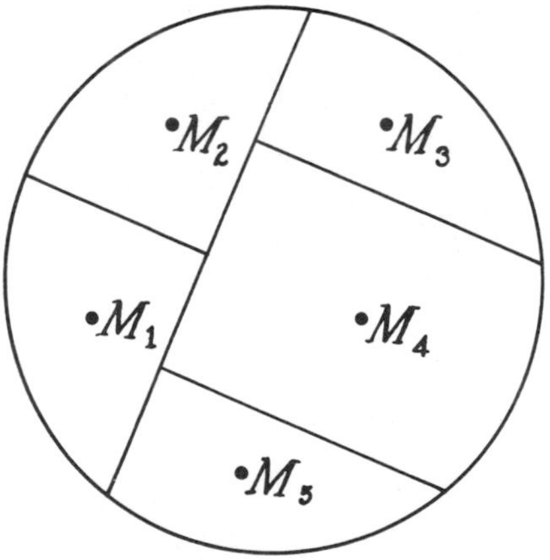

Fig. 24

The proof is suggested by Fig. 24 drawn for the case $n = 5$ and by the additive property of the index discussed in 4.663.

4.6652. THEOREM. *Consider a system*

$$\frac{dx}{dt} = P_n(x, y) + F_n(x, y)$$

$$\frac{dy}{dt} = Q_m(x, y) + G_m(x, y)$$

in which

$$\lim \frac{F_n^2(x, y) + G_m^2(x, y)}{P_n^2(x, y) + Q_m^2(x, y)} = 0 \quad \text{as} \quad x^2 + y^2 \to 0,$$

and in which the origin is an isolated singular point. Let J_0, J_0' *be the indices of the origin relative to the vector fields of the given system and of the truncated system*

$$\frac{dx}{dt} = P_n(x, y)$$

$$\frac{dy}{dt} = Q_m(x, y).$$

Then $J_0 = J_0'$.

Write

$\vec{V}'(M) = (P_n(x, y), Q_m(x, y))$, $\vec{V}''(M) = (F_n(x, y), G_m(x, y))$, then the vector field of our equation is given by

$$\vec{V}(M) = \vec{V}'(M) + \vec{V}''(M).$$

By hypothesis we may choose a neighborhood $S(O, \varepsilon)$ of the origin so small that

$$|\vec{V}''(M)| < \alpha\, |\vec{V}'(M)|, \qquad 0 < \alpha < 1,$$

for every point M inside or on the circumference k of S. Thus along k the vectors $\vec{V}(M)$ and $\vec{V}'(M)$ are never in opposition. Then by 4.662, $J_0 = J_0'$.

In particular if the conditions of our theorem apply to the system

$$\frac{dx}{dt} = cx + dy + F(x, y),$$

$$\frac{dy}{dt} = ax + by + G(x, y),$$

that is, if F and G are infinitesimals of order higher than the first, then the index J_0 of the origin relative to this system is the same as the index J_0' of the approximating truncated linear system

$$\frac{dx}{dt} = cx + dy.$$

$$\frac{dy}{dt} = ax + by.$$

The index $J_M = J_k$ relative to the vector field (4.6601) can be conveniently expressed in the form

$$J_k = \frac{1}{2\pi}\int_k d \arctan \frac{Q}{P} = \frac{1}{2\pi}\int \frac{PdQ - QdP}{P^2 + Q^2}.$$

We shall use this formula to prove.

4.6653. THEOREM. *Consider a linear system*

$$\frac{dx}{dt} = ax + by, \quad \frac{dy}{dt} = cx + dy, \quad \begin{vmatrix} a & b \\ c & d \end{vmatrix} = q \neq 0.$$

If the origin is either a focus, a center, or a node then the index of the origin is equal to $+1$. *If the origin is a saddle point then its index is equal to* -1.

We have

$$J_0 = \frac{1}{2\pi}\int_k \frac{(ax + by)d(cx + dy) - (cx + dy)d(ax + by)}{(ax + by)^2 + (cx + dy)^2}.$$

If we take the ellipse

$$(ax + by)^2 + (cx + dy)^2 = 1$$

as one closed curve k, then

$$J_0 = \frac{q}{2\pi}\int_k (xdy - ydx) = \frac{q}{\pi} A$$

where A is the area of the above ellipse. To calculate A we use the transformation

$$\xi = ax + by, \qquad \eta = cx + dy$$

to map our ellipse into the circle $\xi^2 + \eta^2 = 1$. Its area A_1 is related to A by

$$A_1 = A_1 \left| \frac{D(\xi, \eta)}{D(x, y)} \right| = A \left\| \begin{matrix} a & b \\ c & d \end{matrix} \right\| = A|q|.$$

Thus $A = \pi/|q|$ and hence

$$J_0 = \frac{q}{|q|}.$$

Since the product of the roots of the characteristic equation

$$\begin{vmatrix} a & b - \lambda \\ c - \lambda & d \end{vmatrix} = 0$$

is equal to q, we arrive at the desired result.

A transformation

$$x_1 = \varphi(x, y), \qquad y_1 = \psi(x, y)$$

is called regular at the origin if $\varphi(x, y)$ and $\psi(x, y)$ are holomorphic at the origin and if

$$x_1 = \alpha x + \beta y + P_2(x, y), \quad y_2 = \gamma x + \delta y + Q_2(x, y), \quad \begin{vmatrix} \alpha\beta \\ \gamma\delta \end{vmatrix} \neq 0,$$

and $P_2(x, y)$ and $Q_2(x, y)$ are power series which contain no terms of order smaller than the second.

4.6654. LEMMA. *Let the origin be an isolated singular point of the system*

$$\frac{dx}{dt} = P(x, y), \qquad \frac{dy}{dt} = Q(x, y),$$

and let the index of the origin be J_0. Apply a transformation $x_1 = \varphi(x, y)$, $x_2 = \Psi(x, y)$ regular at the origin. Then the index of the origin relative to the transformed system is also equal to J_0.

By hypothesis the inverse matrix

$$A^{-1} = \begin{bmatrix} \alpha_1 & \beta_1 \\ \gamma_1 & \delta_1 \end{bmatrix} \text{ of } A = \begin{bmatrix} \alpha & \beta \\ \gamma & \delta \end{bmatrix}$$

exists. Since, in view of **4.664**, the linear transformation

$$x_2 = \alpha_1 x_1 + \beta_1 y_1, \qquad y_2 = \gamma_1 x_1 + \delta_1 y_1$$

does not change our index, we may replace the original regular transformation by its product with the above linear transformation of matrix A^{-1}. This new transformation is of the form

$$x_2 = x + \bar{P}_2(x, y), \qquad y_2 = y + \bar{Q}_2(x, y)$$

and when applied to our differential equation it yields the equation of the form

$$\frac{dx_2}{dt} = (1 + \lambda)P + \mu Q,$$

$$\frac{dy_2}{dt} = \nu P + (1 + \varrho)Q,$$

where λ, μ, ν and ϱ are linear combinations of $\partial P_2/\partial x$ and $\partial P_2/\partial y$, and hence are continuous and tend to zero as the point $M(x, y)$ tends to the origin.

Let $\vec{V}'(M) = (P, Q)$ and $\vec{V}''(M) = (\lambda P + \mu Q, \nu P + \varrho Q)$. We notice that for our transformed system the vector field is $\vec{V} = \{\vec{V}(M)\} = \{\vec{V}'(M) + \vec{V}''(M)\}$ and that the conditions of Theorem 4.6642 are satisfied, i.e.,

$$\frac{|\vec{V}''(M)|^2}{|\vec{V}'(M)|^2} = \frac{(\lambda P + \mu Q^2) + (\nu P + \varrho Q)^2}{P^2 + Q^2}$$

$$\leqq \frac{2[\lambda^2 P^2 + \mu^2 Q^2 + \nu^2 P^2 + \varrho^2 Q^2]}{P^2 + Q_2}$$

$$\leqq 2[\lambda^2 + \mu^2 + \nu^2 + \varrho^2]$$

and the last expression tends toward zero as $M(x, y)$ tends toward the origin. This establishes our lemma.

Lemma 4.6654 applies to any isolated singular point. Making use of this one can establish without difficulty the following

4.6655. Corollary. *Let the transformation* $x_1 = \varphi(x, y)$, $y_1 = \psi(x, y)$ *be regular in the interior K of a simply-closed Jordan curve k and let K contain only a finite number of singular points. Then if k′ is the image of k, we have*

$$J_k = J_{k'}.$$

4.6656. Corollary. *If k is a closed trajectory containing only a finite number of singular points in its interior, then* $J_k = 1$.

There exists a conformal mapping of the interior K of k onto the interior K' of the unit circle with the circumference k'. By the preceding lemma we have $J_k = J_{k'}$. But for every $M \in k'$, the vector $\vec{V}(M)$ of our vector field is tangent to the circle k'. Hence $\vec{V}(M)$ turns continuously in the same direction and its total variation is exactly 2π. Thus $J_{k'} = 1$.

4.6657. Corollary. *If for every point M on a closed Jordan curve k of Corollary 4.6655, the vectors $\vec{V}(M)$ all point into the interior (exterior) of k, then $J_k = 1$.*

To prove this we again employ a conformal mapping of K onto the interior K' of the unit circle and compare the index $J_{k'}$ with the index of k' relative to the vector field $\vec{V}' = \{\vec{V}'(M)\}$ with each $\vec{V}'(M)$ pointing toward the origin along the radial ray OM.

4.67. Criteria for the existence of periodic solutions. In this section we shall discuss some criteria for the existence (or the nonexistence) of periodic solutions. We shall express these in terms of some pertinent properties of the vector field $(P(x, y), Q(x, y))$.

4.671. Theorem. (Bendixson [7]). *If in a simply-connected domain G the partial derivatives of $P(x, y)$ and $Q(x, y)$ are continuous and the expression $\partial P/\partial x + \partial Q/\partial y$ does not change sign and does not vanish indentically, then there is no closed trajectory L in G.*

Suppose that the system

$$\frac{dx}{dt} = P(x, y), \qquad \frac{dy}{dt} = Q(x, y),$$

does have in G a periodic solution L: $x = x(t)$, $y = y(t)$ of period l. Then

$$\int_L Pdy - Qdx = \int_0^l (PQ - QP)dt = 0$$

and hence if Γ is the domain bounded by L,

$$\iint_\Gamma \left(\frac{\partial P}{\partial x} + \frac{\partial Q}{\partial y}\right) dxdy = \int_L Pdy - Qdx = 0.$$

However this last equality may hold only if $\partial P/\partial x + \partial Q/\partial y$ either vanishes identically or changes its sign in Γ.

4.6711. The above criteria are particularly useful for systems of the form

$$\frac{dx}{dt} = \varphi(x) + \Psi(y), \qquad \frac{dy}{dt} = \bar{\varphi}(x) + \bar{\Psi}(y).$$

4.6712. We shall apply Bendixon's criterion to the equation

$$\ddot{x} + f(x)\dot{x} + g(x) = 0$$

which plays an important role in the theory of nonlinear oscillations. We write

$$\frac{dx}{dt} = y, \qquad \frac{dy}{dt} = -f(x)y - g(x),$$

and see at once that *if the function* $f(x)$ *does not change its sign in the strip* $a \leqq x \leqq b$, *then there are no closed trajectories in this strip.* For, here

$$\frac{\partial P}{\partial x} = 0, \qquad \frac{\partial Q}{\partial y} = -f(x).$$

4.672. Poincaré [47] **method of contact curves.** Consider a system of nonintersecting, closed, differentiable curves $F(x, y) = C$. Such a family of curves we shall call a *topographical system.*

A *contact curve* consists of points at which the curves of the topographical system are tangent to trajectories of the system

$$\frac{dx}{dt} = P(x, y), \qquad \frac{dy}{dt} = Q(x, y).$$

This curve satisfies the equation

$$\frac{P}{Q} = -\frac{\dfrac{\partial F}{\partial y}}{\dfrac{\partial F}{\partial x}}. \tag{4.6721}$$

If, in particular, we take the system of concentric circles $x^2 + y^2 = R^2$ as our topographic system, then the equation (4.6721) becomes

$$\frac{P}{Q} = -\frac{y}{x}.$$

It is clear that if the contact curve has no real branches in a domain G then $(Px + Qy)/Qx$ does not change its sign in G and in general there exist no periodic trajectories in this domain.

4.6722. *Example* (Andronow and Chaikin [2]). Consider an ordinary pendulum with linear friction under the action of a constant torque P. The equation of motion is of the form

$$I\ddot{\varphi} + h\dot{\varphi} + \text{mg}\, l \sin \varphi = P,$$

where I, g, h, m, l are constants. The corresponding system is given by

$$\frac{d\varphi}{dt} = u, \quad \frac{du}{dt} = -\frac{mgl}{I}\sin\varphi - \frac{h}{I}u + \frac{P}{I}.$$

Since φ takes on values in the interval $0 \leqq \varphi \leqq 2\pi$, our phase space is a cylinder. As our topographical system we choose the parallels of the cylinder, i.e., the family of circles $u = c$. The equation of the contact curve has the form

$$\frac{-\frac{mg\,l}{I}\sin\varphi - \frac{h}{I}u + \frac{P}{I}}{u} = 0,$$

i.e., $u = (P - mgl \sin \varphi)h^{-1}$, and consequently there are no periodic solutions outside of the strip

$$\frac{P - mgl}{h} < u < \frac{P + mgl}{h}.$$

4.673. A symmetry principle. Let the origin be again an isolated singular point of the system

$$\frac{dx}{dt} = P(x, y), \quad \frac{dy}{dt} = Q(x, y).$$

Suppose that $P(x, y)$ is an even function relative to x and that $Q(x, y)$ is odd, relative to x, i.e., suppose that

$$P(-x, y) = P(x, y) \quad \text{and} \quad Q(-x, y) = -Q(x, y).$$

Then all the integral curves of our system are symmetric with respect to the y-axis, and in order to show that a trajectory $f(p, t)$ through a point p on the y-axis is closed, it suffices to prove that $f(p, t)$ cuts the y-axis again (i.e., for $t > 0$).

Symmetry with respect to the x-axis is implied by the condition

$$P(x, -y) = -P(x, y), \quad Q(x, -y) = Q(x, y)$$

and leads to a similar criterion for periodicity.

4.6731. *Example.* Consider an equation of the type

$$\frac{dy}{dx} = \frac{-x + P(x, y)}{y + Q(x, y)}$$

where $P(x, y)$ and $Q(x, y)$ are analytic functions whose expansions begin with terms of order higher than the first. From the study of trajectories of such a system in the neighborhood of the origin, it follows that the origin is either a focus or a center. *In case $P(x, y)$ contains only terms of even degree in x, and $Q(x, y)$ contains only terms of odd degree in x, the symmetry principle implies that the origin is a center.*

We arrive at the same conclusion in case $P(x, y)$ contains only terms of odd degree in y and $Q(x, y)$ contains only terms of even degree in y.

In general, the difficulty in applying the symmetry principle consists in establishing the existence of a trajectory $f(p, t)$ through a point p on the y-axis (x-axis) such that $f(p, t)$ cuts the y-axis (x-axis) again for $t > 0$. This is illustrated by the following

4.6732. Theorem (E. A. McHarg [22]). *Let $f(x)$ and $g(x)$ be odd functions such that $f(x) > 0$ and $g(x) > 0$ for $x > 0$. Let k and x_1 be positive constants such that*

$$f(x) < kg(x) \quad \text{for} \quad 0 < x < x_1.$$

Consider the differential equation

$$\ddot{x} + f(x)\dot{x} + g(x) = 0.$$

Every solution of this differential equation satisfying the initial conditions $x = 0$, $\dot{x} = v_0$, with

$$v_0 \leqq \min \{\frac{1}{k}, \quad [2G(x_1)]^{1/2}\}, \quad G(x) = \int_0^x g(x)dx$$

is periodic.

We replace the given equation by the system

$$\text{(A)} \qquad \frac{dx}{dt} = v, \quad \frac{dv}{dt} = -f(x)v - g(x)$$

in the (x, v)-phase plane, and this, in turn, by

$$\text{(B)} \qquad \frac{dv}{dx} = -f(x) - \frac{g(x)}{v}.$$

Multiplying the first equation (A) by $g(x)$, the second one by v, and adding the results we obtain

$$g(x)\frac{dx}{dt} + v\frac{dv}{dt} = -f(x)v^2,$$

and hence

$$\text{(E)} \qquad \frac{d\lambda(x, v)}{dt} = -f(x)v^2, \quad \text{where} \quad \lambda(x, v) = \tfrac{1}{2}v^2 + G(x).$$

All the curves of the family $\lambda(x, v) = C$ for $C > 0$, are closed curves symmetric with respect to both the x and the v axes.

Consider the trajectory $f(p, t)$ of (B) through a point $p(0, v_0)$ with $v_0 < x_1$.

If $x \geqq 0$ and $v > 0$, then it follows from (A) that x increases with t, whence (B) implies that $dv/dx < 0$, i.e., that v decreases with increasing x. From equation (E) we see that as long as $x > 0$ and $v \neq 0$, we have

$$\frac{d\lambda(x, v)}{dt} < 0.$$

This last inequality means that our trajectory cuts across the curves of our system $\lambda(x, v) = C$ and in doing so passes from the exterior of each such curve into its interior. Thus $f(p, t)$ remains in the interior of the curve $\lambda(x, v) = \lambda(0, v_0) = \frac{1}{2}v_0^2$ and therefore must cut the x-axis in a point $q(\xi, 0), \xi < 0$. Since $\lambda(x, v)$ decreases along $f(p, t)$, we have

$$\lambda(\xi, 0) = G(\xi) < \tfrac{1}{2}v_0^2 = \lambda(0, v_0).$$

In the quadrant $x \geqq 0$, $v < 0$, x decreases with t, and $d\lambda(x, v)/dt$ remains negative. Thus upon cutting across the x-axis our trajectory remains in the domain bounded by the curve $\lambda(x, v) = G(\xi) < \frac{1}{2}v_0^2$ to which it is tangent at q. Also, along $f(p, t)$ in the second quadrant, we have

$$|v| < [2G(\xi)]^{1/2} < v_0 \leqq 1/k.$$

Then it follows from (B) and the hypotheses of our theorem that for $0 < x < x_1$,

$$\frac{dv}{dx} = -f(x) - \frac{g(x)}{v} = \frac{g(x)}{|v|} - f(x) > kg(x) - f(x) > 0.$$

We see that our trajectory $f(p, t)$ cannot tend to the origin, that it must therefore cut the v-axis in a point $(0, \eta)$, $-1/k < \eta < 0$, and hence, in view of the symmetry principle, $f(p, t)$ is a closed trajectory.

4.674. Limit cycles. We should note that the symmetry principle cannot be applied to establish the existence of limit cycles. There exist, however, geometrical considerations which lead to useful criteria.

4.6741. First we observe that if a ring-shaped region Γ is such that all the trajectories which cut across its boundary pass from its exterior into its interior with increasing t (decreasing t), and if, moreover, Γ contains no singular points, then Γ contains a limit cycle. This follows from the discussion in **1.47**. The inner boundary of Γ may in a special case contract to a singular point. The difficulty in applying this last geometrical observation consists in the selection of a suitable ring-shaped region Γ. One may use it in conjunction with a family of topographical curves $F(x, y) = C$, as was frequently done by Lyapunov in the study of stability problems.

Consider, for example, the topographical family consisting of concentric circles

$$F(x, y) = x^2 + y^2 = r^2.$$

Then

$$\frac{dF}{dt} = 2\left(x\frac{dx}{dt} + y\frac{dy}{dt}\right) = 2[xP(x, y) + yQ(x, y)] = \frac{dr^2}{dt},$$

and we have

4.6742. THEOREM. *If there exist two constants r_0, r_1, $r_0 < r_1$, and such that $F = xP + yQ \geqq 0$ for $x^2 + y^2 = r_0^2$ and $F \leqq 0$ for $x^2 + y^2 = r_1^2$, and if there are no singular points in the ring-shaped region Γ: $r_0^2 \leqq x^2 + y^2 \leqq r_1^2$, then Γ contains a stable limit cycle. If F is negative on the inner circle and is positive on the outer circle, the Γ contains an unstable limit cycle.*

4.6743. *Example.* For the system

$$\frac{dx}{dt} = -y + x(x^2 + y^2 - 1),$$

$$\frac{dy}{dt} = x + y(x^2 + y^2 - 1),$$

we have

$$\tfrac{1}{2}F = xP + yQ = (x^2 + y^2)(x^2 + y^2 - 1),$$

and hence $F < 0$ for $x^2 + y^2 = 1 - \varepsilon$, and $F > 0$ if $x^2 + y^2 = 1 + \varepsilon$. Thus, the circle $x^2 + y^2 = 1$ is an unstable limit cycle of our system.

It should be mentioned that the existence of a semi-stable cycle cannot be established by the above simple geometrical considerations.

4.675. Differential equations of the second order. The problem of vibrations in a mechanical or an electrical system in the presence of constant friction and under the influence of a linear restoring force leads to the study of differential equations of the form

$$\ddot{x} + a\dot{x} + bx = 0.$$

In the case of ordinary (dissipative) friction the coefficient a is positive. If a negative, the energy of the system would increase and *friction* would represent a source of energy. This situation may arise only if there is an external supply of energy. Such systems are discussed in the "Theory of Oscillations" by Adronow and Chaikin [2] and in Theodorchik's "Autooscillations".

As long as the coefficient of friction a is constant there may exist no auto-oscillations and no relaxation oscillations. These occur only in the case of variable friction. If we should have a system with variable friction, say depending on x, and if the restoring force depends on x not necessarily in a linear manner, then we are led to an equation of the type

(A) $$\ddot{x} + f(x)\dot{x} + g(x) = 0,$$

or to one of the type

(B) $$\ddot{x} + f(x, \dot{x})\dot{x} + g(x) = 0$$

in case friction depends on $\dot{x}$ as well as on x.

We tacitly assume those properties of $f(x)$ and $g(x)$ which assure the existence and the uniqueness of solutions. Whenever needed we assume the differentiability of $f(x)$ and $g(x)$.

Physical considerations suggest the following requirements for the variable coefficients: $f(x) < 0$ for $|x|$ sufficiently small, and $f(x) > 0$ for $|x|$ sufficiently large with similar restrictions for $f(x, \dot{x})$. Regarding $g(x)$ on physical grounds one may assume that it has the sign of x. Furthermore we usually assume that it is a nondecreasing or slowly decreasing function of x.

In what follows we shall consider conditions under which equations of type (A) or of type (B) possess limit cycles. Levinson and Smith [29] proved a number of theorems establishing the existence of limit cycles under rather general conditions. We state without proof one such result for equations of type (B).

4.6751. THEOREM (Levinson-Smith [29]). *Let* $xg(x) > 0$ *for* $|x| > 0$. *Moreover let*

$$\int_0^{\pm\infty} g(x)dx = \infty.$$

Let $f(0, 0) < 0$ *and let there exist some* $x_0 > 0$ *such that* $f(x, v) \geqq 0$ *for* $|x| \geqq x_0$. *Further, let there exist an* M *such that for* $|x| \leqq x_0$

$$f(x, v) \geqq -M.$$

Finally, let there exist some $x_1 > x_0$ *such that*

$$\int_{x_0}^{x_1} f(x, v)dx \geqq 10Mx_0,[12]$$

where $v < 0$ *is an arbitrary decreasing positive function of* x *in the integration. Under these conditions equation (B) has at least one periodic solution.*

4.6752. Liénard's plane. We return to Equation (A) and set $\dot{x} = v$. Then (A) is replaced by the system

$$\frac{dx}{dt} = v, \qquad \frac{dv}{dt} + f(x)v + g(x) = 0$$

in the phase space. Introducing a new variable $y = v + F(x)$ where

$$F(x) = \int_0^x f(x)dx,$$

we obtain the system

[12] Adamov in proving this result by a different method seems to be able to replace the constant $10M$ by $4M$.

$$\text{(E)} \qquad \frac{dx}{dt} = y - F(x), \qquad \frac{dy}{dt} = -g(x),$$

in Liénard's plane. For system (E) we have

4.6753. THEOREM (Dragilev [14]-Ivanov [24]). *Suppose that*

1. *$g(x)$ satisfies*

$$xg(x) > 0 \quad \text{for} \quad x \neq 0, \qquad \int_0^{\infty} g(x)dx = \infty.$$

2. *$F(x)$ is single valued for $-\infty < x < +\infty$, satisfies a Lipschitz condition in every finite interval, and $xF(x) < 0$ for $x \neq 0$ and $|x|$ sufficiently small.*

3. *There exist constants N, k, k' $(k' < k)$ such that*

$$F(x) \geqq k \quad \text{for} \quad x > N, \qquad F(x) \leqq k' \quad \text{for} \quad x < -N.$$

Then the system (E) *admits at least one limit cycle.*

This theorem implies Theorem 4.6751 for the special case when $f(x, \dot{x}) = f(x)$ does not depend on $\dot{x}$. The function $g(x)$, except for differentiability, is subject to the same requirements in both theorems. Condition 2 is implied by the requirement that $f(0) < 0$ and $f(x)$ be continuous. Finally, if M, x_0, x_1 are chosen as in Theorem 4.6751, then

$$F(x_0) = \int_0^{x_0} f(x)dx \geqq -Mx_0, \quad F(-x_0) = \int_0^{-x_0} f(x)dx \leqq Mx_0,$$

and for $x \geqq x_1 > x_0$

$$F(x) = \int_0^{x_0} f(x)dx + \int_{x_0}^{x_1} f(x)dx + \int_{x_1}^{x} f(x)dx \geqq -Mx_0 + 10Mx_0 = 9Mx_0,$$

whereas for $x < -x_0$ we have

$$F(x) = \int_0^{-x_0} f(x)dx + \int_{-x_0}^{x} f(x)dx \leqq Mx_0.$$

Thus, condition 3 is satisfied for $N = x_1$, $k = 9Mx_0$, and $k' = Mx_0$.

4.6754. Proof of 4.6753 (cf. Levinson-Smith [22]). Consider the family of nonintersecting closed curves

$$\lambda(x, y) = \tfrac{1}{2}y^2 + G(x) = C, \quad G(x) = \int_0^{x} g(x)dx,$$

surrounding the origin. Since $g(x)$ and $F(x)$ are of opposite signs for small $|x|$, then along a trajectory of (E)

$$\frac{d\lambda}{dt} = y\dot{y} + g(x)\dot{x} = -g(x)F(x) > 0$$

near the y-axis. Hence as t increases, every trajectory sufficiently close to the origin cuts across the curves of the family $\lambda(x, y) = C$ and in so doing passes from the interior to the exterior of each such curve. Thus no trajectory tends to the origin (the only singular point of (E)), as $t \to +\infty$.

Let D be an upper bound of the (continuous) function $F(x)$ in the interval $-N \leqq x \leqq +N$. We may assume that $D > k$ and $-D < k'$, where both k and k' are positive, and we consider the trajectory $f(Q, t)$ through a point $Q(\bar{x}, \bar{y})$ with $\bar{x} = \pm N$ and $|\bar{y}| > D$.

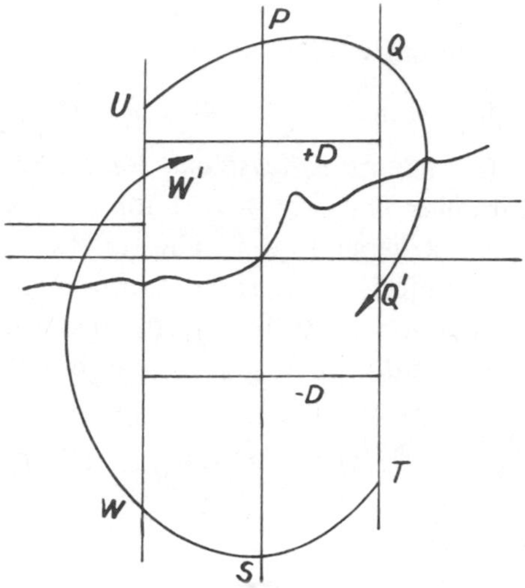

Fig. 25

If Q is in the first quadrant, i.e., if $\bar{x} = +N$ and $\bar{y} > D$ (cf. Fig. 25), then along $f(Q, t)$

$$\frac{dx}{dt} = y - F(x) > y > 0 \quad \text{and} \quad \frac{dy}{dt} = -g(x) < 0$$

as long as $-N \leqq x \leqq +N$. Therefore, as t decreases the trajectory $f(Q, t)$ will move to the left of Q and upward, and since $x > y > 0$, it will cut the y-axis at some point P. As t increases $f(Q, t)$ will move at first to the right of Q and downward. We assert that it will cut the graph of $F(x)$. For along our trajectory

$$0 < \frac{dy}{dx} = \frac{-g(x)}{y - F(x)} < \frac{-g(x)}{y - k}$$

as long as this trajectory remains above the graph $y = F(x)$. Integrating, we see that

$$y < \bar{y} - \frac{1}{\bar{y} - k} \int_N^x g(x)dx.$$

The last inequality cannot hold for points above the curve $y = F(x)$ for arbitrarily large x, since $\int_N^x (x)$ tends to $+\infty$ with x and the right-hand member of the last inequality would become negative. On the other hand, as $t \to +\infty$ our trajectory cannot remain above $y = F(x)$ and within a finite domain $N \leqq x \leqq N_0$, $\bar{y} \leqq y$, since there are no singular points other than the origin.

Since at a point of intersection with the curve $y = F(x)$ the trajectory has a vertical tangent, it cuts this curve but once, proceeds downward and to the left (now $dx/dt < 0$), and, as can easily seen, cuts the line $x = N$ again at some point Q'.

We may assume that Q is so chosen that our trajectory cuts the line $\bar{x} = -M$ at a point U above the line $y = D$. For if we choose a point U_0 on $x = -M$ and above $y = D$, then the trajectory through U_0 when continued in the positive direction, will cut the y-axis (here $dx/dt > \gamma > 0$) at a point P_0. If P_0 is not above P, our original choice of Q will do. If P_0 is above P, then we continue the new trajectory until it cuts $x = N$ at a point Q_0 above Q and choose Q_0 as our new Q.

Next, on the line $x = N$ choose a point T which lies below Q' and also below the line $y = -D$. Reasoning as above we see that T can be so chosen that our trajectory when continued in the positive sense, will move downward and to the left, will cut the y-axis at a point S, then will cut the line $x = -N$ at a point W which lies below the line $y = -D$, and then cutting across the curve $y = F(x)$ will return to the line $x = -N$ at a point W'.

Suppose that W' does not lie above U and consider the domain $\mathscr{D}_0$ bounded by the integral curves $UPQQ'$, $TSWW'$ and the segments $Q'T$ and $W'U$, and in which we delete a sufficiently small (cf. the beginning of this section) neighborhood of the origin. No trajectory can leave this domain with increasing t, and since our domain $\mathscr{D}_0$ contains no singular points, it must contain a limit cycle in view of **1.47**.

Suppose next that W' lies above U (Fig. 26). Continue the trajectory till it intersects the y-axis at P' which lies above P and denote by $\mathscr{D}$ the domain bounded by the integral curves PQQ', $TSWW'P'$ and by the segments $Q'T$ and $P'P$. Consider a trajectory through a point $A(0, y_0)$ lying above P' on the y-axis. As t increases the trajectory through A will move to the right of A and downward, will cut across the line $x = N$ at B, will cut across the graph $y = F(x)$ and across the x-axis (cf. Fig. 26). If this trajectory should enter the domain $\mathscr{D}$ by cutting across $Q'T$, then either it

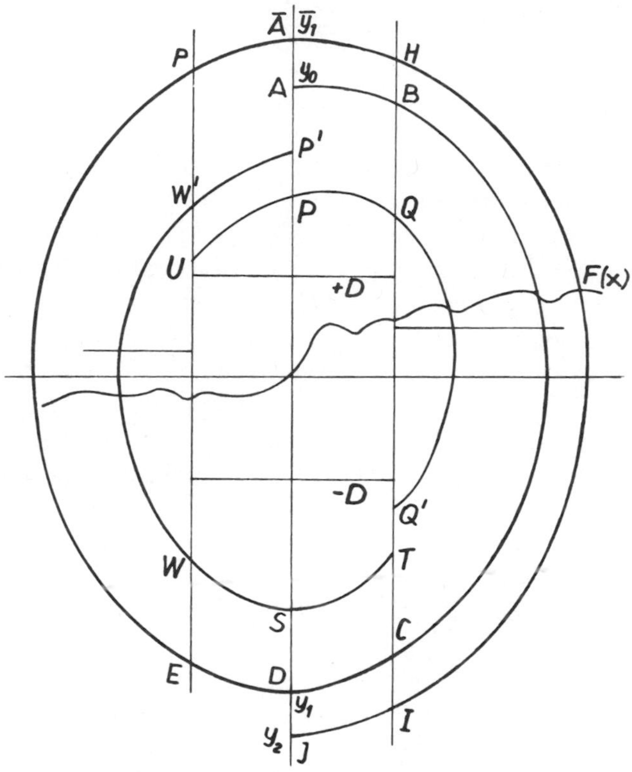

Fig. 26

remains in $\mathscr{D}$ and hence determines a limit cycle (its ω-limit set) or it leaves $\mathscr{D}$ through a point P_1 on the segment $P'P$. The second alternative yields a domain $\mathscr{D}_1$ bounded by the integral curve ABP_1 and the segment P_1A. This domain $\mathscr{D}_1$ must contain a limit cycle by the now familiar argument.

It remains to consider the case when the trajectory through A circles around $\mathscr{D}$ returning to the y-axis in a point $\bar{A}(0, \bar{y}_0)$. We shall show that A can be chosen sufficiently high above the origin on the y-axis so that $\bar{A}$ must fall below A. To accomplish this we prove that

$$\tfrac{1}{2}(\bar{y}^2 - y_1^2) = \lambda(J) - \lambda(D) \leqq 4L_1 + 2Hk - (k - k')\bar{y}_0;$$

where L_1, H are constants not depending on y_0, and $\lambda(Q) = \lambda(x, y)$ if x and y are coordinates of Q. Should $\bar{y}_0 > y_0$ for all y_0 however large then the right-hand member of the last inequality would become negative for sufficiently large values of y_0, whereas the left-hand member $\tfrac{1}{2}(\bar{y}^2 - y_1^2)$ would remain positive, a contradiction.

We proceed to establish the desired inequality. Choosing y as our parameter along the arcs EF and HI, we see that along these arcs

$$\frac{d\lambda}{dy} = \frac{\partial\lambda}{\partial y} + \frac{\partial\lambda}{\partial x}\frac{dx}{dy} = y + g(x)\frac{y - F(x)}{-g(x)} = F(x)$$

and

$$\lambda(I) - \lambda(H) = \int_H^I F(x)dy = -\int_I^H F(x)dy \leqq -k(y_H - y_I),$$

$$\lambda(F) - \lambda(E) = \int_E^F F(x)dy \leqq k'(y_F - y_E).$$

Choosing x as the parameter along the arcs DE, $F\bar{A}$, $\bar{A}H$, and IJ, we see that along these arcs

$$\frac{d\lambda}{dx} = \frac{\partial\lambda}{\partial y}\frac{dy}{dx} + \frac{\partial\lambda}{\partial x} = \frac{-g(x)}{y - F(x)} + g(x) = \frac{-g(x)F(x)}{y - F(x)}.$$

Let

$$L(PQ) = \int_P^Q \frac{|g(x)F(x)|}{|y - F(x)|}dx$$

and let $L(TS)$, $L(SW)$, $L(UP)$ be defined similarly. Let L_1 be the largest of these four values. Then along AB

$$|\lambda(E) - \lambda(D)| = \left|\int_A^B \frac{-g(x)F(x)}{y - F(x)}dx\right| \leqq L_1,$$

since the difference $y - F(x)$ is larger along PQ than along DE. Similar inequalities hold for the arcs $F\bar{A}$, $\bar{A}H$, and IJ. If the variation of the y-coordinate along any arc of trajectory lying outside of $\mathscr{D}$ in the strip $-N \leqq x \leqq N$ does not exceed a constant H, then

$$y_H - y_I \geqq \bar{y}_0 + |y_D| - 2H$$
$$y_F - y_E < \bar{y}_0 + |y_D|$$

and

$$\begin{aligned}\lambda(J) - \lambda(D) &= [\lambda(J) - \lambda(I)] + [\lambda(I) - \lambda(H)] + [\lambda(H) - \lambda(F)] \\ &\quad + [\lambda(F) - \lambda(E)] + [\lambda(E) - \lambda(D)] \\ &\leqq 4L_1 + k'(\bar{y}_0 + |y_D|) - K(\bar{y}_0 + |y_D| - 2H) \\ &= 4L_1 + 2Hk - (k - k')(\bar{y}_0 + |y_D|) \\ &< 4L_1 + 2Hk - (k - k')\bar{y}_0.\end{aligned}$$

It remains to show the existence of H. Indeed, if y_3 and y_4 are two values of y on an arc of trajectory in the strip $-N \leqq x \leqq +N$, and if $\mu = \max_{(-N,\,+N)} |g(x)|$, then

$$\begin{aligned}|y_3^2 - y_2^2| &= |2(\lambda_3 - \lambda_2) + 2\int_{x_3}^{x_2} g(x)dx| \\ &\leqq 2\,|\lambda_3 - \lambda_2| + 4N\mu \\ &\leqq 4L_1 + 4N\mu,\end{aligned}$$

whence

$$|y_3 - y_2| \leqq \frac{4L + 4N\mu}{|y_3 + y_2|} \leqq \frac{2(L+N\mu)}{D} = H.$$

We shall next consider conditions under which there exists but one limit cycle.

We shall illustrate this problem by discussing a result due to Levinson and Smith [29].

4.6755. THEOREM. *Consider the equation*

$$\ddot{x} + f(x)\dot{x} + g(x) = 0.$$

(1) *Let $f(x)$ be an even function such that for the odd function*

$$F(x) = \int_0^x f(x)dx$$

there exists an x_0 with $F(x) < 0$ for $0 < x < x_0$, and $F(x) > 0$ and monotonically increasing for $x > x_0$. (2) *Let $g(x)$ be an odd function such that $g(x) > 0$ for $x > 0$.*
(3) *Next let*

$$\int_0^\infty f(x)dx = \int_0^\infty g(x)dx = \infty.$$

(4) *Finally, we assume that $f(x)$ and $g(x)$ satisfy Lipschitz conditions.*

Under these conditions our differential equation has a unique periodic solution.

For the case $g(x) = x$ this result was proved by Liénard.

We consider again the associated differential equations

$$\text{(A)} \qquad \frac{dx}{dt} = y - F(x), \qquad \frac{dy}{dt} = -g(x)$$

or

$$\text{(B)} \qquad \frac{dy}{dx} = \frac{-g(x)}{y - F(x)}$$

or

$$\text{(E)} \qquad ydy + g(x)dx = F(x)dy$$

in the Liénard plane.

The existence of a limit cycle follows from the theorem of Driagilev-Ivanov. Next we prove its uniqueness.

Since (B) remains unchanged if (x, y) is replaced by $(-x, -y)$ it follows that if a closed integral curve passes through $(0, y_0)$ it must also pass through $(0, -y_0)$. For if this were not the case, the reflection in the origin, that is the replacing of (x, y) by $(-x, -y)$, would give rise to another closed integral curve which intersects the first one.

Again we employ the family of curves

$$\lambda(x, y) = \tfrac{1}{2}y^2 + G(x) = C.$$

Each of these closed curves is symmetric with respect to the x-axis, and hence if one some trajectory $\lambda(0, y)$ assumes the same value for a positive as well as for a negative value of y, then this trajectory is closed.

Consider three trajectory arcs ACB, $A'C'B'$, $A''C''B''$ (Fig. 27) with both the initial and the terminal points on the y-axis. It follows from (B) that for each trajectory $dy/dx < 0$ above the graph $y = F(x)$, that $dy/dx > 0$ below it and that the tangent is vertical at the (unique) point of intersection.

Assume first that the point C of intersection of our trajectory arc and the curve $y = F(x)$, lies in the strip $0 < x < x_0$. Then $F(x) < 0$. Since $dy < 0$ along ACB, we have $F(x)dy > 0$, and along our trajectory

$$\lambda(B) - \lambda(A) = \int_A^B d\lambda(x, y) = \int_A^B F(x)dy < 0$$

by (E). Thus $\overline{OB} < \overline{OA}$.

Next let $A'C'B'$ and $A''C''B''$ be two trajectory arcs which intersect $y = F(x)$ at points C' and C'' to the right of the line $x = x_0$. From (B) and (E) it follows that (cf. 4.6754)

$$d\lambda(x, y) = \frac{-F(x)g(x)}{y - F(x)} dx.$$

Since $-F(x) > 0$ for $0 < x < x_0$ and since $y - F(x)$ is greater along $A''G$ than along $A'E$, we have

$$\int_{A''}^{G} d\lambda(x, y) < \int_{A'}^{E} d\lambda(x, y),$$

where each integral is taken along the proper arc of trajectory. Thus

$$\lambda(G) - \lambda(A'') < \lambda(E) - \lambda(A').$$

Next, along GH we have $d\lambda(x, y) = F(x)dy < 0$, and hence

$$\lambda(H) - \lambda(G) < 0.$$

For the same y, $F(x)$ along HI exceeds $F(x)$ along EF. Then, it follows from $d\lambda(x, y) = F(x)dy$ that

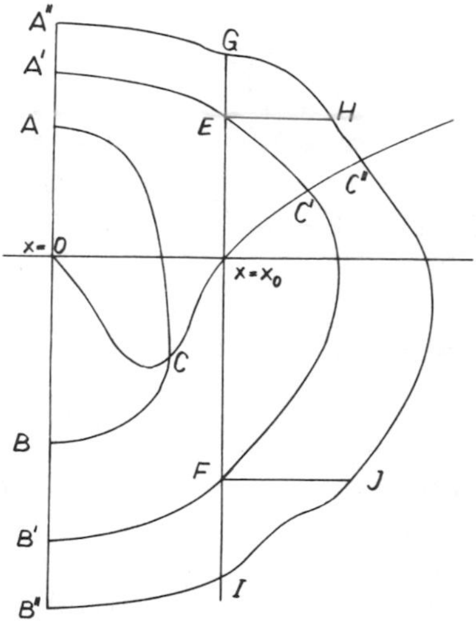

Fig. 27

$$\lambda(I) - \lambda(H) = \int_H^I d\lambda(x, y) < \int_E^F d\lambda(x, y) = \lambda(F) - \lambda(E).$$

Just as along GH, so also along IJ we have

$$\lambda(J) - \lambda(I) < 0.$$

Finally, by the argument in the beginning of this proof,

$$\lambda(B'') - \lambda(J) < \lambda(B') - \lambda(A').$$

Adding all the inequalities for λ, we obtain

$$\lambda(B'') - \lambda(A'') < \lambda(B') - \lambda(A').$$

Thus

$$\overline{OB}'' - \overline{OA}'' \leqq \overline{OB}' - \overline{OA}'.$$

In other words, $\overline{OB} - \overline{OA} > 0$ as long as C lies in $0 < x < x_0$. When C moves outward along $y = F(x)$, then outside of the strip $0 < x < x_0$, the difference $\overline{OB} - \overline{OA}$ is a monotonically decreasing function. Therefore $\overline{OB}$ can equal $\overline{OA}$ at most once, which means that there is at most one closed trajectory. This completes the proof of our theorem.

4.6756. *Example.* Consider Van der Pol's equation

$$\ddot{x} - \mu(1 - x^2)\dot{x} + x = 0.$$

Here

$$F(x) = -\mu \int_0^x (1 - x^2)dx = -\mu\left(x - \frac{x^3}{3}\right).$$

Therefore $F(x)$ is odd, $F(x) < 0$ for $0 < x < 3$, $F(x) \geqq 0$ for $x \geqq 3$, $F'(x) > 0$, and $F(x)$ is monotone for $x \geqq 3$. Finally,

$$\int_0^\infty x dx = \infty \quad \text{and} -\mu \int_0^\infty (1 - x^2)dx = +\infty.$$

Thus Van der Pol's equation has one and only one periodic solution.

CHAPTER III

Systems of n Differential Equations

(The Asymptotic Behavior of Solutions)

1. Introduction

The qualitative theory of differential equations deals with problems which fall into two classes. To the first class belong all the problems which arise from the study of systems of differential equations I, 1.01, whose righthand members are independent of t, the so-called autonomous or stationary systems, and to the second class those associated with the nonautonomous systems I, 1.02, whose righthand members do depend upon the independent variable t.

In the case of a stationary system the geometry of the family of trajectories in the phase space is governed by the laws of the associated dynamical system. This is not true of the nonautonomous systems.

Stationary systems have the following properties:

1.11. Each point $x = (x_1, \ldots, x_n)$ determines, independently of the time t, a unique trajectory passing through the point.

1.12 The unique trajectory of a stationary system determined by an ω-limit point x of an integral curve consists only of ω-limit points of that curve. Thus the limiting behavior of integral curves may be described in terms of integral curves through points of the limit set.

In the case of a nonautonomous system, such a simple description of the limiting behavior is generally impossible.

True enough, in some cases a given nonautonomous system induces a stationary dynamical system, say, through a transformation by means of which the study of the original problem may be reduced to the study of a stationary system.

In general, to obtain a geometrical picture for the behavior of solutions of a nonautonomous system I, 1.02, we pass to the study of the parametric system I, 1.201 in the augmented $(n+1)$-dimensional space. As was shown in Chapter I, the trajectories of I, 1.201 are topologically equivalent to a family of parallel lines; thus, in studying the limiting behavior of a trajectory, we study its asymptotic approach to a curve or, in general, a set. Here asymptotic approach signifies a behavior akin to the approach of a hyperbola to its asymptotes.

A bounded integral curve $x = x(t)$ is one for which

$$||x(t)|| \leqq r.$$

Thus in the n-dimensional space of the autonomous system a bounded trajectory lies within the sphere of radius r about the origin, whereas a bounded trajectory of a nonautonomous system, when considered in the augmented $(n+1)$-dimensional space, lies in the cylinder of radius r with the t-axis as its axis.

If $\lim_{t\to+\infty} x(t) = a$, then in the autonomous case a is a point of equilibrium and a singular point of the system. In the nonautonomous case this condition means that the integral curve $x = x(t)$ approaches the line $(x = a,\ t = \tau)$ asymptotically, but it does not imply that a is in a position of equilibrium of the dynamical system, i.e., that the line $(x = a,\ t = \tau)$ is a trajectory of the system.

One may also compare the behavior of a solution $x(t)$ as t tends to $+\infty$ with the behavior of a monotone function $\psi(t)$. Such questions led Lyapunov to the creation of the theory of characteristic numbers which will be discussed later in this chapter.

1.2. A normalization. In the study of the stability of a solution

$$x_i \equiv x_i(t) \qquad (i = 1, \ldots, n; \quad t \geqq 0).$$

of a nonautonomous system

$$\frac{dx_i}{dt} = f_i(x_1, \ldots, x_n, t) \tag{1.201}$$

it is convenient to make the following normalization:

We observe that if

$$z_i = z_i(t) \qquad (i = 1, 2, \ldots, n, \quad t \geqq 0)$$

is a solution of the system

(1.202) $$\frac{dz_i}{dt} = f_i(z_1 + x_1(t), \ldots, z_n + x_n(t), t) - \frac{dx_i(t)}{dt} = F_i(z_1, \ldots, z_n, t),$$

then

(1.203) $$\bar{x}_i(t) = z_i(t) + x_i(t), \qquad i = 1, 2, \ldots, n, \quad t \geqq 0$$

is a solution of (1.201) and conversely if $x_i = \bar{x}_i(t)$ is a solution of (1.201) then $\{z_i(t)\}$ in (1.203) is a solution of (1.202).

System (1.202) is satisfied by

$$z_1 = 0, \quad z_2 = 0, \ldots, z_n = 0, \quad t \geqq 0.$$

Thus the t-axis is a solution of the corresponding parametric system

(1.204) $$\frac{dz_i}{d\tau} = F_i(z_1, \ldots, z_n, t), \qquad \frac{dt}{d\tau} = 1.$$

Thus, every deviation $z(t)$ of a solution $\bar{x}(t)$ of (1.201) from a fixed solution $x(t)$ of the same system is a solution of (1.204), lying in the neighborhood of the trivial solution $z = 0$, $t = \tau$. So study of the behavior of solutions of (1.201) in the neighborhood of $x(t)$ can be reduced to the study of the behavior of solutions of (1.204) near the trivial solution.

1.21. In the light of the above discussion we may assume that for each f_i in (1.201),

(1.211) $$f_i(0, \ldots, 0, t) \equiv 0$$

identically in t. Then

(1.212) $$x_1 = 0, \ldots, x_n = 0, \quad t \geqq 0$$

is a solution of (1.201).

1.3. Stability in the sense of Lyapunov.

1.31. DEFINITION. *The solution* (1.212) *of* (1.201) *is said to be stable in the sense of Lyapunov, if for every* $\varepsilon > 0$ *there exists an* $\eta > 0$ *and* $t_0 \geqq 0$ *such that every solution*

(1.311) $$x_i = x_i\,(t, t_0, x_{10}, \ldots, x_{n0}),$$

for which

(1.312) $$\sum_{j=1}^{n} x_{j0}^2 \leqq \eta \quad (\text{at } t = t_0)$$

remains in the cylinder

$$\sum_{i=1}^{n} x_i^2 \leqq \varepsilon \quad \text{for} \quad t \geqq t_0.$$

If there exists an $\eta > 0$ such that, in addition,

$$\lim_{t\to\infty} \sum_{i=1}^{n} x_i^2(t) = 0 \tag{1.313}$$

for every solution (1.311) satisfying (1.312), then (1.212) is said to be *asymptotically stable.*

1.32. Definition (Persidski). *The solution* (1.212) *of* (1.201) *is said to be uniformly stable in the sense of Lyapunov, if for every* $\varepsilon > 0$ *there exists a* $\delta > 0$ *such that every solution* (1.311) *with the initial values* $t_0, x_{10}, \ldots, x_{n0}$ *in the cylinder*

$$t \geqq 0, \qquad \sum_{i=1}^{n} x_i^2 \leqq \delta,$$

will remain in the ε*-cylinder*

$$t \geqq 0, \qquad \sum_{i=1}^{n} x_i^2 \leqq \varepsilon$$

for $t \geqq t_0$.

A uniformly stable solution is said to be uniformly asymptotically stable if (1.313) holds

In an autonomous system, stability in the sense of Lyapunov is always uniform.

1.33. Definition. *We speak of an integral curve* $x_i = x_i(t)$ *as an* O^+*-curve if* (1.313) *holds. If on the other hand,*

$$\lim_{t\to-\infty} \sum_{i=1}^{n} x_i^2(t) = 0$$

we say that an integral curve is an O^-*-curve.*

1.4. Systems of linear equations. Consider a system of linear equations

$$\frac{dy_i}{dt} = \sum_{k=1}^{k=n} a_{ik}(t) y_k \qquad (i = 1, 2, \ldots, n). \tag{1.401}$$

Recalling the rules of multiplication and differentiation of matrices, we may write (1.401) in the form

$$\frac{dy}{dt} = A(t)y \tag{1.402}$$

where $A(t) = (a_{ik}(t))$ and $y = (y_1, \ldots, y_n)$ is a matrix of one column — a column vector.

Similarly a nonhomogeneous system

$$\frac{dy_i}{dt} = \sum_{k=1}^{n} a_{ik}(t)y_k + f_i(t) \tag{1.403}$$

may be written in the form

$$\frac{dy}{dt} = A(t)y + F(t) \tag{1.404}$$

where $F(t) = (f_1(t), \ldots, f_n(t))$.

1.41. Homogeneous systems. Let $Y = (y_{ij})$ be a matrix (*a fundamental matrix*) whose columns $(y_{1j}, \ldots, y_{nj})$, $j = 1, \ldots, n$, form a fundamental system of solutions of (1.401). Then

$$\frac{dY}{dt} = A(t)Y \tag{1.411}$$

and the general solution of (1.401) is given by the product $Y \cdot c$, where c is a column vector with constant components c_i, $i = 1, \ldots, n$. That $Y \cdot c$ is a solution of (1.401) follows from

$$\frac{d(Yc)}{dt} = \frac{dY}{dt}c = (AY)c = A(Yc)$$

and (1.402)

If A is a constant matrix, then

$$Y = e^{A(t-t_0)}Y(t_0) \tag{1.412}$$

is a solution of (1.411). Here

$$e^{At} = I + \frac{At}{1!} + \frac{A^2t^2}{2!} + \ldots + \frac{A^nt^n}{n!} + \ldots \tag{1.413}$$

It is easily seen that this series converges for all values of t for every constant matrix A.

To verify that (1.412) is a solution of (1.411), we employ the usual rules of differentiation (which can be easily justified), and obtain

$$\frac{dY}{dt} = A \cdot e^{A(t-t_0)} Y(t_0).$$

1.42. Adjoint systems. The system

$$\frac{dz}{dt} = (-A')z \tag{1.421}$$

is said to be adjoint to (1.402). Here A' is the transpose of A, and z is a column vector.

We note that in terms of matrices of a single row (row vectors) we may rewrite (1.421) as

$$\frac{dz'}{dt} = -z'A. \tag{1.422}$$

A fundamental matrix Z of (1.422) is a matrix whose *rows* form a fundamental system of solutions of (1.422) and hence which satisfies the equation

$$\frac{dZ}{dt} = -ZA. \tag{1.423}$$

Thus Z is the transpose of a fundamental matrix of (1.421).

Next,

$$\frac{d(Z \cdot Y)}{dt} = Z\frac{dY}{dt} + \frac{dZ}{dt}Y. \tag{1.424}$$

If Z and Y satisfy equations (1.423) and (1.411) respectively, then the right-hand member of (1.424) is equal to $Z(AY)+(-ZA)Y=0$, and hence

$$Z \cdot Y = C, \qquad C \text{ a constant matrix.} \tag{1.425}$$

Conversely, if (1.425) holds, then the left-hand member of (1.424) is zero. If, moreover, Y satisfies (1.411) and is nonsingular, then

$$\frac{dZ}{dt}Y = -Z\frac{dY}{dt} = -ZAY$$

and therefore Z satisfies (1.423). Also, if Z satisfies (1.423), and is nonsingular then Y satisfies (1.411). This proves

1.421. THEOREM. *If Z is the transpose of a fundamental matrix of the adjoint system* (1.421), *then for every constant nonsingular*

matrix C, the matrix $Y = Z^{-1}C$ is a fundamental matrix of the given system (1.402). Conversely every fundamental matrix Y of (1.402) is of the form $Z^{-1}C$. In particular, if Y is a matrix of solutions of the given system then Y^{-1} is (the transpose of) a matrix of solutions of the adjoint system.

1.43. Nonhomogeneous systems. We seek a solution of (1.404) in the form

$$y = Y(t) \cdot c(t), \tag{1.431}$$

where $Y(t)$ is a fundamental matrix of the coresponding homogeneous system (1.402), i.e., a matrix whose columns form a fundamental system of solutions of (1.402), and $c(t)$ is a column vector which is to be suitably determined. Substituting (1.431) into (1.404) we get

$$\frac{dy}{dt} = \frac{dY}{dt}c + Y\frac{dc}{dt} = (AY)c + F, \tag{1.432}$$

and since $dY/dt = AY$, it follows that $Y\,dc/dt = F$ or

$$\frac{dc}{dt} = Y^{-1}F. \tag{1.433}$$

Our matrix Y does have an inverse, since

$$|Y(t)| = e^{\int_{t_0}^{t}(a_{11} + a_{22} + \ldots + a_{nn})d\tau}\,|Y(t_0)|. \tag{1.434}$$

Integrating both members of (1.433), we obtain

$$c(t) = c(t_0) + \int_{t_0}^{t} Y^{-1}(\tau)F(\tau)\,d\tau.$$

Thus, in view of (1.431),

$$y(t) = Y(t)\int_{t_0}^{t} Y^{-1}(\tau)F(\tau)d\tau = \int_{t_0}^{t} Y(t)Y^{-1}(\tau)F(\tau)d\tau \tag{1.435}$$

is a particular solution of (1.404). The general solution of (1.404) is given by

$$y(t) = Y(t - t_0)a + \int_{t_0}^{t} Y(t)Y^{-1}(\tau)F(\tau)d\tau, \tag{1.436}$$

where a is a column vector of arbitrary constants (parameters). If $Y(t)$ is a normal fundamental system of solutions, i.e., if $Y(0) = I$, then $a = y^{(0)} = y(t_0)$.

1.44. Formula (1.436) can be considerably simplified if A is a constant matrix. We show first that in this case

$$Y(t)Y^{-1}(\tau) = Y(t-\tau), \tag{1.441}$$

provided

$$Y(0) = I. \tag{1.442}$$

We observe that both the left-hand and the right-hand members of (1.441) are solutions of (1.411) with the same initial values at $t = \tau$, for $Y(\tau)Y^{-1}(\tau) = I$ and $Y(\tau - \tau) = Y(0) = I$ by assumption. The truth of the identity (1.441) follows from the uniqueness property.

Thus if A is constant and Y is normal, i.e., satisfies (1.442), then (1.436) becomes

$$y(t) = Y(t-t_0)y^{(0)} + \int_{t_0}^{t} Y(t-\tau)F(\tau)d\tau. \tag{1.443}$$

Here $y^{(0)} = y(t_0)$. If we remember that $Y(t) = e^{At}$ (cf. (1.412)), we may write (1.443) in the form

$$y(t) = e^{A(t-t_0)}y^{(0)} + \int_{t_0}^{t} e^{A(t-\tau)}F(\tau)d\tau.$$

1.45 An interesting estimate for solutions of a system (1.401) is given by the following

1.451. Theorem. *If*

$$|a_{ik}(t)| \leqq f(t) \quad \textit{for} \quad t \geqq t_0, \tag{1.4511}$$

then

$$|y_i(t)| \leqq \left(\sum_{i=1}^{n}|y_i(t_0)|\right)e^{n\int_{t_0}^{t} f(t)dt} \tag{1.4512}$$

Here, $(y_1(t), \ldots, y_n(t))$ *is a solution of* (1.401).

We have

$$\frac{dy_i(t)}{dt} = a_{i1}(t)y_1(t) + \ldots + a_{in}(t)y_n(t),$$

whence

$$y_i(t) = y_i(t_0) + \int_{t_0}^{t} \left(\sum_{j=1}^{n} a_{ij}(t)y_j(t)\right)dt,$$

and, in view of (1.4511),

$$|y_i(t)| \leqq |y_i(t_0)| + \int_{t_0}^{t} f(t) \sum_{j=1}^{n} |y_j(t)|dt, \quad (i = 1, \ldots, n).$$

Adding these last inequalities together, we obtain

$$\sum_{i=1}^{n} |y_i(t)| \leqq \sum_{i=1}^{n} |y_i(t_0)| + nf(t) \sum_{i=1}^{n} \int_{t_0}^{t} |y_i(t)|dt.$$

Making use of Lemma 4.1 of Chapter I, we obtain (1.4512).

1.452. Corollary. *If all the integrals*

$$\int_{t_0}^{+\infty} |a_{ik}(t)|dt, \quad (i,\ k = 1, 2, \ldots, n)$$

converge, then all the solutions of (1.401) *are bounded for* $t \geqq t_0$.

We may take

$$f(t) = \sum_{i=1}^{n} \sum_{k=1}^{n} |a_{ik}(t)|.$$

1.46. Linear transformations. Consider a linear nonsingular transformation $y = K(t)z$. Substituting this expression into (1.402) we obtain

$$\frac{dy}{dt} = \frac{dK}{dt}z + K\frac{dz}{dt} = AKz.$$

Thus system (1.402) is transformed into a linear homogeneous system

$$\frac{dz}{dt} = Bz, \tag{1.461}$$

where

$$B = K^{-1}AK - K^{-1}\frac{dK}{dt}. \tag{1.462}$$

If K is a constant matrix, then

$$B = K^{-1}AK. \tag{1.463}$$

We see that if K is constant then matrix B of the new system (1.461) is the transform of the original matrix A by K. If, moreover, matrix A of the given system (1.402) is constant, then a matrix K exists such that B in (1.463) is in the Jordan canonical form, i.e.,

$$(1.464)\quad B = \begin{bmatrix} B_1 & 0 & \dots & 0 \\ 0 & B_2 & \dots & 0 \\ \dots & \dots & \dots & \dots \\ 0 & 0 & \dots & B_q \end{bmatrix} \qquad B_i = \begin{bmatrix} \gamma_i & 1 & 0 & \dots & 0 \\ 0 & \gamma_i & 1 & \dots & 0 \\ \dots & \dots & \dots & \dots & \dots \\ 0 & 0 & 0 & \dots & 1 \\ 0 & 0 & 0 & \dots & \gamma_i \end{bmatrix}$$

where γ_i are the roots of the characteristic equation $|A - \gamma I| = 0$. System (1.461) with B as in (1.464) may be written as a system of matrix equations

$$(1.465)\qquad \frac{dz^{(i)}}{dt} = B_i z^{(i)}, \qquad (i = 1, \dots, q).$$

Since each B_i is constant, we may use (1.412) and (1.413) to determine a fundamental matrix Z_i for each of the equations (1.465). Since

$$B_i^r = \begin{bmatrix} \gamma^r & \binom{r}{1}\gamma^{r-1} & \dots & \binom{r}{m-1}\gamma^{r-m+1} \\ 0 & \gamma^r & \dots & \binom{r}{m-2}\gamma^{r-m+2} \\ \cdot & \cdot & \dots & \cdot \\ 0 & 0 & \dots & \gamma^r \end{bmatrix}$$

where $m = m_i$ is the order of B_i, $\gamma = \gamma_i$ is the corresponding characteristic root, and

$$\binom{r}{k} = \frac{r(r-1)\dots(r-k+1)}{1 \cdot 2 \cdot \dots \cdot k},$$

we get $Z_i(t)$ in the familiar explicit form

$$(1.466)\quad Z_i(t) = \begin{bmatrix} e^{\gamma t} & te^{\gamma t} & \dots & \frac{t^{m_i-1}}{(m_i-1)!}e^{\gamma t} \\ 0 & e^{\gamma t} & \dots & \frac{t^{m_i-2}}{(m_i-2)!}e^{\gamma t} \\ \cdot & \cdot & \dots & \cdot \\ 0 & 0 & \dots & e^{\gamma t} \end{bmatrix}$$

1.5. Reducible systems. Following Lyapunov, we introduce the following

1.51. DEFINITION. *System* (1.402) *is called reducible if there*

exists a matrix $K(t)$ *which together with the determinant* $|K^{-1}(t)|$ *of its reciprocal, is bounded on* $t_0 \leqq t < +\infty$, *and for which the system* (1.461), *obtained from* (1.402) *by the linear transformation*

$$y = K(t)z, \tag{1.511}$$

has constant coefficients.

If such a matrix $K(t)$ exists, it is called a Lyapunov matrix.

1.52. THEOREM. (N. P. Erugin [16]). *The system* (1.402) *is reducible if and only if every matrix* Y *of its fundamental system of solutions may be written in the form*

$$Y(t) = M(t)e^{J(t-t_0)}U_0, \qquad Y(t_0) = U_0, \tag{1.521}$$

where $M(t)$ *is a bounded matrix,* M^{-1} *has bounded determinant for* $t \geqq t_0$, *and* J *is a constant matrix in the Jordan canonical form.*

If (1.402) is reducible, then in view of (1.511)

$$Y = K(t)Z, \tag{1.522}$$

where Z is a fundamental matrix of solutions of (1.461) with a constant matrix B. An addiitonal transformation

$$z = Cu \tag{1.523}$$

with a suitably chosen constant matrix C, yields a system

$$\frac{du}{dt} = Ju \tag{1.524}$$

whose matrix J is in the Jordan canonical form. The matrix U in

$$Y = K(t)CU$$

is a fundamental matrix of (1.524) and hence is equal to $e^{J(t-t_0)}U(t_0)$, which proves the second part of our theorem.

Conversely, let (1.402) have a fundamental matrix of the form (1.521). Apply the linear transformation (1.511) with $K(t) = M(t)$. Then $K(t) = YU_0^{-1}e^{-J(t-t_0)}$ by (1.521), and the transformed system has the matrix

$$B = K^{-1}AK - K^{-1}\frac{dK}{dt} = e^{J(t-t_0)}U_0Y^{-1}AYU_0^{-1}e^{-J(t-t_0)}$$

$$- e^{J(t-t_0)}U_0Y^{-1}\left(\frac{dY}{dt}U_0^{-1}e^{-J(t-t_0)} - YU_0^{-1}e^{-J(t-t_0)}J\right)$$

$$= e^{J(t-t_0)}U_0Y^{-1}YU_0^{-1}e^{-J(t-t_0)}J = e^{J(t-t_0)}U_0U_0^{-1}e^{-J(t-t_0)}J = J$$

which is constant and is in the Jordan canonical form.

It is natural to seek criteria for the reducibility of systems. Known criteria are of two kinds. Criteria of the first kind, introduced by Lyapunov, apply to periodic matrices B, and those of the second kind apply to matrices B whose elements approach zero sufficiently rapidly with increasing t.

1.53. THEOREM. *If all the solutions of a system* (1.402) *are bounded and if*

$$\int_{t_0}^{t} (a_{11}(t) + a_{22}(t) + \ldots + a_{nn}(t))dt > d > -\infty \tag{1.531}$$

for all $t \geqq t_0$, *then* (1.402) *reduces to the system*

$$\frac{dz}{dt} = 0. \tag{1.532}$$

Let Y be a fundamental matrix of a system (1.402) subject to the conditions of our theorem. Then Y is a Lyapunov matrix. For,

$$Y^{-1} = \left(\frac{Y_{ik}}{|Y|}\right),$$

where Y_{ik} is the cofactor of the element y_{ik} in the determinant $|Y|$ of Y. Since Y is bounded, Y^{-1} is bounded provided $|Y|$ does not tend to zero as t tends to $+\infty$. But, by (1.434),

$$|Y(t)| = e^{\int_{t_0}^{t}(a_{11} + \ldots + a_{nn})d\tau}\,|Y(t_0)|,$$

whence, in view of (1.531), $|Y| > e^d > 0$.

Substituting $y = Yz$ into (1.402) we obtain

$$\frac{dy}{dt} = \frac{dY}{dt}z + Y\frac{dz}{dt} = AYz.$$

Since Y is nonsingular, this yields (1.532).

When combined with conditions insuring boundedness of solutions Theorem 1.53 yields more direct criteria for reducibility.[1] The following theorem will serve as an example.

1.54. THEOREM. *System* (1.402) *is reducible to* (1.532) *if*

$$\int_0^{\infty} ||A(t)||dt < +\infty. \tag{1.541}$$

Here

[1]For a more complete discussion of reducibility criteria see Erugin [16].

$$||A(t)|| = \sqrt{\sum_{i,k} |a_{ik}|^2}$$

is the modulus (after Wedderburn) of $A(t)$.

Condition (1.541) implies (1.531). Also, in view of Corollary 1.452, all the solutions of (1.402) are bounded.

1.55. THEOREM (Liapunov) [31]. *If $A(t)$ is a periodic matrix, i.e., if $A(t+\omega) = A(t)$, then (1.402) is reducible by means of a periodic matrix.*

1.551. Let $Y(t)$ be a fundamental matrix of solutions of (1.402). Then $Y(t+\omega)$ is also a fundamental matrix.

First, by the periodicity of $A(t)$, $Y(t+\omega)$ is a matrix of solutions of (1.402), and hence

$$Y(t+\omega) = Y(t)C \tag{1.5511}$$

where C is constant.

Replacing t by $t+\omega$ in the identity (1.434), setting $t = t_0$, and making use of the periodicity of $\sum_{i=1}^n a_{ii}(t)$, we get

$$|Y(t_0+\omega)| = |Y(t_0)|e^{\int_0^\omega \sum_{i=1}^n a_{ii}(\tau)d\tau}.$$

Comparing this with (1.5511) we see that

$$|C| = e^{\int_0^\omega \sum_{i=1}^n a_{ii}(\tau)d\tau}$$

Since $|Y(t)| \neq 0$ for all t (cf. (1.434)), we have $|Y(t+\omega)| \neq 0$ and $Y(t+\omega)$ is a fundamental matrix.

1.552. We can choose a fundamental matrix $Y(t)$ so that the matrix C in (1.5511) is in the Jordan canonical form. For, if we replace $Y(t)$ by $Y(t)M$ where M is constant and nonsingular, then $Y(t+\omega)$ is replaced by $Y(t+\omega)M$ and

$$C = Y^{-1}(t)Y(t+\omega)$$

by $M^{-1}Y^{-1}(t)Y(t+\omega)M = M^{-1}CM$. If we choose M so that $M^{-1}CM$ in the Jordan canonical form, then $Y(t)M$ is the desired fundamental matrix.

We write

$$C = \begin{bmatrix} C_1 & 0 & \dots & 0 \\ 0 & C_2 & \dots & 0 \\ \cdot & \cdot & \cdot\cdot\cdot & \\ 0 & 0 & \dots & C_n \end{bmatrix}, \qquad C_i = \begin{bmatrix} \gamma_i & 1 & 0 & \dots & 0 \\ 0 & \gamma_i & 1 & \dots & 0 \\ \cdot & \cdot & \cdot & \cdot\cdot\cdot & 1 \\ 0 & 0 & 0 & \dots & \gamma_i \end{bmatrix}.$$

Since $|C| \neq 0$, each C_i is nonsingular and we can find matrices D_i such that

(1.5521) $$C_i = e^{\omega D_i}$$

and if

$$D = \begin{bmatrix} D_1 & 0 & \dots & 0 \\ 0 & D_2 & \dots & 0 \\ \cdot & & \cdot\cdot\cdot\cdot & \cdot \\ 0 & 0 & \dots & D_n \end{bmatrix}$$

then

(1.5522) $$C = e^{\omega D}.$$

1.553. Let

(1.5531) $$P(t) = e^{Dt} Y^{-1}(t).$$

This matrix is periodic, for

$$\begin{aligned} P(t+\omega) &= e^{D(t+\omega)}\, Y^{-1}(t+\omega) = e^{Dt}\, e^{\omega D}\, (Y(t)C)^{-1} \\ &= e^{Dt}\, e^{\omega D}\, e^{-\omega D}\, Y^{-1}(t) \\ &= e^{Dt}\, Y^{-1}(t) = P(t). \end{aligned}$$

Since $P(t)$ is periodic, it is bounded. Moreover

$$|P(t)| = |e^{Dt}| \cdot |Y^{-1}(t)| \neq 0,$$

and since $|P(t)|$ is periodic, its absolute value is greater than a positive constant d. Thus $P^{-1}(t)$ is also bounded.

The linear transformation

(1.5532) $$Z = P(t)Y = e^{Dt}$$

yields a reduced system

(1.5533) $$\frac{dz}{dt} = Dz, \qquad D = \frac{1}{\omega} \ln C.$$

1.554. The characteristic roots $\mu_1, \dots, \mu_n$ of D are called the *characteristic exponents* of the original periodic system (1.402). In view of (1.466) and (1.5532) each component of $Y(t)$ and hence each component of every solution of a periodic system (1.402) is of the form $\Sigma\alpha_i(t)e^{\mu_i t}$ where each $\alpha_i(t)$ is a polynomial in t with periodic coefficients. In view of (1.5521), the μ's are related to the characteristic roots γ_i of C by the formula

$$\gamma_i = e^{\omega\mu_i}.$$

This shows that the real parts of μ_i are completely determined by C and hence by $A(t)$.

The actual determination of characteristic exponents in general presents a very difficult and still unsolved problem. Some approximate methods are described in the literature. The reader may be referred to T. A. Artemieff's paper "A method of determination of characteristic exponents and its application to two problems of celestial mechanics", Isv. Ak. Nauk. 1944, v. 8, No. 2.

1.6. We shall study systems of the type

$$\frac{dy_i}{dt} = \sum_{k=1}^{n} p_{ik}(t)y_k \quad (i = 1, \ldots, n) \tag{1.601}$$

where the functions $p_{ik}(t)$ are continuous for $0 \leqq t < +\infty$. Our aim is to describe the asymptotic behavior of solutions of such a system in terms of the behavior of its coefficients $p_{ik}(t)$.

Since every solution is a linear combination of the solutions $y^{(k)}(t, t_0) = (y_{ik}(t, t_0))$ in a fundamental system

$$Y(t, t_0) = (y^{(1)}(t, t_0), \ldots, y^{(n)}(t, t_0)) = (y_{ik}(t, t_0)), \tag{1.602}$$

it suffices to describe the asymptotic behavior of these solutions. We may assume that

$$Y(t_0, t_0) = I$$

i.e., $Y(t, t_0)$ is in a normal form.

1.61. We consider first the case when

$$\lim y_{ik}(t, t_0) = b_{ik} \neq \pm\infty$$

for every component in (1.602). We say in this case that *solutions of* (1.601) *behave regularly as t tends to* $+\infty$. Regular behavior implies that every integral curve $(y_1(t), \ldots, y_n(t), t)$ of (1.601) in the augmented $(n + 1)$-dimensional space has a vertical asymptote

$$y_1 = b_1, \; y_2 = b_2, \ldots, y_n = b_n, \; t = \tau,$$

for suitable constants $b_1, \ldots, b_n$.

1.62. THEOREM. *If*

$$\int_{t_0}^{\infty} |p_{ik}(t)|dt < +\infty \tag{1.621}$$

for every i and k, then solutions of (1.601) *behave regularly as t tends to* $+\infty$.

We have

$$y_{ij}(t, t_0) = y_{ij}(t_0, t_0) + \int_{t_0}^{t} \sum_{k=1}^{n} p_{ik}(\tau) y_{kj}(\tau) d\tau.$$

By Corollary 1.452 and (1.621) all the solutions of (1.601) are bounded. Again (1.621) implies the (absolute) convergence of the integral on the right and hence it has a limit as $t \to +\infty$. This in turn implies the of a limit of $y_{ij}(t, t_0)$ as $t \to +\infty$.

The above theorem deals with systems (1.601) all of whose solutions are bounded. The problem of finding criteria for the boundedness of solutions is one of the most difficult problems still unsolved even for the case of one equation of the second order. Actually, the investigation of the asymptotic behavior of solutions has proceeded along a different path pioneered by Lyapunov.

1.7. The theory of characteristic numbers of Lyapunov. Let $f(t)$ be a function continuous in the interval $t_0 \leqq t < +\infty$, and let $\psi(t)$ be a steadily increasing function in this interval, which tends to $+\infty$ with t. We shall say that λ is the *characteristic number of* $f(t)$ *with respect to* $\psi(t)$ if and only if

$$\overline{\lim_{t \to +\infty}} \, (|f(t)| \psi(t)^{\lambda + \varepsilon}) = +\infty, \tag{1.701}$$

$$\lim_{t \to +\infty} (|f(t)| \psi(t)^{\lambda - \varepsilon}) = 0 \tag{1.702}$$

for all $\varepsilon > 0$; and we shall write $\lambda = \lambda(f, \psi)$. If $\lim (|f(t)| \psi(t)^{\nu}) = 0$ for every $\nu > 0$, then we say that $\lambda(f, \psi) = +\infty$ and if $\overline{\lim} \, (|f(t)| \psi(t)^{\nu}) = +\infty$ for every $\nu < 0$, we say that $\lambda(f, \psi) = -\infty$.

The characteristic number of a finite set $\{f_a\}$ of functions f_α is defined by

$$\lambda(\{f_\alpha\}, \psi) = \min_{\alpha} (\lambda(f_\alpha, \psi)). \tag{1.703}$$

Characteristic numbers have the following arithmetic properties which are useful for their estimation.

1.71. LEMMA. $\lambda(f_1 + f_2, \psi) \geqq \min(\lambda(f_1, \psi), \lambda(f_2, \psi))$. *If* $\lambda(f_1, \psi) \neq \lambda(f_2, \psi)$, *then* $\lambda(f_1 + f_2, \psi) = \min(\lambda(f_1, \psi), \lambda(f_2, \psi))$.

Let $\lambda_1 = \lambda(f_1, \psi) \leqq \lambda_2 = \lambda(f_2, \psi)$. Our first assertion follows from

$$\lim (|f_1(t) + f_2(t)|\psi(t)^{\lambda_1-\varepsilon}) \leqq \lim (|f_1(t)|\psi(t)^{\lambda_1-\varepsilon}) + \lim (|f_2(t)|\psi(t)^{\lambda_1-\varepsilon}) = 0 + 0 = 0.$$

If $\lambda_2 \neq \lambda_1$, say $\lambda_2 > \lambda_1$, then for every ε such that $0 < \varepsilon < \lambda_2 - \lambda_1$, and hence for every $\varepsilon > 0$, we have

$$\overline{\lim} (|f_1(t) + f_2(t)|\psi(t)^{\lambda_1+\varepsilon}) = +\infty.$$

This completes the proof of our lemma.

1.711. COROLLARY. $\lambda(\sum_{i=1}^k f_i, \psi) \geqq \min_i (\lambda_i = \lambda(f_i, \psi))$. *If* $\lambda_q = \min_i (\lambda_i)$, *and if* $\lambda_q < \lambda_i$ *for* $i \neq q$, *then* $\lambda(\sum_{i=1}^k f_i, \psi) = \lambda_q$.

Since $\lambda(\sum_{i\neq q} f_i, \psi) \geqq \min_{i\neq q} (\lambda(f_i, \psi)) > \lambda_q$, we may apply the second part of Lemma 1.71 to $\sum_{i\neq q} f_i$ and f_q.

1.72. LEMMA. $\lambda(f_1 \cdot f_2, \psi) \geqq \lambda(f_1, \psi) + \lambda(f_2, \psi)$.

The proof follows at once from the equality

$$\lim (|f_1 \cdot f_2|\psi(t)^{\lambda_1+\lambda_2-\varepsilon}) = \lim (|f_1|\psi(t)^{\lambda_1-\varepsilon/2}) \cdot \lim (|f_2|\psi(t)^{\lambda_1-\varepsilon/2}) = 0.$$

1.721. COROLLARY. *If* $|\varphi_i(t)| \leqq a_i < +\infty$, *then* $\lambda(\sum_{i=1}^n \varphi_i f_i, \psi) \geqq \min (\lambda_i = \lambda(f_i, \psi))$.

This corollary follows at once from Lemmas 1.71—1.72, if we observe that $\lambda(\varphi_i, \psi) \geqq 0$.

1.722. COROLLARY. *For the special case* $f_2(t) = \psi(t)^\eta$ *we get*

$$\lambda(f_1\psi^\eta, \psi) = \lambda(f_1) - \eta. \tag{1.7221}$$

The proof follows directly from (1.701) and (1.702).

1.73. LEMMA. *If* $|f_1(t)| \leqq |f_2(t)|$ *for* $t \geqq t_0$, *then* $\lambda(f_1, \psi) \geqq \lambda(f_2, \psi)$.

Since $|f_1(t)|\psi(t)^\nu \leqq |f_2(t)|\psi(t)^\nu$, $\lim |f_1(t)|\psi(t)^{\lambda_2-\varepsilon} = 0$, and hence $\lambda_1 \geqq \lambda_2$.

1.731. COROLLARY. *If* $f_i(t) \geqq 0$, *then*

$$\lambda(\sum_{i=1}^k f_i, \psi) = \min_i \lambda(f_i, \psi).$$

Since $\Sigma f_i \geqq f_i$, then $\lambda(\Sigma f_i, \psi) \leqq \lambda(f_i, \psi)$ for every f_i, and hence $\lambda(\Sigma f_i, \psi) \leqq \min_i \lambda(f_i, \psi)$. Combining this result with that of Corollary 1.711, we obtain the desired equality.

1.74. For a vector $y(t) = (y_1(t), \ldots, y_n(t))$ we define $\lambda(y, \psi)$ to be the characteristic number of the set of its components (cf. 1.703). That is

(1.741) $$\lambda(y, \psi) = \min_i (\lambda(y_i, \psi)).$$

1.742. LEMMA. *We assert that*

(1.743) $$\lambda(||y||, \psi) = \lambda(y, \psi),$$

whether we set

(1.7431) $$||y|| = \sum_{i=1}^{n} |y_i|,$$

or

(1.7432) $$||y|| = \sqrt{\Sigma |y_i|^2}.$$

Since $\lambda(|y_i|, \psi) = \lambda(y_i, \psi)$, our assertion for the case (1.7431) follows at once from Corollary 1.731.

If there exists a sequence $t_1, t_2, \ldots \to +\infty$ such that $|f(t_i)||\psi(t_i)^{\lambda+\varepsilon} \to +\infty$, then $|f(t_i)^2|\psi(t_i)^{2\lambda+2\varepsilon} \to +\infty$. Thus $2\lambda \geqq \lambda(f^2, \psi)$ Since by Lemma 1.72, $2\lambda \leqq \lambda(f^2, \psi)$ we have

(1.744) $$\lambda(f^2, \psi) = 2\lambda(f, \psi).$$

We use (1.744) to prove

(1.745) $$\lambda(y, \psi) = \lambda(\sqrt{\Sigma y_i^2}, \psi)$$

By (1.731) and (1.744)

$$\lambda(\sqrt{\Sigma y_i^2}, \psi) = \tfrac{1}{2} \min_i \lambda(|y_i^2|, \psi) = \min_i \lambda(|y_i|, \psi)$$

and by definition this last expression is $\min \lambda(y_i, \psi) = \lambda(y, \psi)$, and our assertion follows, proving the lemma.

1.746. LEMMA. *If* $\lambda_i = \lambda_i(y^{(i)}, \psi)$, $i = 1, 2, \ldots, k$, *are all distinct then the k vectors* $y^{(i)}(t) = (y_{1i}(t), \ldots, y_{ni}(t))$ *are linearly independent.*

Suppose that there exists a nontrivial relation

(1.747) $$c_1 y^{(1)}(t) + \ldots + c_k y^{(k)}(t) = 0.$$

We note that $\lambda(c_i y^{(i)}) = \lambda(y^{(i)})$ if $c_i \neq 0$ and is equal to $+\infty$ if $c_i = 0$. Let $\lambda_q = \lambda(c_q y^{(q)}, \psi)$ be the smallest of $\lambda_i = \lambda(c_i y^{(i)})$ with $c_i \neq 0$. We may assume that $\lambda_q = \lambda(y_{1q}, \psi)$. Then, since λ_i are all distinct,

$$\lambda_q < \lambda(c_i y_{1i}, \psi) \quad \text{for} \quad i \neq q.$$

From (1.747) we obtain

$$c_1 y_{11}(t) + \ldots + c_q y_{1q}(t) + \ldots + c_k y_{1k}(t) = 0.$$

This relation together with Corollary (1.711) yields

$$\lambda_q = \lambda(c_q y_{1q}, \psi) = \lambda(\sum_{i=1}^{n} c_i y_{1i}, \psi) = \lambda(0, \psi) = +\infty,$$

a contradiction.

1.75. In the special case of $\psi(t) = e^t$ we shall speak of $\lambda(f) = \lambda(f, e^t)$ as characteristic numbers of Lyapunov. These characteristic numbers may also be defined (after Perron [39]) by

$$\lambda_0 = \lambda(f, e^t) = -\overline{\lim_{t\to\infty}} \frac{\log |f(t)|}{t}. \tag{1.751}$$

To justify this assertion we recall (cf. (1.701) and (1.702)) that if $\lambda = \lambda(f, e^t)$, then

$$\overline{\lim} |f(t)| e^{(\lambda+\varepsilon)t} = +\infty \tag{1.752}$$

$$\lim_{t\to+\infty} |f(t)| e^{(\lambda-\varepsilon)t} = 0 \tag{1.753}$$

for every $\varepsilon > 0$.

1.754. It follows from (1.752) that for every $\varepsilon > 0$ there exists a sequence $t_1, t_2, \ldots, t_n, \ldots$ which tends to $+\infty$ and such that

$$|f(t_n)| e^{(\lambda+\varepsilon)t_n} \to +\infty.$$

Hence for all sufficiently large n,

$$\log |f(t_n)| + (\lambda + \varepsilon)t_n > 0,$$

or

$$\frac{\log |f(t_n)|}{t_n} > -\lambda - \varepsilon.$$

1.755. Equality (1.753) implies that for every $\varepsilon > 0$ and for all sufficiently large t

$$|f(t)| e^{(\lambda-\varepsilon)t} < 1$$

or

$$\frac{\log |f(t)|}{t} < -\lambda + \varepsilon.$$

Thus $\lambda = \lambda(f, e^t)$ satisfies (1.751). Conversely, if λ_0 satisfies

(1.751), then it follows from the discussion in 1.755 and 1.754 that $\lambda_0 = \lambda(f, e^t)$.

A similar argument shows that

$$\lambda(f, e^{q(t)}) = -\overline{\lim_{t\to+\infty}} \frac{\log |f(t)|}{q(t)}. \tag{1.756}$$

1.76. Since at most n solutions of (1.601) can be linearly independent we have

1.761. THEOREM. (Perron) *The set of all the characteristic numbers* $\lambda(y, \psi)$, *of solutions* y *of* (1.601), *contains at most* n *distinct elements.*

1.77. Let $\lambda'_1 < \ldots < \lambda'_r$ be all the distinct characteristic numbers of (1.601). It can be shown that all the λ_i are finite for $\psi(t) = e^t$.

If Y is a fundamental matrix of solutions of this system, i.e., a matrix whose columns $y^{(j)}$ $(j = 1, \ldots, n)$ form a basis for the solutions, we let n_i be the number of the elements $y^{(j)}$ in the basis for which $\lambda(y^{(j)}, \psi) = \lambda'_i$, and we write

$$\nu(Y) = n_1\lambda'_1 + \ldots + n_r\lambda'_r \quad (n_i \geqq 0). \tag{1.771}$$

1.772. A fundamental matrix Y of (1.601) or the corresponding basis $y^{(j)} = (y_{1j}, \ldots, y_{nj})$ $(j = 1, 2, \ldots, n)$ is called *normal* by Lyapunov, if $\nu(Y)$ attains its largest possible value.

1.773. If $\{y^{(j)}\}$ is a normal basis of (1.601), then for any solution

$$y = \sum c_j y^{(j)}, \tag{1.7731}$$

we have

(1.7732) $\lambda(y, \psi) = \min\,(\lambda(y^{(j)}, \psi)$ where j stands for those indices for which $c_j \neq 0)$.

For, if $\lambda(y, \psi) > \min\,(\lambda(y^{(j)}, \psi)$ with $c_j \neq 0)$, we drop one of the $y^{(j)}$'s which occur in (1.7731) with a nonvanishing coefficient c_j and for which moreover $\lambda(y^{(j)}) = \min_h \lambda(y^{(h)}, \psi)$ and replace it by y. Such a replacement leads to a new basis with a larger ν, which contradicts the hypothesis $\{y^{(j)}\}$ is a normal basis.

1.774. All solutions of (1.601) whose characteristic numbers exceed λ'_k form a vector space by Corollary 1.711. Property 1.773 implies that if Y is normal, then the dimension of this vector space is $n_{k+1} + \ldots + n_r$. Thus, the sums $n_{k+1} + \ldots + n_r$ $(k = 1, \ldots r - 1)$, and hence the numbers n_k $(k = 1, \ldots, r)$, are independent of the choice of a normal basis. Each n_k is called the multiplicity of the characteristic number λ_k' of (1.601). The corresponding (maximal) value $\nu(Y)$ in (1.771) is called the *sum of all the characteristic numbers of* (1.601).

1.78. Two systems related by a Lyapunov transformation (1.511) have the same characteristic numbers. For (1.511) implies $\lambda(y) \geqq \lambda(z)$, by Lemma 1.721, and the equality $z = K^{-1}(t)y$, implies $\lambda(z) \geqq \lambda(y)$ by the same lemma.

1.8. Estimates of characteristic numbers. We may assume that Y in (1.434) is normal in the sense of Lyapunov. Let

$$y^{(i)} = (y_{1i}, \ldots, y_{ni}), \qquad \lambda_i = \lambda(y^{(i)}, \psi).$$

Then

$$\nu = \lambda_1 + \ldots + \lambda_n = n_1\lambda_1' + \ldots + n_r\lambda_r'$$

where $\lambda_1', \ldots, \lambda_r'$ are all distinct, ν is maximal and is the sum of the characteristic numbers of (1.601) (cf. 1.774). For each term of $|Y|$ we have

$$\lambda(y_{i_1,1} \ldots y_{i_n n}, \psi) \geqq \sum_{k=1}^{n} \lambda(y_{i_k,k}, \psi) \geqq \sum_{k=1}^{n} \lambda_k = \nu,$$

since $\lambda(y_{jk}, \psi) \geqq \lambda_k = \min_j (\lambda(y_{jk}, \psi))$ (cf. 1.74), and hence

$$\lambda(|Y|, \psi) \geqq \min_{(i_1, \ldots, i_n)} \lambda(y_{i_1,1} \ldots y_{i_n, n}, \psi)$$

Thus, in view of (1.434), we have

1.81. THEOREM (Lyapunov [31]) *If $\nu = \Sigma\lambda_i$ is the sum of all the characteristic numbers (cf.* 1.774) *of* (1.601), *then*

$$\text{(1.811)} \qquad \nu = \sum_{i=1}^{n} \lambda_i \leqq \lambda(e^{\int_{t_0}^{t} \sum_s \operatorname{Re} p_{ss}(\tau)d\tau}, \psi),$$

and if $\psi(t) = e^t$,

$$\text{(1.812)} \qquad \nu = \sum_{i=1}^{n} \lambda(y^{(i)}, e^t) \leqq -\overline{\lim_{t\to\infty}} \frac{1}{t} \int_{t_0}^{t} \sum_{s=1}^{n} \operatorname{Re} p_{ss}(\tau)d\tau$$

in view of (1.751).

1.82. In the case when $|p_{ik}(t)| \leqq b < +\infty$, we can find both lower and upper bounds [7] for the characteristic numbers associated with (1.601).

1.821. If we make the substitution $z_i = y_i e^{\eta t}$ then (1.601) becomes

$$\text{(1.8211)} \qquad \frac{dz_i}{dt} = p_{1i}(t)z_1 + \ldots + (p_{ii}(t) + \eta)z_i + \ldots + p_{ni}(t)z_n.$$

Let $u = z_1^2 + \ldots + z_n^2$. Multiplying the ith equation in (1.8211) by z_i and adding the resulting equations, we obtain

$$\frac{1}{2}\frac{du}{dt} = A(\eta) + B(\eta) = \Gamma(\eta),$$

where

$$A(\eta) = \sum_i (p_{ii}(t) + \eta)z_i^2 \quad \text{and} \quad B(\eta) = \sum_{s \neq i} p_{si}(t)z_s z_i.$$

We note that in general

$$\sum_{s \neq i} |\alpha_s \alpha_i| \leqq 2(n-1)(\alpha_1^2 + \ldots + \alpha_n^2) \qquad (s,\ i = 1, 2, \ldots, n).$$

Hence,

$$\text{(1.8212)} \qquad |B(\eta)| \leqq b \sum_{s \neq i} |z_s z_i| \leqq 2(n-1)bu.$$

Next, let

$$\text{(1.8213)} \qquad \eta_0 = (2n-1)b + \frac{\varepsilon}{2},\ \varepsilon > 0.$$

Then for $\eta \geqq \eta_0$ we have $p_{ii}(t) + \eta \geqq \eta_0 - b > 0$, and therefore

$$\text{(1.8214)} \qquad A(\eta) \geqq (\eta_0 - b)u > 0.$$

Combining (1.8212) and (1.8214), we obtain

$$\Gamma(\eta) = A(\eta) + B(\eta) > (\eta_0 - b)u - 2(n-1)bu = \frac{\varepsilon}{2}u$$

for $\eta \geqq \eta_0$ and $t \geqq t_0$. Thus, for $\eta \geqq \eta_0$

$$\frac{1}{2}\frac{du}{dt} > \frac{\varepsilon}{2}u,$$

that is,

$$\text{(1.8215)} \qquad u \geqq ce^{\varepsilon t} \quad \text{for} \quad \eta \geqq \eta_0, \qquad \varepsilon > 0, \quad t \geqq t_0,$$

and for a suitably chosen constant c.

A similar argument shows that for $\eta \leqq -\eta_0$, we have $\Gamma(\eta) \leqq -\varepsilon_2 u$ whence

(1.8216) $u \leqq c_1 e^{-\varepsilon t}$ for $\eta \leqq -\eta_0, \quad \varepsilon > 0, \quad t \geqq t_0,$

and a constant c_1.

The above estimates yield

1.822. THEOREM. *If* $|p_{ik}(t)| \leqq b < +\infty$, $i, k = 1, \ldots, n$, *then*

$$(1.8221) \quad -2(2n-1)b \leqq \lambda' = \lambda(\sum_{i=1}^{n} y_i^2, e^t) \leqq 2(2n-1)b.$$

Here $y = (y_1, \ldots, y_n)$ *is a solution of* (1.601).

We set $v = y_1^2 + \ldots + y_n^2$. Then

$$u = e^{2\eta t} v.$$

We observe that (1.8216) and the choice of η_0 (cf. (1.8213)) imply that

$$(1.8222) \quad \lim_{t\to+\infty} ve^{2\eta t} = \lim_{t\to+\infty} u = 0 \quad \text{for} \quad \eta < -(2n-1)b,$$

and that from the choice of η_0 and (1.8215), it follows that

$$(1.8223) \quad \lim ve^{2\eta t} = \lim u = +\infty \quad \text{for} \quad \eta > (2n-1)b.$$

By (1.701) and (1.702),

$$\overline{\lim}\, ve^{\lambda'+\varepsilon'} = +\infty, \qquad \lim ve^{\lambda'-\varepsilon'} = 0 \quad \text{for every} \quad \varepsilon' > 0.$$

Hence, in view of (1.8222)

$$\lambda' + \varepsilon' \geqq -2(2n-1)b,$$

and

$$\lambda' - \varepsilon' \leqq 2(2n-1)b,$$

by (1.8223). Since the last two inequalities hold for every $\varepsilon' > 0$, we have

$$-2(2n-1)b \leqq \lambda' \leqq 2(2n-1)b.$$

1.823. COROLLARY. *If* $|p_{ik}(t)| \leqq b < +\infty$, *then*

$$(1.8231) \qquad -(2n-1)b \leqq \lambda(y, e^t) \leqq (2n-1)b,$$

where $y = (y_1, \ldots, y_n)$ *is a solution of* (1.601).

The inequalities (1.8231) follow at once from (1.8221) in view of (1.743)—(1.745).

1.824. COROLLARY. *If* $\lim_{t\to+\infty} p_{ik}(t) = 0$ $(i, k = 1, \ldots, n)$, *then* $\lambda(y, e^t) = 0$ *for every solution* y *of* (1.601).

For every $\varepsilon > 0$ and a sufficiently large t_0, we have $|p_{ik}(t)| \leqq \varepsilon$ for $t \geqq t_0$. Thus the value of b in (1.8231) may be taken arbitrarily small.

The last corollary can be applied to (1.601) to yield a rather general result. We let $\varrho(t)$ be a continuous function such that

$$(1.8241) \qquad \varrho(t) \geqq 0, \ \lim \frac{p_{ij}(t)}{\varrho(t)} = 0.$$

Applying the transformation

$$(1.8242) \qquad \tau = \int_{t_0}^{t} \varrho(s) ds$$

to (1.601) we obtain

$$\frac{dy_i}{d\tau} = q_{i1}(\tau)y_1 + \ldots + q_{in}(\tau)y_n \qquad (i = 1, \ldots, n)$$

with

$$q_{ij}(\tau) = \frac{p_{ij}(t)}{\varrho(t)}.$$

The new system fulfills the conditions of Corollary 1.824, in view of (1.8241). Hence, remembering (1.8242), we get

$$(1.825) \qquad 0 = \lambda(y, e^{\tau}) = \lambda(y, e^{\int_{t_0}^{t} \varrho(s)ds}).$$

As a simple application of this general result we shall prove

1.826. Corollary. *If $a > 0$, $\alpha > 1$ and $|p_{ij}(t)| \leqq at^{-\alpha}$ for $0 < t_0 \leqq t$, then*

$$(1.8261) \qquad \lambda(y, t) = 0$$

for every solution y of (1.601).

We observe that $\varrho(t) = 1/t$ satisfies conditions (1.8241), and that $\int_{t_0}^{t} \varrho(s) ds = \ln t/t_0$.

1.83. We shall consider next the *almost diagonal systems*, i.e., systems (1.601) in which $\lim_{t\to+\infty} p_{ik}(t) = 0$ for $i \neq k$. Here, in some cases, characteristic numbers may be expressed in terms of the coefficients in the main diagonal.

1.831. Theorem (Perron) [39]. *If in* (1.601) *we have*

(1.8311) $\lim p_{ik}(t) = 0$ *for* $i \neq k$, *and* $\mathrm{Re}(p_{k-1,k-1}(t) - p_{kk}(t)) \geqq c > 0$ *for* $t \geqq t_0$, *then*

(1.8312) $$\lambda_k = \overline{\lim_{t\to\infty}} \frac{1}{t} \int_0^t \operatorname{Re}(p_{kk}(\tau))d\tau.$$

The proof of this theorem is based on the following two lemmas which are of interest in themselves.

1.832. LEMMA. *Consider a system (in general with complex coefficients)*

(1.8321) $$\frac{dy_i}{dt} = p_i(t)y_i + \sum_{k=1}^{n} p_{ik}(t)y_k$$

where $p_i(t)$ and $p_{ik}(t)$ are continuous for $t \geqq t_0$, and where

(1.8322) $$\operatorname{Re} p_1(t) > \operatorname{Re} p_i(t) + c, \qquad i \neq 1,\ c > 0,$$

(1.8323) $$\lim_{t\to+\infty} p_{ik}(t) = 0.$$

Such a system has a solution $y = (y_1, \ldots, y_n)$ in which

(1.8324) $$\lim_{t\to\infty} \frac{y_i}{y_1} = 0 \quad \text{for} \quad i \neq 1, \qquad \lim_{t\to\infty} \left(\frac{y_1'}{y_1} - p_1\right) = 0.$$

Consider a solution of (1.8321) satisfying the initial conditions

(1.8325) $$y_i(t_1) = \eta_i, \quad |\eta_1| > |\eta_j| \quad \text{for} \quad j \neq 1, \quad t_1 > t_0.$$

Multiplying (1.8321) by the complex conjugate $\bar{y}_i$ of y_i, we obtain

$$\bar{y}_i y_i' = p_i(t)|y_i|^2 + \sum_{k=1}^{n} p_{ik}(t)\bar{y}_i y_k,$$

and since $1/2\ d/dt\,|y_i^2| = \operatorname{Re}(\bar{y}_i y_i)$, we have

$$\left| \frac{1}{2} \frac{d}{dt} |y_i^2| - \operatorname{Re} p_i(t)|y_i^2| \right| \leqq \sum_{k=1}^{n} |p_{ik}(t)\bar{y}_i y_k|$$

and hence

(1.8326) $$-\sum_{k=1}^{n} |p_{ik}(t)\bar{y}_i y_k| \leqq \frac{1}{2} \frac{d}{dt} |y_i^2| - \operatorname{Re} p_i(t)|y_i^2| \leqq \sum_{k=1}^{n} |p_{ik}(t)\bar{y}_i y_k|.$$

Next, let t_1 be so large that $|p_{ik}(t)| < c/2n$ for $t \geqq t_1$. Then (1.8326) implies that

(1.8327) $$-\frac{c}{2n} \sum_{k=1}^{n} |\bar{y}_i y_k| \leqq \frac{1}{2} \frac{d}{dt} |y_i^2| - \operatorname{Re} p_i(t)|y_i^2| \leqq \frac{c}{2n} \sum_{k=1}^{n} |\bar{y}_i y_k|, \quad (t \geqq t_1).$$

1.833. Employing the above inequalities we shall show that

(1.8331) $$|y_1(t)|^2 > |y_i(t)|^2 \quad \text{for} \quad i \neq 1 \quad \text{and} \quad t \geqq t_1.$$

By (1.8325), this inequality holds for $t = t_1$. If the least upper bound t_2 of the values $t \geqq t_1$ for which (1.8331) holds is finite, then for some q we have

(1.8332) $$|y_1(t_2)|^2 = |y_q(t_2)|^2 \geqq |y_i(t_2)|^2, \quad i \neq 1, q,$$

and

(1.8333) $$\left(\frac{d}{dt}|y_1|^2\right)_{t=t_2} \leqq \left(\frac{d}{dt}|y_q|^2\right)_{t=t_2}.$$

In view of (1.8327) and (1.8332),

(1.8334) $$\operatorname{Re} p_1(t_2)|y_1(t_2)|^2 - \frac{c}{2}|y_1(t_2)|^2 \leqq \left(\frac{1}{2}\frac{d}{dt}|y_1^2|\right)_{t=t_2},$$

(1.8335) $$\left(\frac{1}{2}\frac{d}{dt}|y_q^2|\right)_{t=t_2} \leqq \operatorname{Re} p_q(t_2)|y_q(t_2)|^2 + \frac{c}{2}|y_q(t_2)|^2.$$

We note that $y_1(t_2) \neq 0$ by (1.8332). The inequalities (1.8334) and (1.8335) combined with (1.8333) yield (after division by $|y_1(t_2)|^2$) the following inequality

$$\operatorname{Re} p_1(t_2) - \frac{c}{2} \leqq \operatorname{Re} p_q(t_2) + \frac{c}{2}$$

which contradicts the hypothesis (1.8322) of our lemma.

1.834. We are now ready to prove the first part of our lemma (cf. (1.8324)). Suppose that

(1.8341) $$\overline{\lim} \left|\frac{y_q}{y_1}\right|^2 = \alpha > 0$$

for some q. We shall show that the assumption (1.8341) implies that the inequalities

(1.8342) $$\left|\frac{y_q(t)}{y_1(t)}\right|^2 > \frac{\alpha}{2}, \qquad \frac{d}{dt}\left|\frac{y_q(t)}{y_1(t)}\right|^2 > -\frac{c\alpha}{2}$$

hold simultaneously for arbitrarily large t and that this in turn contradicts (1.8323).

If one can find an arbitrarily large value t_2 of t so that

$$\left|\frac{y_q(t_2)}{y_1(t_2)}\right|^2 \leqq \frac{\alpha}{2},$$

then by (1.8341) there exists $t_3 > t_2$, and indeed a smallest value $t_3 > t_2$, such that

$$\left|\frac{y_q(t_3)}{y_1(t_3)}\right|^2 = \frac{3}{4}\alpha.$$

Then,

$$\left(\frac{d}{dt}\left|\frac{y_q(t)}{y_1(t)}\right|^2\right)_{t=t_3} \geqq 0 > -\frac{c\alpha}{2},$$

and therefore (1.8342) holds true for arbitrarily large values of t.

If the first inequality in (1.8342) holds for all sufficiently large values of t, then our assertion regarding (1.8342) is true, since the second inequality in (1.8342) is satisfied by arbitrarily large values of t. This follows from the fact that the relation

$$\frac{d}{dt}\left|\frac{y_q(t)}{y_1(t)}\right|^2 \leqq -\frac{c\alpha}{2} \quad \text{(for all sufficiently large } t\text{)}$$

would imply that $\lim |y_q(t)/y_1(t)|^2 = -\infty$, a contradiction.

We note that

$$\frac{d}{dt}\left|\frac{y_q}{y_1}\right|^2 = \frac{1}{|y_1|^2}\frac{d}{dt}|y_q|^2 - \frac{|y_q|^2}{|y_1|^4}\frac{d}{dt}|y_1|^2.$$

Applying (1.8326) for $i = 1$ and for $i = q$, we get

$$\frac{1}{2}\frac{d}{dt}\left|\frac{y_q}{y_1}\right|^2 \leqq \sum_{k=1}^{n}\frac{|p_{qk}(t)\bar{y}_q y_k|}{|y_1|^2} + \operatorname{Re} p_q(t)\left|\frac{y_q}{y_1}\right|^2$$
$$+ \sum_{k=1}^{n}\frac{|p_{1k}(t)\bar{y}_1 y_k|\,|y_q|^2}{|y_1|^4} - \operatorname{Re} p_1(t)\left|\frac{y_q}{y_1}\right|^2$$

or

$$\frac{1}{2}\frac{d}{dt}\left|\frac{y_q}{y_1}\right|^2 + \operatorname{Re}\,(p_1(t) - p_q(t))\left|\frac{y_q}{y_1}\right|^2 \leqq \sum_{k=1}^{n}\frac{|p_{qk}(t)\bar{y}_q y_k|}{|y_1|^2}$$
$$+ \sum_{k=1}^{n}\frac{|p_{1k}(t)\bar{y}_1 y_k|\,|y_q|^2}{|y_1|^4},$$

and finally, by (1.8331),

$$\frac{1}{2}\frac{d}{dt}\left|\frac{y_q}{y_1}\right|^2 + \operatorname{Re}\,(p_1(t) - p_q(t))\left|\frac{y_q}{y_1}\right|^2 \leqq \sum_{k=1}^{n} |p_{qk}(t)|$$

$$+ \sum_{k=1}^{n} |p_{1k}(t)|.$$

We have shown that (1.8341) implies that arbitrarily large values of t exist for which (1.8342) holds and for which, therefore,

$$-\frac{c\alpha}{4} + \frac{c\alpha}{2} = \frac{c\alpha}{4} \leqq \sum_{k=1}^{n} |p_{qk}(t)| + \sum_{k=1}^{n} |p_{1k}(t)|.$$

This contradicts (1.8323).

Thus, we have proved the first part of (1.8324). The second conclusion in (1.8324) follows immediately if we apply the first conclusion to the first equation in (1.8321).

1.835. LEMMA. *Consider the system* (1.8321) *which in addition to* (1.8323) *satisfies*

(1.8351) $\operatorname{Re}(p_i(t)) \geqq \operatorname{Re}(p_{i+1}(t)) + c, \qquad c > 0 \quad (i = 1, 2, \ldots, n-1).$

Such a system has n linearly independent solutions

(1.8352) $\quad y^{(k)} = (y_{1k}, y_{2k}, \ldots, y_{nk}) \qquad (k = 1, \ldots, n),$

in which

(1.8353) $$\lim_{t\to\infty} \frac{y_{jk}}{y_{kk}} = 0, \quad (j \neq k)$$

and

(1.8354) $$\lim_{t\to\infty} \left|\frac{y'_{kk}}{y_{kk}} - p_k\right| = 0, \qquad (k = 1, \ldots, n).$$

1.836. The proof will proceed by induction.

By Lemma 1.832, there exists a solution $y^{(1)} = (y_{11}, \ldots, y_{n1})$ such that (cf. 1.8324)

(1.8361) $$\lim_{t\to\infty} \frac{y_{j1}}{y_{11}} = 0 \quad \text{for} \quad j \neq 1, \qquad \lim\left(\frac{y'_{1,1}}{y_{1,1}} - p_1\right) = 0.$$

Let

$$\begin{cases} y_1 = y_{11}\int u\,dt \\ y_j = y_{j1}\int u\,dt + z_{j-1} \end{cases} \qquad (j = 2, \ldots, n). \tag{1.8362}$$

The functions $y_1, \ldots, y_n$ in (1.8362) yield a solution of (1.8321), if and only if

$$\begin{cases} y_{11}\, u = \sum_{k=2}^{n} p_{1k} z_{k-1} \\ y_{j1} u + z'_{j-1} = p_j z_{j-1} + \sum_{k=2}^{n} p_{jk} z_{k-1} \end{cases} \qquad (j = 2, \ldots, n). \tag{1.8363}$$

Substituting the value of u determined by the first equation in (1.8363) into the remaining equations, we obtain

$$z'_{j-1} = p_j z_{j-1} + \sum_{k=2}^{n} \left(p_{jk} - \frac{y_{j1}}{y_{11}} p_{1k}\right) z_{k-1} \quad (j = 2, \ldots, n). \tag{1.8364}$$

Here the coefficients of z_{k-1} tend to zero in view of (1.8323) and (1.8353). Thus the $n-1$ differential equations (1.8364) satisfy the conditions of our lemma. By the hypothesis of the induction, there exist $n-1$ linearly independent solutions

$$z^{(m)} = (z_{1m}, \ldots, z_{n-1,\,m}) \quad (m = 1, \ldots, n-1),$$

such that

$$\lim \frac{z_{im}}{z_{mm}} = 0 \qquad (i \neq m) \tag{1.8365}$$

and

$$\lim \left(\frac{z'_{mm}}{z_{mm}} - p_{m+1}\right) = 0 \qquad (m = 1, \ldots, n-1). \tag{1.8366}$$

Denote by u_j the function u assigned to a solution $(z_{1j}, \ldots, z_{nj})$ by (1.8363). Then

$$y_{11} u_j = \sum_{k=2}^{n} p_{1k} z_{k-1,\,j}$$

and, in view of (1.8323) and (1.8365), we have

$$\lim_{t\to\infty} \frac{y_{11}}{z_{jj}} u_j = 0. \tag{1.8367}$$

Next, by (1.8361) and (1.8366),

$$\frac{y_{11}'}{y_{11}} = p_1 + \varepsilon_1, \qquad \frac{z_{jj}'}{z_{jj}} = p_{j+1} + \varepsilon_{j+1},$$

where $\lim_{t\to\infty} \varepsilon_k = 0$ $(k = 1, \ldots, n)$, and hence

$$y_{11} = c_1 e^{\int_{t_0}^{t}(p_1 + \varepsilon_1)dt} \qquad (c \neq 0)$$

and

$$z_{jj} = c_{j+1} e^{\int_{t_0}^{t}(p_{j+1} + \varepsilon_{j+1})dt} \qquad (c_{j+1} \neq 0).$$

Thus

$$\left|\frac{z_{jj}}{y_{11}}\right| = \left|\frac{c_{j+1}}{c_1}\right| e^{-\int_{t_0}^{t} \mathrm{Re}\,(p_1 + \varepsilon_1 - p_{j+1} - \varepsilon_{j+1})dt}$$
$$< \left|\frac{c_{j+1}}{c_1}\right| e^{-\int_{t_0}^{t} c' + \mathrm{Re}\,(\varepsilon_1 - \varepsilon_{j+1})dt}$$

by (1.8351). The integral $\int_0^\infty |z_{jj}/y_{11}|dt$ exists. This in turn, in view of (1.8367), implies the existence of the integral $\int_0^\infty |u_j|dt$.
Then, by means of (1.8362), our $n - 1$ linearly independent solutions $z^{(j)} = (z_{1j}, \ldots, z_{n-1,\,j})$ of (1.8364) yield the $n - 1$ solutions

$$\text{(1.8368)} \qquad \begin{cases} y_{1,\,j+1} = y_{1,1}\int_{-\infty}^{t} u_j dt \\ y_{k,\,j+1} = y_{k,1}\int_{-\infty}^{t} u_j dt + z_{k-1,\,j} \end{cases}$$

of our original system (1.8321). These solutions satisfy the conditions of our lemma.

First, we observe that

$$\text{(1.8369)} \qquad \lim_{t\to+\infty} \frac{y_{1,\,j+1}}{z_{jj}} = \lim_{t\to\infty} \frac{\int_{-\infty}^{t} u_j dt}{\dfrac{z_{jj}}{y_{11}}} = \lim_{t\to+\infty} \frac{u_j}{\dfrac{d}{dt}\left(\dfrac{z_{jj}}{y_{11}}\right)}$$

$$= \lim \frac{\dfrac{y_{11}}{z_{jj}} u_j}{\dfrac{z_{jj}'}{z_{jj}} - \dfrac{y_{11}'}{y_{11}}} = 0$$

by (1.8367), (1.8366), (1.8361), and (1.8351).

Next, for $k > 1$

$$\lim_{t\to+\infty} \frac{y_{k,\,j+1}}{z_{jj}} = \lim_{t\to+\infty} \frac{y_{k1}\int_{-\infty}^{t} u_j dt}{z_{jj}} + \lim_{t\to+\infty} \frac{z_{k-1,\,j}}{z_{jj}}$$

$$= \lim_{t\to+\infty} \frac{y_{k1}}{y_{11}} \cdot \frac{y_{1,\,j+1}}{z_{jj}} + \lim_{t\to+\infty} \frac{z_{k-1j}}{z_{jj}} = \begin{cases} 0 \text{ for } k-1 \neq j, \\ 1 \text{ for } k-1 = j. \end{cases}$$

Combining these results with those in (1.8369), we see that

$$\lim_{t\to+\infty} \frac{y_{k,\,j+1}}{y_{j+1,\,j+1}} = 0 \quad \text{for} \quad k = 1, 2, \ldots, n \text{ as long as } k \neq j+1.$$

It follows from (1.8321) at once that

$$\lim \left(\frac{y'_{j+1,\,j+1}}{y_{j+1,\,j+1}} - p_{j+1} \right) = 0.$$

The linear independence of the solutions $(y_{1j}, \ldots, y_{nj})$ follows from the linear independence of the solutions $(z_{1j}, \ldots, z_{n-1,\,j})$ in view of the determinant relation

$$\begin{vmatrix} y_{11} & y_{12} & \cdots & y_{1n} \\ y_{21} & y_{22} & \cdots & y_{2n} \\ \cdot & \cdot & \cdots & \cdot \\ y_{n1} & y_{n2} & \cdots & y_{nn} \end{vmatrix} = \begin{vmatrix} y_{11} & 0 & \cdots & 0 \\ y_{21} & z_{11} & \cdots & z_{1,\,n-1} \\ \cdot & \cdot & \cdots & \cdot \\ y_{n1} & z_{n-1,1} & \cdots & z_{n-1,\,n-1} \end{vmatrix} \neq 0.$$

1.837. We are now in a position to prove Theorem 1.831. Condition (1.8311) implies that the solutions of (1.601) have a basis (1.8352) which satisfies (1.8353) and (1.8354). It follows from (1.8354) that for every $\varepsilon > 0$ and for t sufficiently large

$$|\mathrm{Re}\,(\frac{d}{dt} \ln y_{kk} - p_{kk}(t))| \leqq |\frac{d}{dt} \ln y_{kk} - p_{kk}(t)| < \varepsilon,$$

and hence that

$$b_0 e^{\int_{t_0}^{t} \mathrm{Re}\,(p_{kk}(\tau))d\tau - \varepsilon t} \leqq |y_{kk}(t)| \leqq b_0 e^{\int_{t_0}^{t} \mathrm{Re}\,(p_{kk}(\tau))d\tau + \varepsilon t},$$

where $b_0 = |y_{kk}(t_0)|$. Since (1.8353) implies that $|y_{ik}(t)| \leqq |y_{kk}(t)|$ for sufficiently large t, we obtain

(1.8371)
$$b_0 e^{\int_{t_0}^{t} \mathrm{Re}\,(p_{kk}(\tau))d\tau - \varepsilon t} \leqq ||y^{(k)}|| = \sum_{i=1}^{n} |y_{ik}| \leqq n b_0 e^{\int_{t_0}^{t} \mathrm{Re}\,(p_{kk}(\tau))d\tau + \varepsilon t}.$$

Thus, in view of (1.743), (1.701), (1.702) and (1.7221), and by Lemma 1.73

$$\lambda_k = \lambda(y^{(k)}, e^t) = \lambda(e^{\int_{t_0}^{t} \mathrm{Re}(p_{kk}(\tau))d\tau}, e^t).$$

Formula (1.8312) follows at once from (1.751).

1.84. Comparison methods. It is reasonable to inquire whether estimates of characteristic numbers of a system

$$\frac{dy_i}{dt} = \sum_{k=1}^{n} q_{ik}(t)y_k \tag{1.8401}$$

may be deduced from such estimates for (1.601), should it be known that

$$\lim_{t\to\infty} |p_{ik}(t) - q_{ik}(t)| = 0. \tag{1.8402}$$

The limitations inherent in such comparison methods are not completely understood. There exist systems which are related as in (1.8402) and which nevertheless have distinct characteristic numbers.

1.841. *Example.* As can be seen by direct integration, the system

$$\frac{dy_1}{dt} = [\sin\log(t+1) + \cos\log(t+1)]y_1$$

$$\frac{dy_2}{dt} = [-\sin\log(t+1) + \cos\log(t+1)]y_2$$

has characteristic numbers $\lambda_1 = -1$ and $\lambda_2 = -1$. On the other hand, the system

$$\frac{dy_1}{dt} = [\sin\log(t+1) + \cos\log(t+1)]y_1$$

$$\frac{dy_2}{dt} = [-\sin\log(t+1) + \cos\log(t+1)]y_2 + \frac{\alpha}{t+1}y_1$$

has characteristic numbers

$$\lambda_1' = -1, \qquad \lambda_2' = -\left(1 + \frac{1}{e^{4\pi}}\right)$$

for every nonzero value of the parameter α.

1.842. If system (1.8401) has constant coefficients $q_{ik}(t) = a_{ik}$, then it follows from (1.8402) that the limiting system (1.8401) has the same characteristic numbers as the original system (1.601). If (1.601) is reducible, the same conclusion regarding characteristic numbers may be drawn for the associated system with constant coefficients. Both of these classical results will be derived in Chapter IV as a consequence of a more general theorem of A. A. Shestakov.

1.85. Regular systems. The difficulties encountered in the work of estimating characteristic numbers and in the study of the stability of the trivial solution under a nonlinear perturbation of the right-hand members of a linear system, led Lyapunov to the study of *regular systems*, i.e., systems of the following type:

1.851. DEFINITION. *A system* (1.601) *is called regular if and only if*

$$\nu = \sum \lambda_i = -\overline{\lim} \frac{1}{t} \int_{t_0}^{t} \sum_{s=1}^{n} \operatorname{Re} p_{ss}(\tau) d\tau,$$

i.e., if the sum ν *of the characteristic numbers of* (1.601) *has its largest possible value* (*cf.* (1.812)).

The more significant results in the theory of regular systems have been obtained by Lyapunov himself and by Perron [39], Cetaev [12], and Persidski [42]. In this section we shall limit ourselves to an outline of the principal results.

1.852. *Example.* The following is the classical example given by Lyapunov of a system which is not regular. Consider the system

$$\frac{dy_1}{dt} = y_1 \cos \log t + y_2 \sin \log t,$$

$$\frac{dy_2}{dt} = y_1 \sin \log t + y_2 \cos \log t.$$

The function

$$e^{\int_{t_0}^{t} \sum p_{ss}(\tau) d\tau} = ce^{t(\sin \log t + \cos \log t)}$$

has $-\sqrt{2}$ as its characteristic number. The system of solutions

$$y_{11} = e^{t \sin \log t}, \qquad y_{12} = e^{t \sin \log t}$$

$$y_{21} = e^{t \cos \log t}, \qquad y_{22} = -e^{t \cos \log t}$$

is normal in the sense of Lyapunov and the sum of its characteristic numbers is $-2 < -\sqrt{2}$.

1.853. A survey of the principal results. Every system with constant coefficients is regular. For, if $\gamma_1, \ldots, \gamma_n$ are the characteristic roots of the matrix $A = (a_{ij})$ of the system and if $\lambda_1, \ldots, \lambda_n$ are its characteristic numbers, then (cf. 1.412, 1.413)

$$\lambda_i = -\mathrm{Re}\, \gamma_i, \quad \nu = \Sigma \lambda_i = -\mathrm{Re}\,(tr\, A) = -\mathrm{Re} \sum_{s=1}^{n} a_{ss},$$

and

$$\lambda(e^{\int_{t_0}^{t} \Sigma a_{ss}\, d\tau}, e^t) = \lambda(e^{-\nu t}, e^t) = \nu.$$

By 1.78, regularity is a property invariant under linear transformation with a Lyapunov matrix. Thus all reducible systems and in particular all systems with periodic coefficients are regular. For these equations it is again possible to express the sum of characteristic exponents in terms of the coefficients of the system.

Other examples of regular systems exist. In general we have

1.8531. THEOREM.[2] *System* (1.601) *is regular if*

$$\lambda_i + \lambda_i' = 0 \qquad (i = 1, \ldots, n),$$

where $\lambda_1 \leqq \lambda_2 \leqq \ldots \leqq \lambda_n$ *are the characteristic numbers of* (1.601) *and* $\lambda_i' \geqq \ldots \geqq \lambda_n'$ *are those of the system adjoint to* (1.601).

2. Qualitative Study of Systems with Constant Coefficients and of reducible systems.

We consider again a system

$$\frac{dy}{dt} = Ay \tag{2.01}$$

where A is a real constant matrix. By a suitable linear transformation, (2.01) may be transformed into the canonical form

$$\frac{dz}{dt} = Bz, \tag{2.02}$$

where B is given by (1.464). It was shown in (1.46) that we can find a fundamental matrix of solutions of the form

[2]Cf. Perron [39], p. 759.

$$(2.03)\qquad Z(t)=\begin{bmatrix} Z_1(t) & 0 & \dots & 0\\ 0 & Z_2(t) & \dots & 0\\ \cdot & \cdot & \dots & \cdot\\ 0 & 0 & \dots & Z_q(t)\end{bmatrix}=(z_{ij})$$

where each $Z_i(t)$ is given by (1.466). Each $Z_i(t)$ is a matrix of order m_i and the corresponding elementary divisor $\delta_i(t)$ of B is of degree m_i. The jth column of (2.03), for

$$(2.04)\qquad j=m_1+\dots+m_k+l=M_k+l,\quad 1\leqq l\leqq m_{k+1},$$

is given by

$$(2.05)\qquad z^{(j)}=(z_{1j},\dots,z_{nj})=\Big(0,\dots,0,\frac{t^{l-1}}{(l-1)!}e^{\gamma t},\ \dots,e^{\gamma t},0,\dots,0\Big),$$

or more precisely,

$$(2.06)\qquad z_{M_k+1,j}=\frac{t^{l-1}}{(l-1)!}e^{\gamma t},\dots,z_{M_k+l,\,j}=e^{\gamma t},$$

and all other components are zero.

If γ is real, then (2.05) is a real solution of (2.01). If $\gamma=\alpha+\beta i$ is complex, then, since the invariant factors are real, for every elementary divisor $\delta(x)=(x-\delta)^m$ there exists a conjugate elementary divisor $\bar\delta(x)=(x-\bar\delta)$ of the same degree m, and hence conjugate submatrices B_k and $\bar B_k$, if distinct, occur in pairs. We note that the matrices $Z_k(t)$ and $\bar Z_k(t)$ are fundamental matrices of solutions of the subsystems corresponding to B_k and $\bar B_k$ respectively, and we consider the composite subsystem with the matrix

$$(2.07)\qquad \begin{bmatrix} B_k & 0\\ 0 & \bar B_k\end{bmatrix}.$$

Its fundamental matrix is given by

$$(2.08)\qquad \begin{bmatrix} Z_k(t) & 0\\ 0 & \bar Z_k(t)\end{bmatrix}.$$

A simple calculation yields

$$(2.09)\qquad W_k(t)=\begin{bmatrix} I & I\\ iI & -iI\end{bmatrix}\begin{bmatrix} Z_k(t) & 0\\ 0 & \bar Z_k(t)\end{bmatrix}\begin{bmatrix} \tfrac12 I & -\tfrac12 iI\\ \tfrac12 I & \tfrac12 iI\end{bmatrix}$$

$$=\begin{bmatrix} \tfrac12(Z_k+\bar Z_k) & \tfrac12 i(Z_k-\bar Z_k)\\ -\tfrac12 i(Z_k-\bar Z_k) & \tfrac12(Z_k+\bar Z_k)\end{bmatrix}.$$

Here, the left-hand multiplier of (2.08) in (2.09) determines a linear transformation of the subsystem defined by (2.07), and the right-hand multiplier produces a change of basis.

The matrix $W_k(t)$ in (2.09), is a real fundamental matrix of the

$$\frac{dw_k}{dt} = C_k w_k$$

where subsystem

$$C_k = \begin{bmatrix} I & I \\ iI & -I \end{bmatrix} \begin{bmatrix} B_k & 0 \\ 0 & \bar{B}_k \end{bmatrix} \begin{bmatrix} \frac{1}{2}I & -\frac{1}{2}iI \\ \frac{1}{2}I & \frac{1}{2}iI \end{bmatrix}$$
$$= \begin{bmatrix} \frac{1}{2}(B_k + \bar{B}_k) & \frac{1}{2}i(B_k - \bar{B}_k) \\ -\frac{1}{2}i(B_k - \bar{B}_k) & \frac{1}{2}(B_k + \bar{B}_k) \end{bmatrix}.$$

2.1. We may assume that the characteristic roots and the elementary divisors are so numbered that $\bar{B}_k = B_{k+1}$, and that all the elementary divisors corresponding to the real roots precede those corresponding to the complex roots.

If we employ the composite linear transformation suggested by (2.09), we pass from (2.02) to a real canonical form

$$\frac{dw}{dt} = Dw, \tag{2.11}$$

with

$$D = \begin{bmatrix} B_1 & \ldots & 0 & 0 & \ldots & 0 \\ \cdot & \ldots & \cdot & \cdot & \ldots & \cdot \\ 0 & \ldots & B_g & 0 & \ldots & 0 \\ 0 & \ldots & 0 & C_{g+1} & \ldots & 0 \\ \cdot & \ldots & \cdot & \cdot & \ldots & \cdot \\ 0 & \ldots & 0 & 0 & \ldots & C_{g+r} \end{bmatrix} \tag{2.12}$$

where submatrices B_j correspond to real elementary divisors, and submatrices C_j correspond to pairs of conjugate elementary divisors. As our fundamental matrix of solutions we choose

$$W(t) = \begin{bmatrix} Z_1(t) & \ldots & 0 & 0 & \ldots & 0 \\ \cdot & \ldots & \cdot & \cdot & \ldots & \cdot \\ 0 & \ldots & Z_g(t) & 0 & \ldots & 0 \\ 0 & \ldots & 0 & W_{g+1}(t) & \ldots & 0 \\ 0 & \ldots & 0 & 0 & \ldots & W_{g+r}(t) \end{bmatrix}. \tag{2.13}$$

The columns of (2.13) define a basis for the real solutions of

(2.11). We shall refer to it as the real canonical basis. In the canonical basis, solutions corresponding to real roots γ are given by (2.05), and the nonvanishing components of solutions associated with pairs of conjugate complex roots are given by the columns of $W_k(t)$ in (2.09). More precisely, if we write $w^{(j)}(t) = (w_{1j}(t), \ldots, w_{nj}(t))$ then

$$(2.14)\quad \begin{cases} w_{M_h+1,\,j} = \dfrac{t^{l-1}}{(l-1)!}\, e^{\alpha t}\cos\beta t, & w_{M_h+1,\,j+m_{h+1}} = -\dfrac{t^{t-1}}{(l-1)!}\, e^{\alpha t}\sin\beta t, \\ w_{M_h+l,\,j} = e^{\alpha t}\cos\beta t, & w_{M_h+l,\,j+mh_{+1}} = -\,e^{\alpha t}\sin\beta t, \\ w_{M_h+m_{h+1}+1,\,j} = \dfrac{t^{l-1}}{(l-1)!}\, e^{\alpha t}\sin\beta t, & w_{M_h+mh_{+1}+1,\,j+m_{h+1}} = \dfrac{t^{l-1}}{(l-1)!}\, e^{\alpha t}\cos\beta t, \\ w_{M_h+m_{h+1}+l,\,j} = e^{\alpha t}\sin\beta t, & w_{M_h+m_{h+1}+l,\,j+m_{h+1}} = e^{\alpha t}\cos\beta t, \end{cases}$$

and all other components are zero. Here,

$$(2.15)\qquad j = m_1 + m_2 + \cdots + m_h + l = M_h + 1,$$

where h is obtained by adding an even number to the number of real elementary divisions and $1 \leqq l \leqq m_{h+1} = m_{h+2}$.

2.16. In what follows we shall study the behavior of the real integral curves of (2.11) in the neighborhood of the origin.

If the linear system

$$(2.161)\qquad Ac = 0$$

associated with (2.01) has no nontrivial solutions, i.e., if A is nonsingular, then the origin is the only singular point of (2.01).

If, however, A is singular, say of rank r, then there is an $(n-r)$-dimensional subspace S_{n-r} of singular points. For every c in S_{n-r}, and for every solution y of (2.01)

$$x(t) = y(t) + c$$

is also a solution of (2.01) which behaves near c as y behaves near the origin and conversely. It suffices therefore to study in detail merely the behavior of the integral curves of (2.01) near the origin.

2.2. Let $\gamma_1 = \alpha_1 + i\beta_1, \ldots, \gamma_n = \alpha_n + i\beta_n$ be the n characteristic roots of the matrix B in (1.464). We shall distinguish six cases.

$$(2.201)\qquad \alpha_k \neq 0, \quad \alpha_k\alpha_j > 0, \quad k, j = 1, \ldots, n.$$

$$(2.202)\qquad \alpha_k \neq 0, \quad k = 1, \ldots, n; \quad \text{there exist } m, j \text{ such that } \alpha_m\alpha_j < 0.$$

(2.203) $\gamma_k \neq 0, \quad \alpha_k\alpha_j \geqq 0, \quad k, j = 1, \ldots, n.$ There is $\alpha_q = 0$.

(2.204) $\gamma_k \neq 0, \quad k = 1, \ldots, n;$ there exist m, j, q such that $\alpha_m\alpha_j < 0, \ \alpha_q = 0.$

(2.205) $\gamma_k \neq 0, \quad \alpha_k = 0, \quad k = 1, \ldots, n.$

(2.206) At least one of γ_i is zero.

If two differential equations correspond to different cases among (2.201)—(2.206), then the distributions of solutions near the origin are of different topological types. We do not assert that two distributions belonging to one and the same case above, are of the same topological type, i.e., can be transformed into one another by a topological transformation of the n-space into itself. This last question is connected with the problem, still little understood, concerning the manner in which the distribution of integral curves changes with a continuous change of the right-hand members of a linear system.

2.207. Definition. We shall say that almost all integral curves are of a given type relative to a given singular point, if, in a sufficiently small neighborhood of this point, all integral curves except possibly those filling a manifold of dimension less than n, belong to this type.

2.21. Henceforth solutions are written y instead of w. If (2.201) holds, then all functions $y_{jk}(t)$ tend to zero either for $t \to +\infty$ or for $t \to -\infty$. Since every solution $y(t)$ is a linear combination of the solutions $y^{(k)}(t) = (y_{1k}(t), \ldots, y_{nk}(t))$, then $y(t)$ tends to the origin either for $t \to +\infty$ or $t \to -\infty$. In this case we shall say that the origin is a *generalized* node.

22.2. Let $\alpha_i < 0$ for $i = 1, \ldots, k$ and $\alpha_j > 0$ for $j = k + 1, \ldots, n$, and let $y^{(i)}(t)$ be the solution associated with the root $y_i = \alpha_i + \sqrt{-1}\beta_i$ Then, the k-parameter family of solutions $y = \Sigma_{i=1}^{k} c_i y^{(i)}$ fills a hyperplane L_k (this hyperplane may be taken as the set of initial points defining this family) of k dimensions and each integral curve of this family tends to the origin as $t \to +\infty$. Similarly, the $(n-k)$-parameter family $y = \Sigma_{i=k+1}^{n} c_i y^{(i)}$ fills a hyperplane L_{n-k} of $(n - k)$ dimensions perpendicular to L_k. Each integral curve of this family tends to the origin as $t \to -\infty$. The remaining integral curves fill an n-dimensional neighborhood of the origin save for the points of L_k and L_{n-k} in that neighborhood. Each of these integral curves has a positive distance from the origin and leaves this neighborhood both for $t \to +\infty$ and for $t \to -\infty$. These integral curves resemble level curves of a saddle surface. In this case we shall call the origin a *generalized saddle point* (*of the first kind*).

2.23. We recall that in view of (2.14) the components of each of the real canonical solutions associated with a pair of purely imaginary roots $\beta_k i$ and $-\beta_k i$ are of the form $c_{jk}t^q \cos \beta_k t$ and $d_{jk}t^r \sin \beta_k t$. Moreover, if p of the elementary divisors belonging to each of these roots are linear, then p components of the first kind have $q = 0$ and p components of the second kind have $r = 0$. We shall write

$$u_{jk} = c_{jk} \cos \beta_k, \quad v_{jk} = d_{jk} \sin \beta_k \qquad (k = 1, \ldots, p).$$

Let $\gamma_1, \ldots, \gamma_{2p}$ be the $2p$ purely imaginary characteristic roots and let $\alpha_j \alpha_k > 0$ for $j,\ k = 2p + 1, \ldots, n$. We may assume for definiteness that $\alpha_j > 0$ for $j = 2p + 1, \ldots, n$. We suppose that the number of distinct elementary divisors corresponding to the roots $\gamma_1, \ldots, \gamma_{2p}$ is equal to $2q$.

2.231. Case $2p = 2q$. Here all the elementary divisors associated with $\gamma_1, \ldots, \gamma_{2p}$ are linear.

First, consider the family of integral curves given by

$$(2.2311) \quad y_j = c_1 u_{j1} + \ldots + c_q u_{jq} + d_1 v_{j1} + \ldots + d_q v_{jq} \quad (j = 1, \ldots, n).$$

The initial values of these integral curves fill a hyperplane of $2q$ dimensions. The components (2.2311) are almost periodic functions and this family of integral curves fills surfaces of toruses of different dimensions.

Second, consider the family of integral curves defined by

$$y_j = h_1 y_{j2p+1}(t) + \ldots + h_{n-2p} y_{jn}(t) \qquad (j = 1, \ldots, n)$$

where $y_{jk}(t)$ tends toward zero as $t \to -\infty$. These are O^--curves and they fill a hyperplane of $n - 2p$ dimensions.

All other curves, not passing through the above-mentioned perpendicular hyperplanes, are given by the formula

$$y_j = c_1 u_{j1} + \ldots + d_q v_{jq} + h_1 y_{j2p+1} + \ldots + h_{n-2p} y_{jn}$$
$$(j = 1, \ldots, u,\ q + (n - p) = n)$$

in which at least one of h_k and one of c_l or d_l are not equal to zero.

These curves approach asymptotically the curves of the family (2.2311) as t tends to $-\infty$.

Thus, if $2p = 2q$, then almost all curves are asymptotic and we say that the origin is a *generalized focus.*

2.232. Case $2p > 2q$. In this case there is a nonlinear elementary divisor associated with one of the roots $\lambda_1, \ldots, \lambda_{2p}$. Here again we have a family of almost periodic curves whose initial values depend upon $2q$ parameters, a family of O-curves depending upon $n - 2p$ parameters, and finally a family of solutions asymptotic to the almost periodic solutions, whose initial values depend upon $n - 2p + 2q$ parameters. All other solutions are of the form

$$y_j = b_1 y_{j1} + \ldots + b_n y_{jn} \qquad (j = 1, \ldots, n),$$

where every solution has at least one component which is not bounded for $t \to +\infty$ and for $t \to -\infty$.

Thus, if $2p > 2q$, then almost all integral curves are of the saddle type and we shall say that the origin is a *generalized saddle point.* To distinguish this case from the case in 2.12 we shall speak of a generalized saddle point *of the second kind.*

2.24. The analysis of case (2.204) is similar to that of (2.203) in 2.13 and yields the following results.

We suppose that $\alpha_j > 0$ for $j = 1, \ldots, p$, $\alpha_j < 0$ for $j = p + 1, \ldots, p + q$, and that $\alpha_j = 0$ for $j = p + q + 1, \ldots, p + q + s = n$. It follows from (2.104) that $p \neq 0$, $q \neq 0$, and $s \neq 0$. Let s' be the number of the elementary divisors associated with the characteristic roots $\gamma_{p+q+1}, \ldots, \gamma_{p+q+s}$. There exist three mutually perpendicular hyperplanes L_p, L_q, L_s of dimensions p, q, s', respectively, with the properties that through every point of L_p there passes an O-curve approaching the origin as $t \to -\infty$ through every point of L_q there passes an O-curve approaching the origin as $t \to +\infty$, and through every point of $L_{s'}$ there passes an almost periodic integral curve. There exist *two families* of asymptotic curves forming manifolds of $p + s'$ and $q + s'$ dimensions respectively. The two manifolds consist of curves which approach the almost periodic solutions asymptotically as $t \to -\infty$ and $t \to +\infty$ respectively. All other integral curves are of the saddle type. Thus, almost all integral curves are of the saddle type. In this case the origin is called a *composite saddle point.*

2.25. We note that (2.205) can hold only for even n.

2.251. If all the elementary divisors are linear, then all solutions are given by (2.2311) with $n = 2q = 2p$. Thus all the solutions are almost periodic. In this case we say that the origin is a *generalized vortex*. An important special case occurs if $\beta_1, \ldots, \beta_n$ are commensurable with each other and all the solutions are therefore periodic.

2.252. If not all the elementary divisors are linear, then $2q < 2p = n$, and in addition to the solutions of type (2.1311) there will exist solutions whose components contain functions of the type

$$d_{jk} t^r \sin \beta_k t, \qquad r > 0,$$
$$c_{jk} t^q \cos \beta_k t, \qquad q > 0.$$

In fact almost all solutions have this property and are therefore of the saddle type. Here the origin is called *a generalized saddle point* (*of the third kind*).

2.26. There remains to consider the case when some of the characteristic roots are zero. Here, we shall mention briefly only the special case where all the elementary divisors associated with the vanishing roots are linear.

Let $\gamma_1 = \gamma_2 = \ldots = \gamma_s = 0$. Then the Jordan canonical form (cf. (1.461), (1.464)) will appear as

$$\text{(2.261)} \qquad \begin{aligned} &\frac{dz_1}{dt} = 0, \ldots, \frac{dz_s}{dt} = 0 \\ &\frac{dz_{s+1}}{dt} = \lambda_{s+1} z_{s+1} + \ldots, \ldots. \end{aligned}$$

We see at once that the phase space consists of layers of subspaces of $n - s$ dimensions in each of which integral curves are characterized by one and the same system, viz., the subsystem

$$\text{(2.262)} \qquad \frac{dz_{s+1}}{dt} = \lambda_{s+1} z_{s+1} + \ldots, \ldots,$$

of (2.261). Since $\lambda_j \neq 0$ for $j = s + 1, \ldots, n$, the behavior of solutions near the origin in each of the $(n - s)$-dimensional subspaces is completely described in Sections 2.11—2.15.

We should note that in the present case the origin is not an isolated singular point. In fact a whole subspace of s dimensions consists only of singular points.

2.27. In Sections 2.21—2.26 we considered only some of the crude aspects of the behavior of integral curves in the neighborhood of the origin. The types of behavior considered there could be established for the original linear system if the normalizing transformation should be merely one-to-one and bicontinuous either in the whole phase space or in some neighborhood of the origin. Since, however, our normalizing transformation is a linear transformation with constant coefficients, we can give a more precise description of the behavior of solutions near the origin. We shall, for instance, try to characterize *regular O-curves*, i.e., *O*-curves which have asymptotic tangents.

2.271. Let $\alpha + \beta i$ be a complex characteristic root with a linear elementary divisor. The canonical subsystem corresponding to the pair $\alpha \pm \beta i$ of conjugate roots may be written as

$$\frac{du}{dt} = \alpha u - \beta v,$$

$$\frac{dv}{dt} = \beta u + \alpha v.$$

Its solutions are the family of spirals

$$u = ce^{\alpha t} \cos \beta t, \qquad v = de^{\alpha t} \sin \beta t.$$

Thus the projections of the integral curves onto the plane (u, v) are spirals. The curves themselves may lie on a manifold passing through the origin, in which case these curves have the character of screws. The *O*-curves of this type are not regular.

2.272. Consider next the case when all the roots $\gamma_1, \ldots, \gamma_n$ of our characteristic equation are real and are of the same sign. For definiteness we may assume that all $\gamma_i > 0$.

The tangent vector to $z^{(j)}(t)$ at t is given by

$$\text{(2.2721)} \qquad \frac{dz^{(j)}}{dt} = u^j(t), \quad u^j(t) = Bz^{(j)}(t),$$

where $B = D$ is in the form (1.464). Thus if we write

$$u^{(j)} = (u_{1j}, \ldots, u_{nj}),$$

then

$$\text{(2.2722)} \qquad \begin{aligned} u_{m_k+1, j} &= \gamma_k \frac{t^{l-1}}{(l-1)!} e^{\gamma_k t} + \frac{t^{l-2}}{(l-2)!} e^{\gamma_k t}, \\ u_{m_k+l-1, j} &= \gamma_k t e^{\gamma_k t} + e^{\gamma_k t}, \\ u_{m_k+l, j} &= e^{\gamma_k t}, \end{aligned}$$

and all other components of $u^{(j)}$ are zero. Here j has the value (2.04).

Since we are interested in the behavior of the tangents to $z^{(j)}(t)$ at t as t tends to $-\infty$, it will suffice to consider the unit tangent vector

$$\frac{1}{\sqrt{\sum_{i=1}^{n}\left(\frac{dz_{ij}}{dt}\right)^2}}\cdot\frac{dz^{(j)}}{dt}=\frac{u^{(j)}(t)}{\sqrt{\sum_{i=1}^{n}(u_{ij}(t))^2}}. \tag{2.2723}$$

Since, as t tends to $-\infty$,

$$(l-1)!\,|\gamma|^{-1}|t^{-l+1}|\,e^{-\gamma t}\sqrt{\sum_{i=1}^{n}(u_{ij}(t))^2}$$

tends to $+1$, we may replace the unit vector (2.2723) by the vector

$$\frac{(l-1)!}{|\gamma|}\,|t|^{-l+1}\,e^{-\gamma t}\,u^{(j)}(t), \tag{2.2724}$$

parallel to it and of approximately equal length, whose behavior can be studied more easily.
Indeed, one sees at once that as t tends to $-\infty$ the (M_k+1)st component (see 2.04) of this vector approaches $(-1)^{l-1}$ (or $+1$ as t tends to $+\infty$) and that all other components approach zero. Thus the direction of the tangent vector (2.2724), and hence that of (2.2721) approaches the direction parallel to that of the (M_k+1)st axis but possibly opposite in sense, and hence $z^{(j)}$ is asymptotic to this axis as t tends to $-\infty$.

In our case, all of the solutions in the canonical basis (2.13) which are asymptotic to the (M_k+1)-axis are given by (2.06). Every linear combination

$$z=\sum_{j} c_j z^{(j)},\quad J: M_k+1\leqq j\leqq M_k+m_{k+1}, \tag{2.2725}$$

behaves as the term $c_j z^{(j)}$ with the nonzero c_j of highest index. Thus all the solutions of the m_{k+1}-parameter family (2.2725) of the canonical system (2.02) are asymptotic to the (M_k+1)st axis at the origin.

Since $\gamma_k>0$, and hence $z^{(j)}$ in (2.06) tends to zero as t tends to $-\infty$, the integral curve $z^{(j)}$ is a regular (cf. 2.27) O-curve asymptotic to the (M_k+1)st axis.

Each family of solutions (1.466) *belongs* to a characteristic root

γ and to one of the elementary divisors associated with it. Suppose, for simplicity of notation, that successive elementary divisors beginning with the lth are exactly those which are associated with a given root γ and consider a linear combination

$$z = \sum_j c_j z^{(j)}, \qquad M_{l-1} \leqq j \leqq M_{l+q}. \tag{2.2726}$$

Components of z are of the form $P(t)e^{\gamma t}$ where $P(t)$ is a polynomial of degree not exceeding the largest order m_k among the B_k in (1.464). The components of largest degree are contained among the $(M_{k_k} + 1)$st components

$$P_k(t) = \sum_j c_j z^{(j)}, \quad l \leqq k \leqq l + q - 1, \quad M_{k-1} + 1 \leqq j \leqq M_k \tag{2.2727}$$

If there are h polynomials $P_{k_1} \ldots P_{k_k}$ of largest degree among (2.2727), then the direction of z in (2.2726) approaches that of a vector with nonzero components in the $(M_{k_1} + 1)$st, ..., ..., $(M_{k_h} + 1)$st positions and with every other component equal to zero.

If we have a solution which is the sum of solutions $z^{(\gamma_i)}$ belonging to different real characteristic roots γ_i all of which are of the same sign, say all positive, then the tangent vector of this solution will behave as the tangent vector of the part $z^{(\gamma)} = \Sigma_\gamma$ with the largest γ.

It follows at once that when all γ_i are positive, each solution of the canonical system (2.02) as well as each solution of the original system (2.01), is a regular O-curve (as $t \to -\infty$).

The picture is particularly simple for the canonical system (2.02) in the case when all characteristic roots are distinct, say $\lambda_1 > \lambda_2 > \ldots > \lambda_n > 0$. Here, all the solutions except of the $(n - 1)$-parameter family

$$c_2 z^{(2)} + \ldots + c_n z^{(n)}, \tag{2.2728}$$

are asymptotic to the z_1-axis. All those in the family (2.2728) except the curves in the $(n - 2)$-parameter family

$$c_3 z^{(3)} + \ldots + c_n z^{(n)}, \tag{2.2729}$$

are asymptotic to the z_2-axis at the origin, and so on.

The analysis of the case when all λ_i have positive real parts and linear elementary divisors is also quite simple. The more general cases lead to complications which we shall not discuss here.

In what follows, we shall see that in many instances the behavior of solutions of nonlinear systems near a singular point remains essentially the same as that discussed above.

2.3. Non-homogeneous systems. We shall consider a system

$$\frac{dy}{dt} = Ay + b, \tag{2.301}$$

where $A = (a_{ij})$ is a constant matrix, and $b = (b_1, \ldots, b_n)$ is a constant column vector.

2.31. If the associated system

$$Ac + b = 0 \tag{2.311}$$

has a solution c, not necessarily unique, then, if we set

$$y = z + c, \tag{2.312}$$

our system (2.301) will be transformed into a homogeneous system (2.01). Since the substitution (2.312) is a translation, the behavior of the integral curves of (2.301) near c is the same as the behavior of such curves for (2.01) near the origin.

2.32. If (2.311) has no solutions, then the corresponding system of differential equations has no singular points.

2.4. Reducible systems. The solutions $y(t)$ of a reducible system (1.402) are related to the solutions $z(t)$ of a system (1.461) with a constant matrix B, by the identity

$$y(t) = K(t)z(t)$$

or, more explicitly, by

$$y_i(t) = \sum_{j=1}^{n} K_{ij}(t)z_j(t), \tag{2.401}$$

where $K(t)$ is a Lyapunov matrix, and therefore $K_{ij}(t)$ are bounded for $t_0 \leqq t < +\infty$. From the behavior of solutions of (1.461) near the origin one can, in view of (2.401), deduce the behavior of the corresponding solutions of (1.402) near the trivial solution

$$y_1(t) \equiv 0, \ldots, y_n(t) \equiv 0, \qquad t > 0, \tag{2.402}$$

in the augmented $(n + 1)$-dimensional space. We note that (2.401) carries the origin, considered as a singular point (solution) of (1.461), into (2.402).

Thus, if

(2.403) $|z_{ij}(t) - z_{ik}(t)| \to 0$ as $t \to +\infty$, $i = 1, \ldots, n$,

then, since $K_{ij}(t)$ are bounded, say $|K_{ij}(t)| < M < +\infty$, we have

$$|y_{ij}(t) - y_{ik}(t)| < M \sum_{i=1}^{n} |z_{ij}(t) - z_{ik}(t)|$$

and hence •

(2.404) $|y_{ij}(t) - y_{ik}(t)| \to 0$ as $t \to +\infty$, $i = 1, \ldots, n$.

In particular, replacing z_{ik} and y_{ik} in (2.403), (2.404), by zero, we see that if $z^{(j)}$ is an O-curve then the corresponding solution y^j of (1.402) approaches (2.402) as t tends to $+\infty$.

The behavior near the origin of solutions of systems of the type (1.461) with a constant matrix B was classified in Section 2.2 according to the values of the real parts of the characteristic roots of B. From the discussion in 1.78 it follows that the characteristic numbers of a reducible system (1.402) are exactly the real parts of every associated system (1.461) with constant coefficients. Therefore the behavior of solutions of (1.402) near (2.402) is governed by the values of characteristic numbers of this system and is subject to classification similar to that given in Section 2.2.

2.41. If the coefficients of (1.402) are periodic, say, of period 2π, we may identify points $(y, t + 2k\pi)$ and (y, t). Then for a fixed r, the neighborhood $y_1^2 + \ldots + y_n^2 \leqq r^2$ of (2.402) becomes a torus, and the trivial solution (2.402) becomes its axis, i.e., a periodic solution. In the light of this geometric interpretation, our discussion of reducible systems yields information about the behavior of solutions in the neighborhood of a periodic solution. We shall consider this problem in detail in the next chapter.

2.5. Comparable and almost linear systems. Consider a system

(2.501) $$\frac{dy_i}{dt} = \sum_{k=1}^{n} a_{ik} y_k + f_i(t, y_1, \ldots, y_n) \quad (i = 1, 2, \ldots, n),$$

where $A = (a_{ik})$ is a constant matrix,

(2.502) $$f_i(t, 0, \ldots, 0) = 0,$$

and

(2.503) $$|f_i(t, y_1', \ldots, y_n') - f_i(t, y_1'', \ldots, y_n'')| \leqq g(t) \sum_{i=1}^{n} |y_i' - y_i''|.$$

In particular

$$|f_i(t, y_1, \ldots, y_n)| \leqq g(t) \sum_{i=1}^{n} |y_i|. \tag{2.504}$$

If $g(t)$ is continuous and bounded for $t \geqq t_0$, then we shall say that (2.501) is *comparable with the linear system*

$$\frac{dy}{dt} = Ay.$$

It was shown in Chapter I that every solution of such a system (2.501) is defined for all values $t \geqq t_0$.

If $g(t)$ satisfies a stronger condition

$$\int_0^\infty g(t)dt < +\infty \tag{2.505}$$

then we shall say that (2.501) is *an almost linear system.*

We may assume that the constant matrix in (2.501) is in the Jordan canonical form, i.e., that (2.501) is of the form

$$\frac{dz}{dt} = Bz + f(z, t), \tag{2.506}$$

where B is given by (1.464) and $f = (f_1, \ldots, f_n)$. We shall use the notation of Section 2, and set

$$\begin{cases} \alpha = \max\limits_j \alpha_j, \quad m = \max\,(m_j \text{ with } \alpha_j = \alpha), \\ \alpha^* = \text{largest negative } \alpha_j, \quad m^* = \max\,(m_j \text{ with } \alpha_j = \alpha^*), \\ p = \max\,(m_j \text{ with } \alpha_j = 0) \quad \text{or} \quad p = 1 \text{ if } \alpha_j \neq 0 \text{ for} \\ \qquad\qquad\qquad\qquad\qquad\qquad\qquad\qquad \text{every } j. \end{cases} \tag{2.507}$$

We recall that m_j is the order of the canonical submatrix B_j in (1.464) and that $\gamma_j = \alpha_j + i\beta_j$ is the corresponding characteristic root.

In what follows we shall be concerned chiefly with the problem of the comparison of solutions of (2.506) *with those of the associated linear system*

$$\frac{dz}{dt} = Bz. \tag{2.508}$$

2.51. We write

$$B = \begin{bmatrix} \mathscr{M}_1 & 0 \\ 0 & \mathscr{M}_2 \end{bmatrix} = \begin{bmatrix} \mathscr{M}_1 & 0 \\ 0 & 0 \end{bmatrix} + \begin{bmatrix} 0 & 0 \\ 0 & \mathscr{M}_2 \end{bmatrix}, \tag{2.511}$$

where $\alpha_j \geqq 0$ for all the characteristic roots γ_j of $\mathscr{M}_1$, and $\alpha_j < 0$ for all the characteristic roots of $\mathscr{M}_2$. Then

$$z(t) = e^{Bt} = z_1(t) + z_2(t),\ z_1 = \begin{bmatrix} e^{\mathscr{M}_1 t} & 0 \\ 0 & 0 \end{bmatrix}, z_2 = \begin{bmatrix} 0 & 0 \\ 0 & e^{\mathscr{M}_2 t} \end{bmatrix}, \tag{2.512}$$

where $z = z(t)$ is the fundamental matrix of (2.507) such that $z(t_0) = I$.

We let

$$\chi_k(t) = \begin{cases} t^{k-1} & \text{for} \quad t \geqq 1, \\ 1 & \text{for} \quad t < 1. \end{cases} \tag{2.513}$$

We note that $x_k(t)$ *is monotone and bounds* t^{k-1}. Thus

$$\begin{aligned} &(\mu)\ \chi(t_1) \leqq \chi(t_2) \quad \text{for} \quad 0 \leqq t_1 \leqq t_2, \\ &(\beta)\ \ t^{k-1} \leqq \chi_k(t) \quad \text{for all} \quad t. \end{aligned} \tag{2.514}$$

We define the modulus $||A||$ of a matrix $A = (a_{ij})$ by

$$||A|| = \sqrt{\sum |a_{ij}|^2}, \tag{2.515}$$

and we note the following fundamental inequalities:

$$||A + C|| \leqq ||A|| + ||C||, \tag{2.516}$$

$$||AC|| \leqq ||A||\ ||C||, \tag{2.517}$$

$$\left\| \int_{t_0}^{t} A(\tau)d\tau \right\| \leqq \int_{t_0}^{t} ||A(\tau)||d\tau. \tag{2.518}$$

With this notation and the conventions (2.507) and (2.512), we have, in view of (2.514β),

$$\begin{cases} ||z(t)|| \leqq c_0 e^{\alpha t} t^{m-1} \leqq c_0 e^{\alpha t} \chi_m(t), \\ ||z_1(-t)|| \leqq c_1 t^{p-1} \leqq c_1 \chi_p(t), \\ ||z_2(t)|| \leqq c_2 e^{\alpha^* t} t^{m^*-1} \leqq c_2 e^{\alpha^* t} \chi_{m^*}(t) \end{cases} \tag{2.519}$$

for $t \geqq 0$. Here c_0, c_1, c_2 are suitably chosen constants.

2.52. An estimate for the solutions of a comparable system (2.501). In view of (1.436), we may replace (2.506) by a matrix integral equation

$$X(t) = Z(t-t_0)X_0 + \int_{t_0}^{t} Z(t-\tau)\Phi(X(\tau), \tau)d\tau, \tag{2.521}$$

or

$$X(t) = e^{B(t-t_0)} X_0 + \int_{t_0}^{t} e^{B(t-\tau)}\Phi(X(\tau), \tau)d\tau, \tag{2.522}$$

where $\Phi_{jk} = (f_j(x^{(k)}, t))$, and $x^{(k)}$ is the kth column of X.

If we apply the inequalities (2.156)—(2.518), and make use of (2.521), we get

$$||X(t)|| \leqq ||Z(t-t_0)|| \;\; ||X_0|| + n^{\frac{1}{2}}\int_{t_0}^{t} ||Z(t-\tau)|| \;\; g(\tau) \;\; ||X(\tau)||d\tau,$$

since

$$||\Phi(X(t), t)|| = (\sum_{j,k} |f_j(x^{(k)}, t)|^2)^{\frac{1}{2}} \leqq g(t)(\sum_{jk}(\sum_{i}|x_i^{(k)}|)^2)^{\frac{1}{2}} \leqq g(t)n^{\frac{1}{2}}||X||$$

by the hypothesis (2.504). Next, by (2.519),

$$\begin{aligned} ||X(t)|| &\leqq c_0 e^{\alpha(t-t_0)} \chi_m(t-t_0)||X_0|| \\ &+ n^{\frac{1}{2}}\int_{t_0}^{t} c_0 e^{\alpha(t-\tau)} \chi_m(t-\tau)g(\tau) \;\; ||X(\tau)|| \; d\tau. \end{aligned} \tag{2.524}$$

Henceforth we assume that

$$\alpha \geqq 0. \tag{2.525}$$

Then, by the mean value theorem,

$$\begin{aligned} (e^{\alpha(t-t_0)} \chi_m(t-t_0))^{-1} &\int_{t_0}^{t} e^{\alpha(t-\tau)} \chi_m(t-\tau)g(\tau)||X(\tau)||d\tau \\ &\leqq \int_{t_0}^{t} e^{\alpha(t_0-\tau)} \frac{\chi_m(t-\tau)}{\chi_m(t-t_0)} g(\tau) \;\; ||X(\tau)|| \; d\tau. \end{aligned}$$

Thus, since (cf. (2.514 μ))

$$0 < \chi_m(\tau - t_0) \leqq \chi_m(t-t_0) \quad \text{for} \quad t_0 < \tau \leqq t,$$

the relation (2.524) may be written in the form

$$\begin{aligned} \frac{||X(t)||}{e^{\alpha(t-t_0)}\chi(t-t_0)} \leqq c_0||X_0||\Bigg(1 + \frac{n}{||X_0||}\int_{t_0}^{t} \chi_m(t-\tau)g(\tau) \\ \frac{||X(\tau)||}{e^{\alpha(\tau-t_0)}x_m(\tau-t_0)} d\tau.\Bigg) \end{aligned} \tag{2.526}$$

Applying Lemma 2.11 of Chapter I to (2.526) we obtain

$$\frac{||X(t)||}{e^{\alpha(t-t_0)}\chi(t-t_0)} \leqq c_0||X_0||\, e^{c_0 n\int_{t_0}^{t}\chi_m(t-\tau)g(\tau)d\tau},$$

whence, by (2.514 μ),

$$(2.527)\quad ||X(t)|| \leqq c_0 e^{-\alpha t_0}\,||X_0||\, e^{\alpha t}\chi_m(t)e^{c_0 n\int_{t_0}^{t}\chi_m(t-\tau)g(\tau)d\tau}.$$

Combining the constants, we may write

$$(2.528)\qquad ||X(t)|| \leqq c_3||X_0||\, e^{\alpha t}\chi_m(t)e^{c_4\int_{t_0}^{t}\chi_m(t-\tau)g(\tau)d\tau}.$$

If we make use of hypothesis (2.503), then an argument similar to the above yields a related estimate

$$(2.529)\quad ||X(t)-X^{(1)}(t)|| \leqq c_5||X_0-X_0^{(1)}||\, e^{\alpha t}\chi_m(t)e^{c_6\int_{t_0}^{t}\chi_m(t-\tau)g(\tau)d\tau}$$

The estimate in (2.528) and that in (2.529) are both subject to the condition (2.525).

We should note that the right-hand member of (2.528) serves as an estimate for the moduli of individual solutions as well as for the absolute values of the individual components. A similar observation holds true for (2.529).

2.53. Almost linear systems. Under more restrictive conditions of (2.506) than those assumed in Section 2.52, the relationship between the solutions of (2.506) and those of the associated system (2.508), may be described precisely by

2.531. THEOREM (Yakubovic [60]). *If* $\alpha \geqq 0$, *i.e., if not all the solutions of* (2.508) *tend toward zero, if* (2.503) *holds, and if*

$$(2.532)\qquad \int_{t_0}^{+\infty}\tau^{m+p-2}\, e^{\alpha\tau}g(\tau)d\tau < +\infty,$$

then there exists a one-to-one bicontinuous correspondence between the initial values $(z^{(0)}, t_0)$ of (2.508) *and* $(x^{(0)}, t_0)$ *of* (2.506), *such that for every solution* $z(t, z^{(0)}, t_0)$ *of* (2.508) *and the corresponding solution* $x(t, x^{(0)}, t_0)$ *of* (2.506), *we have*

$$(2.533)\quad |x(t, x^{(0)}, t_0) - z(t, z^{(0)}, t_0)| \to 0 \quad \text{as} \quad t \to +\infty.$$

If, however, $\alpha < 0$, *i.e., all the solutions of* (2.508) $\to 0$ *as* $t \to +\infty$, *then all the solutions of* (2.506) $\to 0$, *provided* (2.503) *holds and provided*

$$(2.534)\quad -\eta t + \int_{t_0}^{t} g(\tau)d\tau \to -\infty \quad \textit{as} \quad t \to \infty, \quad \textit{for every } \eta > 0.$$

2.54. We employ the notation in (2.506) and (2.512), note that $Z_1(t-\tau) = Z_1(t)Z_1(-\tau)$, and write (1.436) in the form

$$(2.541)\quad x(t) = Z(t-t_0)x^{(0)} + Z_1(t)\int_{t_0}^{t} Z_1(-\tau)f(x(\tau),\tau)d\tau + \int_{t_0}^{t} Z_2(t-\tau)f(x(\tau),\tau)d\tau.$$

Here $x(t)$ is the solution of (2.506) for which $x(t_0) = x^{(0)}$, i.e., $x(t) = x(t, x^{(0)}, t_0)$.

We shall prove that for every solution $x(t) = x(t, x^0, t_0)$ the integral

$$(2.542)\qquad v_1(x^{(0)}, t^0) = \int_{t_0}^{\infty} Z_1(-\tau)f(x(\tau),\tau)d\tau$$

converges. By the estimate of $Z_1(-\tau)$ in (2.519), and in view of (2.504), we have

$$(2.5421)\qquad \|Z_1(-t)f(x(t), t)\| \leqq nc_1t^{p-1}g(t)\|x(t)\|$$

and hence, by (2.528),

$$\|Z_1(-t)f(x(t),t)\| \leqq nc_1c_3\|X_0\|\, t^{m+p-2}\, e^{\alpha t}\, g(t)e^{c_4\int_{t_0}^{t}\chi_m(t-\tau)g(\tau)d\tau}.$$

The integral in the exponent converges, for

$$\int_{t_0}^{t}\chi_m(t-\tau)g(\tau)d\tau = \int_{t_0}^{t-1}(t-\tau)^{m-1}\, g(\tau)d\tau + \int_{t-1}^{t}g(\tau)d\tau =$$
$$= \sum_{k=0}^{m-1}\binom{m-1}{k}t^{m-1-k}(-1)^k\int_{t_0}^{t-1}\tau^k g(\tau)d\tau + \int_{t-1}^{t}g(\tau)d\tau,$$

where the last term tends to zero as $t\to+\infty$, and

$$\int_{t_0}^{+\infty}\tau^k g(\tau)d\tau < +\infty, \qquad (0 \leqq k \leqq m-1),$$

by the hypothesis (2.532) of our theorem. Thus

$$\|Z_1(-t)f(x(t),t)\| \leqq nc_1c_3\|X_0\|t^{m+p-2}\, e^{\alpha t}\, g(t)c_5,$$

or, combining all the constants,

$$\|Z_1(-t)f(x(t), t)\| \leqq ct^{m+p-2}\, e^{\alpha t}\, g(t) \qquad (\alpha \geqq 0).$$

Integrating this last inequality and using hypothesis (2.532) we obtain

$$(2.543)\quad v_1(x^{(0)}, t_0) \leqq \int_{t_0}^{\infty}\|Z_1(-\tau)f(x(\tau),\tau)\|d\tau \leqq c\int_{t_0}^{\infty}\tau^{m+p-2}e^{\alpha\tau}g(\tau)d\tau < +\infty,$$

by hypothesis (2.532).

Since $v_1(x^{(0)}, t_0)$ in (2.542) is finite, it follows from (2.541) that

$$x(t) - z(t) = -\int_t^{+\infty} Z_1(t-\tau) f(x(\tau), \tau) d\tau + \int_{t_0}^t Z_2(t-\tau) f(x(\tau), \tau) d\tau, \tag{2.544}$$

where

$$z(t) = Z(t - t_0)(x^{(0)} + Z_1(t_0) v_1(x^{(0)}, t_0)) \tag{2.545}$$

is a solution of (2.506).

We observe that for $\tau \geqq t$

$$||Z_1(t - \tau)|| \leqq c_1(\tau - t)^{p-1} \leqq c_1 \tau^{p-1},$$
$$||Z_1(t - \tau) f(x(\tau), \tau)|| \leqq c_{15} \tau^{m+p-2} e^{\alpha t} g(t)$$

and hence

$$\int_t^{+\infty} ||Z_1(t - \tau) f(x(\tau), \tau)|| d\tau \to 0 \quad \text{as} \quad t \to +\infty \tag{2.546}$$

in view of (2.532).

The second integral occurs only if there exist characteristic roots with negative real parts, i.e., if there exist solutions of (2.508) which tend to zero as $t \to +\infty$.

Making use of (2.519) and (2.504), we obtain

$$\begin{cases} \int_{t_0}^t ||Z_2(t - \tau) f(x(\tau), \tau)|| d\tau \leqq c_8 \int_{t_0}^t e^{\alpha^*(t-\tau)} (t - \tau)^{m^*-1} \tau^{m-1} e^{\alpha\tau} g(\tau) d\tau \\ \leqq c_9 e^{\alpha^* \frac{t}{2}} t^{m^*-1} \int_{t_0}^{\frac{t}{2}} \tau^{m-1} e^{\alpha\tau} g(\tau) d\tau + c_{10} \int_{t/2}^t \tau^{m-1} e^{\alpha t} g(\tau) d\tau, \end{cases} \tag{2.547}$$

for sufficiently large values of r. Here

$$c_9 = 2^{1-m^*} c_8$$

and

$$c_{10} = c_8 \max_{t_0 \leqq \tau \leqq t} (e^{\alpha^*(t-\tau)} (t - \tau)^{m^*-1}).$$

This maximum is attained for $\tau = t + (m^* - 1)/a^*$ which lies in the interval $t/2 \leqq \tau \leqq t$, when t is sufficiently large.

Since $\alpha^* < 0$, we see, in view of the hypothesis (2.532), that both terms of the right-hand member of the last inequality $\to 0$ as $t \to +\infty$. This result, together with (2.546), shows that for each solution $x(t)$ of (2.506), given by (2.541), and for the corresponding solution (2.545) of (2.508), we have

$$||x(t) - z(t)|| \to 0 \quad \text{as} \quad t \to +\infty. \tag{2.548}$$

More precisely, it follows from the discussion in this section that

$$(2.549)\qquad \|x(t) - z(t)\| = O\Big(\int_t^\infty \tau^{m+p-2} e^{\alpha\tau} g(\tau)d\tau\Big) + O\Big(\int_{t/2}^t \tau^{m-1} e^{\alpha\tau} g(\tau)d\tau\Big) + O(e^{\frac{\alpha^*}{2}t} t^{m^*-1}),$$

where the last two terms occur only if B has a characteristic root with negative real part.

If all the solutions of (2.508) are bounded, i.e., if $\alpha = 0$ and $m = p = 1$, we have the case considered by Weyl [57].

2.55. We have just shown that to every solution $x(t, x^{(0)}, t_0)$ of the perturbed system (2.506), there corresponds the solution $z(t, x^{(0)} + Z_1(t_0)v_1(x^{(0)}, t_0), t_0)$ of the unperturbed linear system (2.508), and that each pair of corresponding solutions satisfies (2.548).

Next, we shall show that to every solution $z(t, z^{(0)}, t_0)$ of (2.508), there corresponds the solution $x(t, z^{(0)} - Z_1(t_0)v_1(x^{(0)}, t_0), t_0)$ of (2.506) and that these solutions satisfy (2.548). It suffices to show that for every solution $z(t, z^{(0)}, t_0)$ we can construct a solution $x(t)$ of (2.544). Such a solution $x(t)$ of (2.544) will have the desired initial values, will satisfy (2.506), and, together with the given solution of (2.508), will satisfy (2.548).

We shall apply a method of successive approximations. Let

$$(2.551)\qquad \begin{cases} T(y(t)) = z(t) - \int_t^\infty Z_1(t-\tau)f(y(\tau), \tau)d\tau + \int_{t_0}^t Z_2(t-\tau)f(y(\tau), \tau)d\tau, \\ t \geqq t_0 \text{ for a suitably chosen } t_0,\ z(t) \text{ a fixed solution of (2.508).} \end{cases}$$

The transform $y^{(1)}(t) = T(y(t))$ is well defined for every $y(t)$ for which the integral

$$(2.552)\qquad \int_t^\infty Z_1(t-\tau)f(y(\tau), \tau)d\tau$$

converges. We proved that, under the hypotheses of our theorem, this is true for all the solutions of (2.506). Using the estimates in (2.519), we can easily show that (2.552) converges for every solution $y(t)$ of (2.508). In general (2.552) converges for every $y(t)$ for which

$$(2.553)\qquad \|y(t)\| \leqq c_{11} t^{m-1} e^{\alpha t} \qquad (t \geqq t_0),$$

in view of (2.5421) with $x(t)$ replaced by $y(t)$, and the hypothesis (2.532). Also, (2.553) implies that

$$||T(y(t))|| \leqq c_{12} t^{m-1} e^{\alpha t} \qquad (t \geqq t_0),$$

since

$$||T(y(t))|| \leqq 2||z(t)|| \leqq 2c_0 t^{m-1} e^{\alpha t},$$

by (2.551), (2.547), (2.546), and (2.519). Thus T can be iterated indefinitely.

If we have

$$(2.554) \quad \Delta_k(t) = ||T^k(y(t)) - T^{k-1}(y(t))|| < a_k, \qquad a_k \text{— a constant},$$

then

$$\Delta_{k+1}(t) \leqq c_1 \int_t^\infty \tau^{p-1} g(\tau) \Delta_k(\tau) d\tau + c_2 \int_{t_0}^t e^{\alpha^*(t-\tau)} (t-\tau)^{m^*-1} g(\tau) \Delta_k(\tau) d\tau$$

$$\leqq a_k [c_1 \int_t^\infty \tau^{p-1} g(\tau) d\tau + c_2 \int_{t_0}^t e^{\alpha^*(t-\tau)} (t-\tau)^{m^*-1} g(\tau) d\tau] = a_k S(t, t_0).$$

The discussion in Section 2.54 implies that we can choose t_0 large enough, so that

$$(2.555) \qquad S(t, t_0) \leqq \beta < 1 \quad \text{for} \quad t \geqq t_0.$$

The choice of t_0 depends only upon the characterizing constants of the system (2.508) and upon the perturbation $f(z, t)$ in (2.506), through its bounding factor $g(t)$. The choice of t_0 as in (2.555) implies that

$$(2.556) \qquad \Delta_{k+1} \leqq a_k \beta, \quad \Delta_{k+r} \leqq a_k \beta^r \qquad (r = 1, 2, \ldots).$$

In particular if we let $y(t) = z(t)$ in (2.551), then

$$\Delta_1(t) = ||T(z(t)) - z(t)|| \to 0 \quad \text{as} \quad t \to +\infty,$$

and there exists a constant a_1 such that (2.554) holds for $k = 1$. Then

$$\Delta_k \leqq a_1 \beta^{k-1} \quad (k = 1, 2, \ldots),$$

and the series

$$z(t) + [T(z(t)) - z(t)] + [T^2(z(t)) - T(z(t))] + \ldots \qquad (t \geqq t_0),$$

is majorized by a convergent geometric series and hence converges uniformly to a vector $x(t)$ which, in view of (2.551), satisfies (2.544). The solution $x(t)$ constructed in this manner may be extended to values $t < t_0$.

2.56. For a fixed $z(t)$ there is but one $x(t)$ satisfying (2.544). For, let $x(t)$ and $\tilde{x}(t)$ satisfy (2.544). Then

$$\Delta(t) = ||x(t) - \tilde{x}(t)|| \leqq ||x(t) - z(t)|| + ||\tilde{x}(t) - z(t)|| \to 0 \quad \text{as} \quad t \to +\infty$$

and hence $\Delta(t)$ is bounded, say

$$\Delta(t) \leqq a \qquad (t \geqq t_0).$$

Next, by the now familiar argument,

$$\Delta(t) \leqq aS(t, t_0) \leqq a\beta, \qquad \beta < 1, \quad t \geqq t_0.$$

Repeated application of this argument yields

$$\Delta(t) \leqq a\beta^k \qquad (k = 1, 2, \ldots),$$

whence

$$\Delta(t) = 0 \quad \text{or} \quad x(t) = \tilde{x}(t) \quad \text{for} \quad t \geqq t_0,$$

and by the uniqueness of solutions $x(t) = \tilde{x}(t)$ for all t.

2.57. For $t = t_0$ the uniquely determined solution $x(t)$ of (2.506), takes the value

$$\begin{aligned} x^0 = x(t_0) &= z^0 - Z_1(t_0) \int_{t_0}^{+\infty} Z_1(-\tau) f(x(\tau), \tau) d\tau \\ &= z^0 - Z_1(t_0) v_1(x^{(0)}, t_0), \end{aligned} \tag{2.571}$$

in view of (2.544) and (2.542). Comparing this with (2.545), we see that the one-to-one correspondence which we established between the solutions of (2.506) and those of (2.508) may be described in terms of the one-to-one correspondence, given by (2.571), between their values at t_0. We shall show that this correspondence is bicontinuous.

We consider two solutions $x(t, x^{(0)}, t_0)$ and $\tilde{x}(t, \tilde{x}^{(0)}, t_0)$ of (2.506). As in 2.52 and 2.54, we show that

$$||x(t) - \tilde{x}(t)|| \leqq c_{12} t^{m-1} e^{\alpha t} ||x^{(0)} - \tilde{x}^{(0)}||. \tag{2.572}$$

Then, by (2.542) and (2.503),

$$||v_1(x^{(0)}, t_0) - v_1(\tilde{x}^{(0)}, t_0)|| \leqq c_7 || x^{(0)} - \tilde{x}^{(0)}|| \int_{t_0}^{\infty} t^{m+p-2} e^{\alpha t} g(t) dt.$$

This proves the continuity in $x^{(0)}$ of $v_1(x^{(0)}, t_0)$ and hence of $z^{(0)}$.

To prove that $x^{(0)}$ is continuous in $z^{(0)}$, we note that

$$\begin{aligned} x(t_0) - \tilde{x}(t_0) = z(t_0) - \tilde{z}(t_0) &- \int_{t_0}^{+\infty} Z_1(t_0 - \tau)[f(x(\tau), \tau) - f(\tilde{x}(\tau), \tau)] d\tau \\ &+ \int_{t_0}^{t} Z_2(t - \tau)[f(x(\tau), \tau) - f(\tilde{x}(\tau), \tau)] d\tau. \end{aligned}$$

Therefore

$$\|x(t_0) - \tilde{x}(t_0)\| \leqq \|z(t_0) - \tilde{z}(t_0)\| + \delta\|x(t_0) - \tilde{x}(t_0)\|, \tag{2.573}$$

where

$$1 > \delta = c_7 c_{12}\Big[\int_{t_0}^{+\infty} \|Z_1(t_0 - \tau)\| g(\tau)\tau^{m-1} e^{\alpha\tau}\, d\tau + \int_{t_0}^{t} \|Z_2(t_0 - \tau)\| g(\tau)\tau^{m-1} e^{\alpha\tau} d\tau\Big]$$

by a suitable choice of t_0. Then, (2.573) becomes

$$\|x(t_0) - \tilde{x}(t_0)\| \leqq \frac{1}{1-\delta} \|z(t_0) - \tilde{z}(t_0)\|$$

which assures the desired continuity property.

2.58. If $\alpha = \alpha^* < 0$, then, in case $m^* = 1$, (2.528) becomes

$$\|x(t)\| \leqq c_{14}\, e^{\alpha^* t + c'_4 \int_{t_0}^{t} g(\tau)d\tau}. \tag{2.581}$$

And we see at once that (2.534) implies that every solution $x(t)$ of (2.506) also tends to zero as $t \to +\infty$.

In order to reduce the case $m^* > 1$ to that in which $m^* = 1$ we augment the original systems by adjoining to them the equations

$$\frac{dx_{n+1}}{dt} = (\alpha + \varepsilon)x_{n+1}, \quad \frac{dz_{n+1}}{dt} = (\alpha + \varepsilon)z_{n+1},$$

where $\alpha < \alpha + \varepsilon < 0$. The new linear system has the matrix

$$\tilde{A} = \begin{bmatrix} A, & 0 \\ 0, & \lambda + \varepsilon \end{bmatrix}$$

and for it $\tilde{a} = \alpha + \varepsilon$, $\tilde{m} = \tilde{p} = 1$ and every solution $\tilde{x}(t)$ of the augmented system satisfies

$$\|\tilde{x}(t)\| \leqq c_{15} e^{(\alpha+\varepsilon)t + c''_4 \int_{t_0}^{t} g(\tau)d\tau}. \tag{2.582}$$

We observe that again (2.534) implies that $\tilde{x}(t)$ tends to zero as $t \to +\infty$ and hence the same is true for $x(t)$.

2.583. Corollary. *Convergence of*

$$\int_{t_0}^{\infty} g(\tau)d\tau$$

implies (2.534), *and hence assures that every solution* $x(t)$ *of* (2.506) *tends to zero in case* $\alpha < 0$.

2.584. Each of the conditions

$$\alpha^* t + c_4' \int_{t_0}^{t} g(\tau)d\tau \to -\infty \quad \text{as} \quad t \to +\infty \qquad (m = 1) \tag{2.5841}$$

and

$$\left\{ \begin{array}{l} (\alpha^* + \varepsilon)t + c_4'' \int_{t_0}^{t} g(\tau)d\tau \to -\infty \quad \text{as} \quad t \to +\infty \\ \qquad \textit{for some } \varepsilon > 0 \textit{ in case } m > 1 \end{array} \right. \tag{2.5842}$$

may be used in place of (2.534) in the appropriate cases and each of these conditions is weaker than (2.534).

2.585. We shall underline the geometrical meaning of Theorem 2.531. Suppose, for example, that there is a k-dimensional hyperplane P_k of initial values of (2.508), all of whose points determine solutions asymptotic to periodic solutions. If we consider the trajectories $(z(t), t)$ in the $(n + 1)$-dimensional space, defined by solutions $z(t)$ of (2.508), then P_k lies in the n-dimensional hyperplane $t = t_0$ and our theorem asserts that there exists a k-dimensional manifold homeomorphic to P_k in the same hyperplane $t = t_0$, all of whose points determine integral curves of (2.506) which for $t \to +\infty$ tend toward the curves

$$z_1 = \varphi_1(\tau), \ldots, z_n = \varphi_1(\tau), \qquad t = \tau$$

where $\varphi_i(\tau)$ are periodic in τ.

The above statement remains true if the solutions which originate on P_k are themselves periodic.

We should note that if (2.508) has integral curves of the saddle type, then it does not follow from our theorem that (2.506) has integral curves of that type. On the other hand, if (2.506) has a k-dimensional manifold of bounded solutions, then the same is true of (2.508).

It is of interest to consider the special cases when $\alpha < 0$ or $\alpha = 0$. This leads to conditions for the Lyapunov stability of the perturbed system.

2.586. COROLLARY. *Consider a system* (2.506) *for which* (2.502) *holds. The trivial solution of this system is asymptotically stable if we have* (2.503), *if* $\alpha < 0$, *and if* (2.534) *holds true.*

2.587. COROLLARY. *Consider* (2.506) *and assume* (2.502). *If we have* (2.503), *if* $\alpha \leqq 0$, $m = p = 1$, *and if*

$$\int_{t_0}^{t} g(\tau)d\tau$$

converges, then the trivial solution of (2.506) *is stable according to Lyapunov.*

CHAPTER IV

A Study of Neighborhoods of Singluar Points and of Periodic Solutions of Systems of n Differential Equations.

1. Singular Points in the Analytic Case

In this section we shall study systems of the form

$$\frac{dx}{dt} = A(t)x + \varphi(x, t), \ A(t) = (a_{ij}(t)), \tag{1.01}$$

where the modulus $||A(t)||$ is bounded for $t \geqq t_0$ or for $t \leqq t_0$, where each component $\varphi_i(x, t) = \varphi_i(x_1, \ldots, x_n, t)$ of the column vector $\varphi(x, t)$, is a power series in $x_1, \ldots, x_n$ with bounded (for $t \geqq t_0$ or $t \leqq t_0$) functions of t as coefficients. Moreover, we shall assume that the expansions of the $\varphi_i(x, t)$ contain no linear terms.

In view of III, 1.2 we can make these assumptions without loss of generality.

We begin with the case when $A = (a_{ij})$ is a constant matrix and when each of the expansions $\varphi_i(x, t)$ has constant coefficients and contains no linear or constant terms. To indicate this last property explicitly we employ the notation

$$\frac{dx}{dt} = Ax + [x_1, x_2, \ldots, x_n]_2, \qquad A = (a_{ij}). \tag{1.02}$$

We shall study the behavior of integral curves in the neighborhood of a singular point. The basic results here were obtained by H. Poincaré, E. Picard, and A. M. Lyapunov.

In what follows we shall allow the variables $x_1, \ldots, x_n$ as well as the coefficients a_{ij}, to assume complex values.

We seek integrals of our system (1.02), which are expressed in the neighborhood of the origin in the form

$$x_i = g_i(y) = g_i(y_1, \ldots, y_n) \qquad (i = 1, \ldots, n) \tag{1.03}$$

where the g_i are analytic and y is a solution of the associated linear system

$$\frac{dy}{dt} = Ay. \tag{1.04}$$

We shall show that under quite general conditions it is possible to transform a given system by means of an analytic transformation into a system which can be integrated by the method of successive approximations. By studying the behavior of the integral curves of the transformed system we discover all the properties of the original system which remain invariant under an analytic transformation. By this method we can study topological properties of the original family of integral curves as well as the order of contact of integral curves with various axes at the origin. Thus, the method of analytic transformations yields information regarding all those properties of solutions with which the qualitative study of systems of n differential equations concerns itself.

We shall first discuss the existence of formal power series integrals and then study the convergency of the formal power series. The results that follow are due to Poincaré, Picard, and Dulac.

1.1. Formal integration. We assume at first that

1.101. all the elementary divisors of $A - \lambda I$ are linear, and that

1.102. there exists no relation

$$k_1\lambda_1 + \ldots + k_n\lambda_n = 0$$

between the characteristic roots of A, with integral coefficients k_i $(i = 1, \ldots, n)$ not all of which are zero.

1.12. THEOREM. *If the system* (1.02) *satisfies conditions* 1.101 *and* 1.102, *then there exists a formal series*

$$z_i = g_i(x_1, \ldots, x_n) \qquad (i = 1, \ldots, n), \tag{1.121}$$

by means of which (1.02) *is transformed into the form*

$$\frac{dz_i}{dt} = \lambda_i z_i \qquad (i = 1, \ldots, n). \tag{1.122}$$

We may assume at the outset that by means of a linear transformation (in general with complex coefficients) the original system (1.02) has been transformed into the canonical form

$$(1.123)\quad \frac{dx_i}{dt} = \lambda_i x_i + F_i(x_1, \ldots, x_n), \qquad (i = 1, 2, \ldots, n),$$

where $F_i(x_1, \ldots, x_n)$ are again power series in $x_1, \ldots, x_n$ lacking linear and contant terms.

We shall show that by means of a transformation

$$(1.124)\quad z_i = x_i + \varphi_{i2} + \varphi_{i3} + \ldots + \varphi_{iN} \qquad (i = 1, 2, \ldots, n),$$

where φ_{ik} $(i = 1, \ldots, N)$ are polynominals of degree k in $x_1, \ldots, x_n$, we may transform (1.123) into the form

$$(1.125)\quad \frac{dz_i}{dt} = \lambda_i z_i + \Phi_i^{(N)}(z_1, \ldots, z_n) \qquad (i = 1, 2, \ldots, n)$$

where $\Phi_i^N(z_1, \ldots, z_n)$ are power series which begin with terms of degree at least $N + 1$. Let

$$F_i(x_1, \ldots, x_n) = F_{i2} + F_{i3} + \ldots + F_{iN} + \ldots$$

where F_{ik} is the sum of the terms of degree k in F_i. In order that $\Phi_i^{(N)}$ in (1.125) should have the desired property we must have

$$(1.126)\quad \begin{cases} F_{i2} + \sum_{j=1}^{n} \frac{\partial \varphi_{i2}}{\partial x_j} \lambda_j x_j = \lambda_i \varphi_{i2}, \\ F_{i3} + \sum_{j=1}^{n} \left(\frac{\partial \varphi_{i3}}{\partial x_j} \lambda_j x_j + \frac{\partial \varphi_{i2}}{\partial x_j} F_{j2} \right) = \lambda_i \varphi_{i3}, \\ \cdots\cdots\cdots\cdots\cdots\cdots \\ F_{iN} + \sum_{j=1}^{n} \left(\frac{\partial \varphi_{iN}}{\partial x_j} \lambda_j x_j + \sum_{p+q=N+1} \frac{\partial \varphi_{ip}}{\partial x_j} F_{jq} \right) = \lambda_i \varphi_{iN}. \end{cases}$$

If $c_{i2} = c_{i2}(e_1, e_2, \ldots, e_n)$ is the coefficient of $x_1^{e_1} x_2^{e_2} \ldots x_n^{e_n}$ in φ_{i2} (i.e., $e_1 + e_2 + \ldots + e_n = 2$) and d_{i2} is the coefficient of the corresponding term in F_{i2}, then the first equality in (1.126) is equivalent to

$$d_{i2} + (e_1 \lambda_1 + \ldots + (e_i - 1)\lambda_i + \ldots + e_n \lambda_n) c_{i2} = 0.$$

Since the expression in the parentheses is not zero by 1.102, we can solve for c_{i2}.

Having determined the coefficients of φ_{i2} so that the first equation in (1.126) holds, we can determine the coefficients of φ_{i3} so that the second condition in (1.126) is fulfilled. Proceeding in this manner we determine the coefficients of $\varphi_{i2}, \ldots, \varphi_{ik-1}$ so that the first $k - 2$ equalities in (1.126) hold. If we rewrite the $(k - 1)$st equality in the form

$$(1.127) \quad \left(F_{ik} + \sum_{j=1}^{n} \sum_{p+q=k+1} \frac{\partial \varphi_{ip}}{\partial x_j} F_{jq}\right) + \sum_{j=1}^{n} \frac{\partial \varphi_{ik}}{\partial x_j} \lambda_j x_j = \lambda_i \varphi_{ik},$$

and note that since $q \geqq 2$ we have $p \leqq k - 1$ and hence the polynomial in the parentheses is known, we see that (1.127) is equivalent to

$$(1.127\text{a}) \quad d_{ik} + (e_1\lambda_1 + \ldots + (e_i - 1)\lambda_i + \ldots + e_n\lambda_n)c_{ik} = 0,$$

where $c_{ik} = c_{ik}(e_1, \ldots, e_n)$ is the coefficient of the term $x_1^{e_1} x_2^{e_2} \ldots x_n^{e_n}$ in φ_{ik}, i.e. $e_1 + \ldots + e_n = k$, and d_{ik} is the known coefficient of the corresponding term in the polynomial in the parentheses of (1.127). Thus we can determine all the coefficients of φ_{ik} so that (1.127) holds.

Completing this process, we determine the coefficients of $\varphi_{i1}, \ldots, \varphi_{iN}$ so that (1.126) holds, and hence so that the transformation (1.124) yields a system (1.125) with the desired properties.

The value of N in (1.124) and hence in (1.125) may be taken arbitrarily large. The coefficients of the polynomials $\varphi_{i2}, \varphi_{i3}, \ldots, \varphi_{ik}$, for $k < N$, are determined consecutively and are independent of N. Continuing this process indefinitely we obtain a sequence of polynomials φ_{ik} $(k = 2, 3, \ldots)$ such that the transformation

$$(1.128) \qquad z_i = g_i(x_1, \ldots, x_n) = x_i + \sum_{k=2}^{+\infty} \varphi_{ik}(x)$$

carries (1.123) into the desired system (1.122).

Also from adding equations (1.126) it follows that the g_i in (1.128) formally satisfy the partial differential equations

$$(1.129) \quad (\lambda_1 x_1 + F_1)\frac{\partial g_i}{\partial x_1} + \ldots + (\lambda_n x_n + F_n)\frac{\partial g_i}{\partial x_n} = \lambda_i g_i \; (i = 1, \ldots, n).$$

If $x(t)$ is a solution of (1.123), then by (1.129)

$$\frac{dg_i(x)}{dt} = \sum_{j=1}^{n} \frac{\partial g_i(x)}{\partial x_j} \frac{dx_j}{dt} = \sum_{j=1}^{n} \frac{\partial g_i(x)}{\partial x_j} (\lambda_j x_j + F_j) = \lambda_i g_i(x),$$

and hence $z(t) = g(x(t))$ is a solution of (1.122).

1.13. We note that

$$(1.131) \qquad \frac{1}{p_1! \ldots p_n!} \left[\frac{\partial^k g_i(x)}{\partial x_1^{p_1} \partial x_2^{p_2} \ldots \partial x_n^{p_n}} \right]_{x_i=0} = c_{ik}(p_1, \ldots, p_n), \quad k = p_1 + \ldots + p_n,$$

where c_{ik} is the coefficient of the term $x_1^{p_1} x_2^{p_2} \ldots x_n^{p_n}$. Then, we see at once that if $g_i(x_1, \ldots, x_n)$ satisfies (1.129), the constant term and the coefficients of all the linear terms in the expansion of $g_i(x_1, \ldots, x_n)$ must all be zero with the sole exception of the coefficient of x_i. This last is not determined by (1.129) and may be chosen arbitrarily.

1.2. Conditions for convergence. If the formal power series (1.128) converge, then they will yield qualitative information about the behavior of the integral curves near the origin. Certainly if there exists an analytic solution, it will be given by these series in view of uniqueness.

1.21. A basic condition. *In the complex plane, the roots* $\lambda_1, \ldots, \lambda_n$ *of our characteristic equation must lie on one side of some straight line through the origin.*

This condition is assumed to hold in what follows.

If the coefficients of our equation are real then this condition implies that the real parts α of our roots cannot vanish (no pure imaginary roots), and, moreover, that all these real parts α must be of the same sign.

Condition 1.21 can be restated in the following useful form.

1.22. CONDITION. *The convex closure of the characteristic roots* $\lambda_1, \ldots, \lambda_n$ *does not contain the origin.*

1.23. Let $p_1, \ldots, p_n$ be a set of non-negative integers, not all equal to zero, and let

$$(1.231) \quad p = p_1 + \ldots + p_n, \qquad \sigma(\lambda) = p_1\lambda_1 + \ldots + p_n\lambda_n.$$

Then

$$(1.232) \qquad \left| \frac{\sigma(\lambda)}{p} \right| \geqq \delta > 0,$$

for the expression on the left is the modulus of the center of gravity of the masses $p_1, \ldots, p_n$ placed at the points $\lambda_1, \ldots, \lambda_n$ respectively. We assert also that,

$$\left|\frac{\sigma(\lambda)-\lambda_i}{p-1}\right| \geqq \varepsilon > 0 \quad \text{if} \quad p > 1 \quad (i = 1, \ldots, n). \tag{1.233}$$

To see this we write

$$\frac{\sigma(\lambda)-\lambda_i}{p-1} = \frac{\dfrac{\sigma(\lambda)}{p} - \dfrac{\lambda_i}{p}}{1 - \dfrac{1}{p}}. \tag{1.234}$$

As p becomes large the second term in both the numerator and the denominator $\to 0$ and the quotient (1.234) $\to \sigma(\lambda)/p$. Thus in view of (1.232) for all values $p_1, \ldots, p_n$ for which p is greater than a sufficiently large c, the quotient (1.234) is greater in absolute value than a positive number ε_1. On the other hand the smallest one in absolute value among the finite number of values of (1.234) for $1 < p \leqq c$, exceeds by (1.102) some $\varepsilon_2 > 0$. Thus, (1.233) holds with $\varepsilon = \min(\varepsilon_1, \varepsilon_2)$.

1.24. We are now ready to consider the problem of convergence of our formal series in the complex plane.

In view of 1.13, we let

$$\begin{gathered} g_i(x) = bx_i + v(x), \quad v(x) = \sum_{p \geqq 2} c_{p_1 \ldots p_n} x_1^{p_1} \ldots x_n^{p_n} \\ p = p_1 + \ldots + p_n, \quad p_j \geqq 0, \quad j = 1, 2, \ldots, n, \end{gathered} \tag{1.241}$$

be a formal integral of (1.129). Here b is an arbitrary constant and $v(x)$ is a power series without linear or constant terms. Substituting (1.241) into (1.129), we get

$$\begin{aligned} \lambda_1 x_1 \frac{\partial v}{\partial x_1} + \ldots + \lambda_n x_n \frac{\partial v}{\partial x_n} - \lambda_i v = -F_1 \frac{\partial v}{\partial x_1} - \ldots \\ - F_n \frac{\partial v}{\partial x_n} - bF_i, \end{aligned} \tag{1.242}$$

where the F_j are defined in (1.123).

Next let,

$$|F_j(x)| \leqq M \quad \text{when} \quad |x_j| \leqq a, \tag{1.243}$$

and consider an auxiliary equation

$$(1.244)\qquad \varepsilon\left(x_1\frac{\partial V}{\partial x_1}+\ldots+x_n\frac{\partial V}{\partial x_n}-V\right)$$
$$=\left[\frac{M}{1-(X/a)}-M-\frac{MX}{a}\right]\left(\frac{\partial V}{\partial x_1}+\ldots+\frac{\partial V}{\partial x_n}+b\right)$$

where $X=x_1+\ldots+x_n$, and $\varepsilon>0$ satisfies (1.233).

Suppose that (1.244) has an integral V which vanishes at the origin together with its first partial derivatives and write

$$(1.245)\quad V=\sum_{p\geqq 2}\tilde{c}_{p_1\ldots p_n}x_1^{p_1}\ldots x_n^{p_n},\ p=p_1+\ldots+p_n,\ p_i\geqq 0,\quad i=1,2,\ldots,n.$$

If we let

$$(1.246)\quad F_j(x)=\sum_{r\geqq 2}\theta^{(j)}_{r_1\ldots r_n}x_1^{r_1}\ldots x_n^{r_n},\ r=r_1+\ldots+r_n,\quad r_i\geqq 0,\ i=1,2,\ldots,n,$$

then (1.242) implies (cf. (1.231))

$$(1.247)\quad (\sigma(\lambda)-\lambda_i)c_{p_1p_2\ldots p_n}=d_{p_1p_2\ldots p_n}(\theta^{(j)}_{r_1\ldots r_n},c_{q_1\ldots q_n})$$
$$q=q_1+\ldots+q_n\leqq p-2,\quad j=1,\ldots,n,$$

where $d_{p_1\ldots p_n}(\theta^{(j)}_{r_1\ldots r_n},c_{q_1\ldots q_n})$ is linear homogeneous in the (given) coefficients $\theta^{(j)}_{r_1\ldots r_n}$ of the F_j (cf. 1.123) in (1.246)), and is of the first degree in the coefficients $c_{q_1\ldots q_n}$ $(q\leqq p-2)$ of $v(x)$ in (1.241).

On the other hand, from (1.244), it follows that

$$(1.248)\quad \varepsilon(p-1)\tilde{c}_{p_1\ldots p_n}=\tilde{d}_{p_1\ldots p_n}(\mu_{r_1\ldots r_n},\tilde{c}_{q_1\ldots q_n})$$
$$p=p_1+\ldots+p_n\geqq 2,\quad q\leqq p-2$$

where the expression $\tilde{d}_{p_1\ldots p_n}$ may be obtained from $d_{p_1\ldots p_n}$ by replacing each coefficient $\theta^{(j)}_{r_1\ldots r_n}$ of $F_j(j=1,\ldots,n)$ by the coefficient $\mu_{r_1\ldots r_n}$ of the corresponding term in the common majorante (see Picard [45])

$$(1.249)\quad \frac{M}{1-\dfrac{X}{a}}-M-\frac{MX}{a}=M\left(\frac{X^2}{a^2}+\frac{X^3}{a^3}+\ldots\right)$$

of all the F_j, and each coefficient $c_{q_1\ldots q_n}$ of $v(x)$, by the corresponding coefficient $\tilde{c}_{q_1\ldots q_n}$ in (1.245).

$$\frac{dx}{dt} = A(t)x, \qquad A(t) = (a_{ij}(t)) \tag{1.291}$$

of the first approximation, also be real, continuous, and bounded for $t \geqq t_0$. *Let* (1.291) *be regular, and let* $\mu_1, \ldots, \mu_n$ *be its characteristic numbers. Choose* k *of these characteristic numbers, say* $\mu_1, \ldots, \mu_k$. *There exist formal power series solutions of* (1.01)

$$x_s = \sum L_s^{(m_1, \ldots, m_k)} \alpha_1^{m_1} \alpha_2^{m_2} \ldots \alpha_k^{m_k} e^{-\sum_{i=1}^{k} m_i \mu_i t} \tag{1.292}$$
$$(s = 1, 2, \ldots, n)$$

in k *parameters* $\alpha_1, \ldots, \alpha_k$, *in which* $L_s^{m_1, \ldots, m_k}$ *are continuous in* t, *are independent of the parameters* α_i, *and are such that the characteristic numbers* $\mu(L_s^{(m_1, \ldots, m_k)}, e^t) \geqq 0$. *Here, the summation is extended over all sets* $m_1, \ldots, m_k$ *with*

$$m = m_1 + \ldots + m_k > 0, \quad m_i \geqq 0 \quad (i = 1, \ldots, k).$$

If all the selected characteristic numbers $\mu_1, \ldots, \mu_k$ *are positive, then, as long as the absolute values of* $\alpha_1, \ldots, \alpha_k$ *do not exceed certain fixed bounds, the series* (1.292) *are absolutely convergent and represent functions which satisfy* (1.01) *for* $t \geqq t_0$.

We shall not prove this interesting Theorem, since the qualitative properties of (1.01) which follow from it may be deduced from the more general Theorems which are to be considered below.

Lyapunov also investigated the case when A has purely imaginary characteristic roots and established a criterion for the existence of periodic solutions for a rather wide class of analytic differential equations.

1.3. The case of purely imaginary characteristic roots. Theorem of Lyapunov. We shall suppose that our system is given in the canonical form

$$\frac{dx_i}{dt} = -\frac{\partial H}{\partial y_i}, \quad \frac{dy_i}{dt} = \frac{\partial H}{\partial x_i} \quad (i = 1, \ldots, n) \tag{1.301}$$

where $H = H(x_1, \ldots, x_n, y_1, \ldots, y_n)$ does not depend on t, and vanishes together with its first partial derivatives at $x_1 = 0, \ldots, y_n = 0$. Moreover, we shall assume:

1.302. The function H is holomorphic in a neighborhood of the origin, i.e., it can be expanded into a power series in $x_1, \ldots, y_n$, which converges as long as the absolute values of the variables do not exceed certain fixed bounds.

1.303. The above expansion of H lacks linear and constant terms, but the coefficients of the second degree terms do not all vanish.

A function H which has the properties **1.302** and **1.303** is called *nonsingular.*

1.31. THEOREM (Lyapunov). *Consider a canonical system* (1.301) *with a nonsingular* H. *Let the matrix* A *formed by the linear terms of* (1.301), *have* k *pairs* $\pm\nu_1 i, \ldots, \pm\nu_k i$ *of purely imaginary roots and suppose that no ratio* ν_j/ν_s $(j \neq s)$ *is an integer. Then our system* (1.301) *admits* k *distinct one-parameter families of periodic solutions, and all the parameters are real.*

To prove this deep result we first establish a sequence of lemmas which are of interest in themselves for they throw considerable light upon the conditions under which we may forego the requirement that our system should be in the canonical form (1.301).

1.32. LEMMA. *Consider a system*

$$\frac{dr}{d\vartheta} = R, \quad \frac{dz_s}{d\vartheta} = q_{s1}z_1 + \ldots + q_{sn}z_n + \varphi_s(\vartheta)r + Z_s \tag{1.321}$$

$$(s = 1, 2, \ldots, n),$$

where $Q = (q_{sj})$ *is a constant matrix,* $\varphi_s(\vartheta)$ *is a linear form in* $\sin\vartheta$ *and* $\cos\vartheta$, R *and* Z_s *are power series in* $r, z_1, \ldots, z_n$ *which lack terms of degree less than two and whose coefficients are periodic functions of* ϑ *of period* 2π. *We assume that each of the power series* R *and* Z_s *has a majorante power series with constant coefficients, which converges as long as* $|r|$ *and* $|z_s|$ $(s = 1, \ldots, n)$ *do not exceed a fixed bound* b. *Finally, we suppose that* Q *has no purely imaginary roots* μi *with integral* μ. *Then, we have the following two alternatives:*

1.322. *either the system* (1.321) *has a family of periodic solutions of period* 2π, *which are represented by power series*

$$\begin{aligned} z_s &= u_s^{(1)}(\vartheta)c + u_s^{(2)}(\vartheta)c^2 + \ldots \quad (s = 1, \ldots, n), \\ r &= c + u^{(2)}(\vartheta)c^2 + \ldots, \end{aligned} \tag{1.322}$$

which converge for $|c| \leqq \varepsilon$, *and where* $u^{(j)}(\vartheta)$ *and* $u_s^{(j)}(\vartheta)$ *are periodic functions of* ϑ *of period* 2π;

1.323. *or, if we substitute the power series in* **1.322** *into* (1.321), *and compare the coefficients of the same powers of c, we find that for some* $m \geqq 1$,

$$u^{(m)}(\vartheta) = g\vartheta + v(\vartheta),$$

where g is a nonvanishing constant and $v(\vartheta)$ *is a periodic function of period* 2π, *whereas all* $u^{(j)}(\vartheta)$ *and* $u_s^{(j)}(\vartheta)$ *with* $j < m$, *are periodic functions of* ϑ.

1.33. We may assume that by a suitable linear transformation, the system (1.321) has been brought into the form in which its linear approximation

$$\text{(1.331)} \qquad \frac{dz}{d\vartheta} = Qz$$

is in the Jordan canonical form. We may assume, therefore, that all the coefficients q_{sj} in (1.321) are zero with the possible exception of

$$\text{(1.332)} \qquad q_{11} = \lambda_1, \ldots, q_{nn} = \lambda_n, \quad q_{21} = \sigma_1, \ldots, q_{n,\, n-1} = \sigma_{n-1},$$

where the λ_j are the characteristic roots of Q, and $\sigma_h = 1$ or 0. Thus (1.321) may be written in the form

$$\text{(1.333)} \qquad \left\{\begin{aligned} \frac{dr}{d\vartheta} &= R, \\ \frac{dz_1}{d\vartheta} &= \lambda_1 z_1 + \varphi_1(\vartheta)r + Z_1, \\ \frac{dz_2}{d\vartheta} &= \sigma_1 z_1 + \lambda_2 z_2 + \varphi_2(\vartheta)r + Z_2, \\ &\cdots\cdots\cdots\cdots\cdots\cdots \\ \frac{dz_n}{d\vartheta} &= \sigma_{n-1} z_{n-1} + \lambda_n z_n + \varphi_n(\vartheta)r + Z_n. \end{aligned}\right.$$

We seek solutions of this system of the form of power series in 1.322. Substituting these formal power series into (1.333), and equating the coefficients of the like powers of c in both members of the resulting equalities, we obtain the following differential equations for the desired coefficients $u^{(j)}$ and $u_s^{(j)}$ $(s = 1, \ldots, n)$:

$$
(1.334)\qquad \begin{cases} \dfrac{du^{(j)}}{d\vartheta} = U^{(j)}(\vartheta) \\[2mm] \dfrac{du_1^{(j)}}{d\vartheta} = \gamma_1 u_1{}^{(j)} + \varphi_1(\vartheta)u^{(j)} + U_1^j(\vartheta) \\[2mm] \dfrac{du_s^{(j)}}{d\vartheta} = \sigma_{s-1}u_{s-1}^{(j)} + \gamma_s u_s^{(j)} + \varphi_s(\vartheta)u^{(j)} + U_s^{(j)}(\vartheta) \\[2mm] \qquad (s = 2, 3, \ldots, n;\ j = 1, 2, \ldots). \end{cases}
$$

Here the functions $U^{(j)}(\vartheta)$ and $U_s^{(j)}(\vartheta)$are either determined directly by the right-hand members of (1.333) or are expressed in terms of the functions $u^{(h)}$, $u_s^{(h)}$ with $h < j$. It follows from this observation that differential equations (1.334) can be integrated successively, first for $j = 1$, then for $j = 2$, etc.

Suppose that all the functions $u^{(j)}$, $u_s^{(j)}$ $(j < k)$ obtained through $k - 1$ successive integrations for $j = 1, \ldots, j = k - 1$, are periodic in ϑ of period 2π. Then all the functions $U^{(j)}$, $U_s^{(j)}$ $(j \leqq k)$ are also periodic in ϑ of period 2π. To see this we note that $U^{(j)}$ is the coefficient of c^j in the expansion of $R(r, z_1, \ldots, z_n)$ in powers of c. Since R, as a power series in $r, z_1, \ldots, z_n$, lacks linear and constant terms, and since the expansions of $r, z_1, \ldots, z_n$ in 1.322 lack constant terms, it follows that the coefficient of c^j $(j \leqq k)$ is a polynomial in $u^{(h)}$, $u_s^{(h)}$ with $h < j$ $(\leqq k)$. The coefficients of this polynomial are coefficients of R and hence are periodic in ϑ of period 2π. Then, since $u^{(h)}$, $u_s^{(h)}$ $(h < k)$ are assumed to be periodic, the function $U^{(j)}(\vartheta)$ is periodic for $j \leqq k$. A similar argument yields the periodicity of $U_s^{(j)}(\vartheta)$ for $s = 1, \ldots, n$ and $j \leqq k$.

To carry out the kth step of our process, we consider the system obtained from (1.334) by letting $j = k$. The first equation of the resulting system yields

$$
(1.335)\qquad u^{(k)}(\vartheta) = \int_0^\vartheta U^{(k)}(\tau)d\tau = g_k\vartheta + \Phi(\vartheta)
$$

where

$$
g_k = \frac{1}{2\pi}\int_0^{2\pi} U^{(k)}(\tau)d\tau
$$

and $\Phi(\vartheta)$ is periodic of period 2π.

If $g_k = 0$, $u^{(k)}(\vartheta)$ is periodic. Then all the remaining equations of (1.334) with $j = k$, are of the form

$$\frac{du_s^{(k)}}{d\vartheta} = \gamma_s u_s^{(k)} + f_{ks}(\vartheta), \tag{1.336}$$

where $f_{ks}(\vartheta)$ is periodic of period 2π. Since, by hypothesis, $\gamma \neq \mu i$ for integral μ, then

$$u_s^{(k)} = \frac{e^{\gamma_s\vartheta}}{e^{-2\pi\gamma_s} - 1}\int_\vartheta^{\vartheta+2\pi} e^{-\gamma_s\tau} f_{ks}(\tau)\,d\tau \tag{1.337}$$

satisfies (1.336) and is periodic of period 2π.

Let g_k in (1.335) be equal to zero for all k. Then our process yields formal series in **1.322** which satisfy (1.333). We shall show that these series do converge and hence represent a solution of the system (1.333).

1.34. In what follows we shall need information regarding the convergence of the series in **1.322** not only for real values of ϑ but also for $\vartheta = \alpha + \beta i$, where α is an arbitrary real number and β is sufficiently small in absolute value.

Making use of the more precise formulae for $f_{ks}(\vartheta)$ obtained by comparing (1.336) and (1.334), we rewrite (1.337) as

$$(1.341)\quad \begin{cases} u_1^{(j)} = \dfrac{e^{\gamma_1\vartheta}}{e^{-2\pi\gamma_1} - 1}\displaystyle\int_\vartheta^{\vartheta+2\pi} e^{-\gamma_1\tau}[\varphi_1(\tau)u^{(j)}(\tau) + U_1^{(j)}(\tau)]\,d\tau, \\ \cdots\cdots\cdots\cdots\cdots\cdots \\ u_s^{(j)} = \dfrac{e^{\gamma_s\vartheta}}{e^{-2\pi\gamma_s} - 1}\displaystyle\int_\vartheta^{\vartheta+2\pi} e^{-\gamma_s\tau}[\sigma_{s-1}u_{s-1}^{(j)}(\tau) + \varphi_s(\tau)u^{(j)}(\tau) + U_s^{(j)}(\tau)]\,d\tau. \end{cases}$$

We write

$$\gamma_s = \nu_s + \mu_s i.$$

Let

$$(1.342)\quad \begin{aligned} \varrho_s = \varrho(\nu_s, \mu_s) &= \frac{|\nu_s|\cdot|1 - e^{-2\pi\gamma_s}|}{|e^{-2\pi\nu_s} - 1|} = \\ &\frac{|\nu_s|}{|e^{-2\pi\nu_s} - 1|}\sqrt{1 - 2e^{-2\pi\nu_s}\cos 2\pi\mu_s + e^{-4\pi\nu_s}}, \end{aligned}$$

if $\nu_s \neq 0$, and $\varrho(0, \mu_s) = \lim_{\nu_s\to 0}\varrho(\nu_s, \mu_s) = |\sin \pi\mu_s|/\pi$. Since the case $\nu_s = 0$ and $\mu_s =$ an integer is excluded by the hypothesis of our lemma, we see that $\varrho_s = \varrho(\nu_s, \mu_s)$ never vanishes.

Denote by $v^{(j)}$ and $v_s^{(j)}$ the least upper bounds of $|u^{(j)}(\alpha + \beta i)|$ and $|u_s^{(j)}(\alpha + \beta i)|$ respectively, for

(1.343) $\quad -\infty < \alpha < +\infty$ and $|\beta| \leqq b$, b sufficiently small.

We may take $v^{(1)} = 1$. In view of the periodicity of $u^{(j)}(\vartheta)$ and $u_s^{(j)}(\vartheta)$ for real ϑ, the upper bounds $v^{(j)}$ and $v_s^{(j)}$ are finite. Let a_s be the least upper bound of $|\varphi_s(\alpha + \beta i)|$ in the domain (1.343). Finally, let $V^{(j)}$ and $V_t^{(j)}$ be the constants obtained from $U^{(j)}$ and $U_s^{(j)}$ by replacing in these latter the functions $u^{(h)}$, $u_s^{(h)}$ by their bounds $v^{(h)}$, $v_s^{(h)}$ respectively, and by replacing each coefficient in the U's by an upper bound of its absolute value. Then in view of (1.335)

(1.344) $\quad v^{(j)} \leqq qV^{(j)} = \tilde{v}^{(j)} \quad (j \geqq 1,\ q$ an absolute constant),

in view of (1.335), and in view of (1.341)

$$v_s^{(j)} \leqq \frac{e^{\nu_s\alpha - \mu_s\beta}}{|e^{-2\pi\gamma_s} - 1|}\int_\vartheta^{\vartheta+2\pi} |e^{-\gamma_s\tau}|\{|\sigma_{s-1}|v_{s-1}^{(j)} + a_s v^{(j)} + V_s^{(j)}\}d\tau$$
$$(s = 1, 2, \ldots, n;\ \sigma_0 = 0),$$

in view of (1.341). Simple calculations yield

$$(1.345)\quad v_s^{(j)} \leqq \frac{e^{\nu_s\alpha}}{|e^{-2\pi\gamma_s} - 1|}\int_\alpha^{\alpha+2\pi} e^{-\nu_s\tau}\{|\sigma_{s-1}|v_{s-1}^{(j)} + a_s v^{(j)} + V_s^{(j)}\}d\tau$$
$$(s = 1, 2, \ldots, n;\ \sigma_0 = 0),$$

and

$$v_1^{(j)} \leqq \frac{e^{\nu_1\alpha}}{|e^{-2\pi\gamma_1} - 1|}\int_\alpha^{\alpha+2\pi} e^{-\nu_1\tau}\{a_1 v^{(j)} + V_1^{(j)}\}d\tau$$

for the case $s = 1$.

Let $\tilde{v}^{(j)}$ and $\tilde{v}_s^{(j)}$ be defined by the relations

$$(1.346)\quad \begin{cases} \tilde{v}^{(j)} = qV^{(j)}, \\ \tilde{v}_s^{(j)} = \dfrac{e^{\nu_s\alpha}}{|e^{-2\pi\gamma_s} - 1|}\displaystyle\int_\alpha^{\alpha+2\pi} e^{-\nu_s\tau}\{|\sigma_{s-1}|v_{s-1}^{(j)} + a_s v^{(j)} + V_s^{(j)}\}d\tau \\ \qquad (s = 1, 2, \ldots, n;\ \sigma_0 = 0). \end{cases}$$

We note that $\tilde{v}_s^{(j)}$ does not depend on α. For,

$$\tilde{v}_s^{(j)} = \frac{e^{\nu_s\alpha}}{|e^{-2\pi\nu_s} - 1|}\{|\sigma_{s-1}|v_{s-1}^{(j)} + a_s v^{(j)} + V_s^{(j)}\}\int_\alpha^{\alpha+2\pi} e^{-\nu_s\tau}d\tau$$
$$= \frac{|e^{-2\pi\nu_s} - 1|}{|\nu_s||e^{-2\pi\gamma_s} - 1|}\{|\sigma_{s-1}|v_{s-1}^{(j)} + a_s v^{(j)} + V_s^{(j)}\},$$

and hence, in view of (1.342),

$$(1.347)\qquad \tilde{v}_s^{(j)} = \frac{1}{\varrho_s}\{|\sigma_{s-1}|v_{s-1}^{(j)} + a_s v^{(j)} + V_s^{(j)}\}.$$

By (1.344) and (1.345), we have

$$(1.348)\qquad v^{(j)} \leqq \tilde{v}^{(j)} \quad\text{and}\quad v_s^{(j)} \leqq \tilde{v}_s^{(j)}.$$

The series

$$(1.3491)\qquad \begin{aligned} r &= c + \tilde{v}^{(2)}c^2 + \tilde{v}^{(3)}c^3 + \dots \\ z_s &= \tilde{v}_s^{(1)} c + \tilde{v}_s^{(2)} c^2 + \tilde{v}_3^{(3)} c^3 + \dots \quad (s = 1, 2, \dots, n) \end{aligned}$$

formally satisfy the equations

$$(1.3492)\qquad \begin{aligned} r &= c + qF(r, z_1, z_2, \dots, z_n), \\ \varrho_1 z_1 &= a_1 r + F_1(r, z_1, z_2, \dots, z_n) \\ \varrho_s z_s &= \sigma_{s-1} z_{s-1} + a_s r + F_s(r, z_1, \dots, z_n) \\ &\qquad (s = 2, \dots, n), \end{aligned}$$

provided that in the prescription for passing from the U's to the V's one replaces $u^{(h)}$ and $u_s^{(h)}$ by the upper bounds $\tilde{v}^{(h)}$ and $\tilde{v}_s^{(h)}$ respectively and obtains the upper bounds for the moduli of the coefficients of the u's by replacing the coefficients of R and Z_s by the corresponding coefficients of the majorantes F and F_s.

Since all ϱ_s $(s = 1, \dots, n)$ are different from zero, the series (1.3491) converge in the neighborhood of zero and represent there the unique analytic solution of (1.3492). But, each series (1.3491) is a majorante of the corresponding series in (1.322) for all $\vartheta = \alpha + \beta i$ in (1.343). Hence these latter, as well, converge for all ϑ in (1.343) and for c in the neighborhood of zero.

1.35. THEOREM. *Consider a system*

$$(1.351)\qquad \begin{aligned} &\frac{dx}{dt} = -\lambda y + X, \quad \frac{dy}{dt} = \lambda x + Y \quad (\lambda > 0); \\ &\frac{dx_s}{dt} = p_{s1}x_1 + \dots + p_{sn}x_n + X_s \quad (s = 1, \dots, n) \end{aligned}$$

where X, Y, X_s are power series in x, y, $x_1, \dots, x_n$ lacking terms of degree less than two in these variables, and where λ and $P = (p_{ij})$ are constant and the characteristic equation $|P - \mu I| = 0$ has no roots of the form $m\lambda i$ with positive integral m.

If there exist a formal power series solution of (1.351) *of the form*

$$(1.352)\quad \begin{aligned} &x = (c + \tilde{u}^{(2)}(t)c^2 + \ldots)\cos\vartheta \quad y = (c + \tilde{u}^{(2)}(t)c^2 + \ldots)\sin\vartheta, \\ &x_s = \tilde{u}_s^{(1)}(t)c + \tilde{u}_z^{(2)}(t)c^2 + \ldots \quad (s = 1, 2, \ldots, n), \end{aligned}$$

where the coefficients $\tilde{u}^{(j)}(t)$, $\tilde{u}_s^{(j)}(t)$ are periodic functions in t of period T, then these power series converge for values of c sufficiently small in absolute value and thereby define a one-parameter family of periodic solutions of (1.351).

We set

$$(1.353)\quad x = r\cos\vartheta, \quad y = r\sin\vartheta, \quad x_1 = rz_1, \ldots, \quad x_n = rz_n,$$

and choose ϑ as the new independent variable.

First, we have

$$(1.354)\quad \begin{cases} \dfrac{dr}{dt} = X\cos\vartheta + Y\sin\vartheta, \\ \dfrac{d\vartheta}{dt} = \lambda + \Theta, \end{cases}$$

where

$$(1.355)\qquad \Theta = \frac{Y\cos\vartheta - X\sin\vartheta}{r}.$$

Since both X and Y are power series without linear and constant terms,

$$(1.356)\quad \frac{dr}{d\vartheta} = \frac{dr}{dt}\cdot\frac{dt}{d\vartheta} = \frac{1}{\lambda + \Theta}(X\cos\vartheta + Y\sin\vartheta) = R$$

where R is a power series in $r, z_1, \ldots, z_n$ which lacks linear and constant terms, and whose coefficients are periodic functions of ϑ of period 2π. More precisely, the coefficients of R are polynomials in $\sin\vartheta$ and $\cos\vartheta$. As in (1.34), we write $\vartheta = \alpha + \beta i$. For all ϑ subject to (1.343) both $\sin\vartheta$ and $\cos\vartheta$ are bounded and hence each polynomial in $\sin\vartheta$ and $\cos\vartheta$ is bounded. Since, moreover, for all such values of ϑ and for $r, z_1, \ldots, z_n$ sufficiently small in absolute value, the series R converges, we can majorize R by a convergent series in $r, z_1, \ldots, z_n$ with constant coefficients.

Next,

$$\frac{dz_s}{dt} = \frac{1}{r}\frac{dx_s}{dt} - \frac{1}{r^2}x_s\frac{dr}{dt} = p_{s1}z_1 + \ldots + p_{sn}z_n + \frac{1}{r}X_s - \frac{1}{r}z_s\frac{dr}{dt},$$

$$\text{since } z_s = \frac{x_s}{r}.$$

We note that X_s may contain terms of second degree in x, y, and x_s $(s = 1, \ldots, n)$. Among these only the terms containing x and y but not x_s, yield second degree terms in $r, z_1, \ldots, z_n$, following the change of variables (1.353). The expression resulting from the second degree terms containing only x and y, contains only r and ϑ and can be written in the form $r^2 \tilde{\varphi}_s(\vartheta)$ where $\tilde{\varphi}_s(\vartheta)$ is a quadratic form in $\sin\vartheta$ and $\cos\vartheta$. Thus, in view of (1.354),

$$(1.357) \quad \frac{dz_s}{d\vartheta} = q_{s1}z_1 + q_{s2}z_2 + \ldots + q_{sn}z_n + \varphi_s(\vartheta)r + Z_s \qquad (s = 1, \ldots, n),$$

where $Q = (q_{ij}) = (1/\lambda)(p_{ij}) = (1/\lambda)P$, $\varphi_s(\vartheta) = (1/\lambda)\tilde{\varphi}_s(\vartheta)$, and Z_s are power series in $r, z_1, \ldots, z_n$, which lack linear and constant terms, and in which the coefficients are polynomials in $\cos\vartheta$ and $\sin\vartheta$ and hence are periodic functions of period 2π. As in the case of the power series R, we can majorize the power series Z_s for all values of ϑ in (1.343) and for r and z_s sufficiently small in absolute value, by a power series in $r, z_1, \ldots, z_n$ with constant coefficients. Moreover, the characteristic equation $|Q - \mu\nu| = \lambda^{-n}|P - \mu\lambda\nu| = 0$ cannot have roots of the form mi with integral m.

Thus, the transformed system consisting of the equations (1.356) and (1.357) fulfills the conditions of lemma 1.32, and hence, if this system has formal solutions of the form (1.322) with coefficients which are periodic functions of of period 2π, then these formal power series converge and define a one parameter family of solutions periodic in ϑ and also of period 2π.

1.36. If a periodic solution (1.322) of the transformed system is given, then a periodic solution of the original system (1.351), expressed in terms of ϑ, is given by

$$(1.361) \quad \begin{aligned} x &= (c + u^{(2)}(\vartheta)c^2 + \ldots)\cos\vartheta, \quad y = (c + u^{(2)}c^2 + \ldots)\sin\vartheta \\ x_s &= (c + u^{(2)}(\vartheta)c^2 + \ldots)(u_s^{(1)}(\vartheta)c + u_s^{(2)}(\vartheta)c^2 \ldots) \quad (s = 1, 2, \ldots, \end{aligned}$$

where ϑ is defined as a function of t by the second one of the equations (1.354) in which the dependence of Θ upon c and ϑ is determined by (1.355) and (1.352).

We rewrite the second equation of (1.354) in the form

$$(1.362) \qquad \frac{\lambda d\vartheta}{\lambda + \Theta(\vartheta, c)} = \lambda dt,$$

and observe that

$$(1.363)\qquad \frac{\lambda}{\lambda + \Theta(\vartheta,\ c)} = 1 + \Theta_1(\vartheta)c + \Theta_2(\vartheta)c^2 + \dots,$$

where $\Theta_j(\vartheta)$ are periodic functions, in fact, polynomials in $\sin\vartheta$ and $\cos\vartheta$, and the series on the right converges for values of c sufficiently small in absolute value. Integrating (1.362) we have, in view of (1.363),

$$(1.364)\qquad \vartheta + c\int_0^\vartheta \Theta_1(\vartheta)d\vartheta + c^2\int_0^\vartheta \Theta_2(\vartheta)d\vartheta + \dots = \lambda(t - t_0)$$

where t_0 is an arbitrary constant. We note that

$$\int_0^\alpha \Theta_j(\vartheta)d\vartheta = h_j\vartheta + \bar{\Phi}_j(\vartheta), \quad \text{where } h_j = \frac{1}{2\pi}\int_0^{2\pi} \Theta_j(\vartheta)d\vartheta$$

and $\bar{\Phi}_j(\vartheta)$ are periodic of period 2π. Then (1.364) becomes

$$(1.365)\quad (1 + h_1c + \dots)[\vartheta + c\Phi_1(\vartheta) + c^2\Phi_2(\vartheta) + \dots] = \lambda(t - t_0)$$

where each $\Phi_j(\vartheta)$ is a trigonometric polynomial.

We proved that the series (1.322) converges for complex $\vartheta = \alpha + i\beta$. Let

$$\frac{2\pi}{\lambda}(1 + h_1c + h_2c^2 + \dots) = T,$$

$$\frac{2\pi(t - t_0)}{T} = \tau \quad \text{and} \quad \vartheta - \tau = \varphi.$$

Then (1.365) becomes

$$(1.366)\qquad \varphi + c\Phi_1(\varphi + \tau) + c^2\Phi_2(\varphi + \tau) + \dots = 0.$$

We allow those values of the parameter $\tau = \varrho + \sigma\sqrt{-1}$ in which ϱ is arbitrary and $|\sigma|$ is sufficiently small. If the same assumption is made regarding φ then in view of the convergence, for sufficiently small $|c|$, of

$$c|\Phi_1(\vartheta)| + c^2|\Phi_2(\vartheta)| + \dots,$$

the function φ satisfying (1.366) must become arbitrarily small as c tends to zero. Thus our problem reduces to the determination of the function φ satisfying (1.366) and such that the absolute value of φ can be made arbitrarily small for small values of the absolute value of c.

The functions $\Phi_j(\vartheta)$ are trigonometric polynomials in ϑ. Therefore $\Phi_j(\varphi + \tau)$ can be represented as a power series in φ converging for all φ and τ. That is, the left-hand member of (1.366) is a power series in c and φ converging for sufficiently small values of $|c|$ and for arbitrary values of φ. The coefficients of this series depend on τ. There exists, however, a dominant convergent power series in c and φ with constant coefficients.

For $\varphi = c = 0$ the left-hand member of (1.366) vanishes and its partial derivative with respect to φ is $= 1$. Therefore the desired function φ can be represented by a power series in c

$$\varphi = \varphi_1 c + \varphi_2 c^2 + \ldots,$$

where φ_k $(k = 1, 2, \ldots)$ depend on τ but not on c. The coefficients φ_k can be expressed in terms of $\Phi_i(\tau)$ and their derivatives, and therefore are periodic in τ. The last series converges for sufficiently small values of $|c|$. Thus ϑ is given by

$$\vartheta = \tau + \varphi_1 c + \varphi_2 c + \ldots + \varphi_n c^n + \ldots.$$

Substituting this expression into the series defining x, y, and x_s $(s = 1, 2, \ldots, n)$, collecting terms of the same degree in c, and replacing τ by $2\pi(t - t_0)/T$ we obtain the series defining x, y, and x_s as functions of t and the arbitrary constant t_0 and c. These series define a periodic solution of (1.357) of period T, i.e., of period

$$T = \int_0^{2\pi} \frac{d\vartheta}{\lambda + \Theta} = \frac{2\pi}{\lambda}(1 + h_1 c^2 + h_2 c^2 + \ldots).$$

This completes our proof.

1.37. LEMMA. *If in addition to satisfying the conditions of Theorem* 1.35 *the equations* (1.351) *have an integral of the form*

$$x^2 + y^2 + F(x, y, x_1, \ldots, x_n) = C \tag{1.371}$$

where F is a power series beginning with terms of degree greater than the first but not containing x and y in its terms of second order, then the system (1.351) *has a family of periodic solutions depending on a single real parameter.*

As in Theorem 1.35 we let

$$x = r\cos\vartheta, \quad y = r\sin\vartheta, \quad x_1 = rz_1, \ldots, x_n = rz_n.$$

Then the transformed system will have an integral of the form

$$r^2 + r^2\overline{F}(r, \vartheta, z_1, \ldots, z_n) = C,$$

where F is a power series in r, $\cos\vartheta$, $\sin\vartheta$, and in $z_1, z_2, \ldots, z_n$. Extracting the square root we obtain the integral

$$r + r\varphi(r, \vartheta, z_1, \ldots, z_n) = C' \tag{1.372}$$

where φ is a power series which vanishes for $r = z_1 = z_2 = \ldots z_n = 0$ and whose coefficients are periodic functions of ϑ.

Should no periodic solution exist, then in view of Lemma 1.32 for some value of l all the coefficients

$$u^{(2)}, u^{(3)}, \ldots, u^{(l-1)}; \quad u_s^{(1)}, \ldots, u_s^{(l-1)} \quad (s = 1, \ldots, n)$$

are periodic, but the coefficient $u^{(l)}$ has the form

$$u^{(l)} = g\vartheta + V$$

where g is a constant different from 0 and V is a periodic function of ϑ.

We let

$$\left\{\begin{aligned} r &= c + u^{(2)}c^2 + \ldots + u^{(l-1)}c^{l-1} + u^{(l)}c^l, \\ z_s &= u_s^{(1)}c + u_s^{(2)}c^2 + \ldots + u_s^{(l-1)}c^{l-1} \quad (s = 1, \ldots, n) \end{aligned}\right. \tag{1.373}$$

in (1.372), and write the result as a polynomial in c. Since the polynomial expressions for r and z_s in (1.373) are the partial sums of the formal power series satisfying (1.357), then upon substitution into an integral of this system, the resulting coefficients of all the powers of c up to the (l)th will be independent of ϑ. The above substitution, however, will not yield such a result, since the coefficient of the lth power of c in the expansion of $r\varphi$ is periodic and cannot yield a constant when added to the coefficient of the term $(g\vartheta + V)c^l$ in the expansion of r.

Thus we see that there exists a family of periodic solutions depending on c, and that for each c the period is

$$T = \frac{2\pi}{\lambda}(1 + h_1c + h_2c^2 + \ldots).$$

1.38. Proof of Theorem 1.31. We are now ready to prove Lyapunov's theorem. We consider again a canonical system with a nonsingular function H, and we assume that the characteristic equation has k pairs of distinct purely imaginary roots

$$\pm i\beta_1, \ \pm i\beta_2, \ldots, \pm i\beta_k.$$

We shall show that one can find a linear contact transformation, i.e., a transformation preserving the Hamiltonian form of our system, under which the function H will be transformed into

$$(1.381) \quad H = \frac{\beta_1}{2}(x_1^2 + y_1^2) + \ldots + \frac{\beta_k}{2}(x_k^2 + y_k^2) + \bar{H}_2 + \bar{H}^3$$

where H_2 is the sum of those terms of the second degree which do not contain the variables $x_1, x_2, \ldots, x_k, y_1, y_2, \ldots, y_k$, and $\bar{H}_3$ is a function whose expansion in a power series contains no terms of degree smaller than the third. If we fix our attention upon any pair of variables corresponding to a pair of imaginary roots $\pm i\beta_s$, then the transformed canonical system is of type (1.351) and has an integral $H = C$ of the form (1.371); hence, in view of Lemma 1.37, it has a family of periodic solutions of period

$$T_s = \frac{2\pi}{\beta_s}(1 + h_1 c + h_2 c^2 + \ldots).$$

Carrying out this argument for each pair of the variables (x_1, y_1), $(x_2, y_2), \ldots, (x_k, y_k)$ we obtain k one-parameter families of periodic solutions.

It remains to show the existence of the desired linear contact transformation. We first consider a linear system of the form

$$\frac{dx_s}{dt} = -\frac{\partial H_2}{\partial y_s}, \quad \frac{dy_s}{dt} = \frac{\partial H_2}{\partial x_s} \qquad (s = 1, 2, \ldots, n)$$

where H_2 is a quadratic form in $x_1, \ldots, y_n$. By hypothesis, its characteristic roots are

$$\pm i\beta_1, \pm i\beta_2, \ldots, i\beta_n \quad \text{where} \quad i = \sqrt{-1},$$

whence the general solution has the form

$$x_j = \sum (a_{js} C_s e^{i\beta_s t} + \bar{a}_{js} C_{n+s} e^{-i\beta_s t})$$
$$y_j = \sum (b_{js} C_s e^{i\beta_s t} + \bar{b}_{js} C_{n+s} e^{-i\beta_s t})$$

where $\bar{a}_{js}, \bar{b}_{js}$ are complex conjugates of a_{js} and b_{js} $(j = 1, 2, \ldots, n;$ $s = 1, 2, \ldots, n)$.

Solving these equations for

$$C_1 e^{i\beta_1 t}, \ C_2 e^{i\beta_2 t}, \ldots, C_n e^{i\beta_n t}, \ C_{n+1} e^{-i\beta_1 t}, \ldots, C_{2n} e^{-i\beta_n t}$$

we obtain linear forms in x_s, y_s $(s = 1, 2, \ldots, n)$. Here

$$C_s e^{i\beta_s t} = u_s(x, y) + iv_s(x, y),$$
$$C_{s+n} e^{-\beta_s t} = u_s(x, y) - iv_s(x, y),$$

where u_s and v_s are linear forms with real coefficients. Multiplying the last two equalities by $e^{-i\beta_s t}$ and $e^{i\beta_s t}$ respectively, we find $2n$ independent integrals

$$(u_s + iv_s)\, e^{-i\beta_s t} = C_s, \quad (u_s - iv_s)\, e^{i\beta_s t} = C_{s+n} \quad (s = 1, 2, \ldots, n),$$

of our linear system.

Poisson brackets for each pair of these integrals is equal to a constant [1]. Thus

$$e^{\pm i(\beta_s \pm \beta_\sigma)t}\, [u_s \pm iv_s,\ u_\sigma \pm iv_\sigma] = \text{const.}$$

The bracket $[u_s \pm iv_s,\ u_\sigma \pm iv_\sigma]$ is itself a constant, since it is a linear combination of the Poisson brackets $[u_s, u_\sigma]$, $[u_s, v_\sigma]$, $[v_s, v_\sigma]$, $[v_s, u_\sigma]$, which in turn are constant in view of the linearity of u_s and v_s.

Case 1. $\sigma \neq s$. If the bracket $[u_s \pm iv_s,\ u_\sigma \pm iv_\sigma]$ were $\neq 0$ then we would have an integral $e^{\pm i(\beta_s \pm \beta_\sigma)t} = \text{const.}$ which is impossible. Therefore

$$[u_s \pm iv_s,\ u_\sigma \pm iv_\sigma] = 0 \quad \text{for} \quad \sigma \neq s.$$

Case 2a.

$$[u_s + iv_s,\ u_s + iv_s] = 0, \quad [u_s - iv_s,\ u_s - iv_s] = 0.$$

Case 2b. The Poisson bracket

$$\begin{aligned}[(u_s + iv_s)e^{-i\beta_s t},\ (u_s - iv_s)e^{i\beta_s t}] &= [u_s + iv_s,\ u_s - iv_s] \\ &= [u_s, u_s] + i[v_s, u_s] \\ &\quad - i[u_s, v_s] + [v_s, v_s] \\ &= -2i[u_s, v_s]\end{aligned}$$

cannot be $= 0$, for otherwise, for given s, we would have $2n$ equations

$$[u_s + iv_s,\ u_r + iv_r] = 0, \quad [u_s + iv_s,\ u_r - iv_r] = 0 \quad (r = 1, \ldots, n).$$

The determinant of this system considered as a linear system in

$$\frac{\partial(u_s + iv_s)}{\partial x_l}, \ \frac{\partial(u_s + iv_s)}{\partial y_l} \quad (l = 1, 2, \ldots, n),$$

[1]See Whittaker's *Analytical Dynamics*, p. 320.

is the Jacobian

$$\frac{D(u_1 + iv_1, \ldots, u_n + iv_n, u_1 - iv_1, \ldots, u_n - iv_n)}{D(x_1, \ldots, x_n, y_1, \ldots, y_n)}.$$

Since $u_s + iv_s \neq$ a constant identically in x_l, y_l this Jacobian must vanish, which contradicts the independence of our system of integrals $u_s \pm iv_s$.

Thus we have

$$[u_s + iv_s, u_\sigma \pm iv_\sigma] = [u_s, u_\sigma] \pm [v_s, v_\sigma] + i\{[v_s, u_\sigma] \pm [u_s, v_\sigma]\} = 0$$

whence

$$[u_s, u_\sigma] = [v_s, v_\sigma] = 0,$$

and if $s \neq \sigma$

$$[u_s, v_\sigma] = [u_\sigma, v_s] = 0.$$

Next

$$[u_s + iv_s, u_s - iv_s] = -2i[u_s, v_s] = \text{const.} \neq 0.$$

Since $[u_s, v_s]$ are bilinear forms in the coefficients of u_s, v_s we can normalize the integral $u_s + iv_s$ so that $[u_s, v_s] = 1$.

To summarize,

$$[u_s, u_\sigma] = 0, \quad [v_s, v_\sigma] = 0$$
$$[u_s, v_s] = 1, \quad [u_s, v_\sigma] = 0 \quad \text{for} \quad s \neq \sigma.$$

Thus [2] the linear transformation with real coefficients which replaces the variables $x_1, x_2, \ldots, x_n, y_1, y_2, \ldots, y_n$, by the variables $u_1, u_2, \ldots, u_n, v_1, v_2, \ldots, v_n$, is a contact transformation and therefore preserves the Hamiltonian character of our system.

This may also be verified by direct calculation, for on one hand

$$\begin{cases} [u_\sigma, v_s] = \sum_r \frac{\partial u_\sigma}{\partial x_r}\frac{\partial v_s}{\partial y_r} - \sum_r \frac{\partial u_\sigma}{\partial y_r}\frac{\partial v_s}{\partial x_r} = 0, \quad (\sigma \neq s) \\ [v_\sigma, v_s] = \sum_r \frac{\partial v_\sigma}{\partial x_r}\frac{\partial v_s}{\partial y_r} - \sum_r \frac{\partial v_\sigma}{\partial y_r}\frac{\partial v_s}{\partial x_r} = 0, \\ [u_s, v_s] = \sum_r \frac{\partial u_s}{\partial x_r}\frac{\partial v_s}{\partial y_r} - \sum_r \frac{\partial u_s}{\partial y_r}\frac{\partial v_s}{\partial x_r} = 1. \end{cases}$$

On the other hand we have

[2]See Whittaker, *Analytical Dynamics*.

$$\begin{cases} \dfrac{\partial u_\sigma}{\partial u_s} = \sum_r \dfrac{\partial u_\sigma}{\partial x_r}\dfrac{\partial x_r}{\partial u_s} + \sum_r \dfrac{\partial u_\sigma}{\partial y_r}\dfrac{\partial y_r}{\partial u_s} = 0 \qquad (s \neq \sigma) \\ \dfrac{\partial v_\sigma}{\partial u_s} = \sum_r \dfrac{\partial v_\sigma}{\partial x_r}\dfrac{\partial x_r}{\partial u_s} + \sum_r \dfrac{\partial v_\sigma}{\partial y_r}\dfrac{\partial y_r}{\partial u_s} = 0, \\ \dfrac{\partial u_s}{\partial u_s} = \sum_r \dfrac{\partial u_s}{\partial x_r}\dfrac{\partial x_r}{\partial u_s} + \sum_r \dfrac{\partial u_s}{\partial y_r}\dfrac{\partial y_r}{\partial u_s} = 1, \end{cases}$$

$$(\sigma = 1, 2, \ldots, n).$$

Subtracting from the equations of the first system corresponding equations of the second system we obtain a homogeneous system

$$\sum_r \frac{\partial u_\sigma}{\partial x_r}\left(\frac{\partial v_s}{\partial y_r} - \frac{\partial x_r}{\partial u_s}\right) - \sum_r \frac{\partial u_\sigma}{\partial y_r}\left(\frac{\partial v_s}{\partial x_r} + \frac{\partial y_r}{\partial u_s}\right) = 0,$$

$$\sum_r \frac{\partial v_\sigma}{\partial x_r}\left(\frac{\partial v_s}{\partial y_r} - \frac{\partial x_r}{\partial u_s}\right) - \sum_r \frac{\partial v_\sigma}{\partial y_r}\left(\frac{\partial v_s}{\partial x_r} + \frac{\partial y_r}{\partial u_s}\right) = 0,$$

$$(\sigma = 1, 2, \ldots, n).$$

the determinant of which (Jacobian in u_σ and v_σ) is different from zero. Therefore

$$\frac{\partial v_s}{\partial y_r} = \frac{\partial x_r}{\partial u_s}, \quad \frac{\partial v_s}{\partial x_r} = -\frac{\partial y_r}{\partial u_s}.$$

Interchanging u and v we obtain, in a similar manner,

$$\frac{\partial u_s}{\partial x_r} = \frac{\partial y_r}{\partial v_s}, \quad \frac{\partial u_s}{\partial y_r} = -\frac{\partial x_r}{\partial v_s}.$$

Carrying out the change of variables in the canonical system

$$\frac{dx_r}{dt} = -\frac{\partial H}{\partial y_r}, \quad \frac{dy_r}{dt} = \frac{\partial H}{\partial x_r} \quad (r = 1, \ldots, n),$$

we obtain

$$\sum_s \frac{\partial x_r}{\partial u_s}\frac{du_s}{dt} + \sum_s \frac{\partial x_r}{\partial v_s}\frac{dv_s}{dt} = -\sum_s \frac{\partial H}{\partial u_s}\frac{\partial u_s}{\partial y_r} - \sum_s \frac{\partial H}{\partial v_s}\frac{\partial v_s}{\partial y_r},$$

$$\sum_s \frac{\partial y_r}{\partial u_s}\frac{du_s}{dt} + \sum_s \frac{\partial y_r}{\partial v_s}\frac{dv_s}{dt} = \sum_s \frac{\partial H}{\partial u_s}\frac{\partial u_s}{\partial x_r} + \sum_s \frac{\partial H}{\partial v_s}\frac{\partial v_s}{\partial x_r},$$

or

$$\sum_s \frac{\partial v_s}{\partial y_r}\left(\frac{du_s}{dt} + \frac{\partial H}{\partial v_s}\right) - \sum_s \frac{\partial u_s}{\partial y_r}\left(\frac{dv_s}{dt} - \frac{\partial H}{\partial u_s}\right) = 0,$$

$$\sum_s \frac{\partial v_s}{\partial x_r}\left(\frac{du_s}{dt} + \frac{\partial H}{\partial v_s}\right) - \sum_s \frac{\partial u_s}{\partial x_r}\left(\frac{dv_s}{dt} - \frac{\partial H}{\partial u_s}\right) = 0,$$

$$(r = 1, 2, \ldots, n).$$

Since the determinant does not vanish, we have

$$\frac{du_s}{dt} = -\frac{\partial H}{\partial v_s}, \quad \frac{dv_s}{dt} = \frac{\partial H}{\partial u_s} \qquad (s = 1, \ldots, n).$$

For our linear system ($H = H_2$) the transformed equations have the form

$$\frac{d}{dt}[(u_s + iv_s)e^{-i\beta_s t}] = 0, \qquad \frac{d}{dt}[(u_s - iv_s)e^{+i\beta_s t}] = 0$$

$$(s = 1, \ldots, n)$$

or

$$\frac{d}{dt}(u_s + iv_s) = i\beta_s(u_s + iv_s), \qquad \frac{d}{dt}(u_s - iv_s) = -i\beta_s(u_s - iv_s).$$

Equating real and imaginary parts, we obtain

$$\frac{du_s}{dt} = -\beta_s v_s, \quad \frac{dv_s}{dt} = \beta_s u_s \quad (s = 1, \ldots, n).$$

Thus, H_2 as a function of the new variables satisfies the conditions

$$\frac{\partial H_2}{\partial v_s} = \beta_s v_s, \quad \frac{\partial H_2}{\partial u_s} = \beta_s u_s,$$

that is

$$H_2 = \frac{\beta_1}{2}(u_1^2 + v_1^2) + \ldots + \frac{\beta_n}{2}(u_n^2 + v_n^2).$$

If only $2k$ roots of the characteristic equation are purely imaginary ($k < n$) then, applying the above transformation only to the variables

$$x_1, \ldots, x_k, \; y_1, \ldots, y_k$$

(say, by setting $x_r = u_r$, $y_r = v_r$ for $r > k$) we obtain a Hamiltonian system with H in the form (1.381).

This completes the proof of Lyapunov's theorem.

1.4. The neighborhood of a singular point. Consider a system of the form

$$\frac{dx}{dt} = Ax + f(x, t), \tag{1.401}$$

where $f(0, t) = 0$ and

$$\frac{\partial f(x, t)}{\partial x_j} \to 0 \text{ uniformly in } t, \text{ as } ||x|| \to 0. \tag{1.402}$$

Here again $A = (a_{hj})$, a_{hj} are constants, $x = \{x_1, \ldots, x_n\}$ and $f = \{f_1, \ldots, f_n\}$ are column vectors, and $||x|| = \sum_{i=1}^{n} |x_i|$.

In particular, condition (1.402) may hold for stationary systems whose right-hand members do not depend on t.

We shall show that near the origin the distribution of integral curves of a system (1.401) subject to the condition (1.402), is related to the distribution of integral curves near the origin of the associated linear system

$$\frac{dy}{dt} = Ay \tag{1.403}$$

with constant coefficients.

1.41. THEOREM. *We consider a system* (1.401) *subject to the condition* (1.402).

(1) *If all the characteristic roots of A have negative real parts then all the solutions of* (1.401) *the initial values of which are sufficiently near the origin, tend toward zero as $t \to +\infty$.* (2) *If the number of such roots is $k < n$, then there exists a manifold depending on k independent parameters, such that all the integral curves passing through the points of this manifold, tend toward zero as $t \to +\infty$.* (3) *If the number of characteristic roots of A with positive real parts is equal to m, then there is an m-parameter manifold such that every integral curve passing through a point of this manifold tends toward zero as $t \to -\infty$.* [3]

1.42. Let $Y(t)$ be the matrix of the fundamental system of solutions of (1.403), for which $Y(0) = I$, and write $y^{(0)} = \{y_1^{(0)}, \ldots, y_n^{(0)}\}$. Then the solution $y(t) = Y(t)y^0$ has the initial value $y(0) = y^0$.

[3] Lyapunov proved this theorem for systems (1.401) with analytic right-hand members. After Lyapunov this theorem was established under less restrictive conditions by O. Perron [37], [38], [40], I. G. Petrowsky [43], [44], and others.

If all the characteristic roots of A have negative real parts, then

$$D = \int_0^{+\infty} ||Y(\tau)|| d\tau < + \infty. \tag{1.421}$$

Choose $\varepsilon < 1/2D$ and select α_0 so that

$$||f_{x_j}(u, t)|| < \varepsilon \quad \text{for} \quad ||u|| \leqq \alpha_0, \quad \text{uniformly in } t. \tag{1.422}$$

This can be done in view of (1.402). Clearly every component of $Y(t)$ does not exceed $c_1 e^{-\alpha t}$ in absolute value, for suitably chosen constants c_1 and $\alpha > 0$. For instance, $-\alpha$ may be chosen as one half of the largest negative real part of all the characteristic roots of A. Thus

$$||Y(t)y^{(0)}|| \leqq c_0 e^{-\alpha t} \leqq \tfrac{1}{2}\alpha_0 \quad (t \geqq t_1)$$

for suitably chosen c_0, and hence

$$||Y(t)y^0|| \leqq \frac{\alpha_0}{2} \tag{1.4231}$$

for large t $(t \geqq t_1)$. Moreover, in addition to (1.4231), we may assume that

$$||Y(t)|| \cdot ||y^{(0)}|| \leqq \tfrac{1}{2}\alpha_0 \quad \text{for} \quad 0 \leqq t \leqq t_1, \tag{1.4232}$$

if $||y^{(0)}||$ is sufficiently small.

Construct a sequence

$$z^{(0)}(t) = Y(t)y^{(0)}, \quad z^{(j+1)}(t) = Y(t)y^{(0)} + \int_0^t Y(t-\tau) f(z^{(j)}, \tau) d\tau. \tag{1.424}$$

We assert that

$$||z^{(k)}(t)|| \leqq \alpha_0 \qquad (t \geqq 0). \tag{1.425}$$

To prove (1.425) by induction, we note that it holds for $k = 0$, and that

$$\begin{aligned} ||z^{(j+1)}(t)|| &\leqq ||Y(t)y^{(0)}|| + \int_0^t ||Y(t-\tau)|| \cdot ||f(z^{(j)}, \tau)|| d\tau \\ &\leqq ||Y(t)y^{(0)}|| + \varepsilon \int_0^t ||Y(t-\tau)|| \; ||z^{(j)}(\tau)|| d\tau \qquad (t \geqq 0), \end{aligned} \tag{1.426}$$

in view of (1.422) and the hypothesis of induction. Then by the choice of ε and by (1.4231–1.4232)

$$||z^{(j+1)}(t)|| \leqq \tfrac{1}{2}\alpha_0 + \varepsilon\alpha_0 D \leqq \tfrac{1}{2}\alpha_0 + \tfrac{1}{2}\alpha_0 = \alpha_0 \quad (t \geqq 0).$$

The inequalities (1.425) and (1.422) imply that

$$\begin{aligned}||z^{(j+1)}(t) - z^{(j)}(t)|| &= \left|\left| \int_0^t Y(t-\tau)[f(z^{(j)}, \tau) - f(z^{(j-1)}, \tau)]d\tau \right|\right| \\ &\leqq \varepsilon \int_0^t ||Y(t-\tau)|| \; ||z^{(j)}(t) - z^{(j-1)}(t)||d\tau \\ &\leqq \varepsilon D \max_{0 \leqq \tau \leqq t} ||z^{(j)}(\tau) - z^{(j-1)}(\tau)||.\end{aligned}$$

We note that $\varepsilon D < \frac{1}{2}$. Thus,

$$\max_{0 \leqq \tau \leqq t} ||z^{(j+1)}(\tau) - z^{(j)}(\tau)|| \leqq \tfrac{1}{2} (\max_{0 \leqq \tau \leqq t} ||z^{(j)}(\tau) - z^{(j-1)}(\tau)||),$$

therefore

$$\max_{0 \leqq \tau \leqq t} ||z^{(j+1)}(\tau) - z^{(j)}(\tau)|| \leqq (\tfrac{1}{2})^j 2\alpha_0 = (\tfrac{1}{2})^{j-1}\alpha_0,$$

and

$$\begin{aligned}\max_{0 \leqq \tau \leqq t} ||z^{(j+k)}(\tau) - z^{(j)}(\tau)|| &\leqq \sum_{\nu=1}^{k} \max_{0 \leqq \tau \leqq t} ||z^{(j+\nu)}(\tau) - z^{(j+\nu-1)}(\tau)|| \\ &\leqq (\tfrac{1}{2})^{j-1}\alpha_0 \sum_{\nu=0}^{k-1} (\tfrac{1}{2})^{\nu} = (\tfrac{1}{2})^{j-2}\alpha_0(1 - (\tfrac{1}{2})^k) < (\tfrac{1}{2})^{j-2}\alpha_0.\end{aligned}$$

Thus sequence (1.424) converges uniformly to a solution $x(t)$ of the integral equation

$$x(t) = Y(t)y^{(0)} + \int_0^t Y(t-\tau)f(x(\tau), \tau)d\tau. \tag{1.427}$$

We note that $x(0) = y^{(0)}$ and that in view of the uniqueness of solutions with given initial conditions, every solution of (1.427) (and hence of 1.401) whose initial value $y^{(0)}$ satisfies 1.4232, can be constructed as a limit of the sequence (1.424).

To prove that the above construction yields a solution $x(t)$ such that $\lim_{t \to +\infty} x(t) = 0$, it suffices to show that

$$\lim_{t \to +\infty} z^{(j)}(t) = 0, \qquad (j = 0, 1, \ldots). \tag{1.428}$$

To accomplish this, we choose constants α_1, α_2, c_2, c_3 so that $\alpha > \alpha_2 > \alpha_1 > 0$ and

$$||Y(t)y^{(0)}|| \leqq c_2 e^{-\alpha_1 t}, \quad ||Y(t)|| \leqq c_3 e^{-\alpha_2 t} \qquad (t \geqq 0).$$

Then, if c_2 is sufficiently large, we can prove that

$$||z^{(j)}|| \leqq 2c_2 e^{-\alpha_1 t} \qquad (t \geqq 0).$$

This inequality is true for $j = 0$. If it holds true for j, it holds true for $(j + 1)$. For, by (1.426),

$$
\begin{aligned}
||z^{(j+1)}(t)|| &\leqq c_2 e^{-\alpha_1 t} + 2\varepsilon c_3 c_2 e^{-\alpha_2 t} \int_0^t e^{(\alpha_2-\alpha_1)\tau} d\tau \\
&= c_2 e^{-\alpha_1 t} [1 + 2c_3 \varepsilon \frac{1}{\alpha_2 - \alpha_1} (1 - e^{(\alpha_1-\alpha_2)t})] \\
&\leqq 2c_2 e^{-\alpha_1 t}
\end{aligned}
$$

if c_2 is sufficiently large.

This completes the proof of part (1) of our theorem.

1.429. We should note that the conclusions of Section 1.42 hold for every initial point $(y^{(0)}, t)$, $t \geqq 0$, of the tube based on the sufficiently small neighborhood $\{y^{(0)}\}$ of the origin determined by (1.4232) in the phase space x. This can be easily seen if we replace $Y(t)$ by $Y(t - t_0)$, integrate from t_0 to t in the construction (1.424), and make use of the uniformity in t of (1.402).

1.43. Next, let $y^{(1)}(t), \ldots, y^{(k)}(t)$ be k linearly independent solutions of (1.403) such that $||y^{(j)}(t)|| \to 0$ as $t \to +\infty$. Write

$$
(1.431) \qquad y(t) = \sum_{i=1}^{k} a_i y^{(i)}(t);
$$

$||y||$ may be made as small as we wish for all $t > 0$ by taking $\sum_{i=1}^{n} |a_i|$ sufficiently small. The desired k-dimensional manifold is generated by the k constants a_i.

Each element y_{ij} of Y is the sum of terms of the form $e^{\lambda t}(P(t) \cos \mu t + Q(t) \sin \mu t)$; $\lambda + \mu i$ is a characteristic root of A; $P(t)$, $Q(t)$ are polynomials (cf. III, 2). Write u_{ij} for the sum of those terms in y_{ij} for which $\lambda < 0$ and v_{ij} for the sum of the remaining terms with $\lambda \geqq 0$ and let $Y_1 = (u_{ij})$, $Y_2 = (v_{ij})$. Then $Y = Y_1 + Y_2$.

To prove part (2) of our theorem we modify the recurrence relation in (1.424) to construct the sequence

$$
\begin{aligned}
& z^{(0)}(t) = y(t) \\
(1.432) \qquad & z^{(j+1)}(t) = y(t) + \int_0^t Y_1(t-\tau) f(z^{(j)}(\tau), \tau) d\tau \\
& - \int_t^\infty Y_2(t-\tau) f(z^{(j)}(\tau), \tau) d\tau.
\end{aligned}
$$

We can show as before that this sequence converges uniformly for every $y(t)$ in (1.431) for which $||y(0)||$ in (1.431) is sufficiently small, i.e., for every $y(t)$ corresponding to a sufficiently small

$\Sigma|a_i|$. The limit $z(t)$ of the sequence (1.432) satisfies the equation

$$z(t) = y(t) + \int_0^t Y_1(t-\tau)f(z(\tau), \tau)d\tau - \int_t^\infty Y_2(t-\tau)f(z(\tau), \tau)d\tau,$$

and hence it is a solution of (1.401) as well. Also, we have $z(0) = y(0) + \text{const.}$, and $z(t)$ tends toward zero as $t \to +\infty$.

To establish our last assertion we proceed as in 1.42. We let $-\lambda < 0$ be an upper bound of the negative real parts of the characteristic roots of A, and we let μ be an upper bound of the real parts of all the characteristic roots of A. Choose λ_1, λ_2, μ_1, c_4, c_5, c_6 so that $\lambda > \lambda_2 > \lambda_1 > 0$, $\mu_1 > \mu \geqq 0$, $||y(t)|| \leqq c_4 e^{-\lambda_1 t}$, $||Y_1(t)|| \leqq c_5 e^{-\lambda_2 t}$, $||Y_2(t)|| \leqq c_6 e^{\mu_1 t}$, $t \geqq 0$. Here, again, it suffices to show that $\lim z^{(j)} = 0$ for every j. We assert that

$$||z^{(j)}(t)|| \leqq 2c_4 e^{-\lambda_1 t} \qquad (t \geqq 0,\ j = 0, 1, \ldots).$$

This inequality holds true for $j = 0$. If it holds true for j, then, for $z(0)$ sufficiently small,

$$\begin{aligned} ||z^{(j+1)}(t)|| &\leqq ||y(t)|| + \varepsilon \int_0^t ||Y_1(t-\tau)||\ ||z^{(j)}(\tau)||d\tau \\ &\quad + \varepsilon \int_t^{+\infty} ||Y_2(t-\tau)||\ ||z^{(j)}(\tau)||d\tau \\ &\leqq c_4 e^{-\lambda_1 t} + 2c_4c_5\varepsilon e^{-\lambda_2 t}\int_0^t e^{(\lambda_2-\lambda_1)\tau}d\tau + 2c_4c_6\varepsilon e^{\mu_1 t}\int_t^{+\infty} e^{-(\mu_1+\lambda_1)\tau}d\tau \\ &\leqq 2c_4 e^{-\lambda_1 t}. \end{aligned}$$

To prove part (3) of Theorem 1.41, we observe that in replacing t by $-t$ the roles of the negative and the positive real parts of characteristic roots are interchanged.

In later sections we shall study the analogy between the distribution of O-curves in the linear and in the nonlinear systems.

1.44. λ-transformations. We concern ourselves next with the rapidity of the convergence of our integral curves toward the origin. We make use of "λ-transformations", i.e., of transformations of the type

$$x_i = e^{\lambda t}\tilde{x}_i. \tag{1.441}$$

This transformation carries (1.401) into

$$\frac{d\tilde{x}}{dt} = (A - \lambda I)\tilde{x} + e^{-\lambda t}f(e^{\lambda t}\tilde{x}, t) = (A - \lambda I)\tilde{x} + \psi(\tilde{x}, t). \tag{1.442}$$

Thus the real parts of the characteristic roots of the new system are

obtained from those of the original system by subtracting λ.
We observe that

$$(1.443)\qquad \frac{\partial\psi(\tilde{x},\,t)}{\partial\tilde{x}_j} = e^{-\lambda t}\frac{\partial f(e^{\lambda t}\tilde{x},\,t)}{\partial\tilde{x}_j} = e^{-\lambda t}\frac{\partial f(x,\,t)}{\partial x_j}\frac{\partial x_j}{\partial\tilde{x}_j} = \frac{\partial f(x,\,t)}{\partial x_j}.$$

Then, condition (1.402) implies

$$(1.444)\qquad \text{if } \lambda > 0, \text{ then } \frac{\partial\psi(\tilde{x},\,t)}{\partial\tilde{x}_j} \to 0 \text{ as } t \to -\infty \text{ uniformly in } x,$$

for $\tilde{x}$ in a bounded region $\tilde{G}$,

and

$$(1.445)\quad \text{if } \lambda < 0, \text{ then } \frac{\partial\psi(\tilde{x},\,t)}{\partial\tilde{x}_j} \to 0 \text{ as } \|\tilde{x}\| \to 0 \text{ uniformly in } t \geqq 0.$$

1.45. Theorem (A. A. Shestakov [50]). *Let* $0 > \alpha_1 \geqq \alpha_2 \geqq \ldots \geqq \alpha_n$ *be the real parts of the characteristic roots of* A. *Assume that condition* (1.402) *holds for* (1.401). *Then, if* $x(t)$ *is a solution of* (1.401), *with* $x(0)$ *sufficiently small, we have*

1.451. $\lim_{t\to\infty}$ $(\log \|x(t)\|)/t$ *exists and is zero or is equal to one of the* α_i.

1.452. *The manifold of the initial conditions* $x(0)$ *of those O-curves* $x = x(t)$ *for which*

$$(1.4521)\qquad \|x(t)\| \leqq e^{(a+\eta(t))t},$$

where $a < 0$ *and* $\eta(t) \to 0$ *for* $t \to +\infty$, *depends upon* p *independent parameters, where* p *is the number of characteristic roots with* $a_i \leqq a$.

1.46. We shall now prove Theorem 1.45. A nonsingular linear transformation

$$(1.461)\qquad x = Cy$$

with a constant matrix $C = (c_{ij})$ carries (1.401) into

$$(1.462)\qquad \frac{dy}{dt} = By + \varphi(y,\,t)$$

where

$$(1.463)\qquad B = (b_{ij}) = C^{-1}AC,\ \varphi(y,\,t) = C^{-1}f(Cy,\,t).$$

Using (1.461) and its reciprocal, we see that

$$(1.464)\qquad g_1\|y\| \leqq \|x\| \leqq g_2\|y\|.$$

The values of the positive constants g_1 and g_2 depend upon the matrix C.

The nonlinear terms $\varphi(y, t)$ are given by (1.463) and hence fulfill conditions of type (1.402), in view of the second inequality in (1.464). Then, since $||y||$ is small for large t we have

$$|\varphi_i(y, t)| < \varepsilon ||y|| \qquad (t \geqq T_\varepsilon) \tag{1.465}$$

for an arbitrarily small ε. Also, it follows from (1.464) that

$$M = \overline{\lim_{t\to+\infty}} \frac{\log ||y||}{t} = \overline{\lim_{t\to+\infty}} \frac{\log ||x||}{t}$$

and

$$m = \underline{\lim_{t\to+\infty}} \frac{\log ||y||}{t} = \underline{\lim_{t\to+\infty}} \frac{\log ||x||}{t}.$$

Thus M and m remain invariant under linear transformations (1.461).

A suitable choice of the matrix C in (1.461), yields a triangular matrix B in (1.462), say, with zeros above the main diagonal. Then $b_{ii} = \lambda_i$ are the characteristic roots of A, $b_{ij} = 0$ for $i < j$, and we may assume that $|b_{ij}| \leqq b$ for $i > j$ and for an arbitrarily small $b > 0$. The corresponding system (1.462) may be written

$$\frac{dy_i}{dt} = \lambda_i y_i + \sum_{j=1}^{i-1} b_{ij} y_j + \varphi_i(y_1, \ldots, y_n, t). \tag{1.466}$$

Multiplying (1.466) by $\bar{y}_i$, we get

$$\frac{d|y_i|^2}{dt} = 2\mathrm{Re}\left(\bar{y}_i \frac{dy_i}{dt}\right) = 2\alpha_i |y_i|^2 + 2\mathrm{Re}\left(\sum_{j=1}^{i-1} b_{ij}\bar{y}_i y_j + \bar{y}_i \varphi_i(y, t).\right)$$

Hence, if we write $|y_i| = r_i (i = 1, \ldots, n)$, $r_0 = 0$,

$$2\alpha_i r_i^2 - 2b \sum_{j=1}^{i-1} r_i r_j - 2r_i \varphi_i \leqq \frac{dr_i^2}{dt} \leqq 2\alpha_i r_i^2 + 2b \sum_{j=1}^{i-1} r_i r_j + 2r_i \varphi_i.$$

In view of (1.465), we have for $t \geqq T_\varepsilon$

$$2\alpha_i r_i^2 - 2(b + \varepsilon) r_i \sum_{j=1}^{n} r_j \leqq \frac{dr_i^2}{dt} \leqq 2\alpha_i r_i^2 + 2(b + \varepsilon) r_i \sum_{j=1}^{n} r_j. \tag{1.467}$$

Adding together the above inequalities for $i = 1, \ldots, n$, and letting $w = \sum_{i=1}^{n} r_i^2$, we get (cf. III, 1.821)

$$2(\alpha_n - 2nb - 2n\varepsilon)w \leqq \frac{dw}{dt} \leqq 2(\alpha_1 + 2nb + 2n\varepsilon)w.$$

Therefore

$$\frac{d}{dt}\left[we^{-2(\alpha_1+2nb+2n\varepsilon)t}\right] \leqq 0,$$

$$\frac{d}{dt}\left[we^{-2(\alpha_n-2nb-2n\varepsilon)t}\right] \geqq 0,$$

and hence

$$2M = \overline{\lim_{t\to+\infty}} \frac{\log w}{t} \leqq 2\alpha_1 + 4nb,$$

and

$$2m = \underline{\lim_{t\to+\infty}} \frac{\log w}{t} \geqq 2\alpha_n - 4nb.$$

Since M and n are invariant under (1.461) and are, therefore, independent of b, we have

$$\alpha_n \leqq m, \ M \leqq \alpha_1.$$

If $m = \alpha_1$, then $m = M = \alpha_1$.

Let $m < \alpha_1$, then

$$\alpha_k \leqq m < \alpha_{k-1}, \qquad 1 < k \leqq n.$$

If we write

$$w_1(t) = \sum_{i=1}^{k-1} r_i^2 \quad \text{and} \quad w_2(t) = \sum_{i=k}^{n} r_i^2,$$

as before, the inequalities (1.467) imply that

$$\text{(1.468)} \qquad \begin{aligned} \frac{dw_1}{dt} &\geqq 2\alpha_{k-1}w_1 - 4(nb + n\varepsilon)(w_1 + w_2), \\ \frac{dw_2}{dt} &\leqq 2\alpha_k w_2 + 4(nb + n\varepsilon)(w_1 + w_2). \end{aligned}$$

Subtracting the second one of these inequalities from the first, we obtain

$$\frac{d(w_1 - w_2)}{dt} \geqq (2\alpha_{k-1} - 8nb - 8n\varepsilon)w_1 - (2\alpha_k + 8nb + 8n\varepsilon)w_2.$$

Choose τ so that $m < \tau < \alpha_{k-1}$, and let ε and b be so small that

$$\alpha_{k-1} - 4nb - 4n\varepsilon > \tau > \alpha_k + 4nb + 4n\varepsilon.$$

Then,

$$\frac{d(w_1 - w_2)}{dt} \geqq 2\tau(w_1 - w_2)$$

or

$$\frac{d}{dt}[(w_1 - w_2)e^{-2\tau t}] \geqq 0. \tag{1.469}$$

Thus, the product $(w_1 - w_2)e^{-2\tau t}$ increases steadily.

We note that the inequality

$$\frac{\log(w_1 + w_2)}{t} < m + \tau$$

and, hence,

$$(w_1 + w_2)e^{-2\tau t} < e^{(m-\tau)t}$$

holds for infinitely many values t_s of t, such that $t_s \to +\infty$. Since $\lim_{t\to+\infty} e^{(m-\tau)t} = 0$, then for sufficiently large values t_s

$$(w_1 + w_2)e^{-2\tau t_s} < \gamma,$$

where $\gamma > 0$ and is arbitrarily small. Hence, in view of (1.469)

$$w_2(t) \geqq w_1(t) \qquad (t \geqq T_\varepsilon).$$

Since $w_1 + w_2 > 0$, we have $w_2 > 0$ for $t > T_\varepsilon$. Also, from the second inequality in (1.468), it follows that

$$\frac{dw_2}{dt} \leqq (2\alpha_k + 4nb + 4n\varepsilon)w_2,$$

whence

$$\overline{\lim_{t\to+\infty}} \frac{\log w_2}{t} \leqq 2\alpha_k + 8nb.$$

On the other hand,

$$2M = \overline{\lim_{t\to+\infty}} \frac{\log(w_1 + w_2)}{t} \leqq \overline{\lim_{t\to+\infty}} \frac{\log 2w_2}{t} = \overline{\lim} \frac{\log w_2}{t},$$

and hence $M \leqq \alpha_k$.

Thus $m = M = \alpha_k$. This completes the proof of 1.451.

1.47. The truth of 1.452 is obvious for $p = n$. Assume therefore that $p < n$.

Set $n - p = k$. Then $\alpha_i > a$ for $i \leqq k$, and $\alpha_i \leqq a$ for $i > k$. Determine $\lambda_0 > 0$ so that $\lambda_0 > -a$; then

$$(1.471) \qquad \alpha_i + \lambda_0 > 0 \quad \text{for} \quad i \leqq k.$$

Apply λ-transformation (1.491) with $\lambda = -\lambda_0$. The nonlinear part $\psi(x)$ of the resulting system (1.442) fulfills the conditions of Theorem 1.41, in view of (1.445). The characteristic roots of the new system are $\lambda_i + \lambda_0$ $(i = 1, \ldots, n)$.

The choice of λ_0 in (1.471) implies that our new system has exactly $p = n - k$ characteristic roots with negative real parts. By Theorem 1.41, there is a manifold, generated by p independent parameters, of solutions $x(t)$ which tend toward zero as $t \to +\infty$.

If the condition (1.452) of our theorem holds for $x(t)$, then

$$||x|| \leqq e^{(a+\varepsilon)t} = e^{(-\lambda_0-\varepsilon)t} \qquad (t > T_{1\varepsilon}),$$

where $0 < \varepsilon = -\frac{1}{2}(a + \lambda_0)$; hence

$$||\tilde{x}|| = e^{\lambda_0 t}\,||x|| \leqq e^{-\varepsilon t} \qquad (t > T_{1\varepsilon}),$$

that is $\tilde{x}(t)$ is a solution of the above p-parameter family. Conversely, if $\tilde{x}(t)$ is a member of this family, then, by 1.451,

$$\alpha = \lim_{t\to+\infty} \frac{\log ||x||}{t} = -\lambda_0 + \lim_{t\to+\infty} \frac{\log ||\tilde{x}||}{t} \leqq -\lambda_0.$$

Since $\alpha_q \leqq -\lambda_0$, we have $\alpha_q \leqq a$. This completes the proof of our theorem.

1.48. Theorem (I. G. Petrowsky). *If the real parts of all the characteristic roots of A are different from zero, then every trajectory of (1.401) passing through a sufficiently small neighborhood of the origin (the singular point) either tends toward the origin or leaves the neighborhood.*

1.481. We may assume that the matrix A in (1.401) is in the Jordan canonical form. Employing the notation of chapter III we may rewrite the integral form (III, 1.436) of (1.401) as two matrix equations

$$(1.4811) \qquad x^{(1)}(t) = Y_1(t - t_0)\tilde{y}^{(1)} + \int_{t_0}^{t} Y_1(t - \tau)f(x(\tau), \tau)d\tau,$$

$$(1.4812)\qquad x^{(2)}(t) = Y_2(t - t_0)\tilde{y}^{(2)} + \int_{t_0}^{t} Y_2(t-\tau)f(x(\tau), \tau)d\tau,$$

where we number the real parts α_i of the characteristic roots λ_i so that $\alpha_1 \leqq \alpha_2 \leqq \ldots \leqq \alpha_n$, $\alpha_i < 0$ for $i = 1, 2, \ldots, k$, $\alpha_i > 0$ for $i = k + 1, \ldots, n$. Also, $\tilde{y}^{(1)} = (y_{01}, \ldots, y_{0k}, 0, \ldots, 0)$ and $\tilde{y}^{(2)} = (0, \ldots, 0, y_{0k+1}, \ldots, y_{0n})$, $y^{(0)} = \tilde{y}^{(1)} + \tilde{y}^{(2)}$, $x(t) = x^{(1)}(t) + x^{(2)}(t)$, $Y(t) = Y_1(t) + Y_2(t)$.

Functions $f_i(x, t)$ fulfill the condition (1.402) and hence for every $\varepsilon > 0$, there is an $m > 0$ such that

$$|f(x, t)| < \varepsilon m \quad \text{for} \quad |x| < \frac{m}{n}.$$

Here $|x| = \max\{|x_1|, \ldots, |x_n|\}$. Write G_m for the domain of all x for which $|x| < m/n$. We assume that the initial point $y^{(0)} = \tilde{y}^{(1)} + \tilde{y}^{(2)} = x^{(1)}(t_0) + x^{(2)}(t_0)$ is in G_m and we choose ε so that

$$\frac{\varepsilon}{\alpha_{k+1}} < \tfrac{1}{8}.$$

If m is small enough and $x^{(1)}(t_0) = \tilde{x}^{(1)} \in G_m$, then $x^{(1)}(t) \to 0$ as $t \to +\infty$ by 1.43.

Assume that the elementary divisors corresponding to $\lambda_{k+1}, \ldots, \lambda_n$ are all simple.[4] Then

$$x_i(t) = y_{0i}e^{\lambda_i(t-t_0)} + u(t) \qquad (i = k+1, \ldots, n)$$

where

$$(1.4813)\qquad |u(t)| = \left| e^{\lambda_i t}\int_{t_0}^{t} e^{-\lambda_i \tau} f_i(x, \tau)d\tau \right| \leqq \frac{\varepsilon m}{\alpha_i n}(e^{\alpha_i(t-t_0)} - 1)$$

$$< \frac{\varepsilon m}{\alpha_i n} e^{\alpha_i(t-t_0)} \quad \text{for} \quad t \geqq t_0.$$

Suppose that there exists a trajectory $x = x(t)$, $x(t_0) = y^{(0)} \in G_m$, which neither tends toward zero nor leaves the region G_m. Let

$$L = \max_{1 \leqq i \leqq n} \overline{\lim_{t \to +\infty}} |x_i(t)|.$$

Then $0 < L < m/n$. Since $\overline{\lim}_{t\to+\infty} |x_i(t)| = 0$ for $i \leqq k$, the component $x^{(2)}(t)$ of $x(t)$ is not zero, and, for some $h \geqq k + 1$,

[4]Our theorem holds true in general. We give a complete proof only in this simplest case.

$$L = \overline{\lim_{t\to+\infty}} |x_h(t)|. \tag{1.4814}$$

Choose $m_1 < m$ so that $L < m_1/n < 2L$, and if $|y^{(0)}| < \frac{1}{2}L$, then choose $y^{(0)} = y(t_0)$ for a suitable new t_0, as a new initial point on our trajectory so that $|y^0| \geqq \frac{1}{2}L$. Then, by (1.4813) with m replaced by m_1,

$$\begin{aligned}|x_h(t)| \geqq \big| |y_{0h}| e^{\alpha_n(t-t_0)} - |u(t)| \big| &\geq e^{\alpha_h(t-t_0)}\left(|y_{0h}| - \frac{\varepsilon m_1}{\alpha_h n}\right) \\ &\geqq e^{\alpha_h(t-t_0)}\left(\frac{m_1}{4n} - \frac{m_1}{8n}\right) \\ &= e^{\alpha_h(t-t_0)}\frac{m_1}{8n},\end{aligned}$$

which contradicts (1.4814).

1.5. In what follows we shall study the behavior of the tangents to an O-curve near the origin. First we consider a lemma relating to autonomous systems [5].

1.51. Lemma. *Consider systems* (1.401) *and* (1.442). *Here we assume that* (1.442) *is obtained from* (1.401) *by means of a λ-transformation* (1.441) *with $\lambda > 0$. Let O_2 denote the set of integral curves of* (1.442) *which tend toward the origin as $t \to -\infty$ (the O^--curves), and let O_1 be the set of corresponding curves of* (1.401). *Assume moreover that* (1.401) *is an autonomous system (and A, canonical).*

If the real parts $\alpha_1 \leqq \alpha_2 \leqq \ldots \leqq \alpha_k$ of the first k characteristic roots of (1.442) *are negative and the real parts $\alpha_{k+1} \leqq \ldots \leqq \alpha_n$ of the remaining characteristic roots are positive, then the curves in O_2 form an $(n-k)$-dimensional hypersurface S_1, and every manifold of k dimensions normal to the $(n-k)$-dimensional hyperplane $S_{0,k}$ consisting of points $(0, \ldots, 0, x_{k+1}, \ldots, x_n)$, intersects S_1 in a single point.*

Since every trajectory in O_2 is given by

$$\tilde{x} = \tilde{x}(t) = (0, \ldots, 0, \tilde{x}_{k+1}(t), \ldots, \tilde{x}_n(t))$$

and thus lies in the hyperplane $\tilde{S}_{0,k}$, the corresponding trajectory

$$x = x(t) = e^{\lambda t}\tilde{x}(t) = (0, \ldots, 0, e^{\lambda t}\tilde{x}_{k+1}(t), \ldots, e^{\lambda t}\tilde{x}_n(t))$$

of (1.401) lies in the hyperplane $S_{0,k}$. Thus all the trajectories in O_1 lie in $S_{0,k}$. Hence, if two trajectories in O_1 cut the same normal to $S_{0,k}$, then these two trajectories intersect, and since (1.401) is an autonomous system, they coincide. We merely state without proof:

1.52. THEOREM. (I. G. Petrowsky [5]). *If the real parts of all the characteristic roots of the matrix A in* (1.401) *are positive* (*negative*) *and if the nonlinear part* $f(x, t)$ *fulfills the condition* (1.402), *then as* $t \to -\infty$ ($t \to +\infty$) *almost all integral curves are tangent to the hyperplane determined by the leading coordinates.*

2. Lyapunov stability

Consider the system

$$\frac{dy}{dt} = A(t)y + f(y, t), \tag{2.1}$$

where $A(t)$ is a matrix with variable coefficients. As before we shall be concerned with the comparison of the behavior of trajectories of this system near the origin with the behavior of trajectories near the origin of the truncated linear system

$$\frac{dy}{dt} = A(t)y. \tag{2.2}$$

This problem has not been solved completely. Most of the known results are concerned with the Lyapunov stability of the trivial solution. It is this last problem which we shall study in this section.

We replace our nonlinear equation (1) by the integral equation (cf. 1.4)

$$y(t) = Y(t)y_0 + \int_{t_0}^{t} Y(t)Y^{-1}(\tau)f(y(\tau), \tau)d\tau$$

where $Y(t) = (y_{ri}(t, t_0))$ is a fundamental matrix of solutions of the linear system (2.2), and $Y^{-1}(t)$ is the reciprocal of $Y(t)$. For definiteness we may assume that $Y(t_0)$ is the identity matrix, for if this is not so we may replace $Y(t)$ by the fundamental matrix $Y_1(t) = Y(t)Y^{-1}(t_0)$ such that $Y_1(t_0) = I$.

2.1. THEOREM. *If the linear system* (2.2) *has a fundamental matrix of solutions* $(y_{ik}(t, t_0))$, *such that*

$$|y_{ik}| \leqq Be^{-\alpha(t-t_0)}$$

where B and α *are positive constants which do not depend upon the* *initial values t_0, then all the solutions of the system* (1) *are O-curves or, in other words, the trivial solution is stable according to Lyapunov.*

This theorem can be proved by the method of successive approximations.

2.2. Arbitrary nonlinear systems. Lyapunov's second method. Consider a general nonlinear system

$$\frac{dx_i}{dt} = f_i(x_1, x_2, \ldots, x_n, t).$$

In this section we shall discuss the problem of the stability of the trivial solution of this nonlinear system.

We introduce the following definitions.

We say that a continuous function $w(x_1, x_2, \ldots, x_n)$ is positive definite if

$$w(0, \ldots 0) = 0$$

and if

$$w(x_1, \ldots, x_n) > 0 \quad \text{for} \quad \sum_{i=1}^{n} x_i^2 \neq 0.$$

In an analogous manner we define a negative definite function.

Let us consider a level surface $w = c$ of the function $w(x_1, x_2, \ldots, x_n)$ in the $(n+1)$-dimensional $(x_1, x_2, \ldots, x_n, t)$. For sufficiently small values of c as we shall show, these cylindrical surfaces partition our $(n+1)$-dimensional space into two domains. One of these, namely that which contains the t-axis, we call the interior and the other the exterior. Draw a simple arc from a point on the t-axis to a point on the level surface $w = c_0$. Such an arc cuts across every level surface $w = c$ for $c < c_0$. Our assertion that for sufficiently small c the level surface $w = c$ separates the space into two domains, the interior and the exterior, follows at once from the last remark. For, suppose the contrary, namely, that $w(x) = c$ does not separate our space into two domains no matter how small the value of c. Then we could find a point p on the cylinder $\Sigma x_i^2 = 1$ which can be connected with the t-axis by a simple arc. Since $w(p) \leqq c$ and since c is arbitrarily small there would exist in our cylinder a point at which $w = 0$, which contradicts our hypothesis.

We observe that for every $\varrho_0 > 0$ we can find a c_0 such that the level surface $w = c_0$ is contained inside the cylinder $\Sigma x_i^2 = \varrho_0$.

We shall say that a function $v(x_1, x_2, \ldots, x_n, t)$, $v(0, \ldots, 0, t) = 0$, is positive definite if there exists a positive definite function $w(x_1, x_2, \ldots, x_n)$ such that $v \geqq w$ for all t. A level surface of $v(x_1, x_2, \ldots, x_n, t) = c$ is no longer cylindrical. However, for sufficiently small c it still separates the space into two domains.

For every c the cylindrical level surface $w = c$ contains in its interior the surface $v = c$.

Among definite functions a special role is played by functions which tend to zero as $\Sigma x_i^2 \to 0$ uniformly in t for $t \geqq t_0$. Such functions have the property that for every $\varepsilon > 0$ we can find such a ϱ_0 that

$$v(x_1, \ldots, x_n, t) \leqq \varepsilon$$

as long as $\Sigma x_i^2 \leqq \varrho_0$ and $t \geqq t_0$. The above condition implies that for every c we can find ϱ_0 such that the cylinder $\Sigma x_i^2 \leqq \varrho_0$ lies completely in the interior of the level surface $v(x_1, x_2, \ldots, x_n, t) = c$, i.e., the section of our surface by the plane $t = h$ does not contract to a point as $h \to \infty$.

By a derivative of a function $v(x_1, x_2, \ldots, x_n, t)$ along a trajectory of our system we shall mean the expression

$$\frac{dv}{dt} = \frac{\partial v}{\partial t} + \frac{\partial v}{\partial x_1} f_1 + \ldots + \frac{\partial v}{\partial x_n} f_n.$$

2.3. THEOREM (Lyapunov [31]). *In order that the trivial solution of the system*

$$\frac{dx_i}{dt} = f_i(x_1, \ldots, x_n, t)$$

with continuous right-hand members be uniformly stable (cf. Chap. III) it is sufficient that there exists a positive definite function $v(x_1, x_2, \ldots, x_n, t)$ which tends to zero with Σx_i^2 uniformly in t, and whose derivative along any trajectory sufficiently close to the trivial solution should be non-positive.

We can find t_0 and ϱ_0 such that for $\Sigma x_i^2 \leqq \varrho_0$, $t \geqq t_0$, we have $dv/dt \leqq 0$.

Let $\varepsilon < \varrho_0$. Let us find a level surface $v = c$ which lies entirely in the interior of the cylinder $\Sigma x_i^2 = \varepsilon$ and an η such that the cylinder $\Sigma x_i^2 = \eta$ lies entirely in the interior of the level surface $v = c$. Let us consider a trajectory whose initial conditions satisfy the inequalities $\Sigma x_{i1}^2 \leqq \eta$ and $t_1 \geqq t_0$. Set $v(x_{11}, \ldots, x_{n1}, t_1) = v_0$ and note that $v_0 \leqq c$. For positive values of t this trajectory cannot leave the cylinder $\Sigma x_i^2 \leqq \varepsilon$. For along an integral curve lying within the cylinder $\Sigma x_i^2 \leqq \varrho_0$ we have

$$v = v_0 + \int_{t_0}^{t} \frac{dv}{dt}\, dt, \qquad \frac{dv}{dt} \leqq 0$$

and consequently $v \leqq v_0 \leqq c$, whereas if this trajectory should leave the cylinder $\Sigma x_i^2 \leqq \varepsilon$ remaining within the cylinder $\Sigma x_i^2 \leqq \varrho_0$, then it should reach a level surface $w = c_1$ with $c_1 > c$.

This completes the proof of our theorem.

Theorem 2.3 can be extended to the case when level surfaces $v = c$ have lines along which only left or right partial derivatives with respect to x_i exist. In this case we may assume that dv/dt denotes a left (right) derivative.

If the right-hand members of our system do not depend on t (the stationary case) then the existence of a positive definite function v described in Theorem 2.3 would imply that all the solutions in a sufficiently small neighborhood of the origin will remain bounded for large t. More precisely, for every $\varepsilon > 0$ we can find a sufficiently small neighborhood of the origin such that all the solutions through an initial point in the small neighborhood remain in the ε-neighborhood for large values of t. We cannot claim in general that such solutions tend to the origin as $t \to \infty$. Such a behavior would correspond to the asymptotic stability of the nonstationary systems.

2.4. THEOREM. (Lyapunov). *In order that the trivial solution of the system*

$$\frac{dx_i}{dt} = f_i(x_1, x_2, \ldots, x_n, t)$$

be asymptotically and uniformly stable it is sufficient that there exists a positive definite function $v(x_1, x_2, \ldots, x_n, t)$ *which tends to zero with* Σx_i^2 *uniformly in* t *and such that its derivative taken along the trajectories of our system be negative definite.*

Let t_0 be so large and ϱ_0 be so small that for $t \geqq t_0$ and for $\Sigma x_i^2 \leqq \varrho_0$ we have

$$v(x_1, x_2, \ldots, x_n, t) \geqq w(x_1, \ldots, x_n) \geqq 0$$

$$\frac{dv}{dt}(x_1, \ldots, x_n, t) \leqq w_1(x_1, \ldots, x_n) \leqq 0$$

where w and w_1 are a positive definite and a negative definite function respectively.

Choose c_ϱ so that the level surface $v = c$, for $c \leqq c_\varrho$, should lie entirely within the cylinder $\Sigma x_i^2 = \varrho_0$. We assert that every trajectory which passes through a point P_0 on the surface $v = c$ for $t = t_0$, enters the interior of the surface defined by $v = c$, for $t > t_0$.

Let $l = \min dv/dt \neq 0$ in the ring-shaped region between c and c_ϱ. Then in this region we have $dv/dt < -l$. The last inequality implies that our trajectory does not leave the region $v \leqq c_\varrho$ and that in this region

$$v \leqq v(P_0) - l(t - t_0)$$

as long the trajectory remains in the above mentioned ring-shaped region. Consequently our trajectory cuts across the level surface determined by c in finite time and remains in the interior of this surface. This completes the proof of our theorem.

The auxiliary positive definite functions $v(x_1, x_2, \ldots, x_n, t)$ may be used to investigate the stability of the trivial solution for a variety of nonlinear and linear systems of differential equations. For a more complete discussion of this method we refer the reader to the original work of Lyapunov and to the tract of Četaev on the stability of motion.

3. The Behavior of Trajectories in the Neighborhood of a Closed Trajectory

3.1. Formulation of the problem. Consider the system

$$\frac{dy_i}{dt} = f_i(y_1, \ldots, y_{n+1}), \qquad (i = 1, \ldots, n+1), \tag{3.11}$$

and let L: $y_i = \varphi_i(t)$ $(i = 1, \ldots, n + 1)$ be a periodic solution (a closed trajectory)of period ω: Thus $\varphi_i(t + \omega) \equiv \varphi_i(t)$.

The creators of the present qualitative theory, Poincaré, Lyapunov and Birkhoff, have devoted much attention to the study of the behavior of trajectories in the neighborhood of a closed trajectory. In general, in the higher dimensional spaces this behavior is much more complicated than the behavior of such trajectories in the plane.

We consider first the fundamental case in which the structure of the neighborhood of the periodic solution is determined by the character of the linear terms. We assume that (Conditions (A))

(1) the functions $f_i(y_1, \ldots, y_{n+1})$ are continuous in some neighborhood G of the curve $y_i = \varphi_i(t)$ and have there partial derivatives with respect to all their arguments, up to and including the second order, and that

(2) these partial derivatives are continuous on the closed trajectory itself.

For all points of the curve L: $y_i = \varphi_i(t)$, and hence in some neighborhood $D(L)$ of this curve, the sum $\sum_{i=1}^{n+1} f_i^2(y_1, \ldots, y_{n+1})$ does not vanish.

If we take the arc length s as a parameter instead of t, then

$$\frac{dy_i}{ds} = \frac{f_i(y_1, y_2, \ldots, y_{n+1})}{\sqrt{f_1^2 + f_2^2 + \ldots f_{n+1}^2}} = \bar{f}_i(y_1, y_2, \ldots, y_{n+1}) \quad (i = 1, 2, \ldots, n+1). \tag{3.12}$$

The new system (3.12) has no singular points in the neighborhood $D(L)$ of our closed trajectory L.

With each point $P(s)$ of the trajectory L: $y_i = \varphi_i(t) = \bar{\varphi}_i(s)$ we associate the hyperplane of n dimensions normal to L at $P(s)$. In view of conditions (A), the functions $\varphi_i(t) = \bar{\varphi}_i(s)$ have continuous second derivatives, the trajectory L has a continuously varying radius of curvature ϱ given by

$$\varrho^2 = \frac{1}{\sum_{i=1}^{n+1} \left(\frac{d^2 y_i}{ds^2}\right)^2} = \frac{1}{\sum_{i=1}^{n+1} \left(\frac{d\bar{f}_i}{ds}\right)^2},$$

and hence $\varrho(s) > \varepsilon > 0$ on L.

It can be shown that we can find a small enough neighborhood $D(L)$ of L, such that every point of this neighborhood lies on one and only one normal hyperplane to L.

We introduce in each normal hyperplane an orthogonal Cartesian coordinate system with $P(s)$ as the origin. Then each point P in $D(L)$ has coordinates $(s, x_1, \ldots, x_n)$ where s is determined (up to an integral multiple of the length s_1 of L) by the hyperplane in which P lies and $x_1, \ldots, x_n$ are the Cartesian coordinates of P in its hyperplane.

The new and the old coordinates in $D(L)$ are related by the equations

$$y_i = \sum_{j=1}^{n} b_{ij}(s) x_j + \bar{\varphi}_i(s) \qquad (i = 1, 2, \ldots, n + 1),$$

where $b_{ij}(s)$ have continuous first derivatives and where $\bar{\varphi}_i(s)$ have continuous derivatives of the second order as well.

One may choose the x_i so that $B = ||b_{ij}||$ is orthogonal. Then the determinant $|B| = \pm 1$. We set $b_{i,n+1} = \bar{\varphi}_i(s)$.

Substituting the new variables into (3.12), we obtain

$$\sum_{j=1}^{n} b'_{ij}(s)x_j + \sum_{j=1}^{n} b_{ij}(s)\frac{dx_j}{ds} + \overline{\varphi'_i(s)} = \bar{\bar{f}}_i(x_1,\ldots,x_n,s), \ (i = 1,\ldots,n+1).$$

The functions $\bar{\bar{f}}_i(x_1, \ldots, x_n, s)$ are continuous and are periodic in s of period s_1. Rewriting the last system in the form

$$\sum_{j=1}^{n} b_{ij}(s)\frac{dx_j}{ds} + b_{i,\,n+1}(s)\frac{ds}{ds} = \bar{\bar{f}}_i - \sum_{j=1}^{n} b'_{ij}(s)x_j, \quad (i = 1, 2, \ldots, n+1)$$

and remembering that $|b_{ij}| = |B| = \pm 1$, we have

$$\frac{dx_i}{ds} = \pm \begin{vmatrix} b_{11}(s), \ldots, \bar{\bar{f}}_1 - \sum_{j=1}^{n} b'_{1j}(s)x_j, \ldots, b_{1,\,n+1}(s) \\ b_{21}(s), \ldots, \bar{\bar{f}}_2 - \sum_{j=1}^{n} b'_{2j}(s)x_j, \ldots, b_{2,\,n+1}(s) \\ b_{n+1,1}(s), \ldots, \bar{\bar{f}}_{n+1} - \sum_{j=1}^{n} b'_{n+1,j}(s)x_j, \ldots, b_{n+1,\,n+1}(s) \end{vmatrix}.$$

Denote the above determinant by $F_i(x_1, \ldots, x_n, s)$. Since the $\bar{\bar{f}}_i$ are continuous and have continuous derivatives of the first two orders with respect to all variables, and assuming also the same for the b_{ij} relative to s, this must also hold true for F_i. Moreover, since $x_1 = x_2 = \ldots = x_n = 0$ is a solution of our new system, $F_i(0, \ldots, 0, s) = 0$ and

$$F_i(x_1, \ldots, x_n, s) = \sum_{i=1}^{n} a_{ij}(s)x_j + \theta_i(x_1, \ldots, x_n, s),$$

it follows that

$$|\theta_i(x_1, \ldots, x_n, s)| \leqq \varepsilon(|x_1| + |x_2| + \ldots + |x_n|)$$

for small values of $\Sigma_i |x_i|$ and for an arbitrary $\varepsilon > 0$.

Thus we have reduced the study of the behavior of integral curves near a closed trajectory to the study of the behavior of solutions of the system

$$\frac{dx_i}{ds} = \sum_{i=1}^{n} a_{ij}(s)x_j + \theta_i(x_1, \ldots, x_n, s) \tag{3.13}$$

in the neighborhood of the trivial solution. Here a_{ij} and θ_i are periodic in s.

Trajectories lying near a closed trajectory L may be classified according to their behavior as follows:

(a) closed trajectories;

(b) one-sided asymptotic trajectories, i.e., those trajectories for which one but not both of the α- or ω-limit sets lies on L. Here one half-trajectory leaves the neighborhood $D(L)$ of L in finite time;

(c) two-sided asymptotic trajectories, i.e., those for which both the α- and the ω-limit sets lie on L;

(d) trajectories stable in one direction according to Lagrange, i.e., the trajectories for which one (and only one) of the α- or the ω-limit sets lies in a given neighborhood $D(L)$ of L;

(e) trajectories stable in both directions according to Lagrange;

(f) saddle curves, i.e., integral curves which leave the neighborhood $D(L)$ for both positive and negative numerically large values of the parameter.

It is clear that properties (a)—(f) are invariant under the transformation into the space of local coordinates and can therefore be investigated by studying the canonical system (3.13).

3.2. Variation equations. In what follows we again take

$$F_i(x_1, \ldots, x_n, t) = \sum_{j=1}^{n} a_{ij}(t)x_j + \theta_i(x_1, \ldots, x_n; t), \quad (i = 1, \ldots, n)$$

as analytic in $x_1, \ldots, x_n, t$; then $a_{ij}(t) = (\partial F_i/\partial x_j)_0$ and each $\theta_i(x_1, \ldots, x_n; t)$ is a power series in $x_1, \ldots, x_n$ which contains no terms of degree lower than two. In this case the system (3.13) may be written in the form

$$\frac{dx_i}{dt} = \left(\frac{\partial F_i}{\partial x_1}\right)_0 x_1 + \ldots + \left(\frac{\partial F_i}{\partial x_n}\right)_0 x_n + \theta_i(x_1, \ldots, x_n; t), \quad (i = 1, \ldots, n). \tag{3.21}$$

The associated linear system

$$\frac{d\xi_i}{dt} = \left(\frac{\partial F_i}{\partial x_1}\right)_0 \xi_1 + \ldots + \left(\frac{\partial F_i}{\partial x_n}\right)_0 \xi_n \qquad (i = 1, \ldots, n) \tag{3.22}$$

is called the system of *variation equations* for the system (3.21).

The following discussion justifies the name *variation equations* in two different ways.

Let $x_i = \varphi_i(t)$ be a periodic solution of the system

$$\frac{dx_i}{dt} = F_i(x_1, \ldots, x_n;\ t).$$

Set

$$x_i(t) = \varphi_i(t) + \xi_i(t). \tag{3.23}$$

The functions $\xi_i(t)$ (usually small in absolute value) are called variations of the functions $x_i(t)$. We have

$$\frac{d\xi_i}{dt} = -\frac{d\varphi_i}{dt} + F_i(\varphi_1 + \xi_1, \ldots, \varphi_n + \xi_n;\ t)$$

or, since

$$\frac{d\varphi_i}{dt} = F_i(\varphi_1, \ldots, \varphi_n;\ t),$$

we have

$$\frac{d\xi_i}{dt} = \frac{\partial F_i}{\partial x_1}\xi_1 + \ldots + \frac{\partial F_i}{\partial x_n}\xi_n + \psi_i(\xi_1, \ldots, \xi_n;\ t), \tag{3.24}$$

where ψ_i are power series which contain no terms of degree lower than two in $\xi_1, \ldots, \xi_n$, and where the $\partial F_i/\partial x_j$ are evaluated along our periodic solution. Discarding the terms of higher order in the variations ξ_i we obtain variation equations which assume the form (3.22) when the given periodic solution is the trivial solution.

The variation equations may be derived in another (useful) manner as follows. We let the periodic solution $x_i = \varphi_i(t)$ be imbedded in a one-parameter family of solutions $\tilde{x}_i = \tilde{\varphi}_i(t, \mu)$ so that $\tilde{\varphi}_i(t, \mu_0) = \varphi_i(t)$. Differentiating the identities,

$$\frac{d\tilde{\varphi}_i}{dt} = F_i(\tilde{\varphi}_1, \tilde{\varphi}_2, \ldots, \tilde{\varphi}_n; t), \qquad (i = 1, \ldots, n)$$

with respect to μ, setting $\mu = \mu_0$, and writing $(\partial\varphi_i/\partial\mu)_{\mu=\mu_0} = \xi_i$, we obtain (3.24), i.e., the variation equations.

3.21. THEOREM. *If $\Phi(x_1, \ldots, x_n) = C$ is an integral of a system of (nonlinear) differential equations, then $(\partial\Phi/\partial x_1)\xi_1 + \ldots + (\partial\Phi/\partial x_n)\xi_n = const.$ is an integral of the corresponding system of variation equations.*

Substituting into Φ the solutions of our one-parameter family $x_i = \tilde{\varphi}_i(t, \mu)$, we see that

$$\Phi(\tilde{\varphi}_1, \ldots, \tilde{\varphi}_n) = \tilde{C}(\mu),$$

where by the definition of an integral the right hand member $\tilde{C}(\mu)$ does not depend on t. Differentiating the last identity with respect to μ and setting $\mu = \mu_0$, we obtain the identity

$$\left(\frac{\partial \Phi}{\partial x_1}\right)_{x_i=\varphi_i(t)} \xi_1 + \ldots + \left(\frac{\partial \Phi}{\partial x_n}\right)_{x_i=\varphi_i(t)} \xi_n = C',$$

where C' is a constant. This completes the proof of our theorem.

We should note that if a periodic solution $x_i = \varphi_i(t)$ is imbedded in a family of solutions

$$x_1 = \tilde{\varphi}_1(t;\ h_1, \ldots, h_p), \ldots, x_n = \tilde{\varphi}_n(t;\ h_1, \ldots, h_p)$$

depending on p parameters $h_1, \ldots, h_p$, and if $\tilde{\varphi}_i(t; 0, \ldots, 0) = \varphi_i(t)$, then the variation equations have p particular solutions

$$\xi_1^{(k)} = \left(\frac{\partial \tilde{\varphi}_1}{\partial h_k}\right)_0, \ldots, \xi_n^{(k)} = \left(\frac{\partial \tilde{\varphi}_n}{\partial h_k}\right)_0 \qquad (k = 1, \ldots, p).$$

3.3. Linear systems [23]. We consider a linear system

$$\frac{dx_i}{dt} = \sum_{k=1}^{n} a_{ik}(t) x_k, \qquad (i = 1, \ldots, n), \tag{3.31}$$

where $a_{ik}(t)$ are periodic functions of period ω (cf. equations (3.13))

Let the n solutions

$$x_1 = x_{1k}(t),\ x_2 = x_{2k}(t), \ldots, x_n = x_{nk}(t), \quad (k = 1, \ldots, n)$$

form a fundamental system of solutions of (3.31). Then

$$x_1 = x_{1k}(t + \omega), \ldots, x_{nk}(t + \omega) \quad (k = 1, \ldots, n)$$

also form a fundamental system and hence, for $(i = 1, 2, \ldots, n)$,

$$\begin{aligned} x_{i1}(t + \omega) &= b_{11} x_{i1}(t) + \ldots + b_{1n} x_{in}(t), \\ \ldots \quad & \quad \ldots \quad \ldots \quad \ldots \\ x_{in}(t + \omega) &= b_{n1} x_{i1}(t) + \ldots + b_{nn} x_{in}(t), \end{aligned}$$

where the matrix $B = [b_{ij}]$ is nonsingular.

Suppose that B has only the linear elementary divisors

$$s - s_1,\ s - s_2, \ldots, s - s_n,$$

where $s_1, s_2, \ldots, s_n$ are the roots of the characteristic equation

$$|B^{\bullet} - sI| = \begin{vmatrix} b_{11} - s & b_{12} & \ldots b_{1n} \\ b_{21} & b_{22} - s & \ldots b_{2n} \\ b_{n1} & b_{n2} & \ldots b_{nn} - s \end{vmatrix} = 0.$$

Then there exists a nonsingular linear transformation which reduces B to the diagonal form

$$\begin{bmatrix} s_1 & 0 \ldots 0 \\ 0 & s_2 \ldots 0 \\ \cdot & \cdot\ \cdot\ \cdot\ \cdot \\ 0 & 0 \ldots s_n \end{bmatrix}.$$

It follows that there exists a fundamental system of solutions

$$F_{1k}(t),\ F_{2k}(t), \ldots, F_{nk}(t)$$

such that

$$\begin{aligned} F_{i1}(t + \omega) &= s_1 F_{i1}(t) \\ \ldots \quad & \ldots \quad \ldots \\ F_{in}(t + \omega) &= s_n F_{in}(t). \end{aligned}$$

Write[5] $s_i = e^{\omega r_i}$ $(i = 1, \ldots, n)$ and let

$$\begin{aligned} F_{i1}(t) &= e^{r_1 t} \varphi_{i1}(t) = s_1^{\frac{t}{\omega}} \varphi_{i1}(t), \\ \ldots \quad & \ldots \quad \ldots \qquad (i = 1, \ldots, n) \\ F_{in}(t) &= e^{r_n t} \varphi_{in}(t) = s_n^{\frac{t}{\omega}} \varphi_{in}(t). \end{aligned}$$

Then $\varphi_{i1}(t), \ldots, \varphi_{in}(t)$ are periodic functions of t of period ω. For, on the one hand we have

$$F_{ij}(t + \omega) = e^{r_j(t+\omega)} \varphi_{ij}(t + \omega) = e^{\omega r_j} e^{r_j t} \varphi_{ij}(t + \omega) = s_j e^{r_j t} \varphi_{ij}(t + \omega),$$

while on the other

$$F_{ij}(t + \omega) = s_j F_{ij}(t) = s_j e^{r_j t} \varphi_{ij}(t).$$

Therefore

$$\varphi_{ij}(t + \omega) = \varphi_{ij}(t) \qquad (i, j = 1, \ldots, n).$$

[5]Cf. Chap. III.

If the elementary divisors of B are given by

$$(s - s_1)^{\mu_1}, \ldots, (s - s_k)^{\mu_k} \qquad (\mu_1 + \mu_2 + \ldots + \mu_k = n),$$

then B can be reduced to the canonical form

$$\begin{bmatrix} M_1 & 0 & \ldots & 0 \\ 0 & M_2 & \ldots & 0 \\ \cdot & \cdot & \cdots & \cdot \\ 0 & 0 & \ldots & M_k \end{bmatrix}, \quad \text{where } M_i = \begin{bmatrix} s_i & 0 & \ldots & 0 & 0 \\ 1 & s_i & \ldots & 0 & 0 \\ \cdot & \cdot & \cdots & \cdot & \cdot \\ 0 & 0 & & 1 & s_i \end{bmatrix}$$

is a matrix of order μ_i. Therefore there exists a fundamental system of solutions F_{ij} falling into k groups

$$[F_{i,\eta_\alpha+1}(t), \quad F_{i,\eta_\alpha+2}(t) \ldots F_{i,\eta_\alpha+\mu_\alpha}(t)], \quad \eta_\alpha = \mu_1 + \ldots + \mu_{\alpha-1},$$

$$(\alpha = 1, 2, \ldots, k, \quad i = 1, 2, \ldots, n).$$

For each of these groups the replacement of t by $t + \omega$ yields

$$F_{i,\eta_\alpha+1}(t + \omega) = s_\alpha F_{i,\eta_\alpha+1}(t),$$

$$F_{i,\eta_\alpha+2}(t + \omega) = s_\alpha F_{i,\eta_\alpha+2}(t) + F_{i,\eta_\alpha+1}(t),$$

$$\cdot \quad \cdot \quad \cdot \quad \cdot \quad \cdot \quad \cdot \quad \cdot \quad \cdot \quad \cdot \quad \cdot \quad \cdot \quad \cdot \quad \cdot \quad \cdot$$

$$F_{i,\eta_\alpha+\mu_\alpha}(t + \omega) = s_\alpha F_{i,\eta_\alpha+\mu_\alpha}(t) + F_{i,\eta_\alpha+\mu_\alpha-1}(t).$$

We introduce the auxiliary polynomials

$$g_1(t) = \frac{t}{\omega}, \quad g_2(t) = \frac{g_1(g_1 - 1)}{2!}, \ldots, g_q(t) = \frac{g_1(g_1 - 1)\ldots(g_1 - q + 1)}{q!}, \ldots;$$

observe that

$$g_1(t + \omega) = g_1(t) + 1,$$

$$g_2(t + \omega) = g_2(t) + g_1(t),$$

$$\cdot \quad \cdot \quad \cdot \quad \cdot \quad \cdot \quad \cdot \quad \cdot \quad \cdot \quad \cdot \quad \cdot$$

$$g_q(t + \omega) = g_q(t) + g_{q-1}(t),$$

$$\cdot \quad \cdot \quad \cdot \quad \cdot \quad \cdot \quad \cdot \quad \cdot \quad \cdot \quad \cdot \quad \cdot$$

Next we define functions $\varphi_j^{(i,\alpha)}(t)$ by the relations

$$F_{i,\eta_\alpha+1}(t) = e^{r_\alpha t}\varphi_1^{(i,\alpha)}(t),$$

$$F_{i,\eta_\alpha+2}(t) = e^{r_\alpha t}[\varphi_2^{(i,\alpha)}(t) + \frac{1}{s_a}\varphi_1^{(i\ \alpha)}(t)g_1(t)],$$

$$(3.32)\qquad F_{i,\eta_\alpha+3}(t) = e^{r_\alpha t}[\varphi_3^{(i,\alpha)}(t) + \frac{1}{s_\alpha}\varphi_2^{(i,\alpha)}(t)g_1(t) + \frac{1}{s_\alpha^2}\varphi_1^{(i,\alpha)}(t)g_2(t)],$$

$$\cdots\cdots\cdots\cdots\cdots\cdots\cdots\cdots$$

$$F_{i,\eta_\alpha+\mu_\alpha}(t) = e^{r_\alpha t}[\varphi_{\mu_\alpha}^{(i,\alpha)}(t) + \frac{1}{s_\alpha}\varphi_{\mu_\alpha-1}^{(i,\alpha)}(t)g_1(t) + \ldots + \frac{1}{s_\alpha^{\mu_\alpha-1}}\varphi_1^{(i,\alpha)}(t)g_{\mu_\alpha-1}(t)].$$

It is easy to verify that $\varphi_j^{(i,\alpha)}(t)$ are periodic of period ω.

We let

$$F_{i,\eta_\alpha+\mu_\alpha}(t) = e^{r_\alpha t}P_{i,\alpha}(t)$$

where $P_{i,\alpha}$ is a polynomial of degree $\mu_\alpha - 1$ in t with coefficients which are periodic in t of period ω. Then in taking the differences $\Delta P_{i\alpha} = P_{i\alpha}(t+\omega) - P_{i\alpha}(t)$ one may treat the coefficients as constants. With this in mind we write (3.32) in the compact form

$$F_{i,\eta_\alpha+\mu_\alpha-1}(t) = e^{r_\alpha t}s_\alpha \Delta P_{i,\alpha}(t)$$

$$(3.33)\qquad \cdots\cdots\cdots$$

$$F_{i,\eta_\alpha+1}(t) = e^{r_\alpha t}s_\alpha^{\mu_\alpha-1}\Delta^{\mu_\alpha-1}P_{i,\alpha}(t).$$

In studying the properties of closed trajectories it is useful to distinguish the following four cases. Here, with every closed trajectory L we associate a sufficiently small torus-shaped neighborhood $D(L)$.

Case 1. *Asymptotically stable periodic solutions* (*limit cycles*). In this case the closed trajectory L is an α-limit set (ω-limit set) for each of the integral curves which enter the neighborhood $D(L)$ of L. Periodic solutions of a linear system are asymptotically stable if the real parts of all the characteristic exponents are different from zero and have the same sign.

Case 2. *Unstable periodic solutions.* Here all integral curves through $D(L)$ except for L itself and a certain set of curves which fill a manifold of less than n dimensions, are saddle-shaped, i.e., they leave D both with increasing as well as with decreasing t. A periodic solution of a linear system is unstable if at least one pair of the associated characteristic exponents has real parts of opposite signs.

Case 3. *Stable periodic solutions: compound limit cycles.* In this case every trajectory is stable in $D(L)$ (cf. 3.1).

Case 4. *Completely stable periodic solutions.* Here the limit sets of every integral curve through some neighborhood $D' \subset D$ are contained in D. In the linear case, for example, the neighboring integral curves are closed curves surrounding L in case all the elementary divisors are simple, all the characteristic exponents are pure imaginaries $r'\sqrt{-1}$, and the r' are commensurable with $2\pi/\omega$.

3.4. Nonlinear systems. Consider again the nonlinear system

$$\frac{dx_i}{ds} = \sum_{j=1}^{n} a_{ij}(t)x_j + \theta_i(x_1, \ldots, x_n, s), \tag{3.41}$$

in which $a_{ij}(s)$ and $\theta_i(x, s)$ are periodic in s,

$$\frac{\partial \theta_i}{\partial x_j} = O(|x_1| + \ldots + |x_n|)$$

for small $\sum_{j=1}^{n} |x_j|$, and which has the trivial solution $x_1 = x_2 = \ldots = x_n = 0$.

The variation equations

$$\frac{dx_i}{ds} = \sum_{j=1}^{n} a_{ij}(t)x_j \tag{3.42}$$

of (3.41), form a reducible linear system (cf. Chap. III). Therefore there exists a transformation

$$x_i = \sum_{j=1}^{n} p_{ij}(s)z_j \tag{3.43}$$

with periodic coefficients and a nonvanishing determinant, which reduces (3.42) to a linear system with a constant matrix A which may be supposed to be in the Jordan form. The characteristic exponents of (3.42) are the characteristic roots of A.

If we apply (3.43) to the nonlinear system (3.41) we obtain the system

$$\frac{dz}{ds} = Az + f(z, s) \tag{3.44}$$

considered in 1.4.

We observe that since the coefficients $p_{ij}(s)$ in (3.43) are bounded, $\Sigma |x_j|$ tends to zero with $\Sigma |z_j|$ and conversely. Applying Theorem 1.41 to (3.44) we obtain for the (n+1)-dimensional system (3.11).

3.41. THEOREM. *If the real parts of the characteristic exponents of the system of variation equations are all of the same sign then our periodic solution is a limit cycle. If k characteristic exponents* $(0<k<n)$ *have negative real parts and* $n-k$ *of them have positive real parts, then in the neighborhood of the periodic solution there exists a* $(k+1)$-*dimensional manifold of solutions which are asymptotic for* $s \to +\infty$ *and an* $(n-k+1)$-*dimensional manifold of solutions which are asymptotic for* $s \to -\infty$ *and all the remaining solutions of (3.11) in this neighborhood possess two-sided instability.*

3.5. Reduction by a method of formal expansions. Consider the system

$$\frac{dx_i}{dt} = \sum_{k=1}^{n} a_{ik}(t)x_k + \theta_i(x_1, \ldots, x_n; t) \tag{3.51}$$

in which the θ_i are power series in $x_1, \ldots, x_n$ beginning with terms of degree not less than the second. The coefficients $a_{ij}(t)$ as well as the coefficients of the power series θ_i, are periodic functions of t with a common period which, for the sake of simplicity, we assume to be 2π.

We assume further that the matrix of the variation equations has linear elementary divisors and that there exists no linear relation with integral coefficients between the characteristic exponents $r_1, \ldots, r_n$, and $\sqrt{-1}$.

Lyapunov showed that there is a linear transformation with periodic coefficients which reduces the given system (3.51) to the form

$$\frac{dx_i}{dt} = r_i x_i + F_i(x_1, \ldots, x_n; t), \tag{3.52}$$

where r_i are the characteristic exponents and the F_i are power series with the properties of the power series in (3.51).

We seek a transformation of the form

$$z_i = x_i + \psi_i(x_1, \ldots, x_n; t) \qquad (i = 1, \ldots, n)$$

where $\psi_i(x_1, \ldots, x_n; t)$ is a formal power series in $x_1, \ldots, x_n$ with periodic coefficients of period 2π and beginning with terms of degree not lower than two, which would carry the given system into

$$\frac{dz_i}{dt} = r_i z_i \qquad (i = 1, \ldots, n).$$

Write

$$F_i = F_{i2} + F_{i3} + \cdots + F_{in} + \cdots,$$
$$\psi_i = \psi_{i2} + \psi_{i3} + \cdots + \psi_{in} + \cdots,$$

where F_{ik} and ψ_{ik} are homogeneous polynomials of degree k in x_i. Replacing each x_i in (3.52) by the difference $z_i - \psi_i(x_1, \ldots, x_n; t)$ we obtain

$$\frac{dz_i}{dt} - \frac{\partial \psi_i}{\partial t} - \sum_{j=1}^{n} \frac{\partial \psi_i}{\partial x_j}\frac{dx_j}{dt} = r_i z_i - r_i \psi_i + F_i(x_1, \ldots, x_n; t).$$

Since we seek the equality $dz_i/dt = r_i z_i$, we must have

$$-\frac{\partial \psi_i}{\partial t} - \sum_{j=1}^{n} \frac{\partial \psi_i}{\partial x_j}\frac{dx_j}{dt} = -r_i \psi_i + F_i(x_1, \ldots, x_n;\ t),$$

or

$$-r_i\psi_i + \frac{\partial \psi_i}{\partial t} + \sum_{j=1}^{n} \frac{\partial \psi_i}{\partial x_j}(r_j x_j + F_j) + F_i(x_1, \ldots, x_n;\ t) = 0.$$

By equating each homogeneous component to zero we obtain the system of equations

$$r_i \psi_{ik} = \frac{\partial \psi_{ik}}{\partial t} + \sum_{j=1}^{n}\left(\frac{\partial \psi_{ik}}{\partial x_j} r_j x_j + \sum_{p+q=k+1} \frac{\partial \psi_{ip}}{\partial x_j} F_{iq}\right) + F_{ik},$$
$$(k = 2, 3, \ldots).$$

Each term of the function ψ_{i2} is of the form

$c_i(t)x_1^{l_1}x_2^{l_2}\ldots x_n^{l_n}(l_1 \geqq 0, \quad l_2 \geqq 0, \ldots, l_n \geqq 0, \quad l_1 + l_2 + \ldots + l_n = 2)$.

If $d_i(t)$ is the coefficient of the corresponding term in F_{i2}, then $c_i(t)$ must satisfy the equation

$$\frac{dc_i}{dt} + k_i c_i + d_i(t) = 0, \tag{3.53}$$

where

$$k_i = l_1 r_1 + l_2 r_2 + \dots + (l_i - 1) r_i + \dots + l_n r_n.$$

By hypothesis, the coefficient k_i of $c_i(t)$ is not equal to zero and is not an integral multiple of $\sqrt{-1}$. Hence (3.53) has the periodic solution

$$c_i(t) = -\frac{e^{-k_i t}}{e^{2\pi k_i} - 1} \int_t^{t-2\pi} d_i(\tau) e^{k_i \tau} \, d\tau.$$

Having found all the coefficients c_i in ψ_{i2} we can compute the coefficients of ψ_{i3}. To this end we shall have to solve an equation of the same type as the above since the coefficients of the expression $\Sigma_{j=1}^{n} (\partial \psi_{i2} / \partial x_j) F_{j2}$ are already known. Next, we can compute the coefficients of ψ_{i4} in a similar manner, and so on. This shows that the desired formal transformation does exist.

3.6. Canonical systems. We shall study next systems in the canonical form

$$\text{(3.61)} \qquad \frac{dx_s}{dt} = -\frac{\partial H}{\partial y_s}, \quad \frac{dy_s}{dt} = \frac{\partial H}{\partial x_s} \qquad (s = 1, \dots, n).$$

where H is a power series in $x_1, \dots, x_n, y_1, \dots, y_n$ with coefficients which are periodic in t of period T. We write

$$H = H_2 + H_3 + \dots + H_s + \dots$$

where H_k is a homogeneous polynomial in $x_1, \dots, y_n$ of degree k.

3.61. Theorem (Lyapunov [31]). The canonical system (3.61) with periodic coefficients has a reciprocal characteristic equation. In the following proof (due to Wintner) we find it convenient to renumber the x's and the y's by replacing x_i by x_{2i-1} and y_i by y_{2i} (i.e., we introduce essentially new variables z's so that $z_{2i-1} = x_i$ and $z_{2i} = y_i$).

In the new notation the variation equations become

$$\text{(3.6101)} \qquad \frac{dx_{2i-1}}{dt} = -\frac{\partial H_2}{\partial y_{2i}}, \quad \frac{dy_{2i}}{dt} = \frac{\partial H_2}{\partial x_{2i-1}}, \qquad (i = 1, 2, \dots, n).$$

Let $A(t) = [a_{jk}(t)]$ be the matrix of this linear system. This matrix has the following useful symmetries.

$$a_{2i-1,\,2h} = -\frac{\partial H_2}{\partial y_{2i}\,\partial y_{2h}} = a_{2h-1,\,2i}$$

$$(3.6102)\qquad a_{2i,\,2h-1} = \frac{\partial H_2}{\partial x_{2i-1}\partial x_{2h-1}} = a_{2h,2i-1}$$

$$a_{2i-1,\,2h-1} = -\frac{\partial H_2}{\partial y_{2i}\,\partial x_{2h-1}} = -\,a_{2h,\,2i}.$$

We write A in the partitioned form $[A_{\mu\nu}]$ where

$$A_{\mu\nu} = \begin{bmatrix} a_{2\mu-1,\,2\gamma-1} & a_{2\mu-1,\,2\gamma} \\ a_{2\mu,\,2\gamma-1} & a_{2\mu,\,2\gamma} \end{bmatrix}$$

and observe that in view of (3.6102)

$$(3.6103)\qquad A^*_{\nu\mu}U_1 + U_1 A_{\mu\nu} = 0, \quad \text{where } U_1 = \begin{bmatrix} 0 & 1 \\ -1 & 0 \end{bmatrix},$$

and $A^*_{\nu\mu}$ is the transpose of $A_{\nu\mu}$.

Let $X(t)$ be a nonsingular matrix whose rows

$$x^{(j)} = (x_1^{(j)}, y_2^{(j)}, \ldots, x_{2n-1}^{(j)}, y_{2n}^{(j)}) \qquad (j = 1, \ldots, 2n)$$

form a system of fundamental solutions of the variation equations. We already had occasion to observe that if $B(X)$ is a nonsingular constant matrix such that

$$X(t + T) = BX(t)$$

and if K is any nonsingular constant matrix of order $2n$, then

$$(3.6104)\quad B(KX) = [KX(t+T)][KX(t)]^{-1} = K[X(t+T)X^{-1}(t)]K^{-1} = KB(X)K^{-1}.$$

Write

$$C(X) = XUX^*$$

where

$$U = \begin{bmatrix} 0 & 1 & 0 & 0 & 0 & \ldots & 0 & 0 \\ -1 & 0 & 0 & 0 & 0 & \ldots & 0 & 0 \\ 0 & 0 & 0 & 1 & 0 & \ldots & 0 & 0 \\ 0 & 0 & -1 & 0 & 0 & \ldots & 0 & 0 \\ \cdot & \cdot & \cdot & \cdot & \cdot & \cdot & \cdot & \cdot \\ 0 & 0 & 0 & 0 & 0 & \ldots & 0 & 1 \\ 0 & 0 & 0 & 0 & 0 & \ldots & -1 & 0 \end{bmatrix} = \begin{bmatrix} U_1 & 0 & \ldots & 0 \\ 0 & U_1 & \ldots & 0 \\ \cdot & \cdot & \cdots & \cdot \\ 0 & 0 & \ldots & U_1 \end{bmatrix}$$

We see at once that $C(X)$ is nonsingular and that

$$C(KX) = KC(X)K^*. \tag{3.6105}$$

Moreover, the matrix C is constant. For,

$$C_{jk} = x^{(j)}Ux^{(k)*} = \sum_{i=1}^{n} \begin{vmatrix} x_{2i-1}^{(j)}(t) & y_{2i}^{(j)}(t) \\ x_{2i-1}^{(k)}(t) & y_{2\,i}^{(k)}(t) \end{vmatrix},$$

and hence

$$\begin{aligned} \frac{dC_{jk}}{dt} &= \dot{x}^{(j)}Ux^{(k)*} + x^{(j)}U\dot{x}^{(k)*} \\ &= x^{(j)}A^*Ux^{(k)*} + x^{(j)}UAx^{(k)*} \\ &= x^{(j)}(A^*U + UA)x^{(k)*}. \end{aligned}$$

To show that $A^*U + UA$ is a zero matrix it suffices to observe that this is true for its constituent 2×2 blocks $A_{\mu\nu}^*U_1 + U_1A_{\mu\nu}$ by (3.6103).

Since C is independent of t,

$$C(BX(t)) = C(X(t + T)) = C(X(t)). \tag{3.6106}$$

Combining (3.6105) and (3.6106), we get

$$C(X) = C(BX) = BC(X)B^*$$

whence

$$C^{-1}B^{-1}C = B^*.$$

Thus B^{-1} has the same characteristic roots as B^* and hence as B. But the characteristic roots of B^{-1} are the inverses of those of B, since

$$\frac{1}{\lambda}\left|B^{-1}\right| \cdot \left|B - \lambda I\right| = \left|\frac{1}{\lambda}I - B^{-1}\right|.$$

It follows from the last theorem that the characteristic exponents (the logarithms of the characteristic roots) occur in pairs equal in absolute value but opposite in sign. Then, since the total number of characteristic exponents is even, the number of zero exponents is also even.

4. The Method of Surfaces of Section.

The method of characteristic exponents of Poincaré-Lyapunov enabled us to reach definite conclusions regarding the behavior of integral curves when the characteristic exponents are distinct and their real parts do not vanish. However, in general, the method fails to yield definite results. This caused Poincaré to seek subtler methods. One such method employed by Poincaré, but actually developed by Birkhoff, is the method of *surfaces of section.*

Consider a closed trajectory L as in 3.1. As our *surface of section* S we may take the hyperplane normal to L at a point P' on L. The trajectory $f(P_0, t)$ through a point P_0 on S sufficiently close to P', will, as t increases, return to S and will cut S in a uniquely determined point $T(P_0) = P_1$. In this manner we define a one-to-one continuous (or, perhaps, even analytic) mapping T of a neighborhood D' of P' on S into S. The properties of T are closely related to the behavior of trajectories near the closed trajectory L. We saw this in the case of trajectories on the torus, where one of the meridians was taken as a surface of section. In general, we see that P' is a fixed point of T, and that $f(P_0, t)$ is closed if and only if P_0 is a fixed point of the kth iterate T^k of T for some $k \geqq 1$. If P_0 is a fixed point of T^k but not of a smaller positive power of k, then the periodic trajectory $f(P_0, t)$ cuts S in exactly k distinct points

$$P_1 = T(P_0), \quad P_2 = T(P_1), \ldots, P_0 = T(P_{k-1})$$

each of which is a fixed point of T^k. In case $f(P_0, t)$ is not periodic, both sequences

$$P_0, \quad P_1 = T(P_0), \quad P_2 = T^2(P_0), \ldots, \quad P_j = T^j(P_0), \ldots$$

and

$$P_0, \quad P_{-1} = T^{-1}(P_0), \quad P_{-2} = T^{-2}(P_0), \ldots, \quad P_{-j} = T^{-j}(P_0), \ldots$$

consist of distinct points. If the points of either one of the above sequences tend to P', then $f(P_0, t)$ is asymptotic. If these points remain in D', then $f(P_0, t)$ is stable.

A set $\Sigma \subset D'$ is said to be invariant under T if $T(\Sigma) \subset \Sigma$. As we shall see, the presence of invariant curves in D' is quite significant for the behavior of trajectories near L.

We say that the fixed point P' is stable if for every ε-neighborhood Σ_ε of P' there exists a δ-neighborhood Σ_δ ($\delta \leqq \varepsilon$) such that

$$T^k(\Sigma_\delta) \subset \Sigma_\varepsilon \qquad (k = \pm 1, \pm 2, \ldots)$$

4.1. Restricted dynamical systems of two degrees of

freedom. (Birkhoff [8]) We consider dynamical systems of Hamiltonian type and of two degrees of freedom. For such systems the manifold of trajectories is four-dimensional. We shall study the three-dimensional manifold of trajectories which is defined by fixing the value h in the energy integral $H = h$. We give here a resume of the principal results.

The surface of section S introduced above is now two-dimensional and if we choose coordinates u, v properly our transformation T will be defined by the formal power series

$$u_1 = au_0 + bv_0 + \sum_{m+n=2}^{\infty} \varphi_{mn} u_o^m v_o^n = au_0 + bv_0 + \Phi(u_0, v_0),$$

$$v_1 = cu_0 + dv_0 + \sum_{m+n=2}^{\infty} \psi_{mn} u_o^m v_o^n = cu_0 + dv_0 + \Psi(u_0, v_0).$$

Here $(0, 0)$ is the fixed point P' corresponding to our periodic trajectory L, (u_0, v_0) is the point P_0, and (u_k, v_k) denotes the point P_k.

Under a linear transformation the linear part of T is transformed into itself. We may assume therefore that the matrix of linear terms is in a canonical form determined by the roots of its characteristic equation

$$\varrho^2 - (a + d)\varrho + (ad - bc) = 0.$$

Since T is conservative, $ad - bc = 1$, the last equation is reciprocal, and the two characteristic roots are ϱ and $1/\varrho$.

The transformation T may be classified according to the behavior of its linear terms as follows:

(1) If ϱ is real and $\varrho \neq 1$, then T can be reduced to the normal form

$$u_1 = \varrho u_0 + \Phi(u_0, v_0),$$

$$v_1 = \frac{1}{\varrho} v_0 + \Psi(u_0, v_0).$$

We subdivide this case into two subcases: $(1')\varrho > 0$ and $(1'')\varrho < 0$.

(2) If ϱ is complex then $|\varrho| = 1$. In this case and for the case when $\varrho = \pm 1$ and the elementary divisors are linear, T has the normal form

$$u_1 = u_0 \cos \theta - v_0 \sin \theta + \Phi(u_0, v_0), \qquad (\varrho = e^{\theta\sqrt{-1}}),$$
$$v_1 = u_0 \sin \theta + v_0 \cos \theta + \Psi(u_0, v_0).$$

It is convenient to subdivide this case into the *irrational* case (2′) when $\theta/2\pi$ is irrational and the rational cases (2″) when $\theta = 0$, and (2‴) when $\theta/2\pi = p/q \neq$ an integer.

(3) In the remaining case when the elementary divisor is of degree two and hence if $\varrho = \pm 1$, the normal form of T is given by

$$u_1 = \pm u_0 + \Phi(u_0, v_0), \qquad (d \neq 0)$$
$$v_1 = \pm v_0 + du_0 + \Psi(u_0, v_0).$$

If no higher terms in the expansion of T are present, the resulting linear transformation may be regarded as providing a first approximation to the corresponding general types. Here, upon successive applications of T or T^{-1}, the point $P = (u, v)$ moves along a hyperbola $uv =$ const. (we may look upon u and v as rectangular coordinates) in the first case, and it will remain on a pair of parallel lines $u^2 =$ const. in the third case. Unless the point $P = (u_0, v_0)$ lies on the degenerate hyperbola $uv = 0$ in the first case, or on the line $u^2 = 0$ in the third, P will recede to infinity upon successive application of T or T^{-1}. When P lies on the degenerate hyperbola, it will approach the invariant point point $P' = (0, 0)$ upon successive application of T or T^{-1} and will recede to infinity upon application of the other. The points on the line $u^2 = 0$ (the third case) are either fixed or are reflected with respect to the origin according as $\varrho = +1$ or -1.

In the second case the approximating linear transformation is a rotation about $(0, 0)$ through an angle θ, and each point $P = (u, v)$ remains at a fixed distance from $(0, 0)$ upon successive application of either T or T^{-1}. Here, every circle with $(0, 0)$ as the center is an invariant curve.

We see that the linear approximations to T are stable in case (2) and are otherwise unstable even though the line $u^2 = 0$ in the third case contains infinitely many stable points.

The above discussion clarifies the following generally accepted definitions.

In cases (1′), (2′), (2″), (3′) the (general) transformation T is called *hyperbolic* if there exist invariant curves through the origin $(0, 0)$, and *elliptic* in the contrary case. In cases (2‴) or (3″), T

is *hyperbolic* or *elliptic* according as T^q or T^2 (of type 2″) is hyperbolic or elliptic.

Our periodic solution L (fixed point P') is called *elliptic* or *hyperbolic* according as T is elliptic or hyperbolic.

4.2. Hyperbolic fixed points (Birkhoff [8]). It was shown by Poincaré [47], Hadamard [21], and Birkhoff [8] that, for conservative T in case (1′) there exist two analytic invariant curves through the invariant point (0, 0). Actually this can be proved for a more general class of transformations.

If the two invariant curves are taken as the axes by a suitable choice of variables, then every point of the region $u^2 + v^2 \leqq \delta^2$ with $u \neq 0$ is carried out of the region by iteration of T while every point with $v \neq 0$ is carried out of the region by iteration of T^{-1}. This shows that there exist no invariant curves through (0, 0) other than the two above-mentioned analytic curves.

We note that in case (1″), T^2 is of type (1′). It can be shown that for T the two analytic invariant curves through (0, 0) coincide with those for T^2, and that there exist no other invariant curves through. (0, 0). The overall behavior of T in the neighborhood of (0, 0) is the same as that of T^2.

In general, the case (2″) is of hyperbolic type. In case T is of type (2″) and is hyperbolic, then every point of the region $u^2+v^2 \leqq \delta^2$ not on one of the invariant curves is carried out of the region by iteration of T or of T^{-1}, while every point on one of these curves approaches the invariant point (0, 0) by iteration of T and is carried out of the region by iteration of T^{-1} or vice versa. There may then exist either exactly three real invariant curves or just a single one.

If T is hyperbolic and is of type (2‴), $\varrho = 2p\pi/q$, T^q is of type (2″) and its invariant curves through (0, 0) are invariant as a set under T.

If T is hyperbolic and is of type (3″), then T^2 is hyperbolic and of type (2′) and the invariant curves of T^2 passing through (0, 0) are invariant as a set under T.

In case (3), points not on invariant curves behave as above.

4.3. Elliptic fixed points. (Birkhoff [8]). We have the following conditions for stability.

4.31. Theorem (Poincaré). *A fixed point P' is stable if and only if every neighborhood of P' contains an invariant (simply) closed curve surrounding P'.*

It may happen that there exists a neighborhood of P' consisting entirely of invariant closed curves surround P'. In case (2′) we have the following alternative. There exists a neighborhood of P' which, if not filled by invariant closed surrounding P', consists (a) of invariant closed curves and (b) of ring-shaped regions bounded by invariant closed curves but containing no such closed curves in their interior. Both the closed curves and the ring-shaped regions (*rings of instability*) surround P'.

It is of interest to compare the above with the discussion of stable singular points in the plane (cf. Chap. II). There the *lines of section* are the rays from the singular point. The analogy becomes even more striking in view of the following

4.32. THEOREM (Birkhoff). *Let C' and C'' be the two boundary curves of a ring of instability. Then for any $\varepsilon > 0$ there exists a positive integer $\nu(\varepsilon)$ such that for any point P on C' and any point Q on C'' one can find a point P_0 in the ε-neighborhood of P and such that $P_n = T^n(P_0)$ lies within the ε-neighborhood of Q for some n with $n < \nu(\varepsilon)$.*

Without loss of generality we may assume that C' lies in the interior of C''. We take as N_0 the union of an ε-neighborhood of P and of the set of points on C' and in its interior. The set $G = \cup_{k=-\infty}^{+\infty} T^k(N_0)$ lies in the interior of C'', is invariant under T, and is connected. The set O of points *occluded* by G will be defined to be those points of G which can be surrounded by some simple closed curve in G which also surrounds P'. The set O lies in the interior of G and is invariant under T. Moreover, O is *simply connected*. Hence its boundary B is an invariant closed curve surrounding P'. The boundary B of O is different from C' and hence must coincide with C'', for otherwise it would be a closed invariant curve surrounding P' and passing through the interior points of our ring of instability. The last alternative would contradict the definition of such a ring.

To establish the existence of $\nu(\varepsilon)$ we need only to apply twice the finite subcovering argument.

Here again it suffices to assume that T is topological. Therefore the above results apply to the rings of instability, whenever such rings exist, of systems satisfying the Lipschitz conditions.

The above results may be restated in terms of trajectories.

4.33. THEOREM. *Consider a stable closed trajectory L of the*

elliptic type discussed above. Then in a sufficiently small neighborhood of L, either all the trajectories lie on invariant torus-shaped surfaces or there exist zones of instability bounded by successive invariant torus-shaped surfaces. Within each zone of instability there exists a trajectory passing from an arbitrary small neighborhood of any point on one boundary surface to an arbitrary small neighborhood of any point on the other boundary surface. Every trajectory in a zone of instability comes arbitrarily close to either of the boundary surfaces.

The case when every neighborhood of a stable elliptic fixed point P' contains rings of instability is analogous to the case of a composite limit cycle in the plane. It is not known whether such *composite* stable elliptic fixed points exist in the analytic case.

If P' is an unstable elliptic fixed point then there exists a neighborhood D' such that for every neighborhood $N \subset D'$ of P', however small, there are points which leave this neighborhood under indefinite iteration of T or of T^{-1}. However, there also exist points $\alpha(\omega)$ which remain in D' under indefinite iteration of T^{-1} (of T). In fact the set of points $\alpha(\omega)$ has a connected subset $A\,(\Omega)$ extending from $(0, 0)$ to the boundary of D'. Also, $T^{-1}(A) \subset A$ and $T^{-1}(\Omega) \subset \Omega$ are sets of the same type as A and Ω. Under certain additional conditions (Birkhoff's *regularity*, say), $T^{-k}(A)$ and $T^{k}(\Omega)$ contract uniformly to $P' = (0, 0)$ as $k \to +\infty$.

When expressed in terms of the behavior of trajectories the above results yield the following theorem.

4.34. Theorem. *In the unstable elliptic case in every neighborhood of L contained in D there exist connected families of trajectories stable for* $t \to +\infty$ *and such families of trajectories stable for* $t \to -\infty$. *Under suitable restrictions there exist in every neighborhood of L connected families of trajectories asymptotic to L for* $t \to -\infty$ *and such families asymptotic to L for* $t \to +\infty$. *Also, every neighborhood (in D) of L contains infinitely many two-sided asymptotic trajectories (i.e.,* $A \cup \Omega$ *has infinitely many points in every such neighborhood).*

Under certain conditions there exist infinitely many periodic solutions, sometimes even analytic families of such solutions near an unstable periodic solution L.

BIBLIOGRAPHY TO PART ONE

1. Almuhamedov, M. I. On the problem of the center. *Izvestiya Kazan. Mat. Obscestva*, 8, 29–37, 1936–37; On the conditions for the existence of a singular point of the type of a center. *Ibid.* 9, 107–126, 1937. (Russian)
2. Andronov, A. A. and Chaikin, S. E. *Theory of oscillations.* Princeton University Press, 1949. (Edited translation from the Russian.)
3. Barbashin, E. A. On dynamical systems possessing a velocity potential. *Doklady Akad. Nauk* 61, 185–187, 1948. (Russian)
4. Bautin, N. N. Du nombre de cycles limites naissent en cas de variation des coefficients d'un état d'équilibre du type foyer ou centre. *Doklady Akad. Nauk* 24, 669–672, 1939.
5. Bellman, R. The stability of solutions of linear differential equations. *Duke Math. Jour.* 10, 643–647, 1943.
6. Bellman, R. On the boundedness of solutions of nonlinear differential and difference equations. *Trans. Am. Math. Soc.* 62, 357–386, 1947.
7. Bendixson, I. Sur les courbes définies par des équations différentielles. *Acta Math.* 24, 1–88, 1901.
8. Birkhoff, G. D. Surface transformations and their dynamical applications. *Acta Math.* 43, 1–119, 1922.
9. Birkhoff, G. D. *Dynamical systems, Colloquium Publications, No.* 9. New York, American Math. Soc., 1927.
10. Birkhoff, G. D. Sur l'existence de régions d'instabilité en dynamique. *Annales Inst. Poincaré* 2, 369–386, 1932.
11. Brouwer, L. E. J. On continuous vector distributions on surfaces. *Verhandl. d. Konigl. Akad. van Wet.* 11, 850–858, 1909; 12, 716–734, 1910.

    Chaikin, S. E. See Andronov and Chaikin.
12. Cetaev, N. G. *Stability of motion.* Moscow, Gostehizdat, 1946. (Russian)
13. Denjoy, A. Sur les courbes définies par les équations différentielles à la surface du tore. *Journal de Math.* (9), 333–375, 1932.
14. Dragilev, A. V. Periodic solutions of a differential equation of nonlinear oscillations. *Prikladnaya mat. i mek.* 16, 85–88. (Russian)
15. Dulac, H. Solutions d'un système d'équations différentielles dans le voisinage de valeurs singulierès. *Bull. Soc. Math. de France* 40, 324–392, 1912.
16. Erugin, P. P. Reducible systems. *Trudy matematiceskogo instituta imeni Steklova* 13, 1946.
17. Floquet, G. Sur les équations différentielles linéaires à coefficients périodiques *Ann. Éc. Normale* 12, 47–88, 1883.
18. *Filippov*, A. F. A sufficient condition for the existence of a stable limit cycle for an equation of the second order. *Mat. Sbornik, N.S.* 30, 171–180, 1952.
19. Forster, H. Über das Verhalten der Integralkurven einer gewöhnlichen Differentialgleichung erster Ordnung in der Umgebung eines singulären Punktes. *Math. Zeits.* 43, 271–320, 1937.
20. Frommer, M. Die Integralkurven einer gewöhnlichen Differentialgleichung erster Ordnung in der Umgebung rationaler Unbestimmtheitsstellen. *Math. Ann.* 99, 222–272, 1928.

20a. Frommer, M. Über das Auftreten von Wirbeln und Strudeln in der Umgebung rationaler Unbestimmtheitsstellen. *Math. Ann.* 109, 395–424, 1934.

21. Hadamard, Jacques. Sur l'itération et les solutions asymptotiques des équations différentielles. *Bull. Soc. Math. France* 29, 224–228, 1901.

22. McHarg, Elizabeth A. A differential equation. *Jour. London Math. Soc.* 22, 83–85, 1947.

23. Horn, Jakob. *Gewöhnliche Differentialgleichungen.* Berlin, De Gruyter, 1927.

24. Ivanov, V. S. Foundation for a hypothesis of van der Pol in the theory of self oscillations. *Ucenye Zapiski Leningradskogo Universiteta, math. series* 10, 111–119, 1940. (Russian)

25. Kneser, H. Reguläre Kurvenscharen auf Ringflächen. *Math. Ann.* 91, 271–320, 1923.

26. Kukles, I. S. Sur les conditions nécessaires et suffisantes pour l'existence d'un centre. *Doklady Akad. Nauk* 42, 160–163, 1944; Sur quelques cas de distinction entre un foyer et un centre. Ibid., 208–211.

27. Lefschetz, S. Lectures on differential equations. *Annals of Math. studies* 14, Princeton University Press, 1946.

28. Leontovic, E. and Mayer, A. G. General qualitative theory. Complement to the Russian translation of Poincaré [47].

29. Levinson, N. and Smith, O. K. A general equation for relaxation oscillations *Duke Math. Journ.* 9, 382–403, 1942.

30. Lonn, E. R. Über singuläre Punkte gewöhnlicher Differentialgleichungen. *Math. Zeits.* 44, 507–530, 1938.

31. Lyapunov, A. M. Probleme général de la stabilité du mouvement. *Ann. of Math. Studies* 17, Princeton University Press, 1947. [Original Russian, dated 1892. This is a reproduction of the French translation dated 1907].

32. Lyapunov, A. M. Investigation of one of the special cases of the problem of stability of motion. *Mat. Sbornik* 17, 253–333, 1893. (Russian)

33. Maier, A. G. Trajectories on the closed orientable surfaces. *Mat. Sbornik N.S.* 12, 71–84, 1943. See also Leontovic-Maier.

34. Malkin, I. G. Über die Bewegungsstabilität nach der ersten Näherung, *Doklady Akad. Nauk* 18, 159–162, 1938.

35. Nemickii, V. V. Qualitative integration of a system of differential equations. *Mat. Sbornik N.S*, 16, 307–344, 1945. (Russian)

Paivin, A. U. See Shestakov and Paivin.

36. Perron, O. Über lineare Differentialgleichungen, bei denen die unabhängig Variable reel ist. *Journal f. d. reine u. angewandte Math.* 142, 254–270, 1913.

37. Perron, O. Über die Gestalt der Integralkurven einer Differentialgleichung erster Ordnung in der Umgebung eines singulären Punktes. *Math. Zeits.* 15, 121–146, 1922.

38. Perron, O. Über Stabilität und asymptotischer Verhalten der Integrale von Differentialgleichungssystemen. *Math. Zeits.* 19, 129–160, 1928.

39. Perron, O. Die Ordnungszahlen linearer Differentialgleichungssystemen. *Math. Zeits.* 31, 748–766, 1930.

40. Perron, O. Die Stabilitätsfrage bei Differentialgleichungen. *Math. Zeits.* 32, 703–728, 1930.

41. Persidskii, K. P. On the stability of motion in accordance with the first approximation. *Mat. Sbornik*, 40, 284–293, 1933. (Russian)

42. Persidskiy, K. P. On the characteristic numbers of differential equations. *Izvestiya Akad. Nauk Kazakhskoi SSR*, 1947. (Russian)

43 Petrovskii, I. G. Über das Verhalten der Integralkurven eines Systems gewöhnlicher Differentialgleichungen in der Nähe eines singulären Punktes. *Mat. Sbornik* 41, 107–155, 1934.

44. Petrovskii, I. G. *Lectures on ordinary differential equations.* Moscow–Leningrad, 1947. (Russian)

45. Picard, Émile. *Traitè d'Analyse,* Vol. 3, 3rd ed. Paris, Gauthier-Villars, 1928.

46. Poincare, Henri. Sur les propriétés des fonctions définies par les équations aux différences partielles. In *Thèses de Mathématiques,* Paris, Gauthier-Villars, 1879.

47. Poincaré, Henri. Sur les courbes définies par les équations différentielles. *Jour. Math. Pures Appl.* (4) 1, 167–244, 1885.

48. Saharnikov, N. A. On Frommer's conditions for the existence of a center. *Prikladnaya. mat. imek.* 12, 669–670, 1948. (Russian)

49. Siegel, C. L. Note on differential equations on the torus. *Ann. of Math.* 46, 423–428, 1945.

50. Shestakov, A. A. and Paivin, A. U. On the asymptotic behavior of the solutions of a nonlinear system of differential equations. *Doklady Akad. Nauk.* 63, 495–498, 1948.

Smith, O. K. See Levinson and Smith.

51. Solncev, Iu. K. On the limiting behavior of the integral curves of a system of differentual equations. *Izvestiga Akad. Nauk* 9, 233–240, 1945. (Russian)

52. Stepanov, V. V. *A course on differential equations,* 4th ed. 1945, Moscow-Leningrad. (Russian)

53. Trjitzinsky, W. J. Properties of growth for solutions of differential equations of dynamical type. *Trans. Am. Math. Soc.* 50, 252–294, 1941.

54. de la Valleé-Poussin, C. J. *Cours d'analyse infinitésimale.* Louvain, 1921.

55. Vinograd, R. E. On the limiting behavior of an infinite integral curve. *Ucenye Zapiski Moskovskogo Universiteta,* 1949. (Russian)

56. Weil, André. On systems of curves on a ring-shaped surface. *Jour. Indian Math. Soc.* 19, 109–114, 1931–32.

57. Weyl, H. Comment on the preceding paper [by Levinson]. *Amer. Jour. Math.* 68, 7–12, 1946.

58. Whitney, H. Regular families of curves. *Ann. of Math.* 34, 244–270, 1933.

59. Wintner, A. The non-local existence problem of ordinary differential equations. *Amer. Jour. Math.* 67, 277–284, 1945.

60. Wintner, A. The infinities in the non-local existence problem of ordinary differential equations. *Amer. Jour. Math.* 68, 173–178, 1946.

61. Yakubovič, V. A. On the asymptotic behavior of the solutions of systems of differential equations. *Doklady Akad. Nauk* 63, 363–366, 1948. (Russian)

APPENDIX TO PART ONE

Problems of the Qualitative Theory of Differential Equations

by

V. V. NEMICKII

(From the Bulletin of Moscow University; No. 8 (1952), Mathematics)

The purpose of this article is to describe the activity in recent years of the seminar on "The Qualitative Theory of Differential Equations." The seminar was founded in the year 1935-36 by V. V. Stepanov and me and has since then united the efforts of Moscow mathematicians and quite a number of mathematicians from other cities, in solving problems in the development of the qualitative theory of differential equations.

Although for the last two years V. V. Stepanov took no direct part in the work of the seminar, up to the last weeks of his life he took a lively interest in it. The great influence of his scientific personality and scientific ideas is testified to in the work of his students. Most of the participants in the seminar have either been his students — among whom I count myself — or the students of his students.

PART 1. LINEAR PROBLEMS OF SYSTEMS WITH NON-CONSTANT COEFFICIENTS

Consider the system of linear differential equations

$$\frac{dx_i}{dt} = \sum_{k=1}^{n} a_{ik}(t)\, x_k.$$

If the a_{ik} are constant the equations can be completely integrated, and so present no immediate problem for the qualitative theory. However, this does not give directly a topological classification of the possible dispositions of the family of integral curves.

Asymptotic Behavior of Solutions. The Lyapunov Characteristic Numbers [1]

Now let the $a_{ik}(t)$ be functions of t. What can be said about the asymptotic behavior of the solutions, i.e. about the behaviour of the solutions as $t \to +\infty$ and $t \to -\infty$? This is a central problem of the theory.

Referring the reader to the book of Nemickii and Stepanov [2] for the basic facts of this theory, we deal only with questions which have served as subjects for the seminar in recent years.

We shall begin with several questions on which, although they are related to important problems, we have only obtained a few results. The first was raised by Lyapunov. Define the characteristic number of a solution: $x_1(t)$, $x_2(t)$, . . ., $x_n(t)$ to be

$$-\overline{\lim_{t\to+\infty}} \frac{\log \sum_{i=1}^{n} |x_i(t)|}{t}.$$

Lyapunov himself established that for linear systems there can exist no more than n characteristic numbers and raised the question of determining these numbers without integrating the system, directly from the coefficient matrix [1].

Up to now this question has not been solved and there is no visible way of doing so. The difficulties arising here are connected with the fact that for a general linear system the characteristic numbers are unstable, i.e. an arbitrarily small change in the coefficients of the system can lead to a finite change in the characteristic numbers [3].

K. P. Persidskii has established [3] that for linear systems with constant coefficients the characteristic numbers are stable, and consequently, they are stable for the whole class of reducible systems. The characteristic numbers of systems with constant coefficients are equal to the real parts of the roots of the characteristic equations. Lyapunov put to the fore the class of regular systems for which the stability of the trivial solution can be investigated by means of the first approximation.

It seems likely that for regular systems the characteristic numbers are stable, but even this is not yet known.

A number of these important equations were treated in a short paper of B.F. Bylov [4]. In one case Bylov verified the stability of the characteristic numbers of regular systems and gave a method for calculating them.

Bylov's theorem. Let there be given a system of linear differential equations with real coefficients

$$\frac{dx_s}{dt} = \sum_{r=1}^{n} p_{sr}(t)\, x_r,$$

such that

(1) $|p_{sr}| \to 0$ as $t \to +\infty$ for $r > s$;

(2) there exist constants λ_s such that for any $\varepsilon > 0$ there exists A such that for $T_2 - T_1 > A$

$$\left| \frac{1}{T_2 - T_1} \int_{T_1}^{T_2} p_{ss} dt + \lambda_s \right| < \varepsilon.$$

Then the characteristic numbers of the system are

$$-\lim_{t \to +\infty} \frac{1}{t - T_1} \int_{T_1}^{t} p_{ss}\, dt = \lambda_s.$$

Condition (2) is especially interesting and essential, since for triangular systems, as Lyapunov showed, the existence of the limits $\lim (1/T) \int_{T_1}^{T} p_{ss} dt$ is sufficient.

Another advance was made by D. I. Grobman. In the second part of this review we make use of his methods. For linear systems Grobman's result is as follows [5].

Grobman's Theorem. Given two systems of linear equations

$$\frac{dx_i}{dt} = \sum_{k=1}^{n} a_{ik}(t)\, x_k, \qquad \frac{dy_i}{dt} = \sum_{k=1}^{n} b_{ik}(t)\, y_k,$$

let there exist a function $g(t)$ such that

(1) $|a_{ik}(t) - b_{ik}(t)| \leqq g(t); \qquad t \geqq t_0;$

(2) $\int_{t_0}^{\infty} g(t) e^{\alpha t}\, dt < \infty, \qquad \alpha > 0.$

Then the characteristic numbers of the two systems are the same.

If one of the compared systems is a system with constant coefficients then the number α in condition (2) may be set equal to zero.

Lyapunov Stability. The second important problem posed by Lyapunov is that of finding conditions on the coefficients of linear systems from which follow the stability or instability of the trivial solution.

It is important to note that even when one can solve the problem of characteristic numbers and find them to be zero, one does not solve the problem of the boundedness or unboundedness of the solutions, i.e. does not solve the problem of the Lyapunov stability of the trivial solution. Hence this question is attacked by different methods. Let the given system have the form:

$$(\alpha) \qquad \frac{dx}{dt} = (A + B(t))x$$

where A is a constant matrix and $||B(t)|| \to 0$ as $t \to \infty$. As already remarked, in this case the characteristic numbers of the solutions of system (α) coincide with the characteristic numbers of the system

$$(\beta) \qquad \frac{dx}{dt} = Ax.$$

However, if it is shown, for example, that all characteristic numbers of the system are positive except one which is zero we cannot in general assert the stability of the position of equilibrium. On this question we note a result of Demidovič [6].

If the elements of the matrix $B(t)$ are denoted by $b_{ij}(t)$, and if we consider the function $\varrho^0(t)$

$$\varrho^0(t) = \frac{1}{\sum_{i=1}^{n} A_{ij}} \sum_i \sum_j A_{ij} b_{ij}(t)$$

where the A_{ij} are the cofactors of the elements of the matrix A then, under certain restrictions on the function $\varrho^0(t)$, we can determine whether the trivial solution of the system

$$\frac{dx}{dt} = (A + B(t))x$$

is stable.

Lyapunov showed that the trivial solution is stable if we can find a function $V(x_1, x_2, \ldots, x_n, t) \geqq 0$ whose total derivative

dV/dt, substituting from the differential equations, is negative. This result gives a method of analyzing particular systems qualitatively.

For analyzing linear systems one naturally takes as a Lyapunov function a quadratic form

$$G(t, x) = \sum C_{ik}(t)x_i x_k.$$

The total derivative of such a form taking the system into account is also a quadratic form

$$\Gamma(t, x) = \sum_{k=1}^{n} d_{ik}(t)x_i x_k.$$

The coefficients of this last form are easily expressed in terms of the coefficients of the system and the coefficients $C_{ik}(t)$.

A. D. Gorbunov [7] by using these forms has found the following estimates for the solutions of the system:

$$x_s(t) \leqq [g(t_0, x^0) \frac{C_{n-1}^{(s)}(t)}{C_n(t)} \exp \int_{t_0}^{t} N_g(\tau) d\tau]^{\frac{1}{2}},$$

$$s = 1, \ldots, n.$$

where $C_n(t)$ is the discriminant of the quadratic form $G(t, x)$, $C_{n-1}^{(s)}(t)$ is the minor of this discriminant obtained by striking out the sth row and the sth column and

$$N_g(t) = \max_{G(t;\, x)=1} \Gamma(t; x).$$

He showed that there always exists a quadratic form giving the best estimate.

The Spectral Theory of Systems of Linear Equations

Consider the system of linear differential equations

$$\frac{dx_i}{dt} = \sum_{k=1}^{n} a_{ik}(t)\, x_k$$

as a single matrix or matrix-vector equation

$$\frac{dx}{dt} = A(t)x$$

where x may either be regarded as an unknown vector or as an unknown fundamental matrix of solutions (in the latter case it is written X). In particular it is convenient for definiteness to regard it as a matrix normalized so that $X(0) = E$ (the unit matrix). There arises then the problem of the spectral properties of the matrix $A(t)$, i.e. from its proper values and the structure of its elementary divisors or from the spectral properties of matrices directly connected with $A(t)$, to find spectral properties of the normal matrix of solutions and in particular to determine the asymptotic behavior of the solutions as $t \to -\infty$ or $t \to \infty$.

This problem as is well known is completely solved if the matrix A is constant; then $x(t)$ and $A(t)$ satisfy the simple relation

$$X(t) = e^{\int_{t_0}^{t} A\,dt} = e^{At}.$$

For the nonconstant matrices there is in general no such relation, since differentiating the formula:

$$X(t) = e^{\int_0^t A(t)\,dt}$$

gives

$$\frac{dX}{dt} = A + \frac{1}{2!}\left[A\int_{t_0}^{t} A\,dt + \left(\int_{t_0}^{t} A\,dt\right)A\right] + \ldots.$$

Consequently, in order that the formula continue to be valid $A(t)$ and $\int_0^t A(t)dt$ must commute.

It would be interesting to find the structure of matrices for which this commutativity takes place, and to investigate the spectral properties of systems of linear equations with matrices $A(t)$ of this type.[1]

If it is not required to find the spectral properties of the matrices merely from the asymptotic behavior of the solution, then we can use the theory of Lyapunov on reducible systems. He called a system

$$\frac{dX}{dt} = A(t)X$$

reducible if for $t \geqq t_0$ there is a matrix $C(t)$ possessing the following

[1]For systems of two equations the structure of these matrices has been described by N. P. Erugin [8].

properties: it is bounded, differentiable, has a bounded inverse and is such that applying the linear transformation

$$Y = C(t)X$$

transforms the given system into

$$\frac{dY}{dt} = BY, \qquad B = CAC^{-1} + \frac{dC}{dt}C^{-1}$$

where B is a constant matrix.

This shows that the fact that a system is reducible gives little information about the asymptotic behavior of its solutions. Therefore, Erugin was quite right in saying that it is first necessary to solve the problem of reducibility to a fixed system. Given a definite system

$$\frac{dY}{dt} = KY \tag{1}$$

where K is a constant matrix, what are the systems with variable coefficients which can be reduced to it? We note progress of V. A. Yakubovič [9, 10, 11] towards solving this problem. A system

$$\frac{dX}{dt} = A(t)X$$

will be reducible to (1) if

$$\int_0^t t^{m-2}\|A(t) - K\|dt < \infty,$$

where m is the rank of the largest elementary divisor of the Jordan form of the matrix K. The question, of course, is not at all exhausted. In particular, it would be interesting to clarify whether in place of the tending to zero of the integrals or of their differences one should not consider the tending to zero of the integrals of the weighted mean values of these differences.

Lyapunov showed that systems with periodic coefficients are reducible. However, the difficulty indicated above does not permit determining the asymptotic behavior of the solutions of arbitrary systems with periodic coefficients. In particular we cannot give conditions for the boundedness of the solutions of these equations.

As was shown by the very deep investigations of M. G. Krein [12], I. M. Gelfand, V. A. Yakubovič, V. B. Lidskiĭ and M. G.

Neĭgauz [13], when the given system of linear equations is of Hamiltonian type, i.e. is obtained from a quadratic form

$$H(t) = \sum_{i,k}^{m} h_{ik}(t)\, x_i\, x_k,$$

the asymptotic behavior of the solutions can be obtained from the spectral properties of $H(t)$; in particular they found criteria for the boundedness of the solutions. We give as an example one of these criteria [13].

Let $\lambda_{\min}(t)$ be the smallest proper value of the matrix H, and let $\lambda_{\max}(t)$ be the largest proper value, then:

THEOREM (Neĭgauz and Lidskiĭ). *All solutions of* (1) *are bounded from* $-\infty < t < +\infty$ *if*

$$n\pi < \int_0^\omega \lambda_{\min}(t)dt \leqq \int_0^\omega \lambda_{\max}(t)dt < (n+1)\pi.$$

For $n = 2$ this theorem was established by V. A. Yakubovič [10].

If we examine a particular canonical system, namely the system of two vector equations of the first order obtained from the second order equation

$$y'' + P(x)y = 0,$$

where $P(x)$ is a symmetric matrix, all of whose terms are periodic with period ω, then it is possible to get criteria which for a single second order equation are the classical criteria of Lyapunov and Zukovskii [14]. All these results are based on the fact that systems of equations with periodic coefficients satisfy:

$$X(t+\omega) = X(t)X(\omega)$$

where $X(t)$ is a normal fundamental system, and from it follows [2] that the asymptotic behavior of the solutions depends wholly on the spectral properties of $X(\omega)$, and in particular on the location of its characteristic numbers with respect to the unit circle. We remark, by the way, that it follows from this that the investigation of the asymptotic behavior of the solutions of individual systems of equations with periodic coefficients can be carried out computationally, since the matrix $X(\omega)$ can be found to any degree of accuracy by methods of approximate integration.

Investigation of a Second Order System. A Condition for Boundedness of Solutions

One can ask the same questions about equations of the second order

$$y'' + p(t)y = 0 \tag{1}$$

as about systems of equations. We begin by discussing the boundedness of solutions.

It was established by Šepelev [15], that if the coefficient $p(t)$ satisfies the inequality $a^2 \leqq p(t) \leqq b^2$ and has bounded variation on $[t_0, \infty]$ then all solutions of (1) are bounded. Progress when $p(t)$ is not of bounded variation was made by L. A. Gusarov [16]. He showed that if $a^2 \leqq p(t) \leqq b^2$ and the derivative $p'(t)$ is of bounded variation, then all solutions of (1) are bounded. Later this was quite simply proved by I. M. Sobol. However, Gusarov's original methods seem to me to be very interesting and illuminating. He compared (1) with $y'' + q(t)y = 0$ in which $q(t)$ is a piecewise linear function. With Gusarov we call the segment $x_i \leqq x \leqq xi + 1$, determined by two successive zeros of a solution $y(x)$ "distinguished' if $p(x)$ is not constant in this interval and it contains one point where $p(x)$ attains a relative minimum.

Denoting the nth "distinguished" segment by the symbol $(x_1^{(n)}, x_2^{(n)})$, Gusarov established the following interesting inequalities

$$|Y'(x_2^{i+1})|^2 \leqq |Y'(x_2^i)|^2 e^{\frac{3\pi}{a^3}\omega[p'(x);\, x_2^i x_2^{i+1}] \log \frac{b^2}{a^2}} \cdot$$

$$\cdot e^{\frac{1}{2}} \sum_{i=2}^{1i+p+1} \omega[p'(x); \quad x_2^i x_2^{i+1}] \left[\frac{p(x_2^{i+1})}{p(x_2^i)}\right]^{\frac{1}{2}},$$

where $\omega[p'(x);\ x_2^i x_2^{i+1}]$ is the oscillation of $p'(x)$ on the segment $(x_2^i x_2^{i+1})$.

Criteria established for continuous $p'(x)$ have been extended by Gusarov, with the aid of a simple principle of comparison due to Bellman [18], to many cases of discontinuous $p(x)$. If we add to the assumptions of the theorem the condition that

$$q(x) \leqq p(x) \leqq [q(x)]^s; \quad s \geqq 1; \quad \lim_{x\to\infty} q(x) = \infty,$$

then all solutions of (1) will not only be bounded but will also tend to zero as $x \to \infty$ [19]. Other cases leading to bounded solutions were found by Kamynin [20].

However, in this field there remains uninvestigated the quite important case where $p'(x)$ is of bounded variation but may change sign.

The case in which $p(x)$ itself changes sign is not completely analyzed. It may also happen here that the solutions are bounded. For example, using Bellman's principle we find that if $p(x)$ is of bounded variation and if

$$\int_{x_0}^{\infty} \left| p(x) - |p(x)| \right| dx$$

converges then the solutions will be bounded.

We now proceed to examine the more difficult case when neither $p(x)$ nor $p'(x)$ is of bounded variation. In particular, equations with periodic coefficient belong to this class.

Special cases of periodic coefficients were examined by Gusarov and Yakubovič. Gusarov [21], reestablishing the priority of N. E. Zukovskiĭ, extended his method to new cases and generalized the known criteria of Lyapunov. Further investigations were carried out, on the one hand by I. M. Gelfand and M. L. Neĭgauz and on the other hand by V. A. Yakubovič [10]. Gelfand and Neĭgauz reduced the examination of the criteria for boundedness of solutions to certain variational problems. They obtained an infinite sequence of integral criteria of Lyapunov type.

Yakubovič attacked (1) by considering it as a particular case of a canonical system of two equations with periodic coefficients. He analyzed in detail the structure of the space of those matrices $A(t)$ for which the system

$$\frac{dx}{dt} = A(t)X$$

has bounded solutions, and on this basis obtained an infinite series of criteria for boundedness of solutions. An important point in the proof is to show that every matrix $A(t)$ can be put in correspondence with a certain integral number.

Oscillating and Nonoscillating Solutions

The second question examined by the seminar is the distribution of the zeros of the solutions of the equations $y'' + p(x)y = 0$ and $y'' + q(x)y' + p(x)y = 0$.

I. M. Sobol [22] completely classified the nonoscillating solutions of the first. If $p(x)$ is of constant sign, this classification is based on the sign of $p(x)$ and the convergence or divergence of the integral

$$\int_a^\infty xp(x)dx.$$

He gave an interesting criterion for the nonoscillation of solutions. Analogous results can be obtained for equations of the nth order

$$\text{(A)} \qquad Y^{(n)} = \sum_{i=1}^{n} A_i(x)Y^{(n-i)}.$$

Sobol showed that if for $1 \leqq j \leqq n$ the integrals

$$\int_a^\infty |A_j(x)|x^{j-1}\,dx < +\infty,$$

then equation (A) has a fundamental system of the form:

$$y_s(x) = x^s + 0\left(\int_a^x \cdots \int_a^t \psi(t)\,(dt)^s,\right)$$

where

$$\psi(x) = \sum_{i=1}^{n} \int_x^\infty |A_j(x)|x^{j-1}\,dx.$$

Method of Variable Frequency of M. I. El'šin

We conclude with the work of M. I. El'šin [23] repeatedly discussed in our seminar. El'šin examined the equation $y'' + p(t)y' + q(t)y = 0$ and analyzed the space of coefficients $p(t)$, $q(t)$ with the metric defined by: $\max(p(t)) + \max(q(t))$ on the infinite interval. El'šin tried to divide this space into regions in which are found equations having solutions of the same type: bounded, oscillating, nonoscillating, etc. His basic method is the investigation of a certain operator

$$J(\theta; (p, q)) = \left(\theta - \frac{p}{2}\right)' + \theta^2 + q - \frac{p^2}{4}, \qquad (' = d/dt)$$

where the admissible functions are continuous on the interval of investigation and $\theta - (p/2)$ has continuous derivatives.

If we make the transformation

$$y = Ye^{\int_{t_0}^{t}\left(\theta - \frac{p}{2}\right)d\xi}$$

then the given equation goes into the equation

$$Y'' + 2\theta Y' + J(\theta;\,(p, q))Y = 0.$$

Further, according to the ideas of Bohl, the general solution of the initial equation is obtained in the form

$$y = \frac{C_1 e^{-\frac{1}{2}\int_{t_0}^{t} p\,d\xi}}{\sqrt{\omega}} \cos\left[\int_{t^0}^{t} \omega\, d\xi + C_2\right],$$

where $\omega(t)$ is called the variable frequency and satisfies the equation

$$J\left[-\frac{\omega'}{2\omega};\,(p,\,q)\right] = \omega^2,$$

and the quantity

$$\varrho(t) = C_1 \frac{e^{-\frac{1}{2}\int_{t_0}^{t} p\,d\xi}}{\sqrt{\omega}}$$

is called the amplitude. These formulas together with some quite delicate analytic considerations lead to the following results: Let $\lambda = (4\theta J - J')/J$, where J is taken with arguments θ, p, q. Then in order that the amplitude of the solutions be bounded on $[a, b]$, it is necessary and sufficient that there exists a θ such that

$$\lambda \geqq 0;\ \overline{\lim_{t\to c}} \int_{t_0}^{t}\left(\theta - \frac{p}{q}\right)d\xi < \infty,$$

where $c = b$ $(c = a)$. In order that the amplitude be damped it is necessary that

$$\lambda \geqq 0;\ \lim \int_{t_0}^{t}\left(\theta - \frac{p}{q}\right)d\xi = -\infty,$$

where $c = b$ $(c = a)$, etc.

Results of this type are used to establish for various θ a set of sufficient conditions in order that the solutions have one or another type, and in the hands of a clever person this method has unlimited possibilities for investigating particular equations. In this respect El'šin's result is similar to the direct method of Lyapunov in the theory of stability.

Similar methods were applied by El'šin to the qualitative investigation of the stability of solutions of systems of two linear equations with variable coefficients. Unfortunately a detailed exposition of his method is not yet published.

PART 2. NONLINEAR SYSTEMS OF DIFFERENTIAL EQUATIONS.

Consider a system

$$\frac{dx_i}{dt} = F_i(t, x_1, \ldots, x_n).$$

We assume that the hypotheses of the existence and uniqueness theorems are fulfilled for all $t \geqq 0$ (instead of zero we may take, of course, $t_0 \geqq 0$) and for the vector x in a certain open region of n-dimensional space which for convenience of exposition we shall assume contains the origin.

Consider an arbitrary solution

$$x_1(t);\ x_2(t);\ \ldots;\ x_n(t),$$

satisfying the initial conditions

$$x_1(0) = x_{10};\ x_2(0) = x_{20};\ \ldots;\ x_n(0) = x_{n0},$$

where $(x_{10}, x_{20}, \ldots, x_{n0})$ is a point of the region G.

Now let t increase. It may happen that at $t = T$ the solution passes through the boundary of the region G, or else remains in G, but the modulus of the solution, i.e. the expression:

$$|x_1(t)| + |x_2(t)| + \ldots + |x_n(t)|,$$

tends to ∞.

Geometrically this means that the integral curve has a vertical asymptote. It is interesting to note that although it is well known that this case is quite typical for nonlinear equations, nevertheless up to now there have not been stated good sufficient conditions based on the form of the equations for its presence; it is also necessary to investigate the connection between the set of initial values determining solutions with vertical asymptotes and the set of initial conditions which determine solutions indefinitely continuable.

Asymptotic Behavior of Solutions

The Lyapunov characteristic number. Assuming that a certain solution is indefinitely continuable, by analogy with linear systems we can investigate the characteristic number of this solution.

The structure of the set of characteristic numbers for arbitrary nonlinear systems has not yet been established, and so we only consider systems close to linear. This circumstance is partially explained by the fact that solutions of "essentially" nonlinear systems are apt to have, as already mentioned, vertical asymptotes, i.e. not apt to be continuable.

It is convenient to treat only the case when the given system takes the form:

$$\frac{dx_i}{dt} = \sum_{k=1}^{n} a_{ik} x_k + \varphi_i(x_1, x_2, \ldots, x_n, t).$$

The assumptions that

$$|(\varphi_i(x_1', x_2', \ldots, x_n', t) - \varphi_i(x_1'', x_2'', \ldots, x_n'', t)| \leqq \varepsilon \sum_{i=1}^{n} |x_j' - x_j''|,$$

for any

$$\varepsilon > 0; \ |x_1'| \leqq \delta_\varepsilon; \ |x_i''| \leqq \delta_\varepsilon; \ i = 1, 2, \ldots, n; \ t \geqq T$$

are called the "basic conditions."

A. A. Šestakov and A. U. Paĭvin [24] showed that for such systems

$$\lim_{t\to\infty} \frac{\log \sum_{i=1}^{n} |x_i(t)|}{t}$$

always exists and is equal to one of the real parts of the roots of the characteristic equation $|A - \lambda E| = 0$.

Many results in this direction were obtained by D. I. Grobman [25]. He assumed that the "nonlinear perturbation" satisfies the conditions:

$$|\varphi_i(x_1', x_2', \ldots, x_n', t) - \varphi_i(x_1'', x_2'', \ldots, x_n'', t)| \leqq g(t) \sum_{j=1}^{n} |x_j' - x_j''|$$

for $t \geqq t_0 > 0$ and all vectors x in the region G, while the function $g(t)$ for $t \geqq t_0$ satisfies the two conditions:

$$\text{(I)} \qquad \int_{t_0}^{t} e^{-\varepsilon(t-\tau)} g(\tau) d\tau + \int_{t}^{\infty} e^{\varepsilon(t-\tau)} g(\tau) d\tau \leqq \frac{1}{2b} |B|;$$

$$\text{(II)} \qquad \int_{t_0}^{t} e^{-\alpha\tau} g(\tau) d\tau < +\infty \quad \text{for any} \quad \alpha > 0.$$

The constants b and B are defined by the matrix of a fundamental system of solutions of a linear system with constant coefficients. As particular cases one may take the conditions:

$$g(t) \leqq \frac{\varepsilon}{4b|B|} \quad \text{or} \quad \int_{t_0}^{\infty} g(t) dt < \infty$$

and many others. We first of all remark that these assumptions insure the continuability of all solutions.

Now let $\omega_1, \omega_2, \ldots, \omega_n$ be the characteristic numbers of the solutions of the linear system with constant coefficients. As Grobman showed, all solutions of the nonlinear system have characteristic numbers lying in the interval $(\omega_k - 2\varepsilon, \omega_k + 2\varepsilon)$ and have many other properties of the asymptotic behavior of the solutions of a system with constant coefficients. If the conditions of the theorem are not fulfilled for all points of G but only for a small neighborhood of a singular point (the origin) then the same behavior is preserved for all continuable solutions. Grobman's method consists in converting the system of differential equations into a system of integral equations of Volterra type and investigating the behavior of their solutions by a method of successive approximations.

Recently, Grobman obtained new conditions on the nonlinear perturbations, under which for the nonlinear system not only the characteristic numbers of the linear system are preserved, but also more delicate features of the asymptotic behavior of the solutions characterizing the algebraic parts (polynomial multipliers) of the solutions of linear systems with constant coefficients.

Asymptotically Equivalent Systems

We now examine another direction which the investigation of the asymptotic behavior of solutions may take — the method of comparison.

Let two systems be given in the same region G

$$\text{(I)} \qquad \frac{dx_i}{dt} = f_i(x_1, x_2, \ldots, x_n, t),$$

$$\text{(II)} \qquad \frac{d\bar{x}_i}{dt} = \bar{f}_i(\bar{x}_1, \bar{x}_2, \ldots, \bar{x}_n, t).$$

We shall call them asymptotically equivalent if there can be found a 1-1 bicontinuous correspondence between the initial values of the solutions of systems (I) and (II) satisfying the following conditions: Let

$$x_i = \varphi_i(t); \qquad (i = 1, 2, \ldots, n);$$
$$\bar{x}_i = \bar{\varphi}_i(t); \qquad (i = 1, 2, \ldots, n)$$

be the solutions of (I) and (II) satisfying corresponding initial conditions. Then

$$|\varphi_i(t) - \bar{\varphi}_i(t)| \to 0; \qquad (1 = 1, 2, \ldots, n)$$

as $t \to +\infty$.

We can ask what consequences the boundedness of the differences

$$|f_i(x_1, x_2, \ldots, x_n, t) - \bar{f}_i(\bar{x}_1, \bar{x}_2, \ldots, \bar{x}_n, t)|$$

have for the asymptotic equivalence of the solutions of (I) and (II).

On this point the work of V. A. Yakubovič [26] is quite fundamental. For system (I) he took a linear system with constant coefficients:

$$\frac{dx_i}{dt} = \sum_{i=1}^{n} a_{ik} x_k.$$

The basic result is: let $\lambda_1, \ldots, \lambda_n$ be the roots of the characteristic equation $|A - \lambda E| = 0$ and let $m_1, \ldots, m_n$ be the orders of the submatrices of the Jordan form of the matrix A.

Further let $\lambda = \max \operatorname{Re} \lambda_i$; $m = \max m_i$ for that i for which $\operatorname{Re} \lambda_i = \lambda$; $p = \max m_i$, for those i for which $\operatorname{Re} \lambda_i = 0$. Then the systems

$$\frac{dx}{dt} = Ax \quad \text{and} \quad \frac{dx}{dt} = Ax + \varphi(x)$$

are asymptotically equivalent if

$\lambda > 0$ and

$$|\varphi_i(x_1', x_2', \ldots, x_n') - \varphi_i(x_1'', x_2'', \ldots, x_n'')| \leqq g(t) \sum_{j=1}^{n} |x_j' - x_j''|,$$

where

$$\int_{t_0}^{t} t^{m+p-2}\, e^{\lambda t} g(t) dt$$

converges.

When the comparison system $dx/dt = A(t)x$ is linear with variable coefficients, there are only the most preliminary results.

Structure of the Integral Curves in the Neighborhood of a Singular Point

Regarding the structure of the integral curves in the neighborhood of a singular point, we must first ask what properties of the disposition of the integral curves must be studied. A clear answer to this question is not to be found in either the special or the general writings on this topic. However, if we analyze the well-known works of Poincaré, Bendixson, Perron, and Petrovskii we see that they study those properties of the integral curves which are preserved under an affine transformation of coordinates. Of course, certain of these affine transformations do not play an essential role in the disposition of the integral curves; for example, the absolute value of a root of the characteristic equation is of slight importance, so we ordinarily put in one class all those equations which differ from on another only in the absolute value of a root of the characteristic equation. However, this, it seems to me, does not prevent the classification from being considered a classification by affine invariants.

It would be quite interesting to make clear the topological types of dispositions of integral curves, and it is first of all necessary to do this for linear systems with constant coefficients. After this arises the interesting problem of the stability of these topological types when nonlinear terms are added.

It is widely known that for general nonlinear systems there can arise topological forms of behavior which are not observed in linear systems; however, the conditions under which these phenomena arise have not been studied extensively. We return to this question later. We will proceed to expound the basic results obtained by the participants in our seminar. A detailed affine classification of the behavior of the solutions under the "fundamental conditions" with respect to nonlinear perturbation was made by Šestakov. The beginning of this investigations is in the work of Šestakov and Paĭvin [24] and a continuation in as yet unpublished work of Šestakov. It is of course necessary to remember that basic results in this direction already are contained in the work of I. G. Petrovskii [27]. A new feature is the general investigation by Šestakov of the case when the linear part of the system entirely or partially disappears. First of all he investigated from the point of view of giving an affine classification to the structure of the neighborhood of a singular point, systems of the form

$$\frac{dx_1}{dt} = \sum_{i=m}^{\infty} c_i x_1^i; \qquad \frac{dx_i}{dt} = \sum_{j=1}^{n} a_{ij} x_j + X_i, \ (i = 2, \ldots, n).$$

Many other systems with analytic right-hand sides can be transformed to this form.

If the contracted characteristic equation

$$|a_{ij} - \lambda\delta_{ij}| = 0; \quad (\delta_{ij} \text{ Kronecker delta}) \ (i, j = 2, 3, \ldots, n)$$

has roots with real parts different from zero, then the structure of a neighborhood of the singular point is determined by the signs of the numbers c_m, the parity of the number m, and the number of roots of the contracted characteristic equation having positive real parts.

In further work Šestakov [28] investigated the general case

$$(1) \qquad \frac{dx_i}{dt} = X_i(x_1, x_2, \ldots, x_n) = X_i^{(m)} + X_i^{(m+1)} + \ldots$$

where the $X_i^{(m)}$ are homogeneous polynomials of degree m, when all n^2 partial derivatives $(\partial X_i/\partial X_j)$ $(i, j = 1, 2, \ldots, n)$ simultaneously vanish at the origin. In this case the singular point should be considered a singular point of higher order. Here the following method is applied. Consider a system of algebraic equations of the form

$$\frac{x_1}{X_1^{(m)}} = \frac{x_2}{X_2^{(m)}} = \ldots = \frac{x_n}{X_n^{(m)}}.$$

Let $x_i = a_i$ $(i = 1, 2, \ldots, n)$ be real roots of this algebraic equation, different from $(0, 0, \ldots, 0)$ and such that not all

$$X_i(a_1, a_2, \ldots, a_n) = 0, \qquad (i = 1, 2, \ldots, n).$$

The numbers $(a_1, a_2, \ldots, a_n)$ will be considered as the components of a certain vector starting at the origin of coordinates, and we call the direction of this vector critical. Integral curves can only approach the origin in a critical direction if they come in with any definite direction. We propose to study the behavior of the integral curves in the neighborhood of the critical directions. For this purpose the numbers $p_{ij}(i, j = 1, 2, \ldots, n-1)$ are defined by the following formulas:

$$p_{ij} = \begin{vmatrix} X_1^{(m)}; & X_{i+1}^{(m)} \\ \dfrac{\partial X_1^{(m)}}{\partial x_{j+1}}; & \dfrac{\delta X_{j+1}^{(m)}}{\partial x_{j+1}} \end{vmatrix}; \qquad (i, j = 1, 2, \ldots, n-1),$$

$$p_{ii} + 1 = \begin{vmatrix} X_1^m; & X_{i+1}^m \\ \dfrac{\partial X_1^m}{\partial X_{i+1}}; & \dfrac{\partial X_{i+1}^{(m)}}{\partial x_{i+1}} \end{vmatrix}; \quad (i = 1, 2, \ldots, n),$$

$$p_{in} = \begin{vmatrix} X_i^m & X_i^{m+1} \\ X_{i+1}^m & X_{i+1}^{m+1} \end{vmatrix}; \quad (i = 1, 2, \ldots, n-1),$$

where all these determinants are taken for $x_i = a_i$, $i = 1, 2, \ldots, n$.

We now examine the equation

$$D(\lambda) = \begin{vmatrix} p_{11} - \lambda, & p_{12}, & \cdots p_{1,\,n-1} \\ p_{21}, & p_{22} - \lambda, & \cdots p_{2,\,n-1} \\ \cdots & \cdots & \cdots \\ p_{n-1,\,1}, & p_{n-1,\,2}, & \cdots p_{n-1,\,n-1} - \lambda \end{vmatrix} = 0.$$

This equation in the case of a point of higher order plays the role of the characteristic equation in the ordinary case. For example, if the characteristic equation of a system has $n - k$ roots with positive real parts and $n - m$ roots with negative real parts, then the system (1) in the neighborhood of the given critical direction has two k-dimensional surfaces approaching the singular point respectively as $x_1 \to +\infty$ and $x_1 \to -\infty$.

There have already been established a number of theorems about the neighborhood of a critical direction. There are also quite a large number of unsolved problems in this subject. In particular, it is important to find out when and to what extent the neighborhoods of critical directions characterize the behavior of the integral curves in the whole neighborhood of the origin, and also to study the disposition of the integral curves when there are no critical directions.

Qualitative Investigations of Nonlinear Systems in the Large

Under this title we discuss various questions referring to the existence of integral curves of one kind or another, not restricted to lie in the neighborhood of a certain singular point; for example, limit cycles or analytic criteria enabling the characterization in one sense or another of the disposition of the integral curves in the large or in a certain finite region. All these topics are only slightly worked out and belong to a number of difficult problems of the qualitative theory for whose solutions there is as yet no general method.

The Search for Limit Cycles and other Periodic Solutions

We begin with a very classical problem. Consider the two equations

$$\frac{dx}{dt} = P(x, y); \qquad \frac{dy}{dt} = Q(x, y) \tag{1}$$

The problem is to find a method by which, without knowing the general integral of the system, one can establish the existence and location of periodic solutions and in particular, the existence, number, and location of limit cycles. There are two possible ways of solving this problem. The first is to use the special properties of the particular right-hand sides to give sufficient conditions for the existence of periodic solutions. For the second it is necessary to indicate a process which, either in a finite number of steps or in the limit, makes it possible to establish the existence of a periodic solution for any definite system of equations. Both of these approaches have been attacked by the participants in our seminar.

A computational method for finding periodic solutions of systems (1) where the right-hand sides satisfy a Lipschitz condition was developed by the author of this survey [29]. It is based on the following considerations. Let there be given a certain closed region G containing no singular points. Consider the function

$$\Phi(x, y) = P^2 + Q^2$$

and let M and m be respectively the minimum and maximum of this function in $\bar{G}$, and L the Lipschitz constant for $P(x, y)$ and $Q(x, y)$.

We now introduce a number $\eta_\varepsilon(\bar{G})$ characterizing the rate of rotation of the vector field as the point moves in the plane. Namely, if $\varrho(A, B) \leqq \eta_\varepsilon(\bar{G})$ then the angle between the directions of the vector field at A and B is less than ε. It is easy to give a lower bound for this quantity, namely $\eta_\varepsilon(\bar{G}) \geqq M^2/4m^2L$. Further, every closed solution is associated with a certain number called the index. The index is equal to the lower limit of the bound of the distances of points C and D lying on the given closed integral curve at which the tangents are respectively parallel and perpendicular to the x-axis. It is clear that every closed region $\bar{\bar{G}}$ contained together with its boundary in $\bar{G}$ always may be included in a certain canonical region Γ. In the canonical region Γ there can be no closed solution whose index is less than or equal to

$$\frac{1}{\sqrt{3}}\,\eta_{\frac{\pi}{2}}\,(\Gamma).$$

For integral curves of index larger than or equal to a given number we can always set up a countable process leading exactly to those periodic solutions. A further development of these investigations allows one to find a final step to a certain simple process. The number of these steps also depends on the constants M, m, and L.

As a result of this process we construct a finite series of quite narrow rings which alone can contain periodic solutions. If such rings do not appear, then there are no periodic solutions in the canonical region and hence none in $\bar{\bar{G}}$.

Another computational method for discovering periodic solutions based on the fact that periodic solutions are invariant curves under infinitesimal transformations was proposed by Molčanov in his lectures. In particular, he obtained for polynomial right-hand sides an upper estimate for the number of limit cycles.

A curious graphical method of constructing systems of equations having one or several limit cycles was proposed by Stebakov. Consider the system (1). We construct the zero isocline $Q = 0$ and the infinity isocline $P = 0$. Assume that in the given region they intersect only at the point (0, 0); then the region G is divided into four parts by these isoclines. Assume that in neighboring regions the signs of P and Q are different and assume that these signs are respectively in region I (+, +), in region II (—, +), in region III (—, —) and in region IV (+, —) (Fig. 1A). Then we can choose the forms of the isocline so that in the region limit cycles must appear.

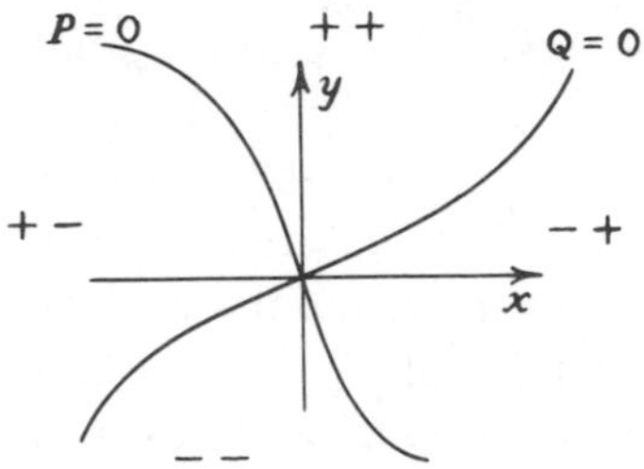

Fig. 1A

For example, in the case illustrated in Fig. 2A limit cycles must arise in the shaded area. It is possible to give forms of isoclines to produce two, three, or many limit cycles.

Stebakov's examples confirm previously known results.

One can indicate geometric disposition of the isoclines $P = 0$ and $Q = 0$ enabling one to describe the topological picture of the distribution of the integral curves in the neighborhood of a singular point.

A special place is occupied by the investigation of the existence of limit cycles for systems of equations obtained from the equation of nonlinear oscillations

$$\ddot{x} + f(x, \dot{x})\dot{x} + g(x) = 0.$$

These investigations begun by physicists have been widely developed especially in recent years. In our seminar they were continued in the work of Dragilev.

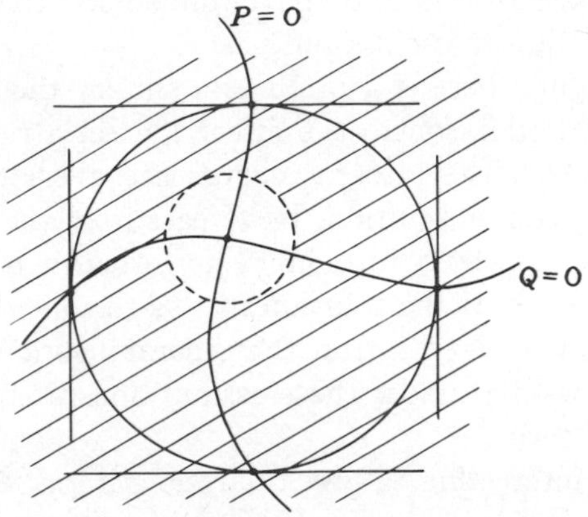

Fig. 2A

A more general result was recently presented by Filippov [30]. He examined the system

$$\frac{dx}{dt} = y - F(x); \qquad \frac{dy}{dt} = -g(x),$$

arising from the second order equation

$$\ddot{x} + f(x)\dot{x} + g(x) = 0.$$

Make a change of variables according to the formulas

$$x > 0; \qquad \int_0^x g(\xi)d\xi = z_1(x); \qquad \int_0^x f(\xi)d\xi = F_1(z_1),$$

$$x < 0; \qquad \int_0^x g(\xi)d\xi = z_2(x); \qquad \int_0^x f(\xi)d\xi = F_2(z_2).$$

Then the system has a stable limit cycle if

(1) $\int_a^{+\infty} g(x)dx = +\infty$; $g(x)$ has the sign of x;

(2) for small z $(|z| < \delta)$; $F_1(z) \leqq F_2(z)$ but they must not always be equal; $F_1(z) < a\sqrt{z}$; $F_2(z) \geqq -a\sqrt{z}$ where $a < \sqrt{8}$;

(3) there exists z_0, such that $\int_0^{z_0} |F_1(z) - F_2(z)| dz > 0$; for $z > z_0$

$F_1(z) \geqq F_2(z)$; $F_1(z) > -a\sqrt{z}$; $F_2(z) < a\sqrt{z}$; $(a < \sqrt{8})$.

Less conclusive results may be obtained for the more general equation of nonlinear oscillations.

As a final result there is a proof of a similar theorem based on the theorem of Bendixson to the effect that in a ring through the boundary of which the integral curves are everywhere entering, there will be a periodic solution. If we pass from the plane to space and replace the ring by a toroidal region, then it is still unknown whether there is a periodic solution in this region when it contains no singular point. However, from the general theorems on dynamical systems it follows that under these assumptions there is a recurrent motion in the torus.

It would be interesting to give analytic criteria for systems of n equations which would establish the existence of such toroidal regions, and also to give analytic criteria for the existence of recurrent motions other than singular points.

Proofs of the existence of periodic solutions for systems with periodic right-hand sides are based on quite different principles. The Brouwer fixed point theorem is applied here.

For systems of two equations of the form

$$\frac{dx}{dt} = X(t, x, y); \qquad \frac{dy}{dt} = Y(t, x, y)$$

B. P. Demidovič [31] obtained interesting results. As they were published earlier in this journal we shall not take time to formulate them. By means of the fixed point theorem applied in cases of two or more dimensions it might be possible to find analytic criteria for the existence of periodic solutions for such systems of equations. However, this has not yet been done.

We now proceed to investigations endeavoring to characterize the disposition of integral curves in the large with few analytic assumptions. We first recall here the recent work of P. N. Papuš. He investigated systems $dx_i/dt = X_i(x_1, \ldots, x_n)$ and considered the first and second derivatives of the functions

$$V = \sum_{i=1}^{n} x_i^2$$

taken with respect to the system. We call neutral a surface on which $dV/dt = 0$. As Papuš proved, if on this surface $d^2V/dt^2 \neq 0$ then it is possible to classify completely the behavior of the integral curves by means of the signs of d^2V/dt^2 and dV/dt in different regions into which the neutral surface divides the region under consideration.

It appears that under these assumptions no limit structures except singular points can arise, i.e. the existence of limit structures excludes constancy of the sign of d^2V/dt^2 on the neutral surface.

It would be quite interesting to study these cases since we do not yet have analytic criteria for finding limit structures, limit cycles, or recurrent motions. For more than two equations this is the analytic problem of the qualitative theory.

On the problem of analytic characterization of systems of differential equations in the large there is the work of E. A. Barbašin [32]. If we look at his contributions merely from this point of view, it appears that he worked on two problems: the analytic characterization of those systems for which the family of integral curves can be mapped on a family of parallel straight lines, and the problem of finding first integrals, i.e. finding conditions for the stratifiability of the family of integral curves. He connects these problems with the existence of continuous solutions of certain linear partial differential equations. The methods he used are based on topological considerations.

I find great difficulty in concluding this survey, since I have left completely unmentioned many quite illuminating investigations in the qualitative theory, and the authors of these investigations may have claims on me.

In this survey I tried to touch only upon those problems, which, essentially, were posed by the classical writers of the qualitative theory — Sturm, Poincaré, Lyapunov — and which have a simple analytic character. However, I shall enumerate other lines of research of the seminar in the post-war years:

1. Theory of dynamical systems (V. V. Nemickiĭ, V. A. Tumarkin, V. I. Grabaĭ, Yu. K. Solnčev, R. E. Vinograd, N. P. Zidkov, E. A. Barbašin, M. A. Al'muhamedov).

2. Problems of the dependence of the solutions on a parameter contained in the equations (A. N. Tihonov, I. M. Gradsteĭn, A. D.

Vasil'eva). Investigation of equations with discontinuous right-hand side (Yu. K. Solnčev).

3. Qualitative theory of equations with delayed time (A. Miškis, L. El'sgolc's).

4. Applied questions of the qualitative theory (G. F. Hil'mi), and also a number of other questions arising from time to time.

Bibliography for Appendix to Part One

1. Lyapunov, A. M. General theory of the stability of motion. *ONTI, L-M.*, 1935. (Also Princeton University Press, 1948 in French).
2. Nemickii, V. V., and Stepanov, V. V. *Qualitative theory of differential equations,* GTTI 2nd edition. Moscow, Leningrad, 1949.
3. Persidskii, K. P. On the characteristic numbers of differential equations. *Izvestiya Akademii Nauk, No.* 1, 1947.
4. Bylov, B. F. On the characteristic numbers of solutions of systems of linear differential equations. *Prikladnaya mat. i meh.* 14, *No.* 4, 1950.
5. Grobman, D. I. Characteristic exponents of the solutions of almost linear systems of differential equations. *Mat. Sbornik.*
6. Demidovic, B. P. On a critical case of stability in the sense of Lyapunov. *Doklady Akad. Nauk* 72, *No.* 6, 1950; On the stability in the sense of Lyapunov of linear systems of ordinary differential equations. *Mat. Sbornik* 28 (3), 1951.
7. Gorbunov, A. D. On a method for obtaining estimates of the solutions of systems of ordinary differential equations. *Vestnik Moskovskogo, Univ. No.* 10, 1950; On certain properties of the solutions of systems of ordinary linear homogeneous equations. *Vestnik Moskovskogo Univ. No.* 6, 1951.
8. Erugin, N. P. Reducible systems. *Trudy Matematiceskogo instituta im. Steklova XIII,* p. 65, 1946.
9. Yakubovic, V. A. Some criteria for the reducibility of systems of differential equations. *Doklady Akad. Nauk* 66, *No.* 4, 1949.
10. Yakubovic, V. A. On the boundedness of the solutions of the equation $y'' = p(x)y$, $p(x+\omega) = p(x)$. *Doklady Akad. Nauk* 74, *No.* 5, 1950.
11. Yakubovic, V. A. Criteria for the stability of systems of two linear differential equations with periodic coefficients. *Uspehi matematiceskich nauk* 6, *issue* 1 (4 1), p. 166–168, 1951.
12. Krein, M. G. A generalization of some investigations of A. M. Lyapunov on differential equations with periodic coefficients. *Doklady Akad. Nauk* 73, *No.* 3, 1950.
13. Neigauz, M. A., and Lidskii, V. B. On the boundedness of the solutions of linear differential equations with periodic coefficients. *Doklady Akad. Nauk* 77, *No.* 2, 1951.
14. Zukovskii, N. E. *Mat. Sbornik XVI,* 1891.
15. Sepelev, V. M. On the question of the stability of motion. *Prikladnaya mat. i meh. II, No.* 1, 1936.
16. Gusarov, L. A. On the boundedness of the solutions of a linear differential equation of the second order. *Doklady Akad. Nauk* 68, *No.* 2, 1949.
17. Sobol', I. M. An investigation of the asymptotic behavior of solutions of a linear equation of the second order using polar coordinates. *Mat. Sbornik* 28, *No.* 3, 1951.
18. Bellman, R. The stability of solutions of linear differential equations. *Duke Math. Journ.* 10, 1943.
19. Gusarov, L. A. On the convergence to zero of the solutions of a linear differential equation of the second order. *Doklady Akad. Nauk* 71, *No.* 1, 1950.
20. Kamynin, L. I. On the boundedness of solutions of differential equations. *Vestnik Moskovskogo Univ. No.* 5, 1951.
21. Gusarova, R. S. On the boundedness of solutions of linear differential equations. *Prikladnaya mat. i meh.*, 13, *No.* 3, 1949.
22. Sobol', I. M. On Ricatti's equation and the second order linear equations reducible to it, *Doklady Akad. Nauk* 65, *No.* 3, 1949; On the asymptotic behavior

of solutions of linear differential equational. *Doklady Akad. Nauk* 61, *No.* 2, 1948.

23. EL'SIN, M. I. The phase method and the classical method of comparison. *Doklady Akad. Nauk* 68, *No.* 5, 1949; on the decremental estimate of the amplitude. *Doklady Akad. Nauk* 63, *No.* 3, 1948.
24. SESTAKOV, A. A. and PAIVAN, A. V. (Full reference not available.)
25. GROBMAN, D. I. Characteristic exponents of almost linear systems. *Mat. Sbormik* 30 (72), *No.* 1, 1962.
26. YAKUBOVIC, V. A. On the asymptotic behavior of systems of differential equations. *Mat. Sbornik* 28, *No.* 1, 1951.
27. PETROVSKII, I. G. Über das Verhalten der Integralkurven eines systems gewohnlicher Differentialgleichungen in der Nähe eines singulären Punktes. *Mat. Sbornik* 41, *issue* 3, 1935.
28. SESTAKOV, A. A. Behavior of the integral curves of systems of the form:
$$dx_1/dt = X_1(x_1); \quad dx_i/dt = \varphi_i(x_1, x_i) + X_i(x_1, x_2, \ldots, x_n).$$
Doklady Akad. Nauk 62, *No.* 5, 1948; On the behavior of integral curves of systems of differential equations in the neighborhood of a singular point. *Doklady Akad. Nauk* 62, *No.* 2, 1948; On the behavior of integral curves of systems of differential equations in the neighborhood of a singular point of higher order. *Doklady Akad. Nauk* 65, *No.* 2, 1948.
29. NEMICKII, V. V. Qualitative integration of systems of differential equations. *Mat. Sbornik* 16, *No.* 3, 1945.
30. FILIPPOV, A.A. Sufficient conditions for the existence of limit cycles for equations of the second order. *Mat. Sbornik* 30 (72) *No.* 1, 1952.
31. DEMIDOVIC, B. P. The existence of periodic solutions of certain non-linear systems of ordinary differential equations. *Vestnik Moskovskogo No* 2, 1949.
32. BARBASIN, E. A. On the existence of continuous solutions of linear partial differential equations, *Doklady Akad. Nauk* 72, *No.* 3, 1950; On homomorphisms of dynamical systems, *Mat. Sbornik* 27 (69), 1950; On homomorphisms of dynamical systems. *Mat. Sbornik Univ.* 29 (71), *No.* 3, 1950.

Index to Part One

PART TWO

CHAPTER V

General Theory of Dynamical Systems

The classical theory of dynamical systems discussed in Section 3, Chapter I considered motions defined by the system of differential equations

$$\text{(A)} \qquad \frac{dx_i}{dt} = X_i(x_1, x_2, \ldots, x_n), \qquad (i = 1, 2, \ldots, n),$$

where the right-hand sides are continuous functions of the point $p(x_1, x_2, \ldots, x_n)$ in some closed domain D of the n-dimensional Euclidean "phase space". Moreover, it was assumed that these functions satisfied a supplementary condition assuring the uniqueness of the solution determined by the initial conditions

$$x_1 = x_1^{(0)}, \; x_2 = x_2^{(0)}, \ldots, x_n = x_n^{(0)} \quad \text{for} \quad t = 0,$$

where $p_0(x_1^{(0)}, x_2^{(0)}, \ldots, x_n^{(0)})$ is the starting point of the motion; such a sufficient condition, for example, is the Lipschitz condition.

In this case there was proved a series of general properties of the motions defined by the system (A): every solution can either be extended without bound as $t \to +\infty$ or else, for a finite value $t = T$, reaches the boundary of the domain D; every solution

$$x_i = f_i(t;\, x_1^{(0)}, x_2^{(0)}, \ldots, x_n^{(0)}), \qquad (i = 1, 2, \ldots, n),$$

is a continuous function of the time t and the coordinates of the initial point; finally, so long as the right-hand sides of the equations (A) do not depend on the time, if a motion beginning at the point p reaches the point p_1 at the time t_1, and if a motion beginning at the point p_1 reaches the point p_2 at the instant t_2, then the first motion reaches the point p_2 at the instant $t_1 + t_2$ (the group property).

For more general investigations of dynamical systems it appears expedient to depart from the definition of a dynamical system by means of the differential equations (A) and introduce an abstract definition of a dynamical system which includes all those properties of it which must be used in the proof of theorems. Finally, if one abstracts further in this direction there is no reason to restrict oneself to the n-dimensional Euclidean space E_n. In fact, for our proofs we will make use of only certain properties of the space; therefore, it is natural to define axiomatically the most general possible space for the dynamical system in which its properties will be the same as in ordinary Euclidean space. All theorems obtained by this course of abstraction will be valid in the special case of systems given by equations (A) in E_n (or in a part of it; for example, in a compact submanifold, if the theorem has been proved for a compact abstract space).

1. Metric spaces

We shall consider dynamical systems defined in metric spaces.[1]

A *metric space* R is a set of elements (points) in which for each pair of points $p, q \in R$ there is defined a non-negative function $\varrho(p, q)$, *distance*, which satisfies the three axioms:

I. $\varrho(p, q) \geqq 0$; moreover $\varrho(p, q) = 0$ if, and only if $p = q$;

II. $\varrho(p, q) = \varrho(q, p)$ — the axiom of symmetry;

III. $\varrho(p, r) \leqq \varrho(p, q) + \varrho(q, r)$ — the triangle axiom.[2]

If A is any set in R, the distance from the point p to the set A is defined as the greatest lower bound of the distances to the points of the set A:

$$\varrho(A, p) = \varrho(p, A) = \inf_{q \in A} \varrho(p, q).$$

The sequence of points $p_1, p_2, \ldots, p_n, \ldots$ *converges* to the point p if $\lim_{n\to\infty} \varrho(p_n, p) = 0$; in such a case we shall write

$$\lim_{n\to\infty} p_n = p \quad \text{or} \quad p_n \to p.$$

There holds the relation: if $p = \lim_{n\to\infty} p_n$ and q is any point, then $\lim_{n\to\infty} \varrho(p_n, q) = \varrho(p, q)$. This follows from Axiom III:

[1]See Hausdorff, *Mengenlehre*, 3. Aufl., Berlin, 1935. §§ 20—26.

[2]We shall say that a metric has been established in a space if distance has been defined in it.

$$|\varrho(p, q) - \varrho(p_n, q)| \leqq \varrho(p, p_n).$$

A point p is called a *limit point* of the set A if there exists a sequence $\{p_n\} \subset A$ such that $\lim_{n\to\infty} p_n = p$. It is obvious that in this case $\varrho(p, A) = 0$. Conversely, if the last equation is fulfilled, then either $p \in A$ or p is a limit point of A.

The set $F \subset R$ is called *closed* if it contains all its limit points. The null set and a set of a finite number of points have no limit points and are closed. The complement of a closed set is called an *open* set.

We introduce the following notation: the set of points $p \in R$ which satisfy the inequality $\varrho(p, p_0) < \varepsilon$, where $p \in R$ and ε is any positive number, we shall denote by $S(p_0, \varepsilon)$ and call a *sphere of radius ε around the point p_0*. Analogously, the set of points p such that $\varrho(p, A) < \varepsilon$ we shall denote by $S(A, \varepsilon)$ and call the ε-neighborhood of the set A.

If F is a closed set and $p \in R - F$, then $\varrho(p, F) > 0$. Otherwise $\varrho(p, F) = 0$; then we have $\inf_{q\in F} \varrho(p, q) = 0$, meaning that there exists a sequence $\{q_n\} \subset F$ such that $\lim_{n\to\infty} \varrho(q_n, p) = 0$. But then p is a limit point of F; and since F is closed we have $p \in F$, contrary to the hypothesis. Hence $\varrho(p, F) > 0$.

From this it follows directly that if G is an open set and $p \in G$ then there exists a positive ε such that $S(p, \varepsilon) \subset G$; *i. e.*, every point of an open set can be surrounded by some sphere likewise in the set. This property is characteristic of open sets. For let G possess the property that for any point $p \in G$ there can be found an $\varepsilon > 0$ such that $S(p, \varepsilon) \subset G$. We shall show that $R - G$ is a closed set. If $R - G$ is empty or contains a finite number of points the theorem is obvious. If $R - G$ contains an infinite number of points and there exists a convergent sequence $\{p_n\}$, $\lim_{n\to\infty} p_n = p_0$, then, by virtue of the definition of convergence, for any positive ε there can be found a p_n such that $\varrho(p_0, p_n) < \varepsilon$, *i.e.*, $p_n \in S(p_0, \varepsilon)$. Thus in any sphere containing the point p_0 there are found points $p_n \in R - G$, meaning that p_0 is not a point of G. Thus $p_0 \in R - G$, *i.e.*, $R - G$ is a closed set and G is open.

From this criterion it follows at once that $S(p_0, \varepsilon)$ is an open set, for if $p \in S(p_0, \varepsilon)$, then $\varrho(p, p_0) = d < \varepsilon$ and, by axiom III, $S(p, \varepsilon - d) \subset S(p_0, \varepsilon)$.

There hold the theorems:

1.01. THEOREM. *The intersection of any aggregate of closed sets is a (possibly empty) closed set.*

Let $A = \Pi_\alpha F_\alpha$, where F_α are closed sets. Suppose that $\{p_n\} \subset A$ and $p_n \to p$; we shall show that $p \in A$. From the condition $p_n \in A$ it follows that for any set F_α we have $\{p_n\} \subset F_\alpha$ and so, because F_α is closed, $p \in F_\alpha$, which means that $p \in \Pi_\alpha F_\alpha = A$.

Passing from the closed sets F_α to their complements, the open sets $G_\alpha = R - F_\alpha$, we obtain

1.02. THEOREM. *The union of any aggregate of open sets is an open set (possibly the whole space R).*

Let $A \subset R$ be any set. The set $\bar{A}$ obtained by annexing to $\bar{A}$ all its limit points is called the *closure* of the set A. For every set A we have $A \subset \bar{A}$; for a closed set F we have

$$\bar{F} = F.$$

1.03. THEOREM. *The closure $\bar{A}$ of any set A is a closed set*; i.e., $\bar{\bar{A}} = \bar{A}$.

Let $\{p_n\} \subset \bar{A}$ be a convergent sequence: $p_n \to p_0$; we shall show that $p_0 \in \bar{A}$. Let an arbitrary positive ε be assigned. From the definition of convergence it follows that there can be found a point p_n at a distance $\varrho(p_0, p_n) < \varepsilon/2$. Since $p_n \in \bar{A}$, it either lies in A or is a limit point of A; in both cases there exists a point $q \in A$ such that $\varrho(q, p_n) < \varepsilon/2$. From axiom III we obtain $\varrho(p_0, q) < \varepsilon$, *i.e.*, p_0 is a limit point of the set A, or in other words $p_0 \in \bar{A}$, which it was required to prove.

It is easily verified that $\varrho(A, p) = \varrho(\bar{A}, p)$.

Sometimes we shall consider the set of points p satisfying the condition $\varrho(p_0, p) \leqq \varepsilon$; we shall call this set the closed sphere of radius ε around the point p_0 and denote it by $S[p_0, \varepsilon]$. It is easy to show that this set is closed (if $\varrho(p_0, p_n) \leqq \varepsilon$ and $p_n \to p$, then $\varrho(p_0, p) \leqq \varepsilon$), but it may not be the closure $\overline{S(p_0, \varepsilon)}$ of the open sphere $S(p_0, \varepsilon)$. For example, let R be the set of numbers x satisfying the conditions $-\infty < x \leqq -1$ or $0 \leqq x < +\infty$; distance is defined as usual on the number line by $\varrho(x, y) = |x - y|$. Then $S(0, 1) = (0, 1)$, *i.e.*, the set $0 \leqq x < 1$; its closure, $\overline{S(0, 1)}$, is $[0, 1]$, but the closed sphere $S[0, 1] = [0, 1] + (-1)$.

A set $E \subset R$ is called *connected* if it is impossible to represent it in the form $E = A + B$, where A and B are not empty and $A\bar{B} + B\bar{A} = 0$, *i.e.*, it is impossible to represent E as a union such that neither term has limit points in the other. If such a decomposition is possible, then the set E is called *disconnected* and A and B, if themselves connected, are called its *components*.

It follows from the definition that a closed connected set cannot be represented as a sum of two closed nonempty sets without common points, for we would have

$$F = A + B,\ AB = 0,\ A = \bar{A},\ B = \bar{B},$$

i.e., F is a disconnected set.

1.04. THEOREM. *The components of a disconnected, closed set are closed sets.*

Let F be a closed set and let $F = A + B$, where $A\bar{B} + B\bar{A} = 0$. Assume, for example, that A is connected but not closed; then there exists a sequence of points $\{p_n\} \subset A$ such that $p_n \to p$ and p is not contained in A. Since $\{p_n\} \subset F$, and F is closed, $p \in F$, that is $p \in B$. On the other hand, $p \in \bar{A}$, that is, $p \in \bar{A} \cdot B$ which, therefore, is not empty. We have arrived at a contradiction and the theorem is proved.

Every open set $U(p)$ containing a given point p we shall call a *neighborhood* of the point p; it follows from the definition that $U(p)$ is a neighborhood for any point $q \in U(p)$. A system of neighborhoods $\{U_\sigma\}$ is called a *base* for the space R if for any point $p \in R$ and any of its neighborhoods $U(p)$ there can be found a neighborhood U_σ of the base such that $p \in U_\sigma \subset U(p)$.

Every open set $G \subset R$ can be represented as a union of neighborhoods belonging to the base. In fact, according to the hypothesis, for each point $p \in G$ there can be found a neighborhood $U(p)$ of the base such that $p \in U(p) \subset G$. But then, obviously

$$\sum_{p \in G} U(p) = G.$$

In the applications we shall be almost exclusively concerned with metric spaces possessing a *countable base*:

$$\{U_1, U_2, \ldots, U_n, \ldots\};$$

such spaces are said to satisfy *the second axiom of countability*.

1.05. DEFINITION. The set $A \subset R$ is called *everywhere dense* in R if $\bar{A} = R$. If in R there exists an everywhere dense set A which is countable, then R is said to be *separable*.

1.06. THEOREM. *A metric space with a countable base is separable.*

In each neighborhood U_n we mark one point $p_n \in U_n$ (among the points p_n there may be repetitions). Then $\{p_n\}$ is the everywhere dense set which is to be found. Indeed, suppose that $p \in R$ is an arbitrary point and $\varepsilon > 0$ is an arbitrary number. For the open set $S(p, \varepsilon)$, by virtue of the definition of a base, there can be found a U_n such that $p \in U_n \subset S(p, \varepsilon)$, but then for $p_n \in U_n$ we have $\varrho(p, p_n) < \varepsilon$, which proves the theorem.

1.07. THEOREM. *If a metric space R is separable, then there exists in it a countable base all of whose neighborhoods are spheres.*

Take the countable set of all the positive rational numbers $r_1, r_2, \ldots, r_n, \ldots$ and the countable, everywhere dense set $\{p_n\}$ and construct the countable set of spheres $S(p_n, r_k)$ $(n = 1, 2, \ldots; k = 1, 2, \ldots)$. This is the base sought. In fact, let $U(p)$ be any neighborhood of a point p. By the property of an open set there exists an $\varepsilon > 0$ such that $S(p, \varepsilon) \subset U(p)$; further, there can be found a point $p_n \in S(p, \varepsilon/2)$ and a rational number r_k, where $\varepsilon/2 > r_k > \varrho(p_n, p)$. Then we have

$$p \in S(p_n, r_k) \subset S(p, \varepsilon) \subset U(p).$$

From Theorems 1.06 and 1.07 there follows:

1.08. COROLLARY. *If a metric space has a countable base, then it also has a countable base consisting of spherical neighborhoods.*

An example of a metric space with a countable base is the Euclidean space $E_n = (x_1, x_2, \ldots, x_n)$ with the distance

$$\varrho[(x_1, x_2, \ldots, x_n), \quad (y_1, y_2, \ldots, y_n)] = [\sum_{i=1}^{n} (x_i - y_i)^2]^{1/2}.$$

Here a countable, everywhere dense set is, for example, the set of rational points $p_k = (r_1^k, r_2^k, \ldots, r_n^k)$, i.e., points all of whose coordinates are rational. A countable base is the family of spheres with rational centers and rational radii.

1.09. THEOREM (Baire). *If in a space with a countable base there is a totally ordered sequence of distinct closed sets, the ordering relation being inclusion, then the sequence is at most countable.*

Suppose we have a sequense of closed sets

$$\text{(F)} \qquad F_1 \supset F_2 \supset \ldots \supset F_n \supset \ldots \supset F_\omega \supset F_{\omega+1} \supset \ldots,$$

wherein by hypothesis $F_{\alpha+1}$ is a proper part of the set F_α. Consider the increasing sequence of open sets

$$G_1 \subset G_2 \subset \ldots \subset G_n \subset \ldots \subset G_\omega \subset G_{\omega+1} \subset \ldots,$$

where $G_a = R - F_a$. Let $\{U_n\}$ be a countable base of the space R. For any α there can be found a point $p_\alpha \in G_{a+1} - G_a$ (this difference is not empty since it is equal to $F_a - F_{a+1}$) and contained in a neighborhood U_{n_α} of the base, $U_{n_\alpha} \subset G_{a+1}$. Obviously U_{n_α} is not contained wholly in G_a. By virtue of this twofold property, if $\alpha \neq \beta$ then $U_{n_\alpha} \neq U_{n_\beta}$. Since the base is countable, the set of indices α in the sequence (F) is not more than countable, whence follows the theorem.

Thus the sequence (F) has the type of a *natural number* or of a *transfinite number of class* II. If one modifies the hypothesis of the theorem so as to allow $F_a = F_{a+1}$ somewhere in (F), then it follows that beginning with a certain number β of class I or II we shall have $F_a = F_{a+1}$ for $\alpha \geqq \beta$ (where F_β may be empty).

1.10. THEOREM. *If a certain system of open sets $\{G_\nu\}$ covers a space R possessing a countable base $\{U_n\}$, $R = \sum_\nu G_\nu$, then from this system there can be chosen a system of not more than a countable number of sets which possesses the same property:*

$$R = \sum_{n=1}^{\infty} G_n.$$

It is not difficult to deduce this theorem from the preceding one. We shall give an independent proof.

From the system U_n we choose those neighborhoods $U_{n'}$ which lie wholly within some set G_ν, and for each such $U_{n'}$ we choose one set $G_\nu \supset U_{n'}$ which we denote by $G_{n'}$. $\{G_{n'}\}$ is obviously at most countable. We shall show that it covers R. Suppose that $p \in R$ is any point; there can be found a G_ν such that $p \in G_\nu$. Further, by the property of a base, there can be found a U_n such that $p \in U_n \subset G_\nu$, that is, this U_n belongs to the system $\{U_{n'}\}$. *Let* $U_n = U_{m'}$; since $U_{m'} \subset G_{m'}$, $p \in G_{m'}$; i.e., $\sum_{n'} G_{n'}$ covers R.

1.11. THEOREM. *In a metric space with a countable base every open set can be represented as the union of a countable number of closed sets.*

Let $G \subset R$ be an open set. To each point $p \in G$ there can be set in correspondence a number $r(p) > 0$ such that $S(p, r(p)) \subset G$; if one takes $r' < r(p)$, then obviously $\overline{S(p, r')} \subset G$. From the covering of the set G by the sets $S(p, r')$ there can be chosen, by Theorem

1.10, a countable covering $S(p_n, r_{n'})$ such that $\sum_{n=1}^{\infty} S(p_n, r_{n'}) \supset G$; but by construction we have

$$\sum_{n=1}^{\infty} S(p_n, r_{n'}) \subset \sum_{n=1}^{\infty} \overline{S(p_n, r_{n'})} \subset G.$$

Consequently, G is represented as a sum of closed sets

$$G = \sum_{n=1}^{\infty} \overline{S(p_n, r_{n'})}.$$

A metric space R is called *compact* if any infinite sequence of its points contains a convergent subsequence.[3] Any closed bounded set of a Euclidean space E_n serves as an example of a compact space (the Bolzano-Weierstrass principal).

1.12. THEOREM. *In a compact space a countable decreasing sequence of (nonempty) closed sets has a nonempty intersection.*

Let these sets be $F_1, F_2, \ldots, F_n, \ldots$. If, beginning with a certain k, we have $F_k = F_{k+1} = F_{k+2} = \ldots$, then $\prod_{n=1}^{\infty} F_n = F_k$ is not empty. If among F_n there is an infinite number of distinct sets, we choose a subsequence of sets

$$F_{n1} \supset F_{n2} \supset \ldots \supset F_{n_k} \supset \ldots$$

such that each succeeding set constitutes a proper part of the preceding one and we choose points $p_k \in F_{n_k} - F_{n_{k+1}}$ $(k = 1, 2, \ldots)$. Because of the compactness, the sequence p_k has a limit point p. Since by construction $p_{k+m} \subset F_{n_k}$ $(m = 1, 2, \ldots)$ and since F_{n_k} is closed, then $p \subset F_{n_k}$ for any k, i.e.,

$$p \in \prod_{k=1}^{\infty} F_{nk} = \prod_{n=1}^{\infty} F_n.$$

The theorem is proved.

1.13. COROLLARY. *If a compact space R is covered by a countable system of open sets $\{G_n\}$ $(n = 1, 2, \ldots)$, then from this system there can be chosen a finite system covering R.*

Suppose that this proposition is false. Construct the sequence of closed sets:

$$F_1 = R - G_1,\ F_2 = R - (G_1 + G_2), \ldots,\ F_n = R - \sum_{i=1}^{n} G_i, \ldots$$

[3]This concept is usually defined in general (not necessarily metric) topological spaces, but this restricted definition will suffice for our purposes. [Ed.]

By our supposition no one of these is empty; moreover, obviously $F_1 \supset F_2 \supset \ldots \supset F_n \supset \ldots$. By 1.12 their intersection $\prod_{n=1}^{\infty} F_n$ is nonempty. But $\prod_{n=1}^{m} F_n = F_m = R - \sum_{i=1}^{m} G_i$; consequently

$$\prod_{n=1}^{\infty} F_n = R - \sum_{i=1}^{\infty} G_i$$

is nonempty, but this contradicts the hypothesis $R = \sum_{i=1}^{\infty} G_i$.

For a compact metric space there holds

1.14. Theorem. *A compact metric space has a countable base (and hence is separable).*

We shall first prove that a compact metric space K, for any positive ε, has an ε-net, *i.e.*, a finite set of points $p_1, p_2, \ldots, p_k$ possessing the property that for any point $p \in K$ there can be found a point p_i such that $\varrho(p, p_i) < \varepsilon$. Indeed, if for some $\varepsilon > 0$ there should not exist an ε-net, then, having chosen one point p_1, a p_2 could be found such that $\varrho(p_1, p_2) > \varepsilon$; in general, for any n, having found the points $p_1, p_2, \ldots, p_n$ such that $\varrho(p_i, p_j) > \varepsilon$ $(i, j = 1, 2, \ldots, n)$ we could find a point p_{n+1} such that $\varrho(p_{n+1}, p_i) > \varepsilon$ $(i = 1, 2, \ldots, n)$. The countable sequence $\{p_n\}$ constructed in this manner would have no limit point. In fact, assuming the existence of such a point p, we would have for $n_1 > N$, $n_2 > N$, $\varrho(p_{n_1}, p) < \varepsilon/2$, $\varrho(p_{n_2}, p) < \varepsilon/2$, *i.e.* $\varrho(p_{n_1}, p_{n_2}) < \varepsilon$ in contradiction with the property of the sequence of points $\{p_n\}$. But the nonexistence of a limit point for p_n contradicts the compactness of the space K. Thus the existence of an ε-net for any $\varepsilon > 0$ has been proved.

On constructing ε-nets for $\varepsilon = 1, 1/2, 1/3, \ldots, 1/n, \ldots$ and taking the union of the corresponding finite sets, we obtain a countable set of points $q_1, q_2, \ldots, q_n, \ldots$ which, as it is easy to see, is everywhere dense in K. Now we can take as a countable base for K the totality of spheres $S(q_n, r_k)$, where $\{r_k\}$ is the set of all positive rational numbers, as in Theorem 1.07.

The *distance between two sets* A and B in a metric space R is defined to be

$$\varrho(A, B) = \inf_{\substack{p \in A \\ q \in B}} \varrho(p, q).$$

1.15. Theorem. *In a compact space K the distance between two disjoint closed sets F_1 and F_2 is positive.*

Suppose that $F_1F_2 = 0$ and $\varrho(F_1, F_2) = 0$. From the latter condition there follows the existence of sequences $\{p_n\} \subset F_1$ and $\{q_n\} \subset F_2$ such that $\lim_{n\to\infty} \varrho(p_n, q_n) = 0$. Because of the compactness of the space K, there can be chosen from $\{p_n\}$ a convergent subsequence $\{p_{n_k}\}$; $\lim_{k\to\infty} p_{n_k} = p$. It is obvious that likewise $\lim_{k\to\infty} q_{n_k} = p$ and, from the fact that F_1 and F_2 are closed, the point p belongs both to F_1 and F_2, which contradicts the hypothesis of the theorem. The theorem is proved.

1.16. REMARK. The condition that F_1 and F_2 lie in a compact space is essential. Consider, in fact, in $E^2(x, y)$ the two closed sets $F_1 = \{y = 0; 1 \leqq x < +\infty\}$ and $F_2 = \{y = 1/x; 1 \leqq x < +\infty\}$; we have $F_1F_2 = 0$ and $\varrho(F_1, F_2) = 0$.

1.17. THEOREM (Heine-Borel). *From any open covering of a compact metric space there can be chosen a finite subcovering.*

This theorem follows directly from Theorems 1.14 and 1.10 and the Corollary 1.13. For, since a compact metric space has a countable base, from any covering by open sets there can be selected a countable covering, and, because of the compactness, from the countable covering there can be selected a finite covering.

We shall be concerned in the sequel with a *locally compact* space. A space R is called *locally compact* if every point $p \in R$ has a neighborhood $U(p)$ such that $\overline{U(p)}$ is a compact set. E^1, the infinite real line, serves as an example of a locally compact space as likewise any space E_n. We shall need to employ only a single property of a locally compact (metric) space.

1.18. THEOREM. *Every locally compact metric space R with a countable base can be represented as the union of a countable increasing sequence of closed compact sets*:

$$R = \sum_{n=1}^{\infty} F_n, \quad F_n \subset F_{n+1} \qquad (n = 1, 2, \ldots).$$

Let $\{U_n\}$ be a countable base of the space R. We note first of all that it can be replaced by a countable base of neighborhoods whose closures are compact. For let $p \in R$ be any point; by hypothesis there can be found a neighborhood $V(p)$ of p such that $V(p)$ is compact. But since $\{U_n\}$ forms a base there can be found a neighborhood $U_{n'}$ such that $p \in U_{n'} \subset V(p)$. Obviously since $\overline{U}_{n'} \subset \overline{V}(p)$, the set $\overline{U}_{n'}$ is compact. The totality of all neighborhoods possessing this property is countable as a part of the system $\{U_n\}$. It constitutes, as it is easy to see, a base for the space R. We shall call this base $\{U_n^*\}$.

There is no difficulty now in proving the theorem. Indeed, we define

$$F_1 = \overline{U}_1^*, \quad F_n = \sum_{i=1}^{n} \overline{U}_i^* \qquad (n, = 2, 3, \ldots).$$

Obviously the F_n are compact sets, $F_{n+1} \supset F_n$ $(n = 1, 2, \ldots)$, and since, according to the definition of a base, $\sum_{n=1}^{\infty} U_n^* = R$, then *a fortiori*

$$\sum_{n=1}^{\infty} \overline{U}_n^* = \sum_{n=1}^{\infty} F_n = R.$$

The theorem is proved.

1.19. Corollary. *Given $p_n \in R - F_n$, then the sequence $\{p_n\}$ has not a single limit point.*

Assume the contrary. Let the subsequence $\{p_{n_k}\}$ converge and let $p_0 = \lim_{k\to\infty} p_{n_k}$; let $\overline{U}_{n_0}^* (p_0)$ be the compact closure of a neighborhood of the point p_0. From the limit concept it follows that $p_{n_k} \in U_{n_0}^*$ for $k > K$; for $k \leqq K$ suppose that the points $p_{n_k} \in \overline{U}_{m_k}^*$ and let $N = \max\,[n_0, m_1, m_2, \ldots, m_K]$. Then obviously $\{p_{n_k}\} \subset F_N$. But from the hypothesis if follows that for $n_k > N$ we have

$$p_{n_k} \in R - F_{n_k} \subset R - F_N.$$

This is a contradiction and proves our assertion.

A sequence $\{p_n\}$ is called *fundamental* if it satisfies the Cauchy criterion, *i.e.*, for any positive ε there can be found an N such that $\varrho(p_n, p_{n+m}) < \varepsilon$ for $n \geqq N$; $m \geqq 1$. A metric space R is called *complete* if any fundamental sequence $\{p_n\} \subset R$ has a limit point. Obviously, a fundamental sequence cannot have more than one limit point; therefore in a complete space every fundamental sequence converges.

Every compact space is complete. The connection between a complete and a compact space is given by the following theorem.

1.20. Theorem. *If a complete space R has a finite ε-net for any $\varepsilon > 0$, then it is compact.*

Let there be given a sequence $\{p_n\} \subset R$. We construct an ε-net for $\varepsilon = 1$; let this be $q_1^{(1)}, q_2^{(1)}, \ldots, q_{N_1}^{(1)}$. Then since $R = \sum_{k=1}^{N_1} S(q_k^{(1)}, 1)$

there can be found an $S(q_{i_1}^{(1)}, 1)$ containing an infinite set of points of the sequence $\{p_n\}$. Let p_{n_1} be a point of this sequence, for example, that point with the smallest index which lies in $S(q_{i_1}^{(1)},1)$.

Further, for $\varepsilon = \frac{1}{2}$ we construct the net $q_1^{(2)}, q_2^{(2)}, \ldots, q_{N_2}^{(2)}$; consider the open sets $S(q_{i_1}^{(1)}, 1) \cdot S(q_k^{(2)}, \frac{1}{2})$, $(k = 1, 2, \ldots, N_2)$. Their union gives $S(q_{i_1}^{(1)}, 1)$; consequently at least one of them, say

$$S(q_{i_1}^{(1)}, 1) \cdot S(q_{i_2}^{(2)}, \tfrac{1}{2}),$$

contains an infinite set of points of $\{p_n\}$; we choose from these the point p_{n_2} with index $n_2 > n_1$.

In general, having chosen the point P_{n_k} from the infinite set of points of the sequence lying in

$$S(q_{i_1}^{(1)}, 1) \cdot S(q_{i_2}^{(2)}, \tfrac{1}{2}) \ldots S\left(q_{i_k}^{(k)}, \frac{1}{k}\right),$$

we consider the intersection of this set with each of the sets

$$S\left(q_l^{(k+1)}, \frac{1}{k+1}\right) \qquad (l = 1, 2, \ldots, N_{k+1}),$$

where $\{q_l^{(k+1)}\}$ is a $1/(k+1)$-net. Then we choose from these intersections that which contains an infinite number of points of the sequence $\{p_n\}$ and of these points we mark the point $p_{n_{k+1}}$, where $n_{k+1} > n_k$.

We shall prove that the subsequence $p_{n_1}, p_{n_2}, \ldots, p_{n_k}, \ldots$ converges. For since $p_{n_{k+m}} \subset S(q_{i_k}^{(k)}, 1/k)$ for $m = 0, 1, 2, \ldots$, then $\varrho(p_{n_m}, p_{n_{m+r}}) < 2/k$ for $m \geqq k$, $r \geqq 1$; *i.e.*, the Cauchy criterion is satisfied and by virtue of the completeness of R the sequence converges.

We have chosen from an arbitrary sequence $\{p_n\}$ a convergent subsequence; *i.e.*, R is compact.

As an example of a complete (noncompact) space one can take E_n. More interesting is the following example. Consider the set of continuous functions $f(x)$ defined, for example, on the segment $0 \leqq x \leqq 1$ (or defined on any compact metric space), as a metric space C. As the distance of two "points" $f_1(x)$ and $f_2(x)$ we take $\sup_{0 \leqq x \leqq 1} |f_1(x) - f_2(x)|$. It is easy to verify that the distance satisfies axioms I—III, and it is obvious that a sequence $\{f_n(x)\}$ will converge to $f(x)$ in the sense of this metric if the sequence of functions $f_n(x)$ converges uniformly to $f(x)$. The completeness of the space follows from the Cauchy criterion: if for any $\varepsilon > 0$ there can be found an $N(\varepsilon)$ such that $\varrho(f_n(x), f_{n+m}(x)) < \varepsilon$ for $n \geqq N(\varepsilon)$, $m \geqq 1$, then the sequence converges uniformly and, consequently, converges to a continuous function. This space, as it is easy to see, is not compact; moreover, the set $\overline{S(O, k)}$ where O represents the function $f(x) \equiv 0$, $0 \leqq x \leqq 1$, and k is any positive constant, is not compact since, for example, from the sequence $\{k \sin nx\}$, $n = 1, 2, \ldots$, it is impossible to select a convergent subsequence. Thus the space is not locally compact. In Chapter V we prove that this space has a countable, everywhere dense set, *i.e.* possesses a countable base.

Another important example of a complete space is the countable-dimensional (or separable) *Hilbert space* E^∞. The points of this space are the countable sequences of numbers $x = (x_1, x_2, \ldots, x_n, \ldots$ such that the series $\sum_{i=1}^{\infty} x_i^2$ converges. The numbers x_i are called the *coordinates* of the point x. The *distance* between two points $x = (x_1, x_2, \ldots, x_n, \ldots)$ and $y = (y_1, y_2, \ldots, y_n, \ldots)$ is defined by the formula

$$\varrho(x, y) = [\sum_{i=1}^{\infty} (x_i - y_i)^2]^{1/2},$$

wherein the series under the radical converges by virtue of the inequality

$$(x_i - y_i)^2 \leqq 2(x_i^2 + y_i^2).$$

The distance defined in such a way satisfies axioms I-III. For axioms I and II it is obvious. For III we have

$$[\varrho(x, y) + \varrho(y, z)]^2$$
$$= \sum_{i=1}^{\infty}(x_i - y_i)^2 + \sum_{i=1}^{\infty}(y_i - z_i)^2 + 2[\sum_{i=1}^{\infty}(x_i - y_i)^2 \cdot \sum_{i=1}^{\infty}(y_i - z_i)^2]^{1/2}.$$

But from the Cauchy-Schwarz inequality there follows

$$[\sum_{i=1}^{\infty}(x_i - y_i)^2 \cdot \sum_{i=1}^{\infty}(y_i - z_i)^2]^{1/2} \geqq \sum_{i=1}^{\infty} |x_i - y_i| \cdot |y_i - z_i|,$$

whence we obtain

$$\varrho(x, y) + \varrho(y, z) \geqq [\sum_{i=1}^{\infty}(x_i - z_i)^2]^{1/2} = \varrho(x, z).$$

If a sequence of points $x^{(1)}, x^{(2)}, \ldots, x^{(n)}, \ldots$ satisfies the Cauchy criterion, *i.e.*, if for $\varepsilon > 0$ there exists an $N(\varepsilon)$ such that $\varrho(x^{(n)}, x^{(n+m)}) < \varepsilon$ for $n \geqq N$, $m \geqq 1$, then this sequence has a limit point. For, on writing the distance in the expanded form, we have

$$\text{(A)} \qquad \sum_{i=1}^{\infty} (x_i^{(n)} - x_i^{(n+m)})^2 < \varepsilon^2,$$

whence for any fixed i $(i = 1, 2, \ldots)$ we obtain

$$|x_i^{(n)} - x_i^{(n+m)}| < \varepsilon,$$

i. e., there exists $\lim_{n\to\infty} x_i^{(n)} = x_i^{(0)}$. Passing to the limit as $m \to \infty$ in the inequality (A), we have

$$\sum_{i=1}^{\infty} (x_i^{(n)} - x_i^{(0)})^2 \leqq \varepsilon^2,$$

whence it is easily obtained that the series $\sum_{i=1}^{\infty} [x_i^{(0)}]^2$ converges. Thus the point $x^{(0)} = (x_1^{(0)}, x_2^{(0)}, \ldots)$ is the limit of the given sequence and, consequently, our assertion is proved.

This space is not compact because the coordinates can increase without bound. Moreover, it is interesting to note that $S[0, \alpha]$, for any $\alpha > 0$, is noncompact. For take the sequence of points $x^{(1)} = (\alpha, 0, 0, \ldots)$, $x^{(2)} = (0, \alpha, 0, \ldots)$ and in general $x^{(i)} = (x_1^{(i)}, x_2^{(i)}, \ldots)$, where $x_k^{(i)} = 0$ for $i \neq k$, $x_i^{(i)} = \alpha$. This sequence does not have a single limit point since $\varrho(x^{(i)}, x^{(j)}) = \sqrt{2}\alpha$ for $i \neq j$. Thus any sphere is noncompact.

From this demonstration, Hilbert space is not locally compact. This space is separable, i.e. it has a countable, everywhere dense subset, the set of points with all coordinates rational. Consequently, by Theorem 1.07, it possesses a countable base.

1.21. THEOREM. *In a complete space R a decreasing sequence of closed, nonempty sets with diameter*[4] *tending toward zero has a nonempty intersection (a point).*

Let $F_1 \supset F_2 \supset \ldots \supset F_n \supset \ldots$ be a sequence of closed sets and let $D(F_n) = d_n \to 0$. If, beginning with a certain k, we have $F_k = F_{k+1} = \ldots$, then $D(F_k) = 0$ and, since F_k is nonempty, F_k consists of a single point p; $\prod_{n=1}^{\infty} F_n = p$ and the theorem is proved.

If the situation above does not hold, then there can be found a sequence $F_{n_1} \supset F_{n_2} \supset \ldots \supset F_{n_k} \supset \ldots$ such that $F_{n_k} - F_{n_{k+1}}$ is nonempty; then in this difference we choose any point and denote

[4]The diameter of a set A is the number $D(A) = \sup_{p, q \in A} \varrho(p, q)$. Obviously, if $A \subset B$, then $D(A) \leqq D(B)$.

it by p_k. The sequence $\{p_k\}$ satisfies the Cauchy criterion since $\sum_{k=m}^{\infty} p_k \subset F_{n_m}$ and, consequently, $\varrho(p_s, p_{s+l}) < d_{n_m}$ for $s \geqq m$, $l > 1$; hence, by the completeness of the space R, the sequence $\{p_n\}$ converges to a point p. This point, being a limit of the sequence $\{p_k, p_{k+1}, \ldots\}$ lies in the closed sets F_{n_k} $(k = 1, 2, \ldots)$. Consequently

$$p \in \prod_{k=1}^{\infty} F_{n_k} = \prod_{n=1}^{\infty} F_n.$$

But the intersection $\prod_{n=1}^{\infty} F_n$ cannot contain more than one point. Suppose that there were two, p_1 and p_2, and let $\varrho(p_1, p_2) = \alpha > 0$. There can be found an F_n such that $D(F_n) = d_n < \alpha$; but, on the other hand, $p_1 + p_2 \subset F_n$ and, therefore, $D(F_n) = d_n \geqq \varrho(p_1, p_2) = \alpha$ and we arrive at a contradiction. The theorem is proved.

1.22. Definition. A set A is *nowhere dense* in R if in any (nonempty) open set $G \subset R$ there can be found a (nonempty) open set $G_1 \subset G$ such that $G_1 \cdot A = 0$.

A finite or countable union of nowhere dense sets in R is called a *set of category* I of Baire in R; the complement of a set of category I is called a *set of category* II of Baire. There holds the theorem:

1.23. Theorem. *In a complete metric space R a set of category* II *is everywhere dense in R.*

Let $E = \sum_{i=1}^{\infty} A_i$, where A_i is nowhere dense in R, be of category I. Let $G \subset R$ be any open set. The theorem will be proved if we show that G contains a point $p \in R - E$. Since A_1 is nondense in R, there can be found a $G_1 \subset G$ such that $G_1 \cdot A_1 = 0$; moreover, G_1 can be taken such that $D(G_1) \leqq 1$. In the set G_1 there can be found a nonempty open set $G^{(1)}$ such that $\overline{G^{(1)}} \subset G_1$. For suppose that $p_1 \in G_1$; then there exists an $S(p_1, \varepsilon_1) \subset G_1$; as $G^{(1)}$ one can take $S(p_1, \varepsilon_1/2)$. Then $\overline{G^{(1)}} \cdot A_1 = 0$. Within $G^{(1)}$, by virtue of the nondenseness of A_2, there can be found an open set G_2 such that $G_2 \cdot A_2 = 0$; moreover, we choose G_2 such that $D(G_2) \leqq \frac{1}{2}$. Further, we find a $G^{(2)}$ such that $\overline{G^{(2)}} \subset G_2$. We have $\overline{G^{(2)}} \subset \overline{G^{(1)}}$ and $\overline{G^{(2)}} \cdot A_i = 0$ $(i = 1, 2)$. In general, after having constructed the set $G^{(k)}$ possessing the properties $D(G^{(k)}) \leqq 1/k$, $\overline{G^{(k)}} \cdot A_i = 0$, $(i = 1, 2, \ldots, k)$, we find an open set $G_{k+1} \subset G^{(k)}$ such that $D(G_{k+1}) \leqq 1/(k+1)$, $G_{k+1} \cdot A_{k+1} = 0$ and then construct an open set $G^{(k+1)} \subset G_{k+1}$ such that $\overline{G^{(k+1)}} \subset G_{k+1}$. Then $\overline{G^{(k+1)}} \subset \overline{G^{(k)}}$ and $\overline{G^{(k+1)}} \cdot A_i = 0$ $(i = 1, 2, \ldots, k+1)$.

Consider the intersection of the closed sets $G^{(k)}$,

$$\prod_{k=1}^{\infty} \overline{G^{(k)}}.$$

By Theorem 1.21 it determines a point $p \in R$. Moreover,

$$p \in R - \sum_{i=1}^{\infty} A_i.$$

For, assuming the contrary, for some i we would have $p \in A_i$. But $p \in \overline{G^{(k)}}$ for any k, in particular $p \in \overline{G^{(i)}}$; meanwhile, by construction, $A_i \cdot \overline{G^{(i)}} = 0$. The theorem is proved.

We note that if A is nowhere dense in R then $\bar{A}$ is also nowhere-dense. Therefore every set of category I, $E = \sum_{i=1}^{\infty} A_i$, is contained in a set of category I, $E_1 = \sum_{i=1}^{\infty} \bar{A}_i$. The set E_1 is a union of closed (nondense) sets. A countable union of closed sets is called a *set of the type* F_σ. Thus every set of category I is contained in a set of the type F_σ of category I. The set complementary to E_1, a set of category II,

$$R - E_1 = R - \sum_{i=1}^{\infty} \bar{A}_i = \prod_{i=1}^{\infty} (R - \bar{A}_i),$$

is the intersection of a countable number of open sets; such a set is called a *set of the type* G_δ and we find that every set of category II contains a set of type G_δ of the second category.

Theorem 1.23 shows that a complete space R as well as every set of category II in R cannot be exhausted by a countable union of nondense sets.

We note that every set $R_1 \subset R$ can be regarded as a metric space which preserves for the points $p \in R_1$ the same metric which they have in R. In order to distinguish such concepts as a closed, an open, or a nowhere dense set in R_1, regarded as a space, we shall speak of $F_1 \subset R_1$ as a *relatively closed* set, of $G_1 \subset R_1$ as a *relatively open* set, etc.

For example, a set F_1 is relatively closed in R_1 if every limit point of F_1 which is contained in R_1 belongs to F_1; then the set $G_1 = R_1 - F_1$ will be relatively open in R_1. It is easy to show that every relatively closed set $F_1 \subset R_1$ is the intersection of an (absolutely) closed set F in R with the space R_1, for example, $F_1 = \bar{F}_1 R_1$, where the closure $\bar{F}_1$ is taken in the space R. Analogously, a relatively open set

$$G_1 = R_1 - F_1 = R_1 - R_1F = R_1(R - F)$$

is the intersection of an absolutely open set $R - F$ with the space R_1.

1.24. Concerning certain properties of continuous functions in a metric space. We begin with the obvious remark: if $\varphi(p)$ $(p \in R)$ is a continuous function, then the set $\{p;\ \varphi(p) \geqq a\}$ is closed.

We next prove a lemma which for a metric space R is completely elementary.

1.25. LEMMA ON INTERPOLATION. *If two closed sets $A \subset R$ and $B \subset R$ $(AB = 0)$ and two real numbers a and b are given, there exists a function, continuous over R, which assumes at all points of the set A the value a and at all points of the set B the value b, and elsewhere values between a and b.*

The desired function $f_{ab}(p)$ can be taken, for example, to be the function

$$f_{ab}(p) = \frac{b\varrho(p, A) + a\varrho(p, B)}{\varrho(p, A) + \varrho(p, B)}.$$

Indeed, f_{ab} is defined everywhere in R since the denominator never becomes zero because of the condition $AB = 0$. Its continuity follows directly from the continuity of $\varrho(p, A)$ and $\varrho(p, B)$ as functions of the point p. If $p \in A$, then $\varrho(p, A) = 0$, $f_{ab}(p) = a$ and, analogously, if $p \in B$, then $f_{ab}(p) = b$. We note, finally, that if $p \in R - A - B$ and if, for example, $a < b$, then $a < f_{ab}(p) < b$.

1.26. THEOREM (Extension Theorem) (Brouwer—Urysohn).

If a continuous bounded function $\varphi(p)$ is defined over a closed set $F \subset R$, then there can be constructed a continuous function $f(p)$ defined for all $p \in R$ and coinciding with $\varphi(p)$ for $p \in F$.

Let $\sup_{p \in F} |\varphi(p)| = \mu_0$. We define $\varphi_0(p) = \varphi(p)$ for $p \in F$. Furthermore, the sets $A_0 = \{p;\ \varphi_0(p) \leqq -\mu_0/3\}$ and $B_0 = \{p;\ \varphi_0(p) \geqq \mu_0/3\}$ contained in F are closed. Employing the lemma we construct for $p \in R$ the continuous function $f_0(p)$ assuming over A_0 the value $-\mu_0/3$ and over B_0 the value $\mu_0/3$. We have $|f_0(p)| \leqq \mu_0/3$.

Next we define for $p \in F$ the function $\varphi_1(p) = \varphi_0(p) - f_0(p)$; it is continuous and $\sup_{p \in F} |\varphi_1(p)| = \mu_1 = 2\mu_0/3$. The sets $A_1 = \{p;\ \varphi_1(p) \leqq -\mu_1/3\}$ and $B_1 = \{p;\ \varphi_1(p) \geqq \mu_1/3\}$ lie in F and are closed. We define for $p \in R$ the continuous function $f_1(p)$ assuming over A_1 the value $-\mu_1/3$ and over B_1 the value $\mu_1/3$. Next, for $p \in F$, we define $\varphi_2(p) = \varphi_1(p) - f_1(p)$ for which

$$\sup_{p \in F} |\varphi_2(p)| = \mu_2 = \tfrac{2}{3}\mu_1.$$

Continuing this construction indefinitely, we obtain for $p \in F$ a sequence of continuous functions $\varphi_0(p), \varphi_1(p), \ldots, \varphi_n(p), \ldots$ and for $p \in R$ a sequence of continuous functions $f_0(p), f_1(p), \ldots, f_n(p) \ldots$ such that

$$\varphi_{n+1}(p) = \varphi_n(p) - f_n(p), \quad \mu_n = \sup_{p \in F} |\varphi_n(p)| = (\tfrac{2}{3})^n \mu_0,$$

$$\sup_{p \in R} |f_n(p)| = (\tfrac{2}{3})^n \mu_0/3.$$

Now, for $p \in R$, set

$$f(p) = \sum f_n(p).$$

On the right-hand side is a uniformly convergent series of continuous functions and thus $f(p)$ is a continuous function. Furthermore,

$$|f(p)| \leqq \sum_{n=0}^{\infty} (\tfrac{2}{3})^n \mu_0/3 = \mu_0,$$

i. e., $|f(p)|$ is bounded by the same number as $|\varphi(p)|$. Finally, for $p \in F$ we have

$$\sum_{i=0}^{n} f_i(p) = \sum_{i=0}^{n} [\varphi_i(p) - \varphi_{i+1}(p)] = \varphi_0(p) - \varphi_{n+1}(p),$$

so that for $p \in F$ we obtain

$$f(p) = \lim_{n \to \infty} [\varphi_0(p) - \varphi_{n+1}(p)] = \varphi_0(p).$$

The theorem is proved.

We remark that if the given function $\varphi(p)$ is bounded by the numbers α and β, for example, $\inf_{p \in F} \varphi(p) = \alpha$, $\sup_{p \in F} \varphi(p) = \beta$ then on applying the construction above to the function $\varphi(p) - (\alpha + \beta)/2$ we obtain for $f(p)$ the inequalities $\alpha \leqq f(p) \leqq \beta$.

For metric spaces R with countable bases the Hilbert space is universal; any such R can be topologically imbedded in it.

1.27. THEOREM (Urysohn). *Any metric space R with a countable base is homeomorphic to some set of the Hilbert space E^∞.*

In Hilbert space the set of points $(x_1, x_2, \ldots, x_m, \ldots)$ for which

$|x_m| \leqq 1/m$ is called a *fundamental parellelopiped* Q^∞. The space E^∞, as we have seen, is not compact (and not locally compact). But $Q^\infty \subset E^\infty$ is compact.

For let there be given a sequence $\{p_n\} \subset Q^\infty$, where $p_n = (x_1^{(n)}, x_2^{(n)} \ldots, x_k^{(n)}, \ldots)$. Because of the boundedness of the set $\{x_1^{(n)}\}$ $(n = 1, 2, \ldots)$, one can select from it a convergent subsequence $x_1^{(n_{1k})}$ $(k = 1, 2, \ldots)$; let $\lim_{k\to\infty} x_1^{(n_{1k})} = x_1^{(0)}$. From the numbers n_{1k} there can be chosen a subsequence n_{2k} such that $\lim_{k\to\infty} x_2^{(n_{2k})} = x_2^{(0)}$, etc. Then, considering the diagonal sequence $n_{11}, n_{22}, \ldots, n_{kk}, \ldots$ we shall have $\lim_{k\to\infty} x_m^{(n_{kk})} = x_m^{(0)}$ for $m = 1, 2, 3, \ldots$, wherein obviously $|x_m^{(0)}| \leqq 1/m$. Denoting by p_0 the point with coordinates $(x_1^{(0)}, x_2^{(0)}, \ldots, x_n^{(0)}, \ldots)$, we shall prove that $\lim_{k\to\infty} p_{n_{kk}} = p_0$, where $p_{n_{kk}} = (x_1^{(n_{kk})}, x_2^{(n_{kk})}, \ldots)$. Let $\varepsilon > 0$ be arbitrary. We can find a P such that

$$\sum_{m=P+1}^{\infty} \frac{1}{m^2} < \frac{\varepsilon^2}{8}.$$

Having chosen P, we define N_i by the condition $|x_i^{(0)} - x_i^{(n_{kk})}| < \varepsilon/\sqrt{2P}$ for $k > N_i$ $(i = 1, 2, \ldots, P)$. Finally, we denote $N = \max_{i=1,2,\ldots,P} N_i$. Then for $k > N$ we obtain

$$[\varrho(p_{n_{kk}}, p_0)]^2 = \sum_{m=1}^{\infty} (x_m^{(n_{kk})} - x_m^{(0)})^2 = \sum_{m=1}^{P} (x_m^{(n_{kk})} - x_m^{(0)})^2$$

$$+ \sum_{m=P+1}^{\infty} (x_m^{(n_{kk})} - x_m^{(0)})^2 \leqq P \cdot \frac{\varepsilon^2}{2P} + 2 \sum_{m=P+1}^{\infty} (x_m^{(n_{kk})})^2 + 2 \sum_{m=P+1}^{\infty} (x_m^{(0)})^2$$

$$\leqq \frac{\varepsilon^2}{2} + 2\frac{\varepsilon^2}{8} + 2\frac{\varepsilon^2}{8} = \varepsilon^2.$$

This proves that $\lim_{k\to\infty} p_{n_{kk}} = p_0$. Thus the compactness of Q^∞ has been proved.

Now we can sharpen the assertion of the theorem: *a metric space R with a countable base can be mapped homeomorphically onto some subset of the space Q^∞.*

We proceed to the proof of the theorem. We remark at the outset that we can always assume that the distance ϱ satisfies the condition $\varrho(p, q) \leqq 1$. For if this should not be so we would introduce the new function

$$\varrho^*(p, q) = \frac{\varrho(p, q)}{1 + \varrho(p, q)};$$

then $\varrho^*(p, q) < 1$ and the topological properties of the space for this new metric are not altered, since if $\varrho(p, p_k) \to 0$ as $k \to \infty$ then $\varrho^*(p, p_k) \to 0$, and conversely.

By Theorem 1.06 there exists in R a countable, everywhere dense set $A = \{a_n\}$. We place in correspondence to the point $p \in R$ the point $\xi \in Q^\infty$ with the coordinates $\xi_m = (1/m)\,\varrho(p, a_m)$ $(m = 1, 2, 3, \ldots)$. This is the desired mapping $\xi = \Phi(p)$.

By construction, to each point $p \in R$ there corresponds a unique point $\xi \in Q^\infty$; for suppose now that $p \neq q$. Since A is everywhere dense in R, there can be found a point a_k such that $\varrho(p, a_k) < \frac{1}{2}\varrho(p, q)$; then $\varrho(q, a_k) > \frac{1}{2}\varrho(p, q)$, and if $\Phi(q) = \eta = (\eta_1, n_2, \ldots)$, then $\eta_k > \xi_k$, *i.e.*, $\eta \neq \xi$.

We shall prove the continuity of the mapping Φ. Suppose that $\varrho(p, q) < \varepsilon$; then $|\varrho(p, a_m) - \varrho(q, a_m)| \leqq \varrho(p, q) < \varepsilon$; therefore, if $\Phi(p) = \xi$, $\Phi(q) = \eta$, then $|\xi_m - \eta_m| < \varepsilon/m$, and, if we denote the distance in the Hilbert space by ϱ_1, we have

$$[\varrho_1(\xi, \eta)]^2 = \sum_{m=1}^{\infty} (\xi_m - \eta_m)^2 \leqq \varepsilon^2 \sum_{m=1}^{\infty} \frac{1}{m^2} = \frac{\varepsilon^2\pi^2}{6}; \quad \varrho_1(\xi, \eta) \leqq \frac{\pi}{\sqrt{6}}\,\varepsilon.$$

Consequently, the mapping Φ is even uniformly continuous.

We shall prove, finally, the continuity of the mapping Φ^{-1}. Let the image of R in Q^∞ be M, $\xi \in M$, $\Phi^{-1}(\xi) = p \in R$ and let $\varepsilon > 0$ be arbitrary. It is required to find a $\delta > 0$ such that from $\varrho_1(\xi, \eta) < \delta$ will follow $\varrho(p, q) < \varepsilon$, where $q = \Phi^{-1}(\eta)$. Let a_n be the first point of A for which $\varrho(p, a_n) < \varepsilon/3$. Then it is sufficient to take $\delta = \varepsilon/3n$, for if $\varrho_1(\xi, \eta) < \delta$ then in particular $|\xi_n - \eta_n| < \delta$; but recalling the definition of ξ_n we find $|\varrho(p, a_n) - \varrho(q, a_n)| < n\delta = \varepsilon/3$. Finally, from the triangle axiom we have

$$\varrho(p, q) \leqq \varrho(p, a_n) + \varrho(q, a_n) < 2(p, a_n) + \frac{\varepsilon}{3} < \varepsilon.$$

Thus Φ^{-1} is continuous, but in general not uniformly continuous. (It will be uniformly continuous if R is compact).

2. General Properties and Local Structure of Dynamical Systems

Let there be given a metric space R and a family of mappings of R onto itself, *i.e.* a function

$$f(p, t)$$

which to any point $p \in R$ for any real number $t(-\infty < t < +\infty)$ puts in correspondence some definite point $f(p, t) \in R$. We shall call the parameter t the *time*. We impose the following conditions on the function $f(p, t)$.

I. *The initial condition*:

$$f(p, 0) = p.$$

II. *The condition of continuity with respect to the set of variables p and t*: if there be given a convergent sequence of numbers $\{t_n\}$, where $\lim_{n\to\infty} t_n = t_0$, and a convergent sequence of points $\{p_n\}$, where $\lim_{n\to\infty} p_n = p_0$, then there holds the relation

$$\lim_{n\to\infty} f(p_n, t_n) = f(p_0, t_0).$$

It is easy to see that this definition of continuity is equivalent to the following: for a given point $p_0 \in R$ and a given number $t_0(-\infty < t_0 < +\infty)$ and for any $\varepsilon > 0$ there can be found a $\delta > 0$ such that if $\varrho(p, p_0) < \delta$ and $|t - t_0| < \delta$ then

$$\varrho[f(p, t), f(p_0, t_0)] < \varepsilon.$$

From Condition II there is obtained as a corollary the property:

II'. *The continuity of $f(p, t)$ as a function of the initial point.* We formulate this property thus: for any point $p \in R$, any number $T > 0$ (arbitrarily large) and any $\varepsilon > 0$ (arbitrarily small) there can be found a $\delta > 0$ such that if $\varrho(p, q) < \delta$ and $|t| \leqq T$ then there holds the inequality

$$\varrho[f(p, t), f(q, t)] < \varepsilon.$$

In other words, if the initial points be chosen sufficiently close, then in the course of a given arbitrarily large time interval $-T < t < T$ the distance between simultaneous positions of the moving points will remain less than an assigned positive quantity ε.

2.01. Proof. If the proposition were false, then there could be found a sequence of points $\{q_n\}$, $\lim_{n\to\infty} q_n = p$, and a corresponding sequence of numbers $\{t_n\}$, $|t_n| \leqq T$, such that

$$\varrho[f(p, t_n), f(q_n, t_n)] > \alpha > 0.$$

According to the theorem of Bolzano-Weierstrass, the sequence of numbers $\{t_n\}$ contains a convergent subsequence; in order not to complicate the notation we assume that $\{t_n\}$ is this subsequence. Thus

$$\lim_{n\to\infty} t_n = t_0; \qquad |t_0| \leqq T.$$

By Axiom III of a metric space (section 1) we have:

$$\varrho[f(p, t_n), f(q_n, t_n)] \leqq \varrho[f(p, t_n), f(p, t_0)] + \varrho[f(p, t_0), f(q_n, t_n)].$$

In view of the continuity of the function f, both distances on the right-hand side can be made less than $\alpha/2$ for sufficiently large n and we obtain the contradiction $\alpha < \alpha$.

III. *The group condition*:

$$f(f(p, t_1), t_2) = f(p, t_1 + t_2)$$

for any $p \in R$ and any real t_1 and t_2. The variable t is the parameter of the group.

From properties I and III there follows the existence of a transformation inverse to $f(p, t)$. Such is the transformation $f(p, -t)$ since it satisfies the relation

$$f(f(p, -t), t) = p.$$

To the value $t = 0$ of the parameter, by virtue of I, there corresponds the identical transformation of the group.

We shall call the group $f(p, t)$ of transformations of the space R into itself possessing the properties introduced above a *dynamical system* and the parameter t the *time*. Thus *a dynamical system is a one-parameter group $f(p, t)$ $(-\infty < t < +\infty)$ of transformations of the space R $(p \in R)$ into itself $(f(p, t) \in R)$ satisfying the conditions*:

I. $f(p, 0) = p$;

II. $f(p, t)$ *is continuous in the pair of variables p and t*;

III. $f(f(p, t_1), t_2) = f(p, t_1 + t_2)$ (*group property*).

We shall call the function $f(p, t)$, for fixed p, a *motion*; we shall call the *set of points*

$$\{f(p, t); -\infty < t < +\infty\}$$

for fixed p, a *trajectory* of this motion and shall denote it by the symbol $f(p; -\infty, +\infty)$, or more briefly, $f(p; \mathrm{I})$. Analogously, we shall call the sets

$$\{f(p, t);\ 0 < t < +\infty\} \quad \text{and} \quad \{f(p, t);\ -\infty < t < 0\}$$

the *positive* and *negative half-trajectories* respectively, with the notations

$$f(p, \mathrm{I}^+) = f(p; 0, +\infty) \quad \text{and} \quad f(p, \mathrm{I}^-) = f(p; -\infty, 0).$$

Finally, we shall call the set of points

$$\{f(p, t);\ T_1 \leqq t \leqq T_2\},$$

where p is fixed and $-\infty < T_1 < T_2 < +\infty$, a *finite arc of the trajectory*; it will be written

$$f(p; T_1, T_2).$$

We shall call the number $T_2 - T_1$ the *time length* of this arc of the trajectory.

In a dynamical system there may exist motions such that for all values t

$$f(p, t) = p. \tag{2.02}$$

We shall call a point p to which there corresponds such a "motion" a *rest point* (or point of equilibrium or critical point).

If for some motion $f(p, t)$ there holds the equality

$$f(p, t_1) = f(p, t_2), \qquad (t_1 \neq t_2),$$

then, writing $t_2 - t_1 = \tau$, we obtain for any t by III

$$\begin{aligned} f(p, t + \tau) &= f(p, t + t_2 - t_1) = f(f(p, t_2), t - t_1) \\ &= f(f(p, t_1), t - t_1) = f(p, t). \end{aligned}$$

A motion which satisfies the condition

$$f(p, t + \tau) = f(p, t) \tag{2.03}$$

for any t we shall call *periodic*, admitting the period τ. It is easily verified, employing property III, that a periodic motion admits as periods along with τ all the numbers $n\tau (n = \pm 1, \pm 2, \ldots)$. The smallest positive number τ satisfying now condition (2.03) is called the *period* of the motion $f(p, t)$.

If a periodic motion does not have such a least period, then $f(p, t)$ reduces to rest. In fact, for any $\varepsilon > 0$, by properties I and

II, there can be found a $\delta > 0$ such that $\varrho(p, f(p, t)) < \varepsilon$ for $|t| < \delta$. But by hypothesis there exists for $f(p, t)$ a period τ smaller than δ; therefore, representing any t in the form $t = n\tau + t'$ (n is an integer, $0 \leqq t' < \tau$), we obtain $\varrho(p, f(p, t)) < \varepsilon$ for all values of t whence, in view of the arbitrariness of ε, there follows $f(p, t) = p$ as we wished to prove.

The trajectory of a periodic motion with the period τ is obviously a simple closed curve, a one-to-one continuous image of the real line $[0, \tau]$, wherein the points 0 and τ are identified. Obviously, it is a closed, compact set.

The following more general assertion is valid:

2.04. THEOREM. *A finite arc of a trajectory, $f(p; T_1, T_2)$, is a closed, compact set.*

Let there be a sequence of points

$$\{p_n\} \subset f(p; T_1, T_2).$$

Let $p_n = f(p, t_n)$, wherein by hypothesis $T_1 \leqq t_n \leqq T_2$ ($n = 1, 2, \ldots$). From the bounded sequence $\{t_n\}$ there can be chosen a convergent subsequence $\{t_{n_k}\}$, $\lim_{k\to\infty} t_{n_k} = \tau$; obviously, $T_1 \leqq \tau \leqq T_2$. By property II we have $\lim_{k\to\infty} f(p, t_{n_k}) = f(p, \tau) = q$; *i.e.*, there exists a limit point $q \in f(p; T_1, T_2)$, and this establishes the theorem.

We shall denote by $f(A, t)$ the image of the set A under the transformation corresponding to a given t.

In the sequel invariant sets will have an important significance. A set A is called *invariant* (with respect to the dynamical system $f(p, t)$) if under all the transformations of the group it goes into itself, *i.e.* satisfies the condition

$$f(A, t) = A \quad (-\infty < t < +\infty). \tag{2.05}$$

We shall explain the meaning of this definition. Suppose that $p \in A$; then by the condition (2.05) we have

$$f(p, t) \subset f(A, t) \subset A;$$

i.e., if a point p belongs to an invariant set, then the entire trajectory determined by p lies in this set.

Obviously, each entire trajectory represents an invariant set. A set which is the sum of any set of trajectories is also an invariant set. In particular, the entire space R is also an invariant set. Thus, an invariant set is a set consisting of entire trajectories, and conversely.

2.06. THEOREM. *The closure of an invariant set is an invariant set.*

Let A be an invariant set and $\bar{A}$ its closure. If $p \in A$, then by the property of an invariant set noted above, $f(p, t) \subset A \subset \bar{A}$. Now suppose $p \in \bar{A} - A$; this means that there exists a sequence of points

$$\{p_n\} \subset A, \lim_{n\to\infty} p_n = p.$$

By property II, for any t we have $\lim_{n\to\infty} f(p_n, t) = f(p, t)$, and since $\{f(p_n, t)\} \subset A$ then $f(p, t) \subset \bar{A}$. That is, $f(\bar{A}, t) \subset \bar{A}$ for any t; from this last inclusion there follows $\bar{A} \subset f(\bar{A}, -t)$ for any t. Consequently, $\bar{A} = f(\bar{A}, t)$.

2.07. REMARK. The system of motions, determined by the differential equations (A) in the introduction to section 1, is a dynamical system in the n-dimensional Euclidean space E_n, if each solution of this system can be extended for all values of $t(-\infty < t < +\infty)$. This same system determines a dynamical system over the set $M \subset E_n$ if for any $p \in M$ a solution $f(p, t) \subset M$ is defined for $-\infty < t < +\infty$. Obviously, such a set is invariant and can be regarded as a space R.

We introduce some theorems concerning rest points.

2.08. THEOREM. *The set of rest points is a closed set.*

Let $p_1, p_2, \ldots, p_n, \ldots$ be rest points and let

$$\lim_{n\to\infty} p_n = p_0.$$

We shall show that p_0 is also a rest point. Taking any value of t, we have

$$f(p_n, t) = p_n.$$

Passing to the limit as $n \to \infty$ and using property II, we obtain

$$f(p_0, t) = p_0$$

which was to be shown.

2.09. THEOREM. *No trajectory enters a rest point for a finite value of t.*

Assume that $f(p, T) = p_0$, where $p \neq p_0$ and p_0 is a rest point. Then by property III we have

$$p = f(p_0, -T),$$

i. e., $f(p_0, -T) \neq p_0$, which contradicts the definition of a rest point.

2.10. THEOREM. *If, for any $\delta > 0$, there exists a point $q \in S(p, \delta)$ such that the half-trajectory $f(q;\ 0, +\infty) \subset S(p, \delta)$, then p is a rest point.*

We assume that p is not a rest point; then for some $t_0 > 0$ we have $f(p, t_0) \neq p$. Let $\varrho(p, f(p, t_0)) = d > 0$. By property II there exists a δ such that

$$\varrho[f(p, t), f(q, t)] < \frac{d}{2}$$

for $|t| \leqq t_0$ and any q satisfying the inequality $\varrho(q, p) < \delta < d/2$. By hypothesis there can be found in $S(p, \delta)$ a point q such that

$$f(q;\ 0, +\infty) \subset S(p, \delta).$$

We obtain by Axiom III of a metric space

$$\varrho(p, f(p, t_0)) < \varrho(p, f(q, t_0)) + \varrho[f(q, t_0), f(p, t_0)] < \delta + \frac{d}{2} < d.$$

The contradiction so obtained proves the theorem.

Obviously, the theorem remains valid if in its hypothesis the positive half-trajectory be replaced by the negative.

2.11. COROLLARY. *If*

$$\lim_{t=\infty} f(q, t) = p$$

exists, then p is a rest point.

For in this case, from the definition of a limit, for any $\delta > 0$ we have

$$\varrho(f(q, t), p) < \delta \quad \text{for} \quad t \geqq t_0.$$

Setting $q_1 = f(q, t_0)$, we deduce directly from property III that

$$f(q_1;\ 0, +\infty) \subset S(p, \delta),$$

i. e., by our theorem p is a rest point.

The study of the local structure of a dynamical system requires the introduction of a series of new concepts.

2.12. DEFINITION. For any set $E \subset R$ we shall call the set

$$\Phi = f(E;\ -T, +T) = \sum_{|t| \leqq T} f(E, t)$$

a *finite tube of time length $2T$.*

2.13. Definition. We shall call a set $F \subset \Phi$, closed in Φ, a *local section* of the finite tube Φ if to each point $q \in \Phi$ there corresponds a unique number t_q such that $f(q, t_q) \in F$, and $|tq| < 2T$.

Otherwise expressed, each segment of a trajectory entering Φ intersects F in one and only one point. A local section for a system defined in E_n by differential equations is easily constructed as a section perpendicular to one of these trajectories. However, for a dynamical system in a metric space the construction of such a section is far from trivial.

2.14. Theorem (Bebutov). *If p is not a critical point of a dynamical system, then for a sufficiently small $\tau_0 > 0$ there can be found a number $\delta > 0$ such that the tube constructed on $S(p, \delta)$ of time length $2\tau_0$ has a local section.*

Since p is not a critical point there exists a θ_0 such that $\varrho(p, f(p, \theta_0)) > 0$. We define a continuous function of q and t:

$$\varphi(q, t) = \int_t^{t+\theta_0} \varrho[f(q, \tau), p] d\tau.$$

From the group property there follows:

$$\varphi(q, t_1 + t_2) = \int_{t_1+t_2}^{t_1+t_2+\theta} \varrho[f(q, \tau), p] d\tau = \int_{t_2}^{t_2+\theta_0} \varrho[f(q, t_1 + \tau), p] d\tau$$
$$= \int_{t_2}^{t_2+\theta_0} \varrho[f(f(q, t_1), \tau), p] d\tau = \varphi[f(q, t_1), t_2].$$

The function $\varphi = \varphi(q, t)$ has the partial derivative

$$\varphi'_t(q, t) = \varrho[f(q, t + \theta_0), p] - \varrho[f(q, t), p].$$

Obviously, the function φ is continuous with respect to q and t.

Since

$$\varphi'_t(p, 0) = \varrho[f(p, \theta_0), p] > 0$$

there can be found an $\varepsilon > 0$ such that

$$\varphi'_t(q, 0) > 0 \quad \text{for} \quad q \in S(p, \varepsilon).$$

We next define τ_0 by the condition that for $|t| \leqq 3\tau_0$ there holds

$$f(p, t) \in S(p, \varepsilon).$$

Then we shall have

$$\varphi(p, \tau_0) > \varphi(p, 0) > \varphi(p, -\tau_0).$$

Next we choose $\eta > 0$ such that

$$\overline{S[f(p, \tau_0), \eta]} \subset S(p, \varepsilon), \ \overline{S[f(p, -\tau_0), \eta]} \subset S(p, \varepsilon)$$

and such that for $q \in S[f(p, \tau_0), \eta]$ there holds $\varphi(q, 0) > \varphi(p, 0)$, and for $q \in S[f(p, -\tau_0), \eta]$ there holds $\varphi(q, 0) < \varphi(p, 0)$.

Finally, we determine $\delta > 0$ such that

$$f[\overline{S(p, \delta)}, \tau_0] \subset S[f(p, \tau_0), \eta], \quad f[\overline{S(p, \delta)}, -\tau_0] \subset S[f(p, -\tau_0), \eta]$$

and such that for $|t| \leqq 3\tau_0$

$$f(S(p, \delta), t) \subset S(p, \varepsilon)$$

(fig. 28).

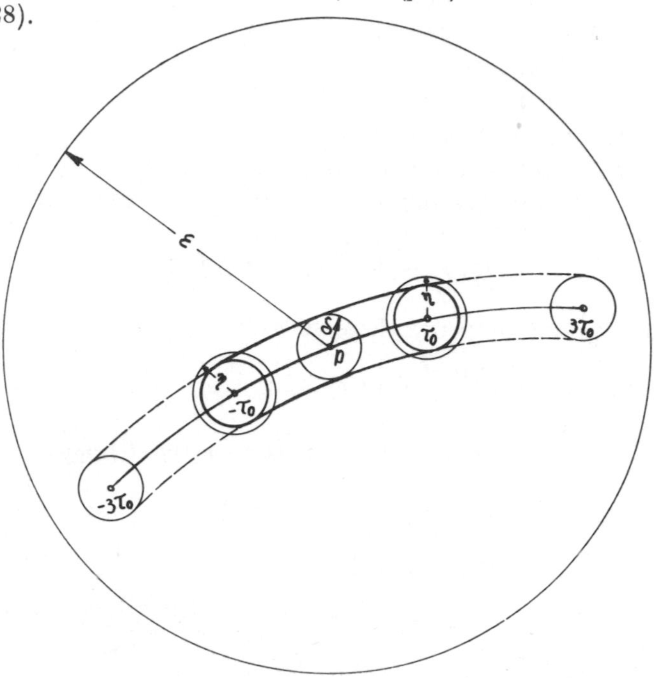

Fig. 28

We shall show that if $q \in \overline{S(p, \delta)}$ there exists one and only one value t_q, $|t_q| < \tau_0$, such that

$$\varphi(q, t_q) = \varphi(p, 0).$$

This follows from the fact that $\varphi(q, t)$ inside $\overline{S(p, \varepsilon)}$ is an increasing, continuous function of t and $\varphi(q, \tau_0) > \varphi(p, 0) > \varphi(q, -\tau_0)$ since

$$f(q, \tau_0) \in \overline{S[f(p, \tau_0), \eta]} \quad \text{and} \quad f(q, -\tau_0) \in \overline{S[f(p, -\tau_0), \eta]}.$$

The desired section F for the tube $\Phi = f[\overline{S(p, \delta)}; -\tau_0, \tau_0]$ is the set Q of points $q \in \Phi$ for which $\varphi(q, 0) = \varphi(p, 0)$. We shall prove this. It is easily seen that the tube Φ constructed on the closed set $\overline{S(p, \delta)}$ is a closed set; from this it follows that F is closed, since if $\{q_n\} \in F$ and $q_n \to q$ then $q \in \Phi$ and $\varphi(q, 0) = \varphi(p, 0)$ because of the continuity of the function φ.

It remains to prove that for any $q \in \Phi$ there exists a unique number t_q, $|t_q| \leqq 2\tau_0$, such that $f(q, t_q) \in F$. For a given point $q \in \Phi$ there can be found, by virtue of the definition of Φ, a certain t', $|t'| \leqq \tau_0$, such that $q' = f(q, t') \in \overline{S(p, \delta)}$, and for q' there can be found, according to what has been proved, a t'', $|t''| \leqq \tau_0$, such that $f(q', t'') \in F$, i. e. $f(q, t' + t'') = f(q, t_q) \in F$, where $t_q = t' + t''$, $|t_q| \leqq 2\tau_0$.

We assume, finally, that there exist two numbers t_q' and t_q'', $|t_q'| \leqq 2\tau_0$, $|t_q''| \leqq 2\tau_0$, such that $f(q, t_q') \in F$, $f(q, t_q'') \in F$; again let $q' = f(q, t') \in \overline{S(p, \delta)}$, $|t'| \leqq \tau_0$. Then $\varphi(q', t_q' - t') = \varphi(q', t_q'' - t') = \varphi(p, 0)$. But, since $|t_q' - t'| \leqq 3\tau_0$, $|t_q'' - t'| \leqq 3\tau_0$, and for t included between $t_q' - t'$ and $t_q'' - t'$, $\varphi_t'(q', t) > 0$, then $t_q' - t' = t_q'' - t'$ or $t_q' = t_q''$. The theorem is proved.

Theorem 2.14 assures the existence of a section in a sufficiently "short" tube. The following theorem establishes the existence of a section in a tube of a preassigned time length at the cost of a decrease in its cross section.

2.15. Theorem. *Let there be given the point $p \in R$, which is not a rest point, and the number $T > 0$, restricted only by the condition $T < \omega/4$ if the motion $f(p, t)$ is periodic of period ω; then there exists a $\delta > 0$ such that the finite tube*

$$\Phi = f(\overline{S(p, \delta)}; -T, +T)$$

has a local section.

We determine, according to Theorem 2.14, the numbers τ_0 and δ. Next we choose $\tau_n = \tau_0/n$ and corresponding to it, according to Theorem 2.14 the number $\delta_n < 1/n$. Then the "short" tube

$$\Phi_n' = f(\overline{S(p, \delta_n)}; -\tau_n, \tau_n),$$

according to what has been proved, has the section F_n. Next we construct the ("long") tube

$$\Phi_n = f(\overline{S(p, \delta_n)}; -T, +T).$$

The sets Φ'_n, Φ_n and F_n are closed, and for each point $q \in \Phi_n$ there can be found a t_q, $|t_q| \leqq T + \tau_n$, such that $f(q, t_q) \in F_n$.

We shall show that there can be found an n_0 such that F_{n_0} will be a local section of the tube Φ_{n_0}. For this it is necessary only to show that the number t_q, for sufficiently large n_0, is unique. We assume the contrary; then for each n there can be found a point $q'_n \in \Phi_n$ such that

$$f(q'_n, t'_n) \in F_n, \; f(q'_n, t''_n) \in F_n; \; |t'_n| \leqq 2T, \; |t''_n| \leqq 2T, \; t'_n \neq t''_n.$$

For definiteness, let $t''_n - t'_n = t_n > 0$; we denote

$$f(q'_n, t'_n) = q_n \in F_n,$$

then

$$f(q'_n, t''_n) = f(q_n, t_n) \in F_n,$$

wherein

$$|t_n| \leqq |t'_n| + |t''_n| \leqq 4T.$$

On the other hand, one can find a point $\bar{q}_n \in \overline{S(p, \delta_n)}$ such that $q_n = f(\bar{q}_n, \bar{t}_n)$, $|\bar{t}_n| < \tau_0$. We note that $\varphi'_t(\bar{q}_n, t) > 0$ for $|t| \leqq 3\tau_0$. Since $f(\bar{q}_n, \bar{t}_n) \in F_n$ and $f(\bar{q}_n, \bar{t}_n + t_n) \in F_n$, then $\varphi(\bar{q}_n, \bar{t}_n) = \varphi(\bar{q}_n, \bar{t}_n + t_n)$; consequently, $|\bar{t}_n + t_n| > 3\tau_0$ and in view of the fact that $t_n > 0$ we have

$$t_n > 3\tau_0 - |\bar{t}_n| > 3\tau_0 - \tau_0 = 2\tau_0.$$

The points q_n converge to p because of the choice of τ_n and δ_n, since $q_n \in F_n \subset f(\overline{S(p, \delta_n)}; -\tau_n, +\tau_n)$; furthermore, there can be chosen from the sequence $\{t_n\}$, by virtue of the inequalities $2\tau_0 < t_n \leqq 4T$, a convergent sequence $\{t_{n_k}\}$, $\lim_{k\to\infty} t_{n_k} = t$, $2\tau_0 \leqq t \leqq 4T$; we note that $f(q_n, t_n) \in F_n$ and therefore $f(q_n, t_n) \to p$. In such a case the identity $f(q_{n_k}, t_{n_k}) = f(q_{n_k}, t_{n_k})$ gives, as $k \to \infty$,

$$p = f(p, t), \qquad 2\tau_0 \leqq t \leqq 4T,$$

i. e., p belongs to a periodic trajectory of period $\omega \leqq 4T$, which contradicts the hypothesis. The theorem is proved.

The theorem of a local character just proved enables one to state the topological character of a dynamical system given in a metric space in the neighborhood of any point distinct from a rest point.

2.16. THEOREM. *If a finite tube Φ of time length $2T$, constructed on* *a set E, has a local section F, then Φ is homeomorphic to a system of parallel segments of a Hilbert space.*

Suppose that we have a mapping Ψ_1 of the set F into the space $E^\infty(\xi_1, \xi_2, \ldots, \xi_n, \ldots)$. Suppose that $q \in \Phi$; according to the definition of a local section there exists a number t_q such that $f(q, -t_q) \in F$, $|t_q| \leqq 2T$. If to the point $f(q, -t_q)$ there corresponds in E^∞ a point with the coordinates $(\xi_1, \xi_2, \ldots, \xi_n, \ldots)$, then we set in correspondence to the point q a point of the new Hilbert space $R^\infty(t, \xi_1, \xi_2, , , ,, \xi_n, \ldots)$ in the following way:

$$\Psi(q) = (t_q, \xi_1, \xi_2, \ldots, \xi_n, \ldots).$$

The mapping Ψ is one-one; that the mapping Ψ is single-valued follows from the definition, that the mapping Ψ^{-1} is single-valued is easily proved. In fact, suppose that $q_1 \neq q_2$ are points of the tube Φ; then, if $f(q_1, -t_1) \in F, f(q_2, -t_2) \in F$, either $f(q_1, -t_1) \neq f(q_2, -t_2)$ or $t_1 \neq t_2$; in both cases $\Psi(q_1) \neq \Psi(q_2)$.

The mapping Ψ is bi-continuous. Suppose that $\{q_n\} \in \Phi$, $\lim_{n\to\infty} q_n = q \in \Phi$. Then, according to the definition of a section, there can be found numbers $\{t_n\}$, t, where $|t_n| \leqq 2T$, $|t| \leqq 2T$, such that $\{f(q_n, -t_n)\} \in F$, $f(q, -t) \in F$. We shall show that $\lim_{n\to\infty} t_n = t$. In the contrary case there could be found a subsequence $\{t_{n_k}\}$, $\lim_{k\to\infty} t_{n_k} = t' \neq t$, $|t'| \leqq 2T$. Then, since F is closed, there follows $\lim_{k\to\infty} f(q_{n_k}, -t_{n_k}) = f(q, -t') \in F$. But this latter inclusion together with $f(q, -t) \in F$ contradicts the definition of a section. Consequently, $\lim_{n\to\infty} t_n = t$ and $\lim_{n\to\infty} f(q_n, -t_n) = f(q, -t)$. But then, denoting distance in the Hilbert space R^∞ by ϱ_1, we have

$$\varrho_1[\Psi(q_n), \Psi(q)] = [(t_n - t)^2 + \sum_{i=1}^{\infty} (\xi_i^{(n)} - \xi_i)^2]^{1/2},$$

where $\xi_i^{(n)}$ are the coordinates of the point $\Psi_1(f(q_n, -t_n))$. Because of the bi-continuity of the mapping Ψ_1 on F, there follows from this

$$\lim_{n\to\infty} \varrho_1[\Psi(q_n), \Psi(q)] = 0.$$

The continuity of the mapping Ψ^{-1} is obvious from this formula for ϱ_1; if it is known that the left side tends to 0, then $t_n \to t$ and $f(q_n, -t_n) \to f(q, -t)$; consequently, $q_n \to q$.

Thus Ψ is a topological mapping. Under this mapping the arc of the trajectory $f(q; -T, +T)$ for any point $q \in E$ is mapped into a segment of the line

$$\xi_i = \xi_i(f(q, -t_q)) = \text{const.}, \quad (i = 1, 2, \ldots), \quad -T \leqq t \leqq T,$$

of R^∞, and these segments are parallel to one another.

2.17. COROLLARY. *If p is distinct from a rest point, and $T > 0$ is arbitrary for a nonperiodic motion, or $T < \omega/4$ for a periodic motion of period ω, then there exists a $\delta > 0$ such that the set $f(\overline{S(p, \delta)};\ -T, +T)$ is homeomorphic to a system of parallel segments in R^∞.*

Thus the local structure of the neighborhood of an ordinary point of a dynamical system in a metric space R is topologically similar to the local structure of the neighborhood of an ordinary point of a system of differential equations.

3. ω- and α-Limit Points

Let there be given in the metric space R a dynamical system $f(p, t)$. We consider a certain positive half-trajectory $f(p; 0, +\infty)$. We take any bounded increasing sequence of values of t:

$$0 \leqq t_1 < t_2 < \ldots < t_n < \ldots, \lim_{n\to\infty} t_n = +\infty;$$

if the sequence of points

$$f(p, t_1), f(p, t_2), \ldots, f(p, t_n), \ldots$$

has a limit point q then we shall call this point an *ω-limit* point of the motion $f(p, t)$. Analogously, any limit point q' of a negative half-trajectory $f(p; -\infty, 0)$ is called an *α-limit* point of the motion $f(p, t)$.

3.01. THEOREM. *Both the set Ω_p and the set A_p of all ω- and α-limit points respectively of the motion $f(p, t)$ are invariant closed sets.*

We shall prove the theorem for ω-limit points. Let q be an ω-limit point for $f(p, t)$. Then there exists a sequence of values $t_1, t_2, \ldots, t_n, \ldots$ $(t_n \to +\infty)$ such that

$$\lim_{n\to\infty} f(p, t_n) = q. \tag{3.02}$$

Let $f(q, \tau)$ be an arbitrary, but definite point of the trajectory passing through q.

By property II′, for any $\varepsilon > 0$ and for $T = |\tau|$ there can be found a δ such that if

$$\varrho(f(p, t_n), q) < \delta \quad \text{then} \quad \varrho[f(p, t_n + \tau), f(q, \tau)] < \varepsilon.$$

But by (3.02) the first inequality is satisfied for $n \geqq N(\delta)$ and thus the second also is satisfied, *i. e.*, $f(q, \tau)$ is a limit point of the sequence $f(p, t_n + \tau)$ and, consequently, an ω-limit point of the motion $f(p, t)$. Thus there lies in Ω_p along with each point q the whole trajectory $f(q, t)$; *i. e.*, Ω_p is an invariant set.

In order to prove that the set Ω_p is closed, we take a sequence of points $q_1, q_2, \ldots, q_n, \ldots$; $q_n \in \Omega_p$ $(n = 1, 2, \ldots)$; $\lim_{n\to\infty} q_n = q$; and we shall show that $q \in \Omega_p$. We assign any $\varepsilon > 0$ and determine n such that $\varrho(q_n, q) < \varepsilon/2$; since q_n is an ω-limit point for $f(q, t)$, there can be found $t = \tau_n$ such that $\varrho(f(p, \tau_n), q_n) < \varepsilon/2$. From this, $\varrho(f(p, \tau_n), q) < \varepsilon$, *i. e.*, q is an ω-limit point for $f(p, t)$. The theorem is proved.

There also hold the relations

$$\Omega_p \subset \overline{f(p;\ 0, +\infty)}; \qquad A_p \subset \overline{f(p;\ -\infty, 0)}, \tag{3.03}$$

since the closure of a half-trajectory contains all its limit points.

We shall consider the structure of the sets Ω_p and A_p for trajectories of the simplest forms.

If p is a rest point then, obviously, $\Omega_p = A_p = p$. If we have

$$\lim_{t\to\infty} f(p, t) = q,$$

then $\Omega_p = q$; moreover, q, as we have seen, is a rest point.

3.04. Theorem. *If $f(p, t)$ is a periodic motion, then*

$$\Omega_p = A_p = f(p;\ I).$$

For if $f(p, t)$ has the period τ and $q = f(p, t_0)$ is any point of the trajectory, then we shall have also $q = f(p, t_0 \pm n\tau)$, $n = 1, 2, \ldots$, *i. e.*, $q = \lim_{n\to\infty} f(p, t_0 \pm n\tau)$; but $\lim_{n\to\infty} (t_0 \pm n\tau) = \pm\infty$. Therefore $q \in \Omega_p$ and $q \in A_p$.

Conversely, suppose that $q \in \Omega_p$, that is, there exists a sequence $\{t_n\}$, $t_n \to +\infty$, such that $q = \lim_{n\to\infty} f(p, t_n)$. Every number t_n can be represented in the form $t_n = k_n\tau + t'_n$, where k_n is an integer and $0 \leqq t'_n < \tau$. From the bounded sequence $\{t'_n\}$ there can be chosen a convergent subsequence $\{t'_{n_k}\}$, $\lim_{k\to\infty} t'_{n_k} = t_0$. In this case

$$q = \lim_{n\to\infty} f(p, t_n) = \lim_{n\to\infty} f(p, t'_n) = \lim_{k\to\infty} f(p, t'_{n_k}) = f(p, t_0),$$

i. e., any ω-limit q lies on the trajectory of the periodic motion, and this proves the theorem.

3.05. Definition. A motion $f(p, t)$ is called *positively stable according to Lagrange* (abbreviated, stable L^+) if the closure of the half-trajectory $f(p; 0, +\infty)$ is a compact set. Analogously, a motion is *negatively stable according to Lagrange* (abbreviated, stable L^-) if $\overline{f(p; -\infty, 0)}$ is compact. A motion which is at the same time positively and negatively stable according to Lagrange is called *stable according to Lagrange* (which we shall denote stable L).

Obviously, if the space R is compact then all motions are stable according to Lagrange. In general, if $f(p; 0, +\infty)$ lies in a compact subset $M \subset R$ then it is *positively* stable according to Lagrange. In the case of a Euclidean space E_n, stability according to Lagrange implies that the trajectory is situated in a bounded portion of the space E^n.

Furthermore, it follows from the definition that *for a motion $f(p, t)$ positively stable according to Lagrange the set Ω_p is not empty and for a negatively stable motion the set A_p is not empty.*

The converse statement is incorrect.

3.06. Example. Consider in the auxiliary plane XOY a family of motions tracing the logarithmic spirals $\varrho = ce^{\theta}$, where ϱ and θ are polar coordinates, and the law of the motion is given by the differential equations

$$\frac{d\varrho}{dt} = \frac{\varrho}{1+\varrho}; \quad \frac{d\theta}{dt} = \frac{1}{1+\varrho}; \qquad \varrho \geqq 0.$$

It is easily verified that all the motions are continuable for $-\infty < t < +\infty$, *i.e.*, we have a dynamical system; moreover, all the motions are negatively stable according to Lagrange, having the origin (rest point) as their α-limit point. All the motions are positively unstable according to Lagrange since as $t \to +\infty$ the radius vector $\varrho \to +\infty$.

We now map the plane XOY on the half-plane $-\infty < y < +\infty$, $-1 < x < +\infty$, by the transformation

$$X = \log(1+x), \qquad Y = y.$$

We shall have

$$\varrho^2 = [\log(1+x)]^2 + y^2, \quad \theta = \arg\{\log(1+x) + iy\}.$$

The integral curves will have the form illustrated in Fig. 29 and the differential equations of the new system will be

$$\dot{x} = \frac{(1 + x)[\log (1 + x) - y]}{1 + \varrho}, \quad \dot{y} = \frac{y}{1 + \varrho} + \frac{\log (1 + x)}{1 + \varrho}$$

We complete our space with the line $x = -1$; moreover, we shall define the motion along it to be the limiting motion for the differential equations we have written as $x \to -1 + 0$

Since, furthermore, $\log(1 + x)/(1 + \varrho) \to -1$, then $\dot{y}$ remains finite and we obtain along the line $x = -1$:

$$\dot{x} = 0, \qquad \dot{y} = -1.$$

In this way the dynamical system is defined for the closed half-plane $x \geqq -1$. Obviously, all the motions are positively unstable according to Lagrange, since as $t \to +\infty$ they do not remain in a

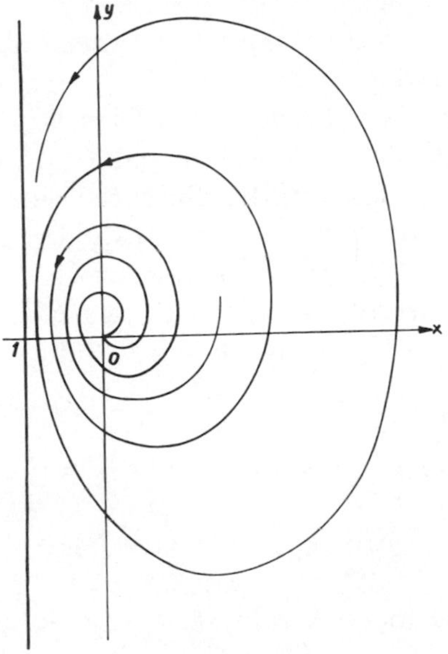

Fig. 29

bounded part of the plane; meanwhile, it is easily seen that for any point $p = (x_0, y_0)$, $x_0 > -1$, $p \neq (0, 0)$, the set Ω_p is the straight line $x = -1$, *i. e.*, Ω_p is not empty.

3.07. THEOREM. *If $f(p, t)$ is positively stable according to Lagrange, then*

$$\lim_{t \to +\infty} \varrho[f(p, t), \Omega_p] = 0.$$

Assume that the statement is false; then there can be found a sequence of positive numbers $\{t_n\}$, $\lim_{n\to\infty} t_n = +\infty$, and a number $\alpha > 0$ such that

$$\varrho[f(p, t_n), \Omega_p] \geqq \alpha. \tag{3.08}$$

The set of points $\{q_n\} = \{f(p, t_n)\}$, belonging to the compact set $\overline{f(p; 0, +\infty)}$, has a limit point q which, according to the definition of the set Ω_p, lies in it; on the other hand the inequality (3.08), on passing to the limit by the corresponding subsequence, gives

$$\varrho(q, \Omega_p) \geqq \alpha.$$

The contradiction proves the theorem.

An analogous statement holds for motions negatively stable according to Lagrange.

3.09. THEOREM. *If $f(p, t)$ is positively stable according to Lagrange, then the set Ω_p is connected.*

Assume that Ω_p is not connected. Then, since it is closed, we would have $\Omega_p = A + B$, where A and B are closed, nonempty disjoint sets and, consequently, since the set Ω_p is obviously compact, $\varrho(A, B) = d > 0$. Since $A \subset \Omega_p$ and $B \subset \Omega_p$, there can be found values t'_n, arbitrarily large, for which $f(p, t'_n) \in S(A, d/3)$ and values t''_n, arbitrarily large, for which $f(p, t'_n) \in S(B, d/3)$. The sequences $\{t'_n\}$ and $\{t''_n\}$ can be chosen to fulfill the inequalities

$$0 < t'_1 < t''_1 < t'_2 < t''_2 < \ldots < t'_n < t''_n < t'_{n+1} < \ldots .$$

Since $\varrho(f(p, t), A)$ is a continuous function of t and we have

$$\varrho(f(p, t'_n), A) < \frac{d}{3}; \quad \varrho(f(p, t''_n), A) > \varrho(A, B) - \varrho(B, f(p, t''_n)) > \frac{2d}{3};$$

then there can be found a value $\tau_n (t'_n < \tau_n < t''_n)$ such that

$$\varrho(f(p, \tau_n), A) = \frac{d}{2}.$$

Because of the compactness of the set $\overline{f(p; 0, +\infty)}$, there can be chosen from the sequence of points $\{f(p, \tau_n)\}$ a subsequence converging to some point q and we shall have

$$q \in \Omega_p, \quad \varrho(q, A) = \frac{d}{2}; \quad \varrho(q, B) \geqq \varrho(A, B) - \varrho(A, q) = \frac{d}{2},$$

i. e., $\Omega_p \neq A + B$. The contradiction proves that Ω_p is connected.

We note that if R is compact, then for any point $p \in R$ the sets Ω_p and A_p are nonempty and connected.

3.10 EXAMPLE. We shall show that in the case of a non-compact space R the set Ω_p may be disconnected. For constructing the example, we take the same auxiliary plane XOY and the same differential equations as in example 3.06 of the present section,

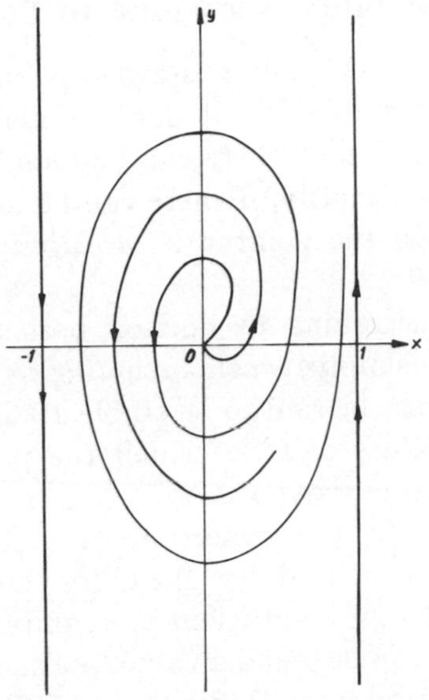

Fig. 30

but this time we map the XOY-plane on the strip $-1 < x < 1$ of the xoy-plane by the transformation

$$X = \frac{x}{1 - x^2}, \qquad Y = y.$$

In the new variables the differential equations take the form

$$\dot{x} = \frac{x(1 - x^2) - y(1 - x^2)^2}{(1 + x^2)(1 + \varrho)}; \quad \dot{y} = \frac{y}{1 + \varrho} + \frac{x}{1 - x^2} \cdot \frac{1}{1 + \varrho}.$$

We complete our space with the lines $x = \pm 1$ with the corresponding limiting equations (we note that, as $x \to \pm 1 \mp 0$, $\lim x/(1 - x^2)(1 + \varrho) = \pm 1$):

$$\dot{x} = 0, \qquad \dot{y} = \pm 1.$$

The dynamical system so obtained (see Fig. 30) is unstable according to Lagrange in the positive direction; but it is readily seen that the set Ω_p, for any point $p = (x_0, y_0)(-1 < x_0 < 1)$, $p \neq (0, 0)$, is disconnected; it consists of the two lines $x = 1$ and $x = -1$.

4. Stability According to Poisson

4.01. DEFINITION. A point p is called *positively stable according to Poisson* (written stable P^+) if, for any neighborhood U of the point p and for any $T > 0$, there can be found a value $t \geqq T$ such that $f(p, t) \in U$. Analogously, if there can be found a $t \leqq -T$ such that $f(p, t) \in U$ then the point p is *negatively stable according to Poisson* (written P^-).

A point stable according to Poisson both as $t \to +\infty$ and as $t \to -\infty$ is called (simply) *stable according to Poisson* (stable P).

Thus a point p may be said to be stable P^+ if there can be found arbitrarily large values of t for which the point appears in any neighborhood of its initial position.

4.02. REMARK. In the condition of stability P^+ one may weaken the requirement that $f(p, t) \in U$ for some t greater than an arbitrarily assigned T; it is sufficient to require that for any neighborhood $U(p)$ there can be found a value $t \geqq 1$ such that $f(p, t) \in U(p)$.

For assume that under the fulfillment of this condition the point p is unstable P^+. This means that there can be found a neighborhood $U_1(p)$ and a number $T > 1$ such that $f(p, t). U_1(p) = 0$ for $t \geqq T$. Consider the trajectory arc $f(p; 1, T)$. If $f(p, \bar{t}) = p$ for some $\bar{t}(1 \leqq \bar{t} \leqq T)$, then the motion is periodic and the assertion has been proved. If $f(p, t) \neq p$ $(1 \leqq t \leqq T)$, then the arc under consideration, being a closed set, is situated at a finite distance from the point p and there can be found a neighborhood $U_2(p) \subset U_1(p)$ having no points in common with this arc. But in such a case the half-trajectory $f(p; 1, +\infty)$ has no points in common with $U_2(p)$, contradicting the new definition.

4.03. THEOREM. *If a point p is stable P^+ then every point of the trajectory $f(p; I)$ is also stable P^+.*

For the proof we remark that the definition of stability P^+ given above is equivalent to the following: there exists a sequence of values $\{t_n\}$, $\lim_{n\to\infty} t_n = +\infty$, such that $\lim_{n\to\infty} f(p, t_n) = p$. Indeed, from this last property the first follows directly; conversely, if the first property fulfilled, then for any sequence $\varepsilon_1 > \varepsilon_2 > \ldots > \varepsilon_n > \ldots$, $\lim_{n\to\infty} \varepsilon_n = 0$, there can be found numbers $t_n > n$ such that $\varrho(p, f(p, t_n)) < \varepsilon_n$. Obviously, $\lim_{n\to\infty} t_n = +\infty$ and $\lim_{n\to\infty} f(p, t_n) = p$, *i. e.*, the second definition is fulfilled.

Consider now an arbitrary point of the trajectory $f(p, t)$. By properties II and III of a dynamical system we have $\lim_{n\to\infty} f(p, t + t_n) = f(p, t)$, *i. e.*, the point $f(p, t)$ is stable P^+. The theorem is proved.

An analogous theorem is valid for stability P^- and stability P. Thus, in the sequel we shall speak of motions and trajectories positively, negatively, and simply stable according to Poisson.

The condition that $f(p, t)$ be stable P^+ can obviously be written thus: $f(p; I) \subset \overline{f(p; 0, +\infty)}$; the condition for stability $P^- : f(p; I) \subset \overline{f(p; -\infty, 0)}$. Alternatively we can say: $p \in \Omega_p$, or $p \in A_p$. The fulfillment of both conditions simultaneously is equivalent to stability P. By Theorem 4.03, a sufficient condition for $f(p, I) \subset A_p$, $f(p, I) \subset \Omega_p$ is: $f(p, I) . A_p \neq 0$ and $f(p, I) . \Omega_p \neq 0$.

It is obvious that a rest point represents a motion stable P. For in such a case $f(p, t) = p$ for $-\infty < t < +\infty$, *i. e.*, $f(p, t) \subset U(p)$ and the condition for stability P is fulfilled.

Another example of motion of stability P is a periodic motion: $f(p, t + \tau) = f(p, t) (-\infty < t < +\infty)$, where τ is a constant. Indeed, we have $f(p, 0) = f(p, n\tau)$ $(n = \pm 1, \pm 2, \ldots)$. Thus, for $t = n\tau$, the point $f(p, t)$ coincides with its initial position, *i. e.*, it falls in any neighborhood $U(p)$. It was shown in Chapter I, Theorem 1.2 that the only trajectories in a plane stable P are rest points and the trajectories of periodic motions.

4.4. Example. The simplest example of a motion stable P, neither a rest point or a periodic motion, is the motion on the surface of a torus $\mathfrak{T}(0 \leqq \varphi < 1,\ 0 \leqq \theta < 1;\ (\varphi + k, \theta + k') \equiv (\varphi, \theta)$ if k and k' are integers) determined by the differential equations

$$\frac{d\varphi}{dt} = 1, \qquad \frac{d\theta}{dt} = \alpha \tag{4.05}$$

where α is a positive irrational number (see Chapter II, section 2). Here the trajectory of each motion is everywhere dense on the torus; every motion is stable P and the sets Ω_p and A_p for any point coincide with the surface of the torus.

4.06. Example. We define motions on the torus by the equations

$$\frac{d\varphi}{dt} = \Phi(\varphi,\theta), \qquad \frac{d\theta}{dt} = \alpha\Phi(\varphi, \theta), \tag{4.07}$$

where $\Phi(\varphi, \theta)$ is a continuous function on the torus (periodic in the arguments φ and θ with period 1) everywhere positive except at the point $(0, 0)$, where $\Phi(0, 0) = 0$, and satisfying a Lipschitz condition. The curves along which the motions take place remain the same as in the system (4.05) since they are determined by the differential equations

$$\frac{d\theta}{\alpha} = \frac{d\varphi}{1},$$

but the character of the motion has been altered. Along the curve $\theta = \alpha\varphi$ there are three motions: (1) $\theta = 0$, $\varphi = 0$; rest. (2) Motions along the positive arc $0 < \varphi < +\infty$; for these motions the positive half-trajectory is everywhere dense on the torus and therefore is stable P^+. The negative half-trajectory tends to the rest point $(0, 0)$ as $t \to -\infty$; it is unstable P^-. (3) Motions along the negative arc $-\infty < \varphi < 0$; they are stable P^- and unstable P^+ since the moving point tends to the rest point as $t \to +\infty$.

All the rest of the trajectories remain the same as in system (4.05), since along them $\Phi(\varphi, \theta) \neq 0$; they are everywhere dense on $\mathfrak{T}$ and, therefore, are stable P in both directions; however, the motions along these trajectories are no longer uniform, the velocity being $\Phi(\varphi, \theta) \cdot \sqrt{1 + \alpha^2}$, and hence the motion is retarded on passing near the point $(0, 0)$.

4.08. Example. By slightly complicating Example 4.06, one can construct a system with motions stable P (including the rest points) and with motions unstable P in both directions. For this we construct on the meridian $\varphi = 0$ of the torus a countable set of rest points which lie on the curve $\theta = \alpha\varphi$ and have $(0, 0)$ as their unique limit point. If, for example, α be expanded as an infinite continued fraction and its consecutive convergents be written p_k/q_k,

$$\frac{p_2}{q_2} < \frac{p_4}{q_4} < \ldots < \alpha < \ldots < \frac{p_3}{q_3} < \frac{p_1}{q_1},$$

then as rest points there can be taken the points with coordinates $\varphi = 0$ and respectively

$$\theta_k = \alpha q_k \pmod 1 \quad (0 < \theta_k < 1),$$
$$\theta'_k = -\alpha q_k \pmod 1 \quad (0 < \theta'_k < 1).$$

We shall have (since p_k and q_k are integers)

$$|\theta_k| = |\alpha q_k - p_k| = q_k \left| \alpha - \frac{p_k}{q_k} \right| < q_k \cdot \frac{1}{q_k^2} = \frac{1}{q_k}$$

and analogously for θ'_k.

We construct the continuous function $\Phi(\varphi, \theta)$ satisfying the Lipschitz condition and positive everywhere except for the set of points $(0, 0)$, $(0, \theta_k)$, $(0, \theta'_k)$ $(k = 1, 2, \ldots)$, where it vanishes. The corresponding equations of the form (4.07) will possess the same trajectories as the equations (4.05) so they will be stable P except for the trajectories lying on the curve $\theta = \alpha\varphi$. This latter breaks up into the countable set of arcs

$$0 < \varphi < q_1; \quad q_1 < \varphi < q_2; \quad \ldots;$$
$$0 > \varphi > -q_1; \quad -q_1 > \varphi > -q_2; \quad \ldots$$

separated by rest points. Along each such arc, for example $q_k < \varphi < q_{k+1}$, the motion will be unstable P (in both directions) since as $t \to +\infty$ it approaches the rest point $\varphi = q_{k+1} \equiv 0 \pmod 1$, $\theta \equiv \alpha q_{k+1} \pmod 1$, and analogously as $t \to -\infty$ it approaches the point $\varphi = q_k \equiv 0 \pmod 1$, $\theta = \alpha q_k \pmod 1$.

We return to the study of the structure of the sets Ω_p and A_p for motions stable P. If $f(p, t)$ is stable P^+, then by Theorem 4.03 of the present section all points of its trajectory are ω-limits for it *i. e.*,

$$f(p; I) \subset \Omega_p.$$

Since Ω_p is a closed set, from this last inclusion there follows

$$\overline{f(p; I)} \subset \Omega_p.$$

Comparing this with the inverse inclusion (3.03), which always holds, we have for a motion stable P^+:

$$\Omega_p = \overline{f(p; I)}.$$

From the same formula (3.03) we have $A_p \subset \overline{f(p; I)}$, whence there follows:

4.09. THEOREM. *For a motion $f(p, t)$ stable P^+ we have*

$$A_p \subset \Omega_p = \overline{f(p; I)}$$

(the set A_p may be empty [5]). Analogously, *for a motion $f(p, t)$ stable P^-*

$$\Omega_p \subset A_p = \overline{f(p; I)}$$

(Ω_p may be empty). Finally, comparing these facts we find:

If $f(p, t)$ is stable P (in both directions) *then*

$$\Omega_p = A_p = \overline{f(p; I)}.$$

We saw in the preceding section that for a rest point and a periodic motion, which are stable P, there holds the relation

$$\overline{f(p; I)} = f(p; I).$$

On the other hand, in Examples 4.04, 4.06 and 4.08 of the present section the closure of a trajectory stable P contained, besides its own points, also other points. This is what generally happens if the trajectory be neither a rest point nor a periodic trajectory. There need only be imposed an additional restriction on the space R. Here it is essential that the space be complete since, for example, the trajectory $\theta = \alpha\varphi$ of the motion defined by the system (4.07) is obviously stable P if there be taken as the space R only points of this trajectory, but its closure contains no other points.

4.10. THEOREM. *For a trajectory of a motion $f(p, t)$, stable P^+, which is not a rest point or a closed curve (the trajectory of a periodic motion) and which is situated in a complete metric space R, there is a set of points not in $f(p; I)$ which is everywhere dense in Ω_p; in fact,*

$$\overline{\overline{f(p; I)} - f(p; I)} = \overline{f(p; I)} = \Omega_p.$$

Since, by the group property, any point p of the trajectory is initial for some motion, it is sufficient for us to prove that in the closure $\overline{S(p, \varepsilon)}$ $(\varepsilon > 0)$ of any spherical neighborhood of p there can be found a point $q \in \overline{f(p; I)}$ not lying on the trajectory $f(p; I)$.

Because of stability P^+ there exists a sequence of points $\{p_n = f(p, t_n)\}$ such that $0 < t_1 < t_2 < \ldots$, $\lim_{n\to\infty} t_n = +\infty$

[5]It is sufficient, for example, in the system (4.07) of Example 4.06 to consider the motions in the space $\mathfrak{T}$ from which the point (0, 0) has been deleted; the motion along the trajectory $\theta = \alpha\varphi$, $0 < \varphi < +\infty$, is stable P^+ and has no α-limit points.

and $\lim_{n\to\infty} p_n = p$. We choose $\tau_1 > t_1$ such that $q_1 = f(p, \tau_1) \epsilon S(p, \varepsilon)$. Obviously, $q_1 \cdot f(p; -t_1, t_1) = 0$, and since the arc $f(p; -t_1, t_1)$ is a closed set, $\varrho[q_1, f(p; -t_1, t_1)] > 0$. Let

$$\varepsilon_1 = \min\left[\frac{\varepsilon}{2}; \ \varepsilon - \varrho(p, q_1); \ \tfrac{1}{2}\varrho(q_1, f(p; -t_1, t_1))\right];$$

then

$$S(q_1, \varepsilon_1) \subset S(p, \varepsilon) \text{ and } \overline{S(q_1, \varepsilon_1)} \cdot f(q; -t_1, t_1) = 0.$$

In general, let $q_{n-1} \epsilon f(p; I)$ and ε_{n-1} be already defined; we choose $\tau_n > t_n$ such that $q_n = f(p, \tau_n) \epsilon S(q_{n-1}, \varepsilon_{n-1})$; this is possible by virtue of the stability P^+ of the trajectory $f(p; I)$. We thereupon define

$$\varepsilon_n = \min\left[\frac{\varepsilon_{n-1}}{2}; \ \varepsilon_{n-1} - \varrho(q_{n-1}, q_n); \ \tfrac{1}{2}\varrho(q_n, f(p; -t_n, t_n))\right].$$

We note that $\varrho(q_n, f(p; -t_n, t_n)) > 0$, since from the inequality $\tau_n > t_n$ the point q_n does not belong to the arc $f(p; -t_n, t_n)$. Obviously, we shall have

$$S(q_n, \varepsilon_n) \subset S(q_{n-1}, \varepsilon_{n-1}); \ \overline{S(q_n, \varepsilon_n)} \cdot f(p; -t_n, t_n) = 0.$$

The sequence $\{q_n\}$, by construction, possesses the property that $\varrho(q_n, q_{n-1}) < \varepsilon_{n-1} \leqq \varepsilon/2^{n-1}$ for $n = 1, 2, 3, \ldots$; because of the completeness of the space, this sequence has a limit point q and $\lim_{n\to\infty} q_n = q$. Since $q_n \epsilon f(p; I)$, then $q \epsilon \overline{f(p; I)}$, and since $\varrho(p, q_n) < \varepsilon$, then $\varrho(p, q) \leqq \varepsilon$. It remains to show that q does not belong to the trajectory $f(p; I)$.

Assume the contrary; let $q = f(p, \tau)$. There can be found an n such that $t_n > |\tau|$; then $q \epsilon f(p; -t_n, t_n)$. But we have $q \epsilon \overline{S(q_n, \varepsilon_n)}$ and by construction $\overline{S(q_n, \varepsilon_n)} \cdot f(p; -t_n, t_n) = 0$, meaning that $q \cdot f(p; -t_n, t_n) = 0$. The contradiction proves our assertion.

It is easily seen that in a complete space for a motion stable P^- or stable P and neither a rest point nor a periodic motion there holds the similar relation

$$\overline{\overline{f(p; I)} - f(p; I)} = \overline{f(p; I)} = A.$$

4.11. Corollary. *Under the conditions of Theorem* 4.10 *every finite arc* $f(p; t_1, t_2)$ *is nowhere dense in* $\overline{f(p; I)}$.

$f(p; t_1, t_2)$ is, in fact, a closed compact set. Whatever relatively open set $U \subset \overline{f(p; I)}$ we may take, the set $U - U \,.\, f(p; t_1, t_2)$ will be a nonempty, relatively open set having no points in common with $f(p; t_1, t_2)$ and this is the condition for being nowhere dense.

4.12. Remark. In many investigations one must consider stability for a discrete set of positions $f(p, n)$ $(n = 1, 2, 3, \ldots)$ of a point. In this case we shall call the point p stable P^+ if the sequence of points

$$\text{(A)} \qquad \{f(p, n)\}$$

has p as its limit point. Every point stable P^+ for this discrete variation of t is obviously P^+ for a continuous variation of t. However, the converse is also valid: if a point p is stable P^+ for a continuous variation $0 \leqq t < +\infty$, then it is a limit for the points of the sequence (A).

We shall prove this. Let $\lim_{n\to\infty} f(p, t_n) = p$, where $\lim_{n\to\infty} t_n = +\infty$. We represent each t_n in the form $t_n = k_n - \tau_n$, where k_n is an integer and $0 \leqq \tau_n < 1$. The set of numbers τ_n has a limit point $\tau (0 \leqq \tau \leqq 1)$ and from the sequence $\{\tau_n\}$ there can be chosen a subsequence converging to τ. For brevity in writing we assume that the sequence $\{\tau_n\}$ itself possesses this property: $\lim_{n\to\infty} \tau_n = \tau$. Thus

$$\lim_{n\to\infty} f(p, k_n - \tau_n) = p.$$

From this, by virtue of the continuity of the function f, we have

$$\lim_{n\to\infty} f(p, k_n) = f(p, +\tau).$$

Thus the point $f(p, +\tau)$ is a limit for the sequence (A). Obviously, the points $f(p, \tau \pm 1), f(p, \tau \pm 2), \ldots, f(p, \tau \pm m), \ldots$ possess the same property since, for l an integer, $f(p, \tau + l) = \lim_{n\to\infty} f(p, k_n + l)$ and $f(p, k_n + l) \in$ (A).

In addition the point $f(p, 2\tau)$ is obviously a limit of the sequence $\{f(p, k_n + \tau)\}$. We shall show that it is a limit of the sequence (A). We assign $\varepsilon > 0$. Let

$$\varrho[f(p, k_\nu + \tau), f(p, 2\tau)] < \frac{\varepsilon}{2}.$$

For the arc $f(p; \tau, k_\nu + \tau)$ and the number $\varepsilon/2$ we find, by property

II′, a δ such that if $\varrho[f(p, \tau), q] < \delta$ then $\varrho[f(p, \tau + t), f(q, t)] < \varepsilon/2$ for $0 \leqq t \leqq k_\nu$. In the δ-neighborhood of the point $f(p, \tau)$ there can be found, according to what has been proved, a point $f(p, k_\mu) = q$; then

$$\varrho[f(p, \tau + k_\nu), f(p, k_\mu + k_\nu)] < \frac{\varepsilon}{2}, i. e., \varrho[f(p, 2\tau), f(p, k_\mu + k_\nu)] < \varepsilon$$

and therefore the point $f(p, 2\tau)$ is a limit of (A).

The same reasoning would show that the points $f(p, 3\tau), \ldots, f(p, k\tau), \ldots$ are also limits of this sequence. Comparing this conclusion with what has been proved earlier, we find that the points $f(p, k\tau - m)$, for k and m integers, are limits of (A).

We consider two cases: (1) the number τ is rational, $\tau = p/q$; setting $k = q$, $m = p$, we find that the point $f(p, 0) = p$ is a limit of the sequence (A); (2) τ is irrational; then the set of numbers $k\tau - m$ $(k, m = 1, 2, \ldots)$ is everywhere dense on the real line; in particular, there exists a sequence converging to zero, and the point p, as a limit of the corresponding sequence $f(p, k\tau - m)$, is a limit of the sequence (A). Our assertion is proved.

4.13. Remark. It has been proved that the point p is always a limit point of (A); if τ is irrational, then all points $f(p; I)$ are also limit points of (A). In the case of rational τ one cannot assert this; in fact, for a periodic motion $f(p, t)$ with an integral period l, only a finite number of points $f(p, k)$ $(k = 0, 1, \ldots, l - 1)$ will be limits of (A).

Using the theorem just proved, for an investigation of points stable P^+ we shall find it expedient to regard this set as the set of limit points of $\{f(p, n)\}$.

We apply this remark to the discovery of all points stable P^+. We shall determine the set of points stable P^+ in the whole space R, the space being assumed to be metric with a countable base.

We consider at first an arbitrary set $A \subset R$ and select from it a set A^* in the following manner

$$A^* = A - A \cdot \sum_{n=1}^{\infty} f(A, -n),$$

i. e., A^* is the set of points $p \in A$ not belonging to one of the sets $f(A, -n)$ $(n = 1, 2, \ldots)$. We note that from the definition of the set A^* we have

$$f(A, -n) \,.\, A^* = 0 \qquad (n = 1, 2, \ldots),$$

whence, since $A^* \subset A$,

$$f(A^*, -n) \,.\, A^* = 0 \qquad (n = 1, 2, \ldots).$$

Taking the images of the sets in both relations at the instant $t = n$, we obtain

$$f(A^*, n) \cdot A = 0 \quad \text{and} \quad f(A^*, n) \cdot A^* = 0 \qquad (n = 1, 2, \ldots).$$

Now let $U_1, U_2, \ldots, U_n, \ldots$ be a definite system of neighborhoods of the space R. We construct for each U_n in the manner described the corresponding set U_n^*. We denote $\Sigma_{n=1}^{\infty} U_n^*$ by V^+ and $R - V^+$ by E^+. Then E^+ is a set of points stable P^+ and V^+ is a set of points unstable P^+.

Indeed, suppose $p \in E^+$ and let U_ν be any neighborhood containing p. According to the definition of the set E^+ the point p does not belong to U_ν^* and therefore for some value of k it belongs to the set $U_\nu \cdot f(U_\nu, -k)$:

$$p \in U_\nu \cdot f(U_\nu, -k).$$

Taking the image of both parts of this relation at the instant $t = k$, we have

$$f(p, k) \in U_\nu \cdot f(U_\nu, k) \subset U_\nu,$$

where $k \geqq 1$ and U_ν is any neighborhood of the point p. From this, by virtue of the second definition of stability P^+ (see the remark at the beginning of this section), there follows the stability P^+ of the point p.

Now suppose that $p \in V^+$; this means that there can be found some neighborhood U_μ of the point p such that $p \in U_\mu^*$. Since it has been proved that

$$f(U_\mu^*, n) \cdot U_\mu = 0 \quad \text{for} \quad n = 1, 2, \ldots,$$

then $f(p, n) \,.\, U_\mu = 0$. Thus p leaves forever its neighborhood U_μ, *i. e.*, it is unstable P^+.

Analogously, constructing the set $U_n^{**} = U_n - U_n \cdot \sum_{k=1}^{\infty} f(U_n, k)$, we obtain $V^- = \sum_{k=1}^{\infty} U_n^{**}$, a set of points unstable P^-, and the set of points $R - V^- = E^-$, stable P^-. Obviously, E^+E^- is a set of points stable P.

From this construction it is easy to determine Baire's class, for example, for the set E^+. In fact, U_n^*, being the difference of the two open sets U_n and $U_n \cdot \sum_{k=1}^{\infty} f(U_n, -k)$, can be represented as the union of a countable number of closed sets, *i, e.*, it is of type F_σ,[6] $V^+ = \sum_{n=1}^{\infty} U_n^*$ is a union of sets F_σ, thus also of type F_σ. Finally, E^+, as the complement of an F_σ, is a G_δ set.

Analogously, E^- is a G_δ set. Finally, the set of points stable P (both as $t \to -\infty$ and as $t \to +\infty$) is E^+E^-, *i. e.*, also a G_δ set.

5. Regional Recurrence. Central Motions

We introduce Birkhoff's concept of regional recurrence.

5.01. Definition. A dynamical system $f(p, t)$ defined in some metric space R possesses in R the property of *regional recurrence* if for any domain $G \subset R$ and any T there can be found a value $t > T$ such that $G \cdot f(G, t) \neq 0$. Applying to this inequality the transformation of the group with parameter $-t$, we have also $G \cdot f(G, -t) \neq 0$, *i. e.*, the definition of recurrence refers simultaneously to positive and to negative values of t. (Systems with an invariant measure, which will be studied in the next chapter, possess this property).

We shall show in the present section that under very general assumptions about a dynamical system there can be formed within R a space M in which regional recurrence will hold.

We shall call a point p *wandering* if there exists a neighborhood $U(p)$ of it and a positive number T such that

$$U(p) \cdot f(U(p), t) = 0 \quad \text{for all } t \geqq T. \tag{5.02}$$

Applying to this equality the transformation with parameter $-t$ as in the preceding definition, we obtain $U(p) \cdot f(U(p), -t) = 0$, *i. e.*, the definition of a wandering point is symmetrical with respect to positive and negative values of t.

The set W of wandering points is invariant since for the point $f(p, t_0)$ we have also from the formula (5.02), applying to it the transformation with parameter t_0,

$$f(U(p), t_0) \cdot f[f(U(p), t_0), t] = 0 \quad \text{for} \quad t \geqq T.$$

Furthermore, this set is open since, by (5.02), along with the point p all points of the neighborhood $U(p)$ are also wandering.

[6]Let U and V be open sets, $U \supset V$; then $U - V = U(R - V)$, where $R - V$ is closed; it was proved in section 1 that $U = \Sigma_i F_i$, consequently $U - V = \Sigma_i F_i (R - V)$.

The set of points nonwandering with respect to R,

$$M_1 = R - W,$$

is thus a closed, invariant set. It may be empty. For example, for the dynamical system defined in E^2 by the equations $dx/dt = 1$, $dy/dt = 0$, all the points are wandering.

A nonwandering point $p \in M_1$ is characterized by the property that, for any neighborhood $U(p)$ containing it, there can be found arbitrarily large values t for which

$$U(p) \cdot f(U(p), t) \neq 0. \tag{5.03}$$

If a point p is stable P^+ or P^-, then, by definition, for any $U(p)$ containing it there can be found values of t arbitrarily large in absolute value for which

$$f(p, t) \cdot U(p) = f(p, t) \neq 0,$$

and consequently (5.03) is fulfilled *a fortiori, i. e.*, every *point stable P^+ or P^- is nonwandering.*

The converse statement is false; in all the examples of the preceding section all the points of the surface of the torus are, as it is easy to verify, nonwandering, but in Example 4.08 there existed points unstable P in both directions.

If a closed set of nonwandering points M_1 contains an open invariant set G, then this domain is regionally recurrent. This follows directly from the definition and the relation (5.03) in which $U(p)$ is chosen subject to the condition $U(p) \subset G$.

5.04. THEOREM. *If a dynamical system possesses at least one motion stable L^+ or L^-, then the set M_1 of nonwandering points is not empty.*

Let $f(p, t)$ be stable L^+; then the set Ω_p is nonempty and compact. Regarding Ω_p as the space R of motions, we shall prove the theorem if we show that in a compact metric space R (which by Theorem 1.14 has a countable base) the set M_1 of nonwandering motions is not empty.

Suppose that W is the set of wandering points and $M_1 = R - W = 0$. Then for every point $p \in R$ there can be found a neighborhood $U(p)$ satisfying for $t > T$ the relation (5.02). Because of the compactness of the space R, we can choose from these neighborhoods a finite number $U_1, U_2, \ldots, U_N$ such that $\sum_{k=1}^{N} U_k = R$; let there correspond to these the numbers $T_1, T_2, \ldots, T_N$.

An arbitrary point $p \in R$ will lie in some U_{n_1}; by (5.02) at the expiration of time $\leqq T_{n_1}$ it leaves this neighborhood forever. Let it fall in U_{n_2}; at the termination of time $\leqq T_{n_2}$ it leaves U_{n_2} forever, etc. Finally, for $t > \sum_{k=1}^{N} T_k$ there will be nowhere for it to go. This contradiction proves the theorem.

In the subsequent material of this section we shall consider a motion $f(p, t)$ in a compact metric space possessing, consequently, a countable base. By the theorem just proved, the set N_1 is nonempty and compact as a closed subset of a compact space. We shall show that any motion tends to the set M_1; namely, there holds the

5.05. THEOREM. *If the space R is compact, then for any positive ε every wandering motion $f(p, t)$ remains for only a finite time not exceeding a certain $T(\varepsilon)$ outside the set $S(M_1, \varepsilon)$.*

In fact, since R is compact and $S(M_1, \varepsilon)$ is an open set, then $R - S(M_1, \varepsilon)$ is compact and consists entirely of wandering points. Therefore, for every point $p \in R - S(M_1, \varepsilon)$ there can be found a neighborhood $U(p)$ satisfying the condition (5.02) for $t > T(p)$.

Repeating the argument of Theorem 5.04 of the present section, we cover $R - S(M_1, \varepsilon)$ with a finite number of these domains $U_1, U_2, \ldots, U_N$ and, denoting the corresponding numbers $T(p)$ by $T_1, T_2, \ldots, T_N$, we verify that the duration of the stay of the point p in $R - S(M_1, \varepsilon)$ cannot exceed $T = \sum_{k=1}^{N} T_k$. The theorem is proved.

Birkhoff's further problem consisted in narrowing down the set in whose neighborhood there pass motions of wandering points. Proceeding in this direction he arrived, in the following manner, at the concept of a *center*.

Consider the set M_1 of nonwandering points of R as the space of a new dynamical system. This space is compact and in it, according to what has preceded, there can be determined a closed, invariant, nonempty set M_2 of points nonwandering with respect to M_1. Continuing this process, we obtain a chain of closed sets included one within the other

$$M_1 \supset M_2 \supset \ldots \supset M_n \supset \ldots.$$

If for some number we obtain $M_k = M_{k+1}$, then $M_k = M_{k+2} = \ldots$ and the set M_k is the required set of *central motions*. If each M_{k+1} is a proper part of M_k, then we define

$$M_\omega = \prod_{k=1}^{\infty} M_k.$$

The set M_ω is again compact and invariant. This process can be continued to all numbers of class II by the method of transfinite induction: if $\alpha + 1$ is a number of the first kind and M_α is already defined, then $M_{\alpha+1} \subset M_\alpha$ is a set of nonwandering points in the space of motions M_α; if β is a transfinite number of the second kind and all $M_\alpha (\alpha < \beta)$ are already defined, then $M_\beta = \prod_{\alpha<\beta} M_\alpha$. We obtain a transfinite sequence of closed sets

$$M_1 \supset M_2 \supset \ldots \supset M_n \supset \ldots \supset M_\omega \supset \ldots \supset M_\alpha \supset \ldots$$

By the theorem of Baire 1.09 for some value α not greater than a transfinite number of class II we obtain $M_\alpha = M_{\alpha+1} = \ldots$. *The set M_α is the set of central motions.* We shall denote it by the letter M. Obviously, M is a compact, invariant set.

5.06. Example. We show a case where $M = M_2$. The motions are defined in the region $x^2 + y^2 \leqq 1$ of the plane E^2. They are traced along curves determined by the differential equations

$$\frac{dy}{dx} = \frac{x + y(1 - x^2 - y^2)}{-y + x(1 - x^2 - y^2)} \qquad (x^2 + y^2 \leqq 1),$$

or, in polar coordinates,

$$\frac{dr}{d\theta} = r(1 - r^2) \qquad (0 \leqq r \leqq 1).$$

The system of integral curves has a critical point (focus) at the origin and also a closed curve $r = 1$; all the remaining curves are spirals approaching the critical point as $\theta \to -\infty$ and winding to the limit cycle as $\theta \to +\infty$.

Passing to the dynamical system, we construct it such that the point $x = 1$, $y = 0$ should be a rest point along with the point $x = 0$ $y = 0$. We attain this on defining, for example, the motions by the system of equations

$$\frac{dx}{dt} = [-y + x(1 - x^2 - y^2)][(x - 1)^2 + y^2],$$

$$\frac{dy}{dt} = [x + y(1 - x^2 - y^2)][(x - 1)^2 + y^2],$$

or in polar coordinates,

$$r = r(1 - r)(1 + r^2 - 2r \cos \theta),$$
$$\dot{\theta} = 1 + r^2 - 2r \cos \theta.$$

Along the curve $r = 1$ there exist two trajectories of motions, the rest point $r = 1$, $\theta = 0$, and the motion along the arc $r = 1$, $0 < \theta < 2\pi$ determined by the equation

$$\theta(t) = 2 \operatorname{arc\,ctg} \left(\operatorname{ctg} \frac{\theta_0}{2} - 2t\right);$$

and

$$\lim_{t \to -\infty} \theta(t) = 0, \qquad \lim_{t \to +\infty} \theta(t) = 2\pi.$$

The points of the domain $G = \{0 < r < 1\}$ are all wandering since they approach $r = 0$ and $r = 1$ as $t \to -\infty$ and $t \to +\infty$ respectively, *i. e.*, every point leaves a sufficiently small neighborhood $U(p)$ around it forever. The point $r = 0$, as a rest point, is nonwandering. All the points of the circle $r = 1$ are also nonwandering, since in any neighborhood $U(p)$ of such a point there can be found points not lying on the circle $r = 1$ and, consequently, for an increase in t, when the polar angle θ increases by a multiple of 2π, these points approach still more closely the arc $r = 1$ and will intersect $U(p)$ again and again. Thus M_1 consists of the points $r = 0$ and the circle $r = 1$.

We consider now motions only in the set M_1. The rest points $r = 0$ and $r = 1$, $\theta = 0$ are obviously nonwandering; every other point p with coordinates $r = 1$, $\theta = \theta_0 \neq 0 \pmod{2\pi}$ is wandering since it has limiting positions as $t \to -\infty$ and $t \to +\infty$ and forever leaves its relative neighborhood $U(p)$ if the latter does not contain a rest point.

The same result is obviously obtained under the subsequent process of apportioning with respect to the nonwandering points. Thus $M = M_2$ consists only of the two rest points.

5.07. Remark. There remains open the question as to whether the chain of sets $M_1 \supset M_2 \supset \ldots$ terminates after a finite number of steps in the case of a system defined in E^n.

We have seen that each trajectory stable P^+ or P^- belongs to M_1. Since all its points are nonwandering with respect to the space of the trajectory itself, it lies in M_2. By the method of transfinite induction it is easily proved that *every trajectory stable P, even though only on one side, lies in the set of central motions M* which may be defined as *the greatest closed set whose points are all* *nonwandering with respect to this set, or,* what is the same, as *the greatest closed set in which recurrence holds for any relative domain.*

The structure of the set M is made clear by the following theorem.

5.08. THEOREM. *Points lying on trajectories stable P are everywhere dense in the set of central motions M.*

Consider a given dynamical system in the set M. Let $p \in M$ be any point and $\varepsilon > 0$ be an arbitrary number. It is required to prove that in $S(p, \varepsilon) = S$ there can be found a point stable P. We take a sequence of increasing, positive numbers $\{T_n\}$, where $\lim_{n\to\infty} T_n = +\infty$. Because of the regional recurrence, there can be found $t_1 > T_1$ such that the intersection $S \cdot f(S, t_1)$ is not empty. Since the intersection of two open sets is an open set, there can be found a point p_1 and a number $\varepsilon_1 > 0$ such that $S(p_1, \varepsilon_1) \subset S \cdot f(S, t_1)$. We let $S_1 = S(p, \varepsilon_1/2)$. By virtue of the same recurrence there can be found a $-t_2 < -T_2$ such that $S_1 \cdot f(S_1, -t_2)$ is nonempty and there can be found a point p_2 and a number $\varepsilon_2 > 0$ such that

$$S(p_2, \varepsilon_2) \subset S_1 \cdot f(S_1, -t_2).$$

Obviously, $\varepsilon_2 \leqq \varepsilon_1/2$. We let $S_2 = S(p_2, \varepsilon_2/2)$. Next there can be found a point p_3 and a number $\varepsilon_3 > 0$ such that $S(p_3, \varepsilon_3) \subset S_2 \cdot f(S_2, t_3)$, where $t_3 > T_3$ and $\varepsilon_3 \leqq \varepsilon_2/2$. Let $S(p_3, \varepsilon_3/2) = S_3$. Next we determine a point p_4 and a number $\varepsilon_4 > 0$ for which $S(p_4, \varepsilon_4) \subset S_3 \cdot f(S_3, -t_4)$, where $-t_4 < -T_4$ and $\varepsilon_4 \leqq \varepsilon_3/2$, etc.

Continuing this process without end and noting that $\bar{S}_n \subset S_{n-1}$ $(n = 2, 3, \ldots)$ and, besides, that $D(\bar{S}_n) < \varepsilon_n < \varepsilon/2^{n-1}$, we obtain because of the compactness of the space M a point q as the intersection of the sets S_n:

$$q = \prod_{n=1}^{\infty} S_n.$$

We shall show that the point q is stable P^-. Let there be given an arbitrarily large number $T > 0$ and an arbitrarily small number $\delta > 0$. We determine a natural number n such that simultaneously $T_{2n+1} > T$ and $\varepsilon_{2n} < \delta$. By construction, $q \in S(p_{2n+1}, \varepsilon_{2n+1})$; on the other hand $S_{2n} = S(p_{2n}, \varepsilon_{2n}/2) \subset S(q, \delta)$, since $\varrho(q, p_{2n}) < \varepsilon_{2n}/2$ and $\delta > \varepsilon_{2n}$. We thus obtain the inclusions

$$q \in S(p_{2n+1}, \varepsilon_{2n+1}) \subset S_{2n} \cdot f(S_{2n}, t_{2n+1});$$

whence, applying the transformation of the group with the parameter $-t_{2n+1}$, we obtain

$$f(q, -t_{2n+1}) \in S_{2n} \cdot f(S_{2n}, -t_{2n+1}) \subset S(q, \delta),$$

wherein $-t_{2n+1} < -T_{2n+1} < -T$. This proves the stability P^- of the point q. Its stability P^+ is proved analogously.

5.09. REMARK. In the proof of Theorem 5.08 we employed only the properties of compactness and regional recurrence of the set M. The theorem is therefore valid if, instead of M, any compact set be taken which possesses the property of regional recurrence.

On the basis of Theorem 5.08 and the remark preceding it, the structure of the set M is made fully clear. Namely, *the set of central motions in a compact space is the closure of the set of points lying on all the trajectories stable P.*

5.10 THEOREM. *In the set M of central motions the points situated on the trajectories stable P form a set of the type G_δ of the second category, i. e., its complement can be represented as the union of a countable number of closed (possibly empty) sets nowhere dense in M.*

We assign an unbounded sequence of increasing positive numbers $\{T_n\}$, $\lim_{n\to\infty} T_n = +\infty$, and a sequence of decreasing positive numbers $\{\varepsilon_n\}$ such that $\lim_{n\to\infty} \varepsilon_n = 0$. We denote by F_k the set of points $p \in M$ for which there holds the relation

$$f(p, t) \cdot S(p, \varepsilon_k) = 0 \quad \text{for all } t > T_k;$$

F_k may be empty. Obviously, all the points $p \in F_k$ are unstable P^+, and it is easy to prove that every point unstable P^+ lies in some F_k.

We shall show that F_k is closed. Assuming the contrary, we would have a sequence $\{p_n\} \subset F_k$ for which $\lim_{n\to\infty} p_n = p_0$, where $p_0 \notin F_k$. From this $f(p_0, t_0) \in S(p_0, \varepsilon_k)$ for some $t_0 \geqq T_k$; consequently, there could be found a number $\varepsilon > 0$ such that $S(f(p_0, t_0), \varepsilon) \subset S(p_0, \varepsilon_k)$. By property II′ of section 2, for the point p_0 and the numbers t_0 and ε there could then be found a $\delta > 0$ such that, if $q \in S(p_0, \delta)$, then $f(q, t_0) \in S(f(p_0, t_0), \varepsilon)$, *i. e.*, $f(q, t_0) \in S(p_0, \varepsilon_k)$ for $t_0 \geqq T_k$; consequently, the points q would not lie in F_k. From the condition $p_n \to p_0$ it follows that $p_n \varepsilon S(p_0, \delta)$ for sufficiently large n, *i. e.*, $p_n \cdot F_k = 0$. The contradiction proves that F_k is closed.

Furthermore, F_k is nowhere dense in M; for if it were dense in some domain $G \subset M$, then because the set F_k is closed it would contain $\bar{G}$, which contradicts Theorem 5.08. Thus the set of points $p \in M$ unstable P^+ is $\sum_{k=1}^{\infty} F_k$.

Analogously, we construct the sets F_k^* of points unstable P^-: $p \in F_k^*$ if $f(p, t) \cdot S(p, \varepsilon_k) = 0$ for all $t < -T_k$. The set of all points unstable P^- is $\sum_{k=1}^{\infty} F_k^*$.

It is now clear that the set of points $p \in M$ stable P is

$$M - \sum_{k=1}^{\infty} F_k - \sum_{k=1}^{\infty} F_k^*,$$

i. e., it is of the type G_δ of the second category in M.

5.11. REMARK. Like Theorem 5.08 the present theorem also remains valid if in its hypothesis M be replaced by any compact, invariant set with regional recurrence.

A special kind of invariant set in which regional recurrence holds is the *quasi-minimal set* introduced by H. F. Hilmy. A quasi-minimal set Θ may be defined as the closure of the trajectory, contained in a compact set, of a motion stable P; if $f(p_0, t)$ is stable P and $f(p_0; I) \subset R_1$, where R_1 is compact (*i. e.*, $f(p_0; I)$ is stable L), then

$$\Theta = \overline{f(p_0; I)}.$$

Theorems 5.08 and 5.10 are applicable to these sets because of the property of regional recurrence. But a more precise theorem is also valid.

5.12. THEOREM. *In a quasi-minimal set Θ, the points situated on trajectories stable P and everywhere dense in Θ form a set of the second category of type G_δ.*

The compact metric space R_1 has a countable base $U_1, U_2, \ldots, U_n, \ldots$. We denote by F_1 and F_2 the set of points $p \in R_1$ in which the half-trajectories $f(p; 0, +\infty)$ and $f(p; -\infty, 0)$, respectively, are nowhere dense in Θ. If $f(p; 0, +\infty)$ is nowhere dense in Θ, then there can be found a neighborhood U_k and a number T such that $f(p, t) \cdot U_k = 0$ for $t > T$.

Let there be assigned an increasing sequence of numbers $\{T_n\}$, $\lim_{n\to\infty} T_n = +\infty$, and let us denote by F'_{kn} the set of points $p \in \Theta$ for which

$$f(p, t) \cdot U_k = 0 \quad \text{for} \quad t > T_n.$$

The same arguments as in Theorem 5.10 show that each F'_{kn} is closed; it cannot be dense anywhere in Θ since then it would contain a domain consisting of points for which $f(p; 0, +\infty)$ is nowhere dense in Θ, which is contradicted by the existence of $f(p_0, t)$ for which $f(p_0; 0, +\infty)$ is everywhere dense in Θ by definition. Obviously

$$F_1 = \sum_{k=1}^{\infty} \sum_{n=1}^{\infty} F'_{kn}.$$

We obtain an analogous representation for F_2 in the form of a union of closed sets F''_{kn} nowhere dense in Θ:

$$F_2 = \sum_{k=1}^{\infty} \sum_{n=1}^{\infty} F''_{kn}.$$

Then the set of points in Θ both of whose half-trajectories are everywhere dense in Θ, and which are therefore stable P, is $\Theta - F_1 - F_2$, *i. e.*, is a G_δ set of the second category.

5.13. Corollary.[7] *If a quasi-minimal set Θ is neither a rest point nor the trajectory of a periodic motion, then it contains an uncountable set of motions everywhere dense and stable P.*

In fact, for each motion $f(p, t)$ stable P and everywhere dense in Θ we have $\overline{f(p; I)} = \Theta$; by the Corollary 4.11, each finite arc $f(p; t_1, t_2)$ is nowhere dense in Θ. Assuming that the set of motions dense in Θ and stable P is countable, we might represent the totality of points of their trajectories as a countable sum of nowhere-dense sets $\sum_{i=1}^{\infty} \sum_{k=1}^{\infty} f(p_i; k, k+1)$ which cannot be, since this set is of the second category in a complete space.

6. Minimal Center of Attraction

In this section we shall have to do with the concept of "the probability of finding the point $f(p, t)$ in a set E" as $t \to +\infty$ or as $t \to -\infty$. By this we shall understand the following. Consider the trajectory arc $f(p; 0, T)$ and the set of those values of $t \in [0, T]$ for which $f(p, t) \in E$; let the measure of this set be $\tau = \tau(p; T, E) = \int_0^T \varphi_E\,(f(p, t))dt$, where φ_E is the characteristic function of the set E; *i. e.*,

$$\varphi_E\,(p) = 1 \quad \text{if} \quad p \in E, \quad \text{and}$$
$$\varphi_E\,(p) = 0 \quad \text{if} \quad p \in R - E.$$[8]

[7]All three examples of section 4 may serve as illustrations of quasi-minimal sets which are not rest points or trajectories of periodic motions. The difference between Example 4.04 on the one hand and Examples 4.06 and 4.08 on the other will be explained in section 7.

[8]The sets E which we shall consider here are closed or open sets; therefore, as is easily seen, the sets of values of t for which $f(p, t) \in E$ will be measurable.

It is natural to call the ratio τ/T the relative time of stay of the point p in the set E in the course of the time interval $[0, T]$. Obviously

$$0 \leqq \frac{\tau}{T} \leqq 1.$$

If there exists

$$\lim_{T\to+\infty} \frac{1}{T}\int_0^T \varphi_E\ (f(p,\ t))dt = \lim_{T\to+\infty} \frac{\tau}{T} = \mathbf{P}^+(f(p,\ t) \in E), \tag{6.01}$$

then we shall call this limit the *probability of finding the point p in the set E as $t \to +\infty$.*

Analogously, there is defined the probability of the stay of p in E as $t \to -\infty$: $\mathbf{P}^-(f(p, t) \in E)$. In the sequel we shall, for definiteness, consider only the case $t \to +\infty$ and for simplicity of writing we shall omit the sign $^+$ on the $\mathbf{P}$.

If $\mathbf{P}^+$ does not exist, then there exists a lower probability

$$\underline{\mathbf{P}}^+(f(p,\ t) \in E) = \liminf_{T\to\infty} \frac{\tau}{T} \tag{6.02}$$

and an upper probability

$$\overline{\mathbf{P}}^+(f(p,\ t) \in E) = \limsup_{T\to\infty} \frac{\tau}{T}, \tag{6.03}$$

wherein

$$0 \leqq \underline{\mathbf{P}}^+ < \overline{\mathbf{P}}^+ \leqq 1.$$

Noting that the numerator $\tau = \tau(p;\ T,\ E)$ in the expression (6.01) is a measure, we easily obtain the following equalities and inequalities:

(1) If $A \subset B$, then $\mathbf{P}(f(p,\ t) \in A) \leqq \mathbf{P}(f(p,\ t) \in B)$, and there are analogous inequalities for $\overline{\mathbf{P}}$ and $\underline{\mathbf{P}}$.

(2) $\mathbf{P}(f(p, t) \in A + B) \leqq \mathbf{P}(f(p, t) \in A) + \mathbf{P}(f(p, t) \in B)$; if $AB=0$, then the equality sign holds.

6.04. Definition. An invariant closed set V is called the *center of attraction of the motion $f(p,\ t)$* as $t \to +\infty$ $(t \to -\infty)$ (H. F. Hilmy) if $\mathbf{P}^+(\mathbf{P}^-)$ of the stay of the point p in $S(V,\ \varepsilon)$ for any $\varepsilon > 0$ is equal to 1:

$$\mathbf{P}(f(p,\ t) \in S(V,\ \varepsilon)) = 1. \tag{6.05}$$

If the set V does not admit a proper subset which is likewise a center of attraction, then V is called a *minimal center of attraction.*

6.06. THEOREM. *If the motion $f(p, t)$ is stable $L^+(L^-)$, then there exists a minimal center of attraction for $f(p, t)$ as $t \to +\infty$ $(t \to -\infty)$.*

We shall prove the theorem for a motion $f(p, t)$ stable L^+. By the definition of stability according to Lagrange, there exists a compact set F such that

$$f(p;\ 0,\ +\infty) \subset F$$

(as F one may take $\overline{f(p;\ 0,\ +\infty)}$). Because of its compactness, the set F can be covered by a finite number of relatively open sets $U_k^{(1)}$ of diameter < 1:

$$F = \sum_{k=1}^{n_1} U_k^{(1)}.$$

Since, obviously, $\mathbf{P}(f(p,\ t) \in F) = 1$, there exist closed sets $\overline{U}_k^{(1)}$ for which

$$\overline{\mathbf{P}}(f(p,\ t) \in \overline{U}_k^{(1)}) > 0, \tag{6.07}$$

because, if for all k there should be $\mathbf{P}(f(p,\ t) \in \overline{U}_k^{(1)}) = 0$, we would obtain a contradiction to property (2). We denote by V_1 the union of the sets $\overline{U}_k^{(1)}$ for which (6.07) holds; this set is closed. The probability of the stay of the point p in $F - V_1$ is equal to zero by properties (2) and (1); therefore, on the basis of property (2) we have

$$\mathbf{P}(f(p,\ t) \in V_1) = 1.$$

We cover the compact set V_1 by a finite system of sets $U_k^{(2)}$ open relative to V_1 and of diameter $< \frac{1}{2}$:

$$V_1 \subset \sum_{k=1}^{n_2} U_k^{(2)},$$

and among $\overline{U}_k^{(2)}$ we select those for which

$$\overline{\mathbf{P}}(f(p, t) \in \overline{U}_k^{(2)}) > 0. \tag{6.08}$$

Denoting their union by V_2, we verify, as in the case of V_1, that V_2 is nonempty, compact and

$$\mathbf{P}(f(p,\ t) \in V_2) = 1;$$

moreover, $V_2 \subset V_1$.

If a set V_m with the indicated properties has already been determined, then we cover it by a finite number of relatively open sets of diameter $< 1/2^m$:

$$V_m \subset \sum_{k=1}^{n_{m+1}} U_k^{(m+1)}$$

and set

$$V_{m+1} = \sum_k{}' \, \overline{U}_k^{(m+1)},$$

where the summation extends over those $\overline{U}_k^{(m+1)}$ for which

$$\overline{\mathbf{P}}(f(p, t) \in \overline{U}_k^{(m+1)}) > 0.$$

We obtain in this way the countable sequence of closed compact sets

$$F \supset V_1 \supset V_2 \supset \ldots \supset V_n \supset \ldots$$

We denote their intersection (nonempty, compact) by V (or by V_p if it is necessary to show its dependence on the point p):

$$V = \prod_{n=1}^{\infty} V_n.$$

We shall prove that V is a minimal center of attraction.

First of all it is easy to show that the set V satisfies the condition (6.05). Indeed, for an assigned $\varepsilon > 0$, there can be found an n such that $V_n \subset S(V, \varepsilon)$. Since by construction $\mathbf{P}(f(p, t) \in V_n) = 1$, then by property (1), for any $\varepsilon > 0$, we obtain

$$\mathbf{P}(f(p, t) \in S(V, \varepsilon)) = 1.$$

We consider further certain properties of the set V. If for $q \in R$ there exists an $\eta > 0$ such that

$$\mathbf{P}(f(p, t) \in S(q, \eta)) = 0,$$

then $q \in R - V$. In fact, we determine n such that $1/2^n < \eta$. If V_{n-1} does not contain q, then the assertion is proved; if $q \in V_{n-1}$, then each of the sets $\overline{U}_k^{(n)}$ containing the point q lies within $S(q, \eta)$; by (1) $\mathbf{P}(f(p, t) \in \overline{U}_k^{(n)}) = 0$, i. e. such a $\overline{U}_k^{(n)}$ does not enter in V_n, which means that $q \in R - V_n \subset R - V$.

Conversely, if for any $\varepsilon > 0$ we have

$$\overline{\mathbf{P}}(f(p, t) \in S(q, \varepsilon)) > 0, \tag{6.09}$$

then $q \in V$. Indeed, there exists a $U_{k_1}^{(1)}$ such that $q \in U_{k_1}^{(1)}$. We choose ε_1 such that $S(q, \varepsilon_1) \subset U_{k_1}^{(1)} \subset \overline{U}_{k_1}^{(1)}$. By (1) it follows from (6.09) that

$$\overline{\mathbf{P}}(f(p, t) \in \overline{U}_{k_1}^{(1)}) > 0,$$

i. e. $\overline{U}_{k_1}^{(1)} \subset V_1$, and this means that $q \in V_1$. Next we take $U_{k_2}^{(2)}$ containing q and choose ε_2 such that $S(q, \varepsilon_2) \subset U_{k_2}^{(2)}$. Again $\overline{U}_{k_2}^{(2)} \subset V_2$ and $q \in V_2$.

By induction we can prove that $q \in V_n$ for any n, i. e. $q \in V$. In this manner the set V may be defined as the set of points $q \in R$ such that relation (6.09) holds for any ε, $\varepsilon > 0$.

This proves the independence of V from the choice of $U_k^{(n)}$. We shall show that V is an invariant set. Suppose $q \in V$; we shall show that for any t_0 we have also $f(q, t_0) \in V$. Fixing t_0, we choose arbitrarily $\varepsilon > 0$. For ε and t_0 there can be found a δ by property II′ of section 2 such that

$$f(S(q, \delta), t_0) \subset S(f(q, t_0), \varepsilon).$$

From property (6.09) of the point q and the formula (6.03) for $\overline{\mathbf{P}}$, we have

$$\limsup_{T\to\infty} \frac{\tau(p; T, S(q, \delta))}{T} > 0.$$

Obviously

$$\tau[p; T, S(f(q, t_0), \varepsilon)] \geqq \tau[p; T, f(S(q, \delta), t_0)].$$

Furthermore, if $f(p, t) \in S(q, \delta)$, then $f(p, t + t_0) \in f(S(q, \delta), t_0)$, and therefore

$$\tau[p; T, f(S(q, \delta), t_0)] \geqq \tau[p; T, S(q, \delta)] - |t_0|,$$

and we obtain

$$\limsup_{T\to\infty} \frac{\tau[p; T, S(f(q, t_0), \varepsilon)]}{T} \geqq \limsup_{T\to\infty} \frac{\tau[p; T, S(q, \delta)] - |t_0|}{T} > 0,$$

i. e. the point $f(q, t_0)$ satisfies the condition (6.09) and consequently $f(q, t_0) \in V$. This proves the invariance of the set V.

Thus it has been established that V is a center of attraction. It remains to prove that V is a minimal center of attraction. Suppose that there exists a V', a proper part of the set V, which is a center of attraction. The set $V - V'$ is not empty and for a point $q \in V - V'$ we have $\varrho(q, V') = \alpha > 0$. Choose $\varepsilon < \alpha/2$. The sets

$S(V', \varepsilon)$ and $S(q, \varepsilon)$ have no common points. By assumption $\mathbf{P}(f(p, t) \in S(V', \varepsilon)) = 1$, therefore by property (2), $\mathbf{P}(f(p, t) \in S(q, \varepsilon)) = 0$, which contradicts the inequality (6.09), since $q \in V$. The theorem is proved.

6.10. THEOREM. *In the minimal center of attraction of the individual motion $f(p, t)$ regional recurrence holds.*

Assume that the theorem is not true. In such a case there can be found in the minimal center of attraction V a relative domain U such that $U \cdot f(U, t) = 0$ for $t \geqq t_0 > 0$. Since U is a relative domain, for each point $q \in U$ there can be found an α such that $S(q, \alpha) \cdot V \subset U$. We choose $\varepsilon < \alpha/2$ and set $S(q, \varepsilon) \cdot V = U_1^*$. Next we assign an arbitrarily small positive number η and choose a positive number T_1 such that $2t_0/T_1 < \eta < 1$.

For the numbers ε and T_1 we determine a δ such that for every point $x \in \overline{U}_1^*$ and any point y satisfying the inequality $\varrho(x, y) < \delta$ there is fulfilled for $0 \leqq t \leqq T_1$ the inequality $\varrho[f(x, t), f(y, t)] < \varepsilon$.

Finally, we take a spherical neighborhood U_1' of radius δ of the set U_1^*:

$$U_1' = S(U_1^*, \delta).$$

If at the instant t_1 the point $f(p, t) \in U_1'$, then there exists a point $r \in U_1^* \subset U \subset V$ such that $\varrho(f(p, t_1), r) < \delta$.

By the assumption regarding U, the point $f(r, t)$, belonging to V, will be found for $t \geqq t_0$ outside U and, consequently, outside $S(q, \alpha)$. From the choice of the number δ, for $0 \leqq t \leqq T_1$ we obtain

$$\varrho[f(p, t_1 + t), f(r, t)] < \varepsilon.$$

Thus for each value of t in the time interval $t_0 \leqq t \leqq T_1$ we shall have

$$\varrho(f(p, t_1 + t), q) > \varrho(f(r, t), q) - \varrho[f(r, t), f(p, t_1 + t)] > \alpha - \varepsilon > \varepsilon,$$

i. e. after each stay in the domain U_1' of a duration not exceeding t_0, the point $f(p, t)$ in the course of time $\geqq T_1 - t_0$ is found outside $S(q, \varepsilon)$. Consequently,

$$\mathbf{P}(f(p, t) \in S(q, \varepsilon)) < \frac{t_0}{T_1 - t_0} < \frac{2t_0}{T_1} < \eta.$$

Because $\eta > 0$ is an arbitrary number, we obtain

$$\mathbf{P}(f(p, t) \in S(q, \varepsilon)) = 0,$$

and this contradicts property (6.09) for the point $q \in U \subset V$. The contradiction so obtained proves the theorem.

6.11. DEFINITION. For any invariant set $E \subset R$, a closed invariant set V_E such that

$$\mathbf{P}(f(p, t) \in S(V_E, \varepsilon)) = 1$$

for any $\varepsilon > 0$ if $p \in E$ is called a *center of attraction* as $t \to +\infty$ *of the motions of the set* E.

If no proper subset of the set V_E is a center of attraction for E, then V_E is a *minimal center of attraction for the motions of* E.

The minimal center of attraction as $t \to -\infty$ is defined analogously. We shall be concerned only with the case $t \to +\infty$.

6.12. THEOREM. *If all the motions of an invariant set E are stable L^+, then there exists a minimal center of attraction V_E.*

We define the set V_E as the closure of the sum of the minimal centers of attraction V_p of all the motions $f(p, t)$ lying in E. Obviously this is an invariant closed set. It is easily verified that it is a center of attraction for E.

Indeed, consider any motion $f(p, t)$, $p \in E$. Since $V_p \subset V_E$, then $S(V_p, \varepsilon) \subset S(V_E, \varepsilon)$, but from the definition of V_p we have $\mathbf{P}(f(p, t) \in S(V_p, \varepsilon)) = 1$ and hence from property (1) we obtain $\mathbf{P}(f(p, t) \in S(V_E, \varepsilon)) = 1$.

We shall show that V_E is a minimal center of attraction for the set E. Suppose that V'_E is also a center of attraction for E and that V'_E is a proper part of V_E. In the set $V_E - V'_E$ there can be found a point q lying in V_p for some $p \in E$ and there can be found an $\alpha > 0$ such that $\varrho(q, V'_E) = \alpha > 0$. On repeating the arguments at the end of the proof of Theorem 6.06 we obtain that $\mathbf{P}(f(p, t) \in S(q, \varepsilon)) = 0$ for $\varepsilon < \alpha/2$, but this contradicts the condition $q \in V_p$.

6.13. THEOREM. *In the minimal center of attraction V_E of the set E regional recurrence holds.*

Assume the contrary. Then there exists a relative domain $U \subset V_E$ such that $U \cdot f(U, t) = 0$ for $t \geqq t_0$. Furthermore, a point $p \in E$ can be found such that its center of attraction V_p intersects U, i. e. $V_p \cdot U = U_p \neq 0$. U_p is a relative domain of the set V_p and, since $U_p \subset U$ there holds the relation $U_p \cdot f(U_p, t) = 0$ for $t \geqq t_0$, but this contradicts Theorem 6.10. The theorem is proved.

We shall compare the theory of minimal centers of attraction with the theory of central motions. Let the space R be compact. Then its minimal center of attraction V_R, both as $t \to +\infty$ and as $t \to -\infty$, is, by Theorem 6.12, not empty, and by Theorem 6.13 it possesses regional recurrence. Since the set of central motions M is the largest set in which regional recurrence exists, then V_R, as $t \to +\infty$, lies in M. It is obvious that also V_R, as $t \to -\infty$, lies in M. From this observation we obtain as a corollary a theorem announced by Birkhoff.

6.14. Corollary. *The probability of the stay of any motion of a dynamical system in an ε-neighborhood of the set of central motions for any* $\epsilon > 0$ *is equal to* 1, i. e. $\mathbf{P}(f(p, t) \in S(M, \varepsilon)) = 1$, *where* $\varepsilon > 0$ *and* $p \in R$ *are arbitrary.*

Since V_p and V_E possess recurrence of domains, Theorem 5.10 is applicable to them and we obtain

6.15. Corollary. *In the minimal centers of attraction* V_p *and* V_E, *the set of points lying on trajectories stable* P *is a* G_δ *of the second category.*

The question arises: will not the set of central motions always be exhausted by the sum of the sets V_R for $t \to +\infty$ and $t \to -\infty$. A negative answer to this question is given by the following example.

6.16. Example. We take as the compact space R the surface of the torus $\mathfrak{T}(\varphi, \theta)$: $0 \leqq \varphi < 1$, $0 \leqq \theta < 1$, $(\varphi + k, \theta + k') \equiv (\varphi, \theta)$ if k and k' are integers. We define the motions by the same differential equations as in example 4.06

$$\frac{d\varphi}{dt} = \Phi(\varphi, \theta), \qquad \frac{d\theta}{dt} = \alpha\Phi(\varphi, \theta);$$

$\alpha > 0$ is irrational, $\Phi(0, 0) = 0$, and $\Phi > 0$ for $|\varphi| + |\theta| \neq 0$, in addition Φ is continuous on the torus and satisfies a Lipschitz condition. We make the supplementary assumption that

$$\iint_{\mathfrak{T}} \frac{d\varphi d\theta}{\Phi(\varphi, \theta)} = +\infty.$$

In the present case the motions along the trajectories $\theta = \alpha\varphi + \theta_0$, $\theta_0 \neq k\alpha \pmod 1$ for each integer k are stable P and, therefore, the set of central motions coincides with the entire surface of the torus. Moreover, we shall show that for any $p \in \mathfrak{T}$ and for any $\varepsilon > 0$

$$\mathbf{P}(f(p,\ t) \in S(O,\ \varepsilon)) = 1, \tag{6.17}$$

where the point $O = (0,\ 0)$. Thus the minimal center of attraction both as $t \to +\infty$ and as $t \to -\infty$ consists only of the single point O.

We first prove the LEMMA: *if $f(x)$ is a Riemann integrable periodic function of period* 1 *and* α *is an irrational number, then for any* x_0

$$\lim_{N\to\infty} \frac{1}{N} \sum_{k=0}^{N-1} f(x_0 + k\alpha) = \int_0^1 f(x)\,dx.$$

In fact, we choose at the start an m such that for an assigned $\varepsilon > 0$ we have

$$\left| \int_0^1 f(x)\,dx - \frac{1}{m} \sum_{s=0}^{m=1} f_s \right| < \frac{\varepsilon}{2},$$

where f_s is any number between the upper and lower bounds of $f(x)$ on the segment $[s/m,\ (s+1)/m]$; and we fix this m.

We next compute how many points $(k\alpha)$, $k = 0, 1, \ldots, N$ (where $(k\alpha)$ denotes the fractional part of the number $k\alpha$, i. e. $(k\alpha) = k\alpha - [k\alpha]$) fall along the half-segment $[s/m,\ (s+1)/m]$ of length $1/m$. For an irrational number α there exist rational fractions p/q with an arbitrarily large denominator q such that $|\alpha - p/q| < 1/q^2$. We choose such a q, to be defined more precisely later, and let $N = nq + r$, $0 \leqq r < q$.

We take the series of q points

$$0,\ (\alpha),\ (2\alpha), \ldots, ((q-1)\alpha) \tag{0*}$$

and replace them with the points

$$0, \left(\frac{p}{q}\right),\ \left(\frac{2p}{q}\right), \ldots \left(\frac{(q-1)p}{q}\right). \tag{0**}$$

The terms of the series (0**) differ from the corresponding terms of the series (0*) by less than $(q-1)|\alpha - (p/q)| < 1/q$. The numbers of the series (0**) are distributed along (0, 1) at equal intervals of $1/q$; the number of them falling in the half-segment of length $1/m$ will be $q/m + \theta$, where $|\theta| \leqq 1$. Conversely, replacing the series (0**) by the series (0*) we see that along the same half-segment there fall or depart from it not more than two points of the series (0*); thus the number of points of the series (0*) along any half-segment of length $1/m$ is equal to $q/m + 3\theta$ ($|\theta| \leqq 1$).

Analogously, in place of the series

$$(l^*) \qquad (lq\alpha),\ ((lq+1)\alpha), \ldots, ((lq+q-1)\alpha)$$

we take the series

$$(l^{**}) \qquad (lq\alpha),\ \left(lq\alpha+\frac{p}{q}\right), \ldots, \left(lq\alpha+\frac{(q-1)p}{q}\right).$$

Again the greatest difference between corresponding terms of the series (l^*) and (l^{**}) is less than $1/q$; the terms of the series (l^{**}) are distributed with a spacing of $1/q$ and the number of the terms of the series (l^*) falling along the half-segment of length $1/m$ is $q/m + 3\theta$, $|\theta| \leqq 1$.

We denote by $N_{1/m}$ the number of points falling in the half-segment of length $1/m$. Assigning to l the values $0, 1, \ldots, n-1$ and summing the estimates found, we obtain:

$$N_{1/m} = \frac{nq}{m} + \theta(3n+r), \qquad |\theta| \leqq 1,$$

whence

$$\frac{N_{1/m}}{N} = \frac{nq}{mN} + \frac{\theta(3n+r)}{N},$$

and further

$$\left|\frac{N_{1/m}}{N} - \frac{1}{m}\right| < \frac{2q}{N} + \frac{3n}{N} < \frac{2q}{N} + \frac{3}{q}.$$

Setting $\sup |f(x)| = M$, we can make

$$\left|\frac{N_{1/m}}{N} - \frac{1}{m}\right| < \frac{\varepsilon}{2Mm}$$

if at the beginning we choose q such that $3/q < \varepsilon/4Mm$, and then N so as to have $2q/N < \varepsilon/4Mm$.

It is obvious that the estimate obtained for the number of points $M_{1/m}$ from $(k\alpha)$ will be correct also for points of the form $x_0 + k\alpha$, where x_0 is any real number.

Now in the sum

$$\frac{1}{N}\sum_{k=0}^{N-1} f(x_0 + k\alpha)$$

we select the terms for which the fractional remainders of the argument of the function fall along some half-segment $[s/m, (s+1)/m]$ and we sum these. Denoting the corresponding sum by $\sum_s'$, we obtain

$$\sum_s{}' = \frac{N_{1/m}}{N} f_s',$$

where f_s' is some definite number contained between the upper and the lower bounds of the modulus of the function $f(x)$ along $[s/m, (s+1)/m]$. Then

$$\left|\sum_s{}' - \frac{1}{m} f_s'\right| < M \left|\frac{N_{1/m}}{N} - \frac{1}{m}\right| < \frac{\varepsilon}{2m}.$$

Summing these inequalities for $s = 0, 1, \ldots, m-1$, we find

$$\left|\frac{1}{N}\sum_{k=0}^{N-1} f(x_0 + k\alpha) - \frac{1}{m}\sum_{s=0}^{m-1} f_s'\right| < \frac{\varepsilon}{2}$$

i. e. for sufficiently large N

$$\left|\frac{1}{N}\sum_{n=0}^{N-1} f(x_0 + n\alpha) - \int_0^1 f(x)dx\right| < \varepsilon.$$

Thus the lemma is proved.

We pass to the proof of our assertion regarding the motions on $\mathfrak{T}$.

Formula (6.17), obviously, is true if $p = O$, or as $t \to +\infty$ if p lies on the trajectory $\theta = \alpha\varphi$, $\varphi < 0$, or as $t \to -\infty$ if p belongs to the trajectory $\theta = \alpha\varphi$, $\varphi > 0$.

Restricting ourselves for definiteness to the case $t \to +\infty$, we consider now the motion along the trajectory $\theta = \theta_0 + \alpha\varphi$, $\theta_0 \not\equiv -k\alpha$ (mod 1) $(k = 0, 1, \ldots)$. The corresponding motions are stable P^+. We choose an arbitrary positive number $\delta < 1/2\sqrt{1+\alpha^2}$ and let $\mathfrak{C} = S(0, \delta)$; in addition we define the distance between the points (φ_1, θ_1) and (φ_2, θ_2) as

$$[\{\varphi_1 - \varphi_2\}^2 + \{\theta_1 - \theta_2\}^2]^{1/2}$$

(where the symbol $\{x\}$ is defined by $-\frac{1}{2} < \{x\} \leqq \frac{1}{2}$ and $x = \{x\} + k$, k an integer).

We shall compute $\tau = \tau(\theta_0; T, \mathfrak{C})$, the measure of the time of the interval $[0, T]$ in the course of which the point moving along the trajectory $\theta = \theta_0 + \alpha\varphi$ is found in $\mathfrak{C}$.

Let $m(\delta) > 0$ be the minimum of the function $\Phi(\varphi, \theta)$ in $\mathfrak{T} - \mathfrak{C}$. We define the function

$$\nu(\varphi, \theta) = \begin{cases} 1 \text{ in } \mathfrak{C} \\ 0 \text{ in } \mathfrak{T} - \mathfrak{C}. \end{cases}$$

Next we introduce the function

$$F(\theta_0) = \int_0^1 \frac{\nu(\varphi, \theta_0 + \alpha\varphi)}{\Phi(\varphi, \theta_0 + \alpha\varphi)} d\varphi.$$

This function is defined and continuous for all $\theta_0 \not\equiv 0 \pmod 1$; it is of period 1 and is equal to zero outside the interval $-\delta\sqrt{1+\alpha^2} < \theta_0 < \delta\sqrt{1+\alpha^2}$. In the neighborhood of the point $\theta_0 = 0$ it is unbounded, since

$$\int_{-1/2}^{1/2} F(\theta_0) d\theta_0 = \iint_{\mathfrak{T}} \frac{\nu d\varphi d\theta}{\Phi(\varphi, \theta)} = \iint_{\mathfrak{C}} \frac{d\varphi d\theta}{\Phi} = +\infty.$$

We shall estimate the quantity $T - \tau$ for some motion beginning at the point $p(\varphi = 0, \theta_0 = \bar{\theta}_0)$, wherein φ varies from 0 to N (N a natural number). We have

$$\begin{aligned} T - \tau(\bar{\theta}_0) &= \int_0^T [1 - \nu(\varphi(t), \theta(t))] dt \\ &= \int_0^N [1 - \nu(\varphi, \bar{\theta}_0 + \alpha\varphi)] \frac{d\varphi}{\Phi(\varphi, \bar{\theta}_0 + \alpha\varphi)} \leqq \frac{N}{m(\delta)}, \end{aligned}$$

where $m(\delta)$ is the minimum of the function Φ in $\mathfrak{T} - \mathfrak{C}$; moreover, since

$$dt = \frac{d\varphi}{\Phi(\varphi, \bar{\theta}_0 + \alpha\varphi)} \geqq \frac{d\varphi}{\max \Phi},$$

T tends to ∞ together with N.

Next we shall estimate between the same limits $\tau(\bar{\theta}_0)$:

$$\tau(\bar{\theta}_0) = \int_0^N \frac{\nu(\varphi, \bar{\theta}_0 + \alpha\varphi)}{\Phi(\varphi, \bar{\theta}_0 + \alpha\varphi)} d\varphi = \sum_{k=0}^{N-1} F(\bar{\theta}_0 + k\alpha).$$

We assign an arbitrarily small number $\sigma > 0$. Since $\int_{-\beta}^{\beta} F(\theta_0) d\theta_0$ diverges for any $\beta > 0$, there can be chosen a positive quantity

$\delta_1 < \delta\sqrt{1+\alpha^2}$ such that

$$\int_{-\delta\sqrt{1+\alpha^2}}^{-\delta_1} F(\theta_0)d\theta_0 + \int_{\delta_1}^{\delta\sqrt{1+\alpha^2}} F(\theta_0)d\theta_0 > \frac{1-\sigma}{\sigma m(\delta)} + 1.$$

We denote by $F^*(\theta_0)$ a function equal to $F(\theta_0)$ outside the intervals $(n-\delta_1, n+\delta_1)(n = 0, \pm 1, \pm 2, \ldots)$ and equal to zero along these intervals. Obviously, $F^*(\theta_0)$ is a bounded Riemann integrable function of period 1. From the lemma, for any $\varepsilon < 1$, there can be found an N_0 such that for $N > N_0$ we have

$$\left| \frac{1}{N} \sum_{k=0}^{N-1} F^*(\bar{\theta}_0 + k\alpha) - \int_0^1 F^*(\theta_0)d\theta_0 \right| < \varepsilon.$$

Hence for $\tau(\bar{\theta}_0)$ we obtain the estimate

$$\tau(\bar{\theta}_0) = \sum_{k=0}^{N-1} F(\bar{\theta}_0 + k\alpha) \geqq \sum_{k=0}^{N-1} F^*(\bar{\theta}_0 + k\alpha) >$$

$$> N\left[\int_0^1 F^*(\theta_0)d\theta_0 - \varepsilon\right] > \frac{1-\sigma}{\sigma} \cdot \frac{N}{m(\delta)}.$$

Comparing with the estimate for $T - \tau(\bar{\theta}_0)$, we find

$$\frac{T-\tau}{\tau} < \frac{\sigma}{1-\sigma} \quad \text{or} \quad \frac{\tau}{T} > 1 - \sigma.$$

Passing to the limit as $T \to +\infty$, we obtain

$$\mathbf{P}^+(f(p, t) \in \mathfrak{C}) \geqq 1 - \sigma,$$

or, because of the arbitrariness of the number σ,

$$\mathbf{P}^+(f(p, t) \in \mathfrak{C}) = 1.$$

Analogously, it can be shown that

$$\mathbf{P}^-(f(p, t) \in \mathfrak{C}) = 1.$$

Thus, for the system considered, the minimal center of attraction consists of the single point O.

7. Minimal Sets and Recurrent Motions.

Let a dynamical system $f(p, t)$ be defined in a space R.

7.01. Definition. A set $\Sigma \subset R$ is called *minimal* if it is nonempty, closed and invariant, and has no proper subset possessing these three properties.

A rest point and the trajectory of a periodic motion represent the simplest examples of minimal sets. Motions on the surface of a torus, each of which is everywhere dense on it (example 4.04), present a more complicated example. Here the minimal set is the whole space. On the other hand, in example 4.06, where there exists a rest point on the surface of the torus, the whole surface of the torus no longer forms a minimal set, only the rest point appearing as such. All these minimal sets are compact.

The trajectory of a rectilinear uniform motion in Euclidean space affords an example of a minimal set which is not compact.

The significance of minimal sets is to be found in the fact that a very extensive class of dynamical systems possesses minimal sets, wherein the compact minimal sets present the greatest interest.

7.02. THEOREM. *Every invariant, closed, compact set F contains some minimal set.*

If F itself is a minimal set the theorem is proved. If not, then this means that there exists a closed, invariant set F_1 which is a proper part of F. If F_1 is not minimal, there exists a closed, invariant set $F_2 \subset F_1$ etc. If in a finite number of steps we have not obtained a minimal set we then obtain a countable sequence of invariant sets

$$F \supset F_1 \supset \ldots \supset F_n \supset \ldots$$

Their intersection F_ω, being obviously closed, compact and non-empty, will be likewise an invariant set.

Indeed, if $p \in F_\omega$, then $p \in F_n$ for any n; because of the invariance of F_n we have $f(p; I) \subset F_n$ for any n, whence $f(p; I) \subset F_\omega$.

If F_α is not minimal, we choose a closed, invariant $F_{\omega+1} \subset F_\omega$ etc. If β is a limiting transfinite number and if F_α is constructed for $\alpha < \beta$, then $F_\beta = \prod_{\alpha<\beta} F_\alpha$. We obtain a transfinite sequence of sets, the one within the other,

$$F \supset F_1 \supset \ldots \supset F_r \supset \ldots \supset F_\omega \supset F_{\omega+1} \supset \ldots \supset F_\beta \supset \ldots.$$

By the theorem of Baire there can be found a transfinite number of the second class β such that $F_\beta = F_{\beta+1}$, i. e. the set F_β has no proper subset which is closed and invariant. Thus F_β is minimal. Moreover, it is compact. The theorem is proved.

7.03. Corollary. *If the space of motions R is compact, it contains a minimal set.*

7.04. Corollary. *If a motion $f(p, t)$ is stable L^+ then the set Ω_p of its ω-limit points contains a minimal set.* This follows from the compactness of the set Ω_p.

From the definition of a minimal set there follows directly its *characteristic property*: if Σ is a minimal set and $p \in \Sigma$ is any point of it, then $\overline{f(p; I)} = \Sigma$, i. e. every trajectory contained in the invariant set Σ is everywhere dense in Σ, and conversely.

Indeed, if the property of being everywhere dense is fulfilled, then every nonempty, closed, invariant subset of the set Σ containing the point p contains, by virtue of its invariance, $\overline{f(p; I)}$, i. e. coincides with Σ which is thus minimal. If this property is not fulfilled, i. e. if there exists a $p_0 \in \Sigma$ such that $\overline{f(p_0; I)}$ forms a proper part of the set Σ, then obviously Σ is not a minimal set.

7.05. Definition. The motion $f(p, t)$ is called *recurrent* if for any $\varepsilon > 0$ there can be found a $T(\varepsilon) > 0$ such that any arc of the trajectory of this motion of time length T approximates the entire trajectory with a precision to within ε. This may be written thus: for an assigned $\varepsilon > 0$ there exists a $T(\varepsilon)$ such that for any t_0 we have

$$f(p; I) \subset S(f(p; t_0, t_0, + T), \varepsilon);$$

or, in other words, whatever may be the numbers u and v there can be found a number w such that $v < w < v + T$ and

$$\varrho[f(p, u), f(p, w)] < \varepsilon.$$

It is easy to show that *every recurrent motion is stable according to Poisson.* For, however small may be the number $\varepsilon > 0$ and however large the number $t_0 > 0$, from the recurrence of the motion $f(p, t)$ there can be found for the point p values t_1 and t_2, $t_0 \leqq t_1 \leqq t_0 + T$, $-t_0 - T_2 \leqq t_2 \leqq -t_0$ such that $\varrho(p, f(p, t_i)) < \varepsilon$ $(i = 1, 2)$, which proves both stability P^+ and P^-.

The connection between recurrent motions and minimal sets is established by the following two theorems of Birkhoff.

7.06. Theorem. (Birkhoff). *Every trajectory of a compact minimal set is recurrent.*

Let Σ be a minimal compact set, $p \in \Sigma$, and assume that the motion $f(p, t)$ is not recurrent. Then there can be found a number $\alpha > 0$ and a sequence of unboundedly increasing time intervals

$(t_\nu - T_\nu, t_\nu + T_\nu)$, $T_\nu \to +\infty$, such that each of the corresponding arcs $f(p; t_\nu - T_\nu, t_\nu + T_\nu)$ is situated at a distance $\geqq \alpha$ from some point $q_\nu = f(p, \tau_\nu)$ on the trajectory $f(p; I)$. From the compactness of the set Σ every subsequence of points $\{q_\nu\}$ has a limit point.

Consider, on the other hand, the sequence of points $\{f(p, t_\nu) = p_\nu\}$; again any subsequence of it has a limit point p^*. We shall assume, in order not to complicate the notation, that $\{q_\nu\}$ and $\{p_\nu\}$ have been chosen as convergent subsequences so that $\lim_{\nu\to\infty} q_\nu = q \in \Sigma$ and $\lim_{\nu\to\infty} p_\nu = p^*$.

Consider the motion $f(p^*, t)$. Take any arc $f(p^*; -T, T)$ of its trajectory, where T is an arbitrarily large fixed number. By property II′ of section 2 one can choose $\delta(\alpha/3, T) > 0$ such that from the inequality $\varrho(p^*, r) < \delta$ there will follow $\varrho[f(p^*, t), f(r, t)] < \alpha/3$ for $|t| \leqq T$. Next one can find a ν such that there will be fulfilled simultaneously the inequalities

$$T_\nu > T, \quad \varrho(p^*, p_\nu) < \delta \quad \text{and} \quad \varrho(q_\nu, q) < \frac{\alpha}{3}.$$

We obtain for any fixed $t \in (-T; T)$:

$$\varrho[f(p^*, t), f(p_\nu, t)] < \frac{\alpha}{3}.$$

But from the choice of the point q_ν, taking into consideration that $|t| < T < T_\nu$, we have

$$\varrho(f(p_\nu, t), q_\nu) = \varrho(f(p, t_\nu + t), q_\nu) \geqq \alpha.$$

Comparing these inequalities with the inequality $\varrho(q_\nu, q) < \alpha/3$, we obtain

$$\varrho(f(p^*, t), q) > \frac{\alpha}{3} \qquad \text{for} \quad |t| < T.$$

Because of the arbitrary choice of the number T, this inequality holds for every t $(-\infty < t < +\infty)$, i. e. $\varrho(f(p^*; I), q) \geqq \alpha/3$. But because the set Σ is closed we have

$$p^* \in \Sigma, \qquad q \in \Sigma,$$

whence, by virtue of the invariance of the set Σ, we deduce

$$f(p^*; I) \subset \Sigma.$$

But in such a case the invariant closed set $\overline{f(p^*; I)} \subset \Sigma$ is a proper part of Σ since it does not contain the point q. We have obtained a contradiction to the assumption that Σ is a minimal set. This contradiction proves the theorem.

7.07. THEOREM (Birkhoff). *If a recurrent motion $f(p, t)$ is situated in a complete space, then the closure $\overline{f(p; I)}$ of its trajectory is a compact, minimal set.*

First we shall prove that $\overline{f(p; I)}$ is compact. We assign an arbitrary number $\varepsilon > 0$. From the recurrence of the trajectory $f(p, t)$ there can be found a $T > 0$ such that the arc of the trajectory $f(p; 0, T)$ approximates $f(p; I)$ with a precision to within $\varepsilon/2$, i. e. for any point $f(p, t)$ we have $\varrho[f(p, t), f(p; 0, T)] < \varepsilon/2$. Suppose that the point $q \in \overline{f(p, I)}$; then there exists a sequence of points $p_n = f(p, t_n)$ such that $\lim_{n\to\infty} p_n = q$. Since $\varrho(p_n, f(p; 0, T)) < \varepsilon/2$, then in the limit we have $\varrho(q, f(p; 0, T)) \leqq \varepsilon/2$.

Because of the compactness of the arc $f(p; 0, T)$ there exists on it a finite $\varepsilon/2$-net, i. e. a finite set of points $p^{(1)}, p^{(2)}, \ldots, p^{(N)}$ such that for any point $r \in f(p; 0, T)$ there can be found a $p^{(\nu)}$ with $\varrho(p^{(\nu)}, r) < \varepsilon/2$. Obviously this set $p^{(1)}, p^{(2)}, \ldots, p^{(N)}$ is an ε-net for $\overline{f(p; I)}$ since for any $q \in \overline{f(p; I)}$, by what has been proved, there can be found an $r \in f(p; 0, T)$ such that $\varrho(r, q) < \varepsilon/2$ and, consequently, $\varrho(q, p^{(\nu)}) < \varepsilon$. From this there follows the compactness of the set $\overline{f(p; I)}$.

We shall prove that the set $\overline{f(p; I)} = \Sigma$ is minimal. Assume the contrary; then there can be found a closed, invariant set A forming a proper part of the set Σ. Obviously the point p does not lie in A since otherwise we would have, from the invariance of the set A, $f(p; I) \subset A$, and since A is closed, $\overline{f(p, I)} = \Sigma = A$. Therefore, $\varrho(p, A) = d > 0$. We choose $\varepsilon < d/2$ and determine the number $T(\varepsilon) > 0$ entering in the definition of a recurrent trajectory $f(p, t)$. Suppose that $q \in A$. For the numbers ε and T and the point q there exists, by virtue of the condition II′ of section 2 a δ such that from the inequality $\varrho(q, r) < \delta$ there follows: $\varrho[f(q, t), f(r, t)] < \varepsilon$ for $|t| \leqq T$. Since q lies in the closure of the trajectory $f(p; I)$, there can be found a point of this trajectory within $S(q, \delta)$; let it correspond to the value of time t_1:

$$\varrho(q, f(p, t_1)) < \delta.$$

Then $\varrho[f(q, t), f(p, t + t_1)] < \varepsilon$ for $|t| \leqq T$, or, since $f(q, t) \subset A$, $\varrho(A, f(p, t + t_1)) < \varepsilon$ for $|t| \leqq T$. From this, $\varrho(p, f(p, t_1 + t)) >$

$d - \varepsilon > \varepsilon$. Consequently, the point p does not lie in an ε-neighborhood of the arc of time length $2T$ with its center at the point $f(p, t_1)$, which contradicts the hypothesis of the recurrence of the motion $f(p, t)$. The theorem is proved.

We notice that, just as for stability P, so in particular also for recurrence of the motion $f(p, t)$, the point returns to the neighborhood of its initial position for arbitrarily large values of t. However, the set of those values of t for which this return is valid possesses in the case of a recurrent motion one characteristic property.

7.08. DEFINITION. A set of numbers is called *relatively dense* if there exists an $L > 0$ such that any interval $(\alpha, \alpha + L)$ of length L contains at least one element of this set.

7.09. THEOREM. *A necessary and sufficient condition that a motion stable L be recurrent is that for any $\varepsilon > 0$ the set of values of t for which*

$$\text{(A)} \qquad \varrho(p, f(p, t)) < \varepsilon$$

be relatively dense.

If the motion is recurrent, then there exists a $T(\varepsilon)$ such that any arc of time length T approximates the whole trajectory, and in particular, the point p. From this there follows at once that the set of values of t fulfilling the condition (A) is relatively dense, wherein $L(\varepsilon) = T(\varepsilon)$.

Conversely, for any ε let there exist an $L(\varepsilon)$ such that the inequality (A) is fulfilled for at least one value of t in each interval $(t_0, t_0 + L)$. We shall prove that $f(p, t)$ is recurrent. Assume the contrary. Since the set $\overline{f(p; I)}$ is compact, by theorem 1 it contains a minimal set Σ, moreover p is not contained in Σ. Reasoning as in Theorem 7.07 we verify that $\varrho(p, \Sigma) = d > 0$. We choose $\varepsilon < d/2$. Suppose that $q \in \Sigma$. For the point p, the interval L and the number ε we find, by II′ of section 2, a δ such that the inequality $\varrho(q, r) < \delta$ implies the inequality $\varrho[f(q, t), f(r, t)] < \varepsilon$ for $0 \leqq t \leqq L$. Since $q \in \Sigma \subset \overline{f(p; I)}$, there can be found a t_1 such that $\varrho(f(p, t_1), q) < \delta$, and then

$$\varrho[f(p, t_1 + t), f(q, t)] < \varepsilon \quad \text{for} \quad 0 \leqq t \leqq L.$$

Hence, for $t_1 \leqq t \leqq t_1 + L$ we have

$$\varrho(p, f(p, t)) \geqq \varrho(p, \Sigma) - \varrho(f(p, t), \Sigma) > d - \varepsilon > \varepsilon.$$

Thus the point $f(p, t)$ in the course of the entire time interval $(t_1, t_1 + L)$ does not return to the ε-neighborhood of the point p, which contradicts the inequality (A) and the definition of the number $L(\varepsilon)$. The theorem is proved.

We denote by D_f the totality of all minimal sets belonging to the invariant set $\overline{f(p; I)}$.

7.10. THEOREM. *If $f(p, t)$ is stable L, then for every $\varepsilon > 0$ there exists an $L(\varepsilon)$ such that for any t_1 we have*

$$f(p;\ t_1,\ t_1 + L) \cdot S(D_f,\ \varepsilon) \neq 0.$$

Otherwise expressed, *the set of values t for which $\varrho(f(p, t), D_f) < \varepsilon$ is relatively dense.*

Assume the contrary; then for an increasing sequence $L_1, L_2, \ldots, L_n \ldots$, $\lim_{n\to\infty} L_n = +\infty$, there can be found values $t_1, t_2, \ldots, t_n, \ldots$ and a number $\alpha > 0$ such that

$$\varrho(f(p;\ t_n,\ t_n + L_n),\ D_f) > \alpha.$$

Because p is stable according to Lagrange, from the sequence of points $\{f(p, t_n)\}$ there can be chosen a convergent subsequence. In order not to complicate the notation we shall assume that $\lim_{n\to\infty} f(p, t_n) = \bar{p}$; we obtain for the point $\bar{p}$ by passing to the limit, $\varrho(f(\bar{p};\ 0,\ +\infty),\ D_f) \geqq \alpha$, whence $\varrho(\overline{f(\bar{p};\ 0,\ +\infty)},\ D_f) \geqq \alpha$.

On the other hand, since $\bar{p} \in \overline{f(p; I)}$, then $f(\bar{p}; I) \subset \overline{f(p; I)}$, and since this last set is compact, then the set $\Omega_{\bar{p}}$ of the ω-limit points of the motion $f(\bar{p}, t)$, by corollary 7.04 of the present section, contains a minimal set $\Sigma \subset D_f$. But since

$$\Sigma \subset \Omega_{\bar{p}} \subset \overline{f(\bar{p};\ 0,\ +\infty)},$$

we obtain

$$\varrho(\Sigma,\ D_f) \geqq \alpha > 0.$$

The contradiction so obtained proves the theorem.

The set $\bar{D}_f$ is closed, compact, and invariant; it possesses the property that for any $\varepsilon > 0$ there holds the inequality

$$\underline{\mathbf{P}}(f(p,\ t) \in S(\bar{D}_f,\ \varepsilon)) > 0.$$

In fact, we determine for an assigned ε the number $L(\varepsilon/2)$ of Theorem 7.10. In the course of any time interval of length L the point $f(p, t)$ enters $S(\bar{D}_f, \varepsilon/2)$ and therefore spends in $S(\bar{D}_f, \varepsilon)$

a time exceeding some $\tau > 0$ which is independent of the interval considered.[9] From this

$$\underline{\mathbf{P}}(f(p, t) \in S(\bar{D}_f, \varepsilon)) \geqq \frac{\tau}{L}.$$

Thus the set D_f has the property similar to the property of a center of attraction for $f(p, t)$; but here $\underline{\mathbf{P}} > 0$ and there we had $\mathbf{P} = 1$.

If we have an invariant, compact set E and denote by D_E the sum of all the minimal sets $\Sigma \subset E$, then analogously to Theorem 5 it can be shown that for any $\varepsilon > 0$ the set of values of t fulfilling the condition

$$\varrho(f(p, t), D_E) < \varepsilon$$

is *uniformly relatively dense* for all $p \in E$, since one and the same number $L(\varepsilon)$ of Theorem 7.10 can be found for all p. The proof of this proposition we leave to the reader. As a corollary we obtain

$$\underline{\mathbf{P}}(f(p, t) \in S(\bar{D}_E, \varepsilon)) > 0; \quad p \in E.$$

The D_E is again analogous to the center of attraction of the set.

However, we shall show in the next chapter, making use of Birkhoff's ergodic theorem, that the sets D_f and D_E in the general case constitute a proper part of the corresponding minimal centers of attraction.

7.11. Definition. A set A is called *locally connected* if for any point $p \in A$ and for any neighborhood $V(p)$ there can be found a neighborhood $U(p) \subset V(p)$ such that $U(p)A$ is connected.

The continuum C in the plane $x0y$ consisting of the curve $y = \sin 1/x$, $0 < x \leqq 1$ and the segment $x = 0$, $-1 \leqq y \leqq 1$ is not locally connected since, for example, any sufficiently small

[9]The distance from the point of entry of $f(p, t)$ into the set $\overline{S(D_f, \varepsilon)}$ to the point reached in $\overline{S(D_f, \varepsilon/2)}$ is obviously $> \varepsilon/2$. Therefore, the time length of the corresponding arc will be greater than some positive number τ. For assume the contrary; let there exist a sequence of point pairs $(p_n, q_n) \subset f(p;\ 0, +\infty)$ such that $\varrho(p_n, q_n) > \varepsilon/2$, $q_n = f(p_n, \tau_n)$, wherein $\lim_{n\to\infty} \tau_n = 0$. Since $\overline{f(p;\ 0, +\infty)}$ is by hypothesis compact, there can be chosen from the sequences $\{p_n\}$, $\{q_n\}$ convergent subsequences; in order not to complicate the notation we shall assume that $p_n \to p^*$, $q_n \to q^*$. Passing to the limit we obtain on the one hand $\varrho(p^*, q^*) \geqq \varepsilon/2$ and on the other $q^* = f(p^*, 0) = p^*$. The contradiction proves our assertion.

relative neighborhood of the point (0, 0) consists of a countable number of components, i. e. is a disconnected set.

The examples of minimal sets that we have considered — a point, a simple closed curve, the surface of a torus in the case of everywhere dense trajectories — were locally connected. We introduce an example due to Poincaré of a minimal set which is not locally connected.

7.12. Example. The space of motions is the surface of a torus $\mathfrak{T}(\varphi, \theta)$ with coordinates (φ, θ) taken mod 1. On the circle $\varphi = 0$ with the angular coordinate $\theta_0 (0 \leqq \theta_0 < 1)$ let there be given a perfect, nowhere dense set F and let $\{(\alpha_n, \beta_n)\}$ $(n = 1, 2, \ldots)$ be a system of its adjacent intervals, where α_n precedes β_n in the cyclic order established by means of the coordinate θ_0. Next let there be given an irrational number γ. On an auxiliary circle Γ of length 1 we consider the set of points $\psi = k\gamma$ $(k = 0, \pm 1, \pm 2, \ldots)$, where the cyclic coordinate $\psi(-\infty < \psi < +\infty;\ \psi + k = \psi$ for k an integer) is the length of arc from a certain reference point O in an established positive direction. Since γ is irrational this set is everywhere dense on Γ. We shall establish a one-one correspondence, with a preservation of the cyclic order, between the set of intervals $\{(\alpha_n, \beta_n)\}$ on the circle $\varphi = 0$ and the set of points $\{k\gamma\}$ on Γ. We order the points $\{k\gamma\}$ thus

$$(*) \qquad O, \gamma, -\gamma, 2\gamma, -2\gamma, \ldots, k\gamma, -k\gamma, (k+1)\gamma, \ldots$$

To the point O on Γ we set in correspondence $(\alpha_1, \beta_1) \equiv (\alpha^{(0)}, \beta^{(0)})$; to the point γ the interval $(\alpha_2, \beta_2) \equiv (\alpha^{(1)}, \beta^{(1)})$; to the point $-\gamma$ we shall set in correspondence $(\alpha^{(-1)}, \beta^{(-1)})$ representing the interval (α_n, β_n) with least subscript n lying on that one of the two arcs intercepted between the intervals just chosen such that $(\alpha^{(0)}, \beta^{(0)})$, $(\alpha^{(1)}, \beta^{(1)})$, $(\alpha^{(-1)}, \beta^{(-1)})$ should have the same cyclic order on the circle $\varphi = 0$ as the points $0, \gamma, -\gamma$ on the circle Γ.

Suppose that to the first N points of the sequence (*) there have already been set in correspondence intervals of the set $\{(\alpha_n, \beta_n)\}$; then the $(N + 1)$st point of this sequence in the cyclic order on Γ will occupy a position between two points $k\gamma$ and $k'\gamma$ (k and k' are integers) already taken; we set in correspondence to it the interval (α_n, β_n) of lowest index not yet used and lying in cyclic order on $\varphi = 0$ between $(\alpha^{(k)}, \beta^{(k)})$ and $(\alpha^{(k')}, \beta^{(k')})$. Continuing this process indefinitely, we obtain the desired correspondence.

We shall define now a mapping $\Phi(\theta_0) = \psi$ of the entire circle $\varphi = 0$ onto the circle Γ in the following manner: to the entire closed interval $[\alpha^{(k)}, \beta^{(k)}]$ there corresponds the one point $k\gamma \in \Gamma$. If θ_0 be a point of the second class of the set F, $0 \leqq \theta_0 \leqq 1$, then on the circle $\varphi = 0$ from which $(\alpha_1, \beta_1) = (\alpha^{(0)}, \beta^{(0)})$ has been deleted it forms a cut in the set of intervals $\{(\alpha^{(k)}, \beta^{(k)})\}$, $k \neq 0$. To this cut there corresponds, because of the coincidence of the cyclic order, a cut in the set of points $\{k\gamma\}$, $k \neq 0$, defining some point $\psi_0 \in \Gamma$. Then $\Phi(\theta_0) = \psi_0$; moreover, for a point of the second class the transformation Φ is one-to-one: $\theta_0 = \Phi^{-1}(\psi_0)$.

Let the circle Γ turn through an angle corresponding to the arc γ; then the point $\psi \in \Gamma$ goes into the point $\psi + \gamma$ (mod 1). For this mapping $T_1(\Gamma)$ of the circle Γ on itself we shall have $T_1(k\gamma) = (k + 1)\gamma$ $(k = 0, \pm 1, \pm 2, \ldots)$. On the circle $\varphi = 0$ the adjacent intervals of the set F are subjected to the transformation T_1, wherein $T_1(\alpha^{(k)}, \beta^{(k)}) = (\alpha^{(k+1)}, \beta^{(k+1)})$. Since under the transformation T_1 the cyclic order of the intervals is preserved, it may be extended to points $\theta_0 \in F$ of the second class and we shall have: if $\theta_0 = \Phi^{-1}(\psi_0)$, then $T_1(\theta_0) = \Phi^{-1}(\psi_0 + \gamma)$.

We shall extend the mapping $T_1(\theta_0)$ to points belonging to the closed adjacent intervals: if $\theta_0 \in (\alpha^{(n)}, \beta^{(n)})$, let $\theta_0 = \alpha^{(n)} + \lambda(\beta^{(n)} - \alpha^{(n)})$, $0 \leqq \lambda \leqq 1$; then we set $T_1(\theta_0) = \alpha^{(n+1)} + \lambda(\beta^{(n+1)} - \alpha^{(n+1)})$. This correspondence is one-to-one and continuous (linearly) within each $(\alpha^{(n)}, \beta^{(n)})$. Furthermore, $T_1(\theta_0)$ is obviously one-to-one for all points of the circle $\varphi = 0$ and preserves the cyclic order on it. It is not difficult to show that it is continuous, and therefore one-to-one bi-continuous, on the circle $\varphi = 0$.

We proceed now to the construction of a dynamical system $f(p, t)$ on the torus $\mathfrak{T}(\varphi, \theta)$.

First we define motions issuing from the point $\varphi = 0$, $\theta = \theta_0$ for $0 \leqq t \leqq 1$. For the point $(0, 0) \in \mathfrak{T}$ we define $f(p, t)$ thus: $\varphi = t$, $\theta = tT_1(0)$, where we choose the value of the coordinate $T_1(0)$ for definiteness so that $0 < T_1(0) < 1$. Next, for any point $(0, \theta_0)$ $(0 < \theta_0 < 1)$ we set $\varphi = t$, $\theta(t, \theta_0) = t[T_1(\theta_0) - \theta_0] + \theta_0$, where the value of $T_1(\theta_0)$ is chosen thus: $T_1(0) < T_1(\theta_0) < T_1(0) + 1 = T_1(1)$. By such a choice the trajectories do not intersect among themselves and fill the entire torus $\mathfrak{T}$, since, for example, if $0 \leqq \theta_0' < \theta_0'' < 1$, then $T_1(0) \leqq T_1(\theta_0') < T_1(\theta_0'') < T_1(0) + 1$, and these same inequalities hold for all values of θ for $0 \leqq t \leqq 1$.

Next we define $f(p, t)$ for $p \in \{\varphi = 0\}$ and any t: if $t = n + \tau$, n an integer, $0 \leqq \tau < 1$, then we set

$$\varphi(t) = t \equiv \tau \pmod 1, \quad \theta(t) = T_1^n(\theta_0) + \theta(\tau;\ T_1^n(\theta_0)).$$

Finally, for any initial point (φ_0, θ_0) we set

$$\varphi(t) = \varphi_0 + t,\ \theta(t) = \theta(t + \varphi_0,\ \theta_0'),$$

where θ_0' is the coordinate on $\varphi = 0$ of the point of intersection of the trajectory passing through (φ_0, θ_0) for $t = \varphi_0$. The dynamical system has been constructed.

The space of motions $\mathfrak{T}$ is a compact set. The set P of points lying on trajectories issuing from F is a closed invariant set since the set of ends of the adjacent intervals and of points of the second class of the set F go into themselves under the transformation T_1. Furthermore, for any point $\psi \in F$ the set $\{T_1^k(\psi)\}$ $(k = 0, \pm 1, \pm 2, \ldots)$ is everywhere dense in F; for the ends of adjacent intervals this follows from the construction and for points of the second class, from the fact that on the circle Γ the set of points $\gamma_0 + k\gamma$ is everywhere dense, where $k = 0, \pm 1, \pm 2, \ldots$ and γ_0 is not a multiple of γ. Therefore, for every motion $f(p, t)$, where $p \in P$, we have $\overline{f(p; I)} = P$, i. e. P is a minimal set. Finally, P is locally disconnected since, for example, any relative (with respect to P) neighborhood of the point $p \in F$ contains an infinite number of components — arcs of trajectories issuing from points of F near p.

Regarding the disposition of the minimal sets in the space R we have the following theorem.

7.13. THEOREM. (G. Ts. Tumarkin) *If a compact minimal set Σ has an interior point, then all its points are interior points.*

Let $p \in \Sigma$ and $S(p, \alpha) \subset \Sigma$. Let t be arbitrary and consider the point $q = f(p, t)$. In view of the continuity of the dependence on initial conditions, there will be a $\delta > 0$ for which $\varrho(q, r) < \delta$ implies

$$\varrho[f(q, -t), f(r, -t)] < \alpha.$$

Hence $f[S(q, \delta), -t] \subset S(p, \alpha)$, and so, in view of the invariance of Σ, $S(q, \delta) \subset \Sigma$. In other words, if Σ has a point on some trajectory as an interior point, then all points on that trajectory are interior points. But every trajectory of a minimal set is dense in the set and so has a point $r \in S(p, \alpha)$. Therefore one can find a $\beta > 0$ such that $S(r, \beta) \subset S(p, \alpha) \subset \Sigma$. Then r is an interior point of Σ, which proves our assertion.

In view of what has just been proved, a compact minimal set dense in R is an open set. Since it is also closed we have the following theorems.

7.14. Corollary 1. *If a minimal set is dense in R then it coincides with some component of R* (or all of R if R is connected).

7.15. Corollary 2. Letting $R = E^n$, one obtains: *a compact minimal set is nowhere dense in E^n.*

From Urysohn's theorem the dimension of a compact subset of E^n which has no interior points does not exceed $n - 1$. Hence we have

7.16. Theorem. (G. F. Hilmy) *The dimension of a compact minimal set in E^n does not exceed $n - 1$.*

In addition, A. A. Markov showed that a compact minimal set is a *Cantor manifold* (for the definition, cf. Hurewicz-Wallman, *Dimension Theory,* Princeton University Press). This theorem is proved in Gottschalk-Hedlund, *Topological Dynamics* [37], but we shall not prove it here.

8. Almost Periodic Motions

Let there be given a dynamical system $f(p, t)$ in a complete metric space R. We introduce the following

8.01. Definition. A motion $f(p, t)$ is called *almost periodic* if for any $\varepsilon > 0$ there exists a number $L(\varepsilon)$ defining a relatively dense set of numbers $\{\tau_n\}$ (displacements)[10] which possess the following property:

$$\varrho[f(p, t), f(p, t + \tau_n)] < \varepsilon \quad \text{for} \quad -\infty < t < +\infty.$$

Periodic motions are a special case of the almost periodic; in fact, if a motion admits the period τ, then its multiples $n\tau$ $(n=0, \pm 1, \pm 2, \ldots)$ form a relatively dense set; moreover, $\varrho[f(p, t), f(p, t + n\tau)] = 0$.

Almost periodic motions, in their turn, are a special case of recurrent motions.

8.02. Theorem. *Every almost periodic motion is recurrent.*

Indeed, a motion is recurrent if for any $\varepsilon > 0$ there can be found an interval $T(\varepsilon)$ such that the arc of the trajectory $f(p, t)$,

[10]The definition of a relatively dense set was given in section 7.

$\alpha \leqq t \leqq \alpha + T$, for any α, approximates every point q of the trajectory with a precision to within ε, i. e. there can be found a $t_0 \in (\alpha, \alpha + T)$ such that $\varrho(q, f(p, t_0)) < \varepsilon$.

If we write $q = f(p, t)$, where t is arbitrarily fixed, then, taking for an almost periodic motion $f(p, t)$ the number $T(\varepsilon) = L(\varepsilon)$ and the displacement τ corresponding to ε and lying in the interval $(\alpha - t, \alpha - t + T)$, we obtain by virtue of the definition of almost periodicity, $\varrho[f(p, t), f(p, t + \tau)] < \varepsilon$, or, denoting $t + \tau = t_0$, we obtain $\varrho[f(p, t), f(p, t_0)] < \varepsilon$, where $\alpha < t_0 < \alpha + T$, i. e. the condition of recurrence.

The converse theorem is false: a recurrent motion need not be almost periodic (cf. example 8.14 below).

If an almost periodic motion is situated in a complete space the closure of its trajectory $\overline{f(p; I)}$ is compact and forms a minimal set (by Theorem 7.07).

The property of almost periodicity is closely connected with Lyapunov stability. We define Lyapunov stability for abstract dynamical systems (according to A. A. Markov) in the following manner.

8.03. DEFINITION. A point $p \in R$ or a motion $f(p, t)$ is *positively Lyapunov stable* (or negatively stable, or stable in both directions)[11] *with respect to the set* $B \subset R$, if for each $\varepsilon > 0$ there exists a $\delta > 0$ such that for every $q \in B$ satisfying the condition $\varrho(p, q) < \delta$, the inequality

$$\varrho[f(p, t), f(q, t)] < \varepsilon$$

is fulfilled for all positive (or all negative, or all) values of t.

8.04. THEOREM. *If all points of a compact set* $A \subset B$ *are positively (or negatively, or in both directions) Lyapunov stable with respect to* B, *then the Lyapunov stability is uniform.*

As usual, this last means that for $\varepsilon > 0$ there exists a $\delta > 0$ such that for

$$p \in A, \quad q \in B, \quad \varrho(p, q) < \delta$$

we have:

$$\varrho[f(p, t), f(q, t)] < \varepsilon \quad \text{for} \quad t > 0$$

(or $t < 0$, or $-\infty < t < +\infty$).

[11]Two-sided Lyapunov stability is called the S-property in the first (Russian) edition of this book and elsewhere.

PROOF. From the condition of Lyapunov stability, for a given $\varepsilon > 0$ and for each point $p \in A$ there exists $\delta'(p) > 0$ such that from $r \in B$, $\varrho(p, r') < \delta'(p)$ follows:

$$\varrho[f(p, t), f(r, t)] < \frac{\varepsilon}{2}$$

for $t > 0$. By the Heine-Borel theorem, on account of the compactness of A, there must exist a finite number of points $p_1, p_2, \ldots, p_N$, such that

$$A \subset \sum_{j=1}^{N} S(p_j, \tfrac{1}{2}\delta'(p_j)).$$

We put

$$\delta = \min [\tfrac{1}{2}\delta'(p_1), \tfrac{1}{2}\delta'(p_2), \ldots, \tfrac{1}{2}\delta'(p_N)]$$

and prove that δ satisfies the conditions of the theorem.

For any point $p \in A$ there is a point $p_i \in A$ such that

$$\varrho(p, p_i) < \tfrac{1}{2}\delta'(p_i);$$

if $q \in B$ and $\varrho(p, q) < \delta$, then $\varrho(q, p_i) \leqq \varrho(q, p) + \varrho(p, p_i) < \delta'(p_i)$.

By the definition of $\delta'(p_i)$ we have:

$$\varrho[f(p_i, t), f(q, t)] < \frac{\varepsilon}{2} \quad \text{for} \quad t > 0.$$

Since $p \in A \subset B$, we also have:

$$\varrho[f(p_i, t), f(p, t)] < \frac{\varepsilon}{2},$$

from which follows $\varrho[f(p, t), f(q, t)] < \varepsilon$ for $t > 0$, which was to be proved.

The proofs in the cases of negative and two-sided stability are analogous.

The following theorem establishes the Lyapunov stability of almost periodic motion.

8.05. THEOREM. *An almost periodic motion $f(p, t)$ in a complete space R is Lyapunov stable in both directions uniformly with respect to the set of points $f(p; I)$ of its trajectory.*

Let $f(p, t)$ be the almost periodic motion and $\varepsilon > 0$ an arbitrary number. On account of the almost periodicity there exists an $L > 0$

such that every interval of length L contains an $\varepsilon/3$-displacement. The continuous dependence on the initial conditions is uniform in the compact set $\overline{f(p; I)}$ (cf. Theorem 7.07), and therefore there exists $\delta > 0$ such that for any two points q and $r \subset \overline{f(p; I)}$ it will follow from the condition $\varrho(q, r) < \delta$ that

$$\varrho[f(q, t), f(r, t)] < \frac{\varepsilon}{3}$$

for $0 \leqq t < L$. We shall show that if

$$\varrho[f(p, t_1), f(p, t_2)] < \delta$$

then

$$\varrho[f(p, t_1 + t), f(p, t_2 + t)] < \varepsilon$$

for $-\infty < t < \infty$. We choose an arbitrary fixed t and find a displacement $\tau = \tau(\varepsilon/3)$, satisfying the inequality:

$$-t \leqq \tau < -t + L, \quad \text{that is} \quad 0 \leqq t + \tau < L.$$

By the properties of an $\varepsilon/3$-displacement we have:

$$\varrho[f(p, t_1 + t), f(p, t_1 + t + \tau)] < \frac{\varepsilon}{3},$$

$$\varrho[f(p, t_2 + t), f(p, t_2 + t + \tau)] < \frac{\varepsilon}{3};$$

and, because of the choice of δ, since $0 \leqq t + \tau < L$, we have:

$$\varrho[f(p, t_1 + t + \tau), f(p, t_2 + t + \tau)] < \frac{\varepsilon}{3}.$$

From these three inequalities we get for any t

$$\varrho[f(p, t_1 + t), f(p, t_2 + t)] < \varepsilon,$$

which was to be proved.

8.06. Theorem. (Bochner). *If $f(p, t)$ is an almost periodic motion in a complete space, then from any sequence $\{f(p, t_n + t)\}$ of motions it is possible to choose a subsequence $\{f(p, t_{n_k} + t)\}$, which for $-\infty < t < +\infty$ converges uniformly to a certain motion $f(q, t)$; this last is almost periodic (with the same function $L(\varepsilon)$ as $f(p, t)$, but with the inequality $\leqq \varepsilon$).*

On account of Theorem 7.07 the set $\overline{f(p; I)} = \Sigma$ is compact, since $f(p, t)$ is recurrent by Theorem 8.02 of the present paragraph. Thus, a convergent subsequence $\{f(p, t_{n_k})\}$ can be chosen from the sequence of points $\{f(p, t_n)\}$, where $\lim_{k=\infty} f(p, t_{n_k}) = q \in \Sigma$. For simplicity we denote the corresponding subsequence by $\{t_n\}$ so that

$$p_n = f(p, t_n) \to q.$$

For any $\varepsilon > 0$ we define $\delta(\varepsilon)$ by the S-property,[12] i.e. such that from $\varrho[f(p, t'), f(p, t'')] < \delta$ follows

$$(8.07) \quad \varrho[f(p, t' + t), f(p, t'' + t)] < \varepsilon \quad (-\infty < t < +\infty).$$

Choosing n_0 so large that $\varrho[f(p, t_n), f(p, t_{n+m})] < \delta$ for $n \geqq n_0$, $m \geqq 1$, and putting in the inequality (8.07) $t' = t_n$ and $t'' = t_{n+m}$, we find:

$$\varrho[f(p, t_n + t), f(p, t_{n+m} + t)) = \varrho[f(p_n, t), f(p_{n+m}, t)] < \varepsilon$$
$$(-\infty < t < +\infty).$$

But this is the criterion for uniform convergence; hence, obviously,

$$\lim_{n\to\infty} f(p_n, t) = f(q, t)$$

It remains to be shown that $f(q, t)$ is almost periodic. For a given $\varepsilon > 0$, let τ be a displacement of the function $f(p, t)$ such that for $p_n = f(p, t_n)$ we have

$$\varrho[f(p_n, t + \tau), f(p_n, t)] < \varepsilon \qquad (-\infty < t < +\infty).$$

Passing to the limit as $n \to \infty$ and fixing t we get

$$\varrho[f(q, t + \tau), f(q, t)] \leqq \varepsilon.$$

The theorem is proved.

8.08. Corollary. *If $f(p, t)$ is almost periodic and is neither a rest point nor a periodic motion (and is situated in a complete space), then each of the uncountable set of motions originating in the minimal set $\Sigma = \overline{f(p; I)}$ is almost periodic.*

This follows from the fact for every point $q \in \Sigma$ there is a sequence of points $p_n = f(p, t_n)$ such that $q = \lim_{n\to\infty} p_n$. Also cf. Corollary 5.13.

8.09. Corollary. *In a complete space every point of the minimal set Σ of an almost periodic motion is uniformly Lyapunov stable with respect to Σ.*

[12]Cf. footnote 11 and Theorem 8.05.

Given $\varepsilon > 0$, we define $\delta = \delta(\varepsilon/2)$ from the Lyapunov stability of $f(p, t)$. Let $q \in \Sigma = \overline{f(p; I)}$, $\varrho(p, q) < \delta/2$. There exists $p_n = f(p, t_n)$ such that $\varrho(q, p_n) < \delta/2$. Then $\varrho(p, p_n) < \delta$ and from the Lyapunov stability of the point p with respect to $f(p; I)$ we have $\varrho[f(p, t), f(p_n, t)] < \varepsilon/2$. By Bochner's theorem, $\varrho[f(p_n, t), f(q, t)] \leqq \varepsilon/2$ follows from the uniform convergence of $f(p_n, t)$ to $f(q, t)$. From these two inequalities we get $\varrho[f(p, t), f(q, t)] < \varepsilon$, if $q \in \Sigma$ and $\varrho(p, q) < \delta$. This proves our assertion.

We now investigate the question of when almost periodicity follows from Lyapunov stability. The condition of Lyapunov stability alone is clearly insufficient, since uniform motion along parallel lines in Euclidean space is obviously Lyapunov stable. The following theorem holds:

8.10. THEOREM. *If the motion $f(p, t)$ is recurrent and Lyapunov stable with respect to $f(p; I)$, then $f(p, t)$ is almost periodic.*

Indeed, it follows from the Lyapunov stability that for any $\varepsilon > 0$ there exists a $\delta > 0$ such that if $q \in f(p; I)$ and $\varrho(p, q) < \delta$, then $\varrho[f(p, t), f(q, t)] < \varepsilon$ for $-\infty < t < \infty$.

But by the first part of Theorem 7.09, the recurrence implies the existence of a relatively dense set of numbers $\{\tau\}$, for which the inequality $\varrho[p, f(p, \tau)] < \delta$ is fulfilled. For each δ we find (putting $q = f(p, \tau)$)

$$\varrho[f(p, t), f(p, t + \tau)] < \varepsilon, \quad (-\infty < t < \infty),$$

which proves the almost periodicity of the motion.

A. A. Markov has proved a stronger theorem.

8.11. THEOREM. *If the motion $f(p, t)$ is recurrent and positively Lyapunov stable with respect to $f(p; I)$ then it is almost periodic.*

PROOF. Given $\varepsilon > 0$ and the point p, we determine $\delta(\varepsilon/2)$ from the positive Lyapunov stability. The recurrence implies the existence of a relatively dense set of numbers $\{\tau\}$, satisfying the condition $\varrho[p, f(p, \tau)] < \delta/2$.

We will prove that each τ is an ε-displacement for $f(p, t)$. Because of continuity, for each τ there will be $\sigma > 0$ such that the inequality $\varrho(p, q) < \sigma$ implies $\varrho[f(p, \tau), f(q, \tau)] < \delta/2$. Let t $(-\infty < t < \infty)$ be any number. From the recurrence of the motion there exists $t_1 < t$ such that

$$(*) \qquad \varrho[p, f(p, t_1)] < \min [\sigma, \delta],$$

and hence by the definition of σ we have

$$\varrho[f(p, \tau), f(p, t_1 + \tau)] < \frac{\delta}{2}$$

from which, by the definition of τ we get

$$(**) \qquad \varrho[p, f(p, t_1 + \tau)] < \delta.$$

The positive Lyapunov stability ($t - t_1 > 0$) and the inequalities (*) and (**) give

$$\varrho[f(p, t - t_1), f(p, t)] < \frac{\varepsilon}{2}, \quad \varrho[f(p, t - t_1), f(p, t + \tau)] < \frac{\varepsilon}{2}$$

and finally

$$\varrho[f(p, t), f(p, t + \tau)] < \varepsilon,$$

which was to be proved.

It is possible to give another condition for almost periodic motion which does not *a priori* demand its recurrence; then it is necessary to require Lagrange stability and uniform Lyapunov stability. Franklin has proved such a theorem. A. A. Markov has strengthened it, demanding only one-sided Lagrange and Lyapunov stability (in different directions).

8.12. Theorem (A. A. Markov). *If the motion $f(p, t)$ is uniformly positively Lyapunov stable with respect to $f(p; I)$ and negatively Lagrange stable, then it is almost periodic.*

On account of Theorem 8.11 it is sufficient to prove that under the conditions of the the present theorem $f(p, t)$ is recurrent. If we assumed the contrary, we would find a minimal set Σ in the compact set $A_p \subset \overline{f(p; I)}$ contained properly in $\overline{f(p, I)}$ while, evidently, the point p does not belong to Σ and therefore $\varrho(p, \Sigma) = \alpha > 0$.

We will show that every point $q \in A_p$ is positively Lyapunov stable with respect to $f(p; I)$. Given $\varepsilon > 0$ we determine the number $\delta(\varepsilon)$ from the positive Lyapunov stability of $f(p, t)$. Because of the way q was defined there is a sequence $\{p_n\}$, $p_n = f(p, -t_n)$, $t_n \to \infty$, and $\lim_{n\to\infty} p_n = q$. We define N such that for $n \geqq N$, $\varrho(p_n, q) < \delta/2$, and hence $\varrho(p_n, p_{n+m}) < \delta$, $m = 1, 2, \ldots$. But then the inequality

$$\varrho[f(p_n, t), f(p_{n+m}, t)] < \varepsilon,$$

for any $t > 0$, follows from the uniform positive Lyapunov stability.

We fix t and n and let m become infinite. From property II′ of section 2 we have $\lim_{m\to\infty} f(p_{n+m}, t) = f(q, t)$, and we get

$$\varrho[f(p_n, t), f(q, t)] \leqq \varepsilon$$

for any $t > 0$ if $\varrho(q, p_n) < \delta/2$.

Now let $q \in \Sigma \subset A_p$. We put $\varepsilon = \alpha/2$ and determine $\delta(\alpha/2)$ from the positive Lyapunov stability. There will be points $p_n = f(p, -t_n)$, $t_n > 0$, such that $\varrho(p_n, q) < \delta/2$, and hence by what was already proved,

$$\varrho[f(q, t_n), f(p_n, t_n)] = \varrho[f(q, t_n), p] \leqq \frac{\alpha}{2}.$$

But $f(q, t_n) \in \Sigma$, and by assumption $\varrho(p, \Sigma) = \alpha$. The contradiction shows that $f(p, t)$ is recurrent. The theorem is proved.

8.13. Example. As an example of almost periodic motion we take the motion on the torus $\mathfrak{T}$: $p = (\varphi, \theta)$; $\varphi = \varphi_0 + t$, $\theta = \theta_0 + \mu t$, μ irrational and

$$\varrho[(\varphi_1, \theta_1), (\varphi_2, \theta_2)] = \sqrt{(\varphi_1 - \varphi_2)^2 + (\theta_1 - \theta_2)^2},$$

where the values of $\varphi_1 - \varphi_2$ and $\theta_1 - \theta_2$ are taken as the smallest in absolute value of the differences (mod 1) (cf. the example 6.16). We saw in section 7 that the surface of the torus is here a minimal set: and at the same time

$$\varrho[f(p_1, t), f(p_2, t)] = \varrho(p_1, p_2),$$

that is, we have uniform Lyapunov stability. From this and Theorem 8.12 follows the almost periodicity of the motion.

We may introduce as another example of almost periodic motion, motion on the n-dimensional torus $\mathfrak{T}^{(n)}$: $p = (\varphi_1, \varphi_2, \ldots, \varphi_n)$, $\varphi_i + k^{(i)} \equiv \varphi$ $(i = 1, 2, \ldots, n)$ for integers $k^{(i)}$; the motion is defined by the equations $\varphi_i = \alpha_i t + \varphi_i^{(0)}$ $(i = 1, 2, \ldots, n)$, where $\alpha_i \neq 0$ are given numbers such that there do not exist integers m_i, not all zero, such that $\sum_{i=1}^n m_i \alpha_i = 0$. The whole surface $\mathfrak{T}^{(n)}$ will be a minimal set.

8.14. Example. The motion in the example at the end of section 7, taking place in the minimal locally disconnected set $P \subset \mathfrak{T}$ is recurrent though not almost periodic.

Indeed, let the length of the longest adjacent interval $(\alpha_{n_0}, \beta_{n_0})$ of the set F on $\varphi = 0$ be d_0. However close we may take on the circle $\varphi = 0$ the two points $p_1 = (0, \theta_1)$, $p_2 = (0, \theta_2)$ of the second kind of the set F, between them there can be found an adjacent interval $\{\alpha_{n_1}, \beta_{n_1}\}$; and by construction there exists a transformation T_1^k such that $T_1^k(\alpha_{n_1}, \beta_{n_1}) = (\alpha_{n_0}, \beta_{n_0})$. Correspondingly, $\varrho[f(p_1, k), f(p_2, k)] \geqq d_0$, i.e. the condition of Lyapunov stability is not fulfilled, and hence the motion is not almost periodic.

8.15. Example. We now give an example of a dynamical system having a locally disconnected minimal set of almost periodic motions (the solenoid of Vietoris and van Dantzig).

In the three dimensional space (x, y, z) we define the torus with center line K_1:

$$x = \varrho \cos \varphi, \quad y = \varrho \sin \varphi, \quad z = 0$$

where $\varrho > 0$ is a constant; the torus T_1 is a closed neighborhood of the circumference K_1, whose intersection with each plane $\varphi = \text{const.}$ is a circle with center at the point $x = \varrho \cos \varphi$, $y = \varrho \sin \varphi$, $z = 0$, and radius α_1.

Inside T_1 we construct a simple closed curve K_2 which encircles T_1 twice; let its equations be, for example:

$$x = \left(\varrho + \frac{\alpha_1}{2} \cos \frac{\varphi}{2}\right) \cos \varphi, \; y = \left(\varrho + \frac{\alpha_1}{2} \cos \frac{\varphi}{2}\right) \sin \varphi, \quad z = \frac{\alpha_1}{2} \sin \frac{\varphi}{2}.$$

We obtain the (topological) torus T_2 by taking in each plane $\varphi = \text{const.}$ a closed disk of radius α_2 with center at the point of intersection of K_2 with the plane, where $\alpha_2 < \alpha_1/2$. Evidently, $T_2 \subset T_1$.

Inside T_2 we construct a simple closed curve K_3 which encircles T_2 twice, i.e. an increase of φ by 8π; for example, K_3 is defined by the equations

$$x = \left(\varrho + \frac{\alpha_1}{2} \cos \frac{\varphi}{2} + \frac{\alpha_2}{2} \cos \frac{\varphi}{4}\right) \cos \varphi$$

$$y = \left(\varrho + \frac{\alpha_1}{2} \cos \frac{\varphi}{2} + \frac{\alpha_2}{2} \cos \frac{\varphi}{4}\right) \sin \varphi$$

$$z = \frac{\alpha_1}{2} \sin \frac{\varphi}{2} + \frac{\alpha_2}{2} \sin \frac{\varphi}{4}.$$

For T_3 we take the totality of disks in each plane $\varphi = \text{const.}$ of radius α_3, with center at the point of intersection of the curve K_3 with the plane; if $\alpha_3 < \alpha_2/2$, then $T_3 \subset T_2$.

The further construction proceeds in the same way: K_{n+1} lies inside T_n and makes two turns around T_n; T_{n+1} is a closed neighborhood of K_{n+1} so small that $T_{n+1} \subset T_n$.

Finally, we define

$$\Sigma = \prod_{n=1}^{\infty} T_n.$$

The set Σ is the solenoid. It is easily shown that it is locally disconnected.

We shall show that it is possible to define a motion $f(p, t)$ on the solenoid such that each trajectory is everywhere dense in Σ.

For definiteness, we consider a certain trajectory, starting from a point $p \in \Sigma$, for which $\varphi = 0$. The intersection of the torus T_1 with the plane $\varphi = 0$ is a circle $\Gamma_1^{(1)}$ of radius α_1 with center at the point: $x = \varrho + \frac{1}{2}$, $y = 0$, $z = 0$. The intersection of the torus T_2 with the same plane is two circles $\Gamma_2^{(1)}$ and $\Gamma_2^{(2)}$ of radius α_2, where $\Gamma_2^{(1)} + \Gamma_2^{(2)} \subset \Gamma_1^{(1)}$. The torus T_3 intersects $\varphi = 0$ in four circles: $\Gamma_3^{(i)}$ $(i = 1, 2, 3, 4)$ of radius α_3 lying pairwise in $\Gamma_2^{(1)}$ and $\Gamma_2^{(2)}$. More generally, each $\Gamma_n^{(i)}$ contains two $\Gamma_{n+1}^{(j)}$.

The product $\prod_{n=1}^{\infty} \sum_{j=1}^{2^{n-1}} \Gamma_n^{(j)} = F$ is a closed nowhere dense set, consisting of the intersection of the set Σ with the plane $\varphi = 0$. Let $p \in F$. Then there exists a sequence $\{\Gamma_n^{(i_n)}\}$ of circles such that $p = \prod_{n=1}^{\infty} \Gamma_n^{(i_n)}$. Taking those segments $-\Phi \leqq \varphi \leqq \Phi$, of the tori T_n, whose intersections with $\varphi = 0$ are the disks $\Gamma_n^{(i_n)}$ (Φ is a fixed positive number), we see that through each point $p \in F$ passes a unique open arc $L_p \subset \Sigma(-\infty < \varphi < +\infty)$.

On L_p we define the motion $f(p, t)$ by the law $\varphi = t$. We will show that $f(p; I)$ is everywhere dense in Σ and that $f(p, t)$ is almost periodic.

Let $\varepsilon > 0$ be a positive number. We choose n such that $\alpha_n < \varepsilon/2$. Let $p \in \Gamma_n^{(i_n)}$; by the construction of the torus T_n this point again enters $\Gamma_n^{(i_n)}$ after φ increases by $2\pi \cdot 2^{n-1}$; for the values $2\pi k$, k integral, $0 \leqq k \leqq 2^{n-1}$, it is in all the circles $\Gamma_n^{(i)}$ $(i = 1, 2, \ldots, 2^{n-1})$. Clearly, the arc $f(p; 0, 2\pi \cdot 2^{n-1})$ ε-approximates the whole set Σ, from which follows its minimality.

Moreover, p and $f(p, 2\pi m \cdot 2^{n-1})$, where m is an arbitrary integer, are in the same circle $\Gamma_n^{(i_n)}$; as t varies $(-\infty < t < +\infty)$ they are in the same plane $\varphi = t$ and in the same circle $f(\Gamma_n^{(i_n)}, t)$ of radius α_n; therefore

$$\varrho[f(p, t), f(p, t + 2\pi m \cdot 2^{n-1})] < 2\alpha_n < \varepsilon$$

for $-\infty < t < +\infty$. Consequently $f(p, t)$ has a relatively dense set of ε-displacements $2\pi m \cdot 2^{n-1}$ ($m = 0, \pm 1, \pm 2, \ldots$).

The almost periodicity of any motion $f(p, t)$ where $p \in \Sigma$ is proved in the same way.

The dynamical system which we have defined on Σ could be extended, for example, to the whole of the torus T_1. It is clear that in this extension the set Σ is also minimal.

The structure of a minimal set consisting of almost periodic motions is characterized by the following theorem.

8.16. THEOREM. (Nemickii). *For a compact set to be the closure of an almost periodic trajectory it is necessary and sufficient that it be the space of a compact connected commutative group.*

Let the motion $f(p, t)$ be almost periodic, and let $M = \overline{f(p; I)}$ be the closure of its trajectory. We define in M a commutative group of operations (we will write it additively). Firstly, let q, $r \in f(p; I)$, i.e. $q = f(p, t_q)$ and $r = f(p, t_r)$. The zero of the group will be the point p; we define the sum $q + r = f(p, t_q + t_r)$ and the inverse element $-q = f(p, -t_q)$; these operations clearly satisfy the group axioms and are continuous. We now extend these operations by continuity to the whole of M. First we prove a lemma.

8.17. LEMMA. *Let $f(p, t)$ be an almost periodic motion. If the sequences $f(p, t_n)$ and $f(p, \tau_n)$ are fundamental, then the sequence $f(p, t_n - \tau_n)$ is also fundamental.*

Let $\varepsilon > 0$ be given. We define $\delta(\varepsilon/2)$ from the uniform Lyapunov stability of the motion $f(p, t)$; on account of the hypothesis of the lemma there exists N such that for $n \geqq N$ and $m \geqq N$ we have

$$\varrho[f(p, t_n), f(p, t_m)] < \delta, \quad \varrho[f(p, \tau_n), f(p, \tau_m)] < \delta.$$

From the first inequality we get

$$\varrho[f(p, t_n - \tau_n), f(p, t_m - \tau_n)] < \frac{\varepsilon}{2},$$

and from the second by a time displacement of $t_m - \tau_n - \tau_m$

$$\varrho[f(p, t_m - \tau_m), f(p, t_m - \tau_n)] < \frac{\varepsilon}{2};$$

together these inequalities give

$$\varrho[f(p,\ t_n - \tau_n),\ f(p,\ t_m - \tau_m)] < \varepsilon$$

for $n \geqq N$ and $m \geqq N$, which proves the lemma.

We can now define the group operation on the whole of M.

Let $a \in M$, $a = \lim_{n\to\infty} f(p,\ t_n^a)$, $b \in M$, and $b = \lim_{n\to\infty} f(p,\ t_n^b)$. Then, by definition,

$$a + b = \lim_{n\to\infty} f(p,\ t_n^a + t_n^b), \qquad -a = \lim_{n\to\infty} f(p,\ -t_n^a).$$

These limits exist, since by the lemma the sequences

$$f(p,\ -t_n^a) = f(p,\ 0 - t_n^a) \text{ and } f(p,\ t_n^a + t_n^b) = f(p,\ t_n^b - (-t_n^a))$$

are fundamental; they satisfy the group axioms.

We now prove the continuity of these operations.

Let $a = \lim_{n\to\infty} f(p,\ t_n^a)$ and $b = \lim_{n\to\infty} f(p,\ t_n^b)$, and let there be given $\varepsilon > 0$. We define $\delta(\varepsilon/3)$ from the Lyapunov stability. Let

$$\varrho(a,\ a') < \delta/3 \quad \text{and} \quad a' = \lim f(p,\ t_n').$$

There exists $N > 0$ such that for $n \geqq N$ there hold the inequalities

$$\text{(A)} \qquad \varrho[a,\ f(p,\ t_n^a)] < \frac{\delta}{3}, \quad \varrho[a',\ f(p,\ t_n')] < \frac{\delta}{3}.$$

In order to prove the continuity of the inverse element, we recall that $-a = \lim_{n\to\infty} f(p, -t_n^a)$ and subject N to the additional condition that for $n \geqq N$, there hold the inequalities

$$\text{(B)} \quad \varrho[-a,\ f(p,\ -t_n^a)] < \frac{\varepsilon}{3}, \quad \varrho[-a',\ f(p,\ -t_n')] < \frac{\varepsilon}{3}.$$

Then we get

$$\text{(C)} \qquad \varrho[f(p,\ t_n^a),\ f(p,\ t_n')] < \delta$$

from which, on displacing the time by $-t_n^a - t_n'$, follows

$$\varrho[f(p,\ -t_n'),\ f(p,\ -t_n^a)] < \frac{\varepsilon}{3}.$$

Combining this with the inequalities (B) gives

$$\varrho(-a,\ -a') < \varepsilon \quad \text{if} \quad \varrho(a,\ a') < \frac{\delta}{3}.$$

In order to prove the continuity of the sum we subject N to the condition (A) and to the condition that for $n \geqq N$

(D) $\varrho[a+b,\ f(p,\ t_n^a+t_n^b)] < \frac{\varepsilon}{3}, \quad \varrho[a'+b,\ f(p,\ t_n'+t_n^b)] < \frac{\varepsilon}{3}.$

From (C), translated by t_n^b, we get

$$\varrho[f(p,\ t_n^a+t_n^b),\ f(p,\ t_n'+t_n^b)] < \frac{\varepsilon}{3}$$

and combining this with the inequalities (D), we get

$$\varrho[a+b,\ a'+b] < \varepsilon \quad \text{if} \quad \varrho(a,\ a') < \frac{\delta}{3}.$$

The single-valuedness of the operations we have defined follows from the continuity.

We shall prove the sufficiency of the condition. Let G be a compact connected commutative group. There exists a one parameter subgroup which is everywhere dense in G (see footnote 13). We

[13]We give a proof of the above-mentioned theorem which was communicated to us by N. Ya. Vilenkin; it assumes that the reader is familiar with Chapter 5 of the book "Topological Groups" by L. S. Pontrjagin (to be referred to as [P]).

8.18. THEOREM. *Let G be a connected compact group satisfying the second axiom of countability. Then there exists a continuous homomorphic mapping of the additive group L of real numbers with the ordinary topology onto an everywhere dense subgroup of G.*

PROOF. We consider the character group X of G. It follows from Example 49 [P] that X is a countable Abelian group without torsion, and, therefore, there exists an isomorphic mapping of the group X into the group L, $\varphi(X) \subset L$.

Let M be the character group of L; we see that M itself is the additive group of real numbers ([P], § 32, remark H).

We consider a character $m \in M$; denote by $[m,\ \varphi(\chi)]$ the value in K (the group of rotations of the circle) into which m maps $\varphi(\psi) \in$ L. If m is fixed, then $[m,\ \varphi(\chi)]$ is a homomorphic mapping of the character group X into the group K, that is, it is a character of the group X; we denote it by $\psi(m)$ and it follows from the definition that

$$(*) \qquad [\psi(m),\ \chi] = [m,\ \varphi(\chi)].$$

On account of the duality theorem ([P], Theorem 32) $\psi(m)$ is an element of the group G. In this way there is defined a mapping of the group M (the additive group of real numbers) into G. We will show that $\overline{\psi(M)} = G$.

Assuming the contrary, we denote by θ the annihilator of the subgroup $\overline{\psi(M)}$, i.e. the set of characters of G, mapping $\overline{\psi(M)}$ into zero. By Theorem 33 [P], θ contains an element a different from zero. But, on account of the relation (*), we have

$$[m,\ \varphi(a)] = [\psi(m),\ a] = [a,\ \psi(m)] = 0$$

for any $m \in M$ since $a \in \theta$. From this it follows that $\varphi(a) = 0$ for $a \neq 0$. But this contradicts the assumption that the mapping of X into L is an isomorphism.

The theorem is proved.

denote it by $\{\xi_t\}$; ξ_0 is the zero of the group, and $\xi_{t_1} + \xi_{t_2} = \xi_{t_1+t_2}$.

Let $p \in G$; we define the dynamical system by the equation

$$f(p, t) = p + \xi_t.$$

The conditions for a dynamical system are fulfilled: $f(p, 0) = p$,

$$f[f(p, t_1), t_2] = (p + \xi_{t_1}) + \xi_{t_2} = p + \xi_{t_1+t_2} = f(p, t_1 + t_2);$$

continuity with respect to p and t follows from the continuity of the group operations. We note that because of the compactness of the group this continuity takes place uniformly, therefore, for $\varepsilon > 0$ there exists $\delta > 0$ such that if $\varrho(p, q) < \delta$, then $\varrho(p + \xi_t, q + \xi_t) < \varepsilon$, or, going over to the dynamical system notation, $\varrho[f(p, t), f(q, t)] < \varepsilon$ for $-\infty < t < +\infty$. In this way uniform Lyapunov stability has been established. Let the zero of the group $\xi_0 = p_0$. The trajectory $f(p_0, t) = \xi_t$ is, by the hypothesis, everywhere dense in G. On account of Theorem 8.12 the motion $f(p, t)$ is almost periodic, and the set G is the minimal set $\overline{f(p; I)}$. The theorem is proved.

Theorem 8.16 gives the structure of the minimal sets on which almost periodic motion is possible.

There arises the question of whether a motion defined on a group space and having this space as a minimal set consists of almost periodic motions. A. A. Markov gave a negative answer to this question. Following Markov's idea we construct an example of a non-almost periodic motion taking place on the trajectories of Example 8.13.

8.19. Example. The motion of the dynamical system is given by the equations

$$\frac{d\varphi}{dt} = \frac{1}{\Phi(\varphi, \theta)}, \quad \frac{d\theta}{dt} = \frac{\mu}{\Phi(\varphi, \theta)}.$$

Here μ is an irrational number which we will define later. $\Phi(\varphi, \theta)$

We will investigate when the mapping $\psi(M)$ is an isomorphism. Assume that $\psi(m) = \psi(m')$ for $m \neq m'$, i.e. $[m, \varphi(\chi)] = [m', \varphi(\chi)]$ for all $\chi \in X$, or $m\varphi(\chi) = m'\varphi(\chi)$ (mod 1). Denoting $m - m'$ by $\alpha \neq 0$, we have $\alpha\varphi(\chi) = 0$ (mod 1). This is only possible if $\varphi(\chi) = n/\alpha$, $n = 0, \pm 1, \pm 2, \ldots$, that is, if the character group is cyclic with one generator. In this case $G = K$ and $\psi(M) = G$ (we have a periodic motion). In all the remaining cases (almost periodic motions) $\psi(M)$ is isomorphic to M.

is a periodic function of period 2π, nonvanishing, and expandable in a uniformly convergent double Fourier series, which we will write in the complex form for simplicity:

$$\Phi = \sum_{m,\, n=-\infty}^{+\infty}{}' a_{mn} e^{im\varphi} e^{in\theta}, \quad a_{-m,\, n} = a_{m,\, -n} = \overline{a_{mn}}.$$

For definiteness we put

$$a_{mn} = \frac{1}{(|m|+1)^2(|n|+1)^2} \text{ for } |m|+|n|>0;\ a_{00} > \sum{}' a_{mn} = \left(1+\frac{\pi^2}{3}\right)^2 - 1$$

(the symbol $\sum{}'$ in the following means that the term $m = n = 0$ is omitted.)

The equations of the trajectory are determined by the differential equation $d\theta/d\varphi = \mu$ and give $\theta = \theta_0 + \mu\varphi$. The time on the trajectory beginning at the point $\varphi = 0$, $\theta = \theta_0$, is determined by the formula

$$t(\varphi,\ \theta_0) = \int_0^{\varphi} \Phi(\tau,\ \mu\tau + \theta_0) d\tau.$$

The function under the integral sign is almost periodic in τ. We will use the following facts from the theory of almost periodic functions: if the indefinite integral of an almost periodic function is almost periodic, then its Fourier series may be obtained by formal integration of the series of the function under the integral sign; if the indefinite integral is not almost periodic, then it is unbounded (Bohr, *Almost Periodic Functions*, Chelsea Pub. Co. pp. 56—60); the sum of the squares of the moduli of the Fourier coefficients of an almost periodic function converges (Bohr, p. 52). We form the difference

$$(8.20) \quad t(\varphi,\ \theta_0) - t(\varphi,\ 0) = \int_0^{\varphi} [\Phi(\tau,\ \mu\tau + \theta_0) - \Phi(\tau,\ \mu\tau)] d\tau.$$

Putting the series for the function Φ in the expression under the integral sign and integrating, we get:

$$(8.21) \quad t(\varphi,\ \theta_0) - t(\varphi,\ 0) = \sum_{m,\, n=-\infty}^{\infty} \frac{a_{mn}(e^{in\theta_0} - 1)}{i(m + n\mu)} e^{i(m+n\mu)\varphi}.$$

The sum of the squares of the moduli of the coefficients of the series on the right-hand side is

$$\sum_{m,\,n=-\infty}^{\infty}{}' \sin^2 \frac{n\theta_0}{2} \cdot \frac{|a_{mn}|^2}{(m+n\mu)^2}. \tag{8.22}$$

We shall choose the number μ such that the series

$$\sum_{m,\,n=-\infty}^{\infty}{}' \frac{|a_{mn}|^2}{(m+n\mu)^2} \tag{8.23}$$

diverges. Putting, for definiteness, $\mu > 0$, it is sufficient to determine it such that for an infinite set of pairs (m, n) of integers, the inequality $|m - n\mu| < |a_{mn}|$ is satisfied. This is always possible. We shall prove the existence of such a set for the above-mentioned choice of the coefficients a_{mn}. Express μ as the infinite continued fraction

$$\mu = 0, \qquad (a_1 a_2 a_3 \ldots).$$

Denoting the ν-th convergent by p_ν/q_ν, we have the known inequality:

$$\left|\mu - \frac{p_\nu}{q_\nu}\right| < \frac{1}{q_\nu q_{\nu+1}} < \frac{1}{a_\nu q_\nu^2}$$

from which follows $|p_\nu - q_\nu \mu| < 1/a_\nu q_\nu$. We determine the non-integral parts by the relations: $a_1 = 1$, and $a_\nu = (q_\nu + 1)^3$; then recalling that $p_\nu \leqq q_\nu - 1$ for $\nu > 2$, we have

$$|p_\nu - q_\nu \mu| < \frac{1}{q_\nu (q_\nu + 1)^3} < \frac{1}{(p_\nu + 1)^2 (q_\nu + 1)^2} = a_{p_\nu q_\nu}$$

Thus, for every pair (p_ν, q_ν) the term of the series (8.23) is greater than 1, that is, the series diverges. We now consider the series (8.22); it can only converge for θ_0 in a set E of measure zero. Otherwise, there would be a set $E_0 \subset E$ with mes $E_0 > 0$; integrating over E_0, we would get the convergence of the series (8.23). Thus, for almost all values of θ_0 the function (8.21) is not almost periodic.

Hence, there are arbitrarily small values of θ_0 for which the difference (8.20) is an unbounded function of φ. From this it follows that the motion of the system (8.20) with the initial data: $t = 0$,

$\varphi = 0$, and $\theta = 0$, is not Lyapunov stable. Indeed, for $\varphi = \varphi_0$ let there hold the inequality:

$$|t(\varphi_0, \theta_0) - t(\varphi_0, 0)| > m \geqq 0$$

where m is an arbitrary positive number. Then, at the moment $t(\varphi_0, 0)$ the value of the φ-coordinate on the trajectory through $(0, \theta_0)$ differs from the value of φ_0 on the trajectory through $(0, 0)$ by more that m, and this is true for arbitrarily small values of θ_0.

Thus, the motion described by the system (8.20) is not almost periodic.

In conclusion, we make the following observation. Theorems 7.06 and 7.07 and Theorem 8.16 show that though the properties of recurrence and almost periodicity have not been formulated in a topologically invariant manner, nevertheless, the closures of recurrent and almost periodic motions have invariant characteristics. These invariant characteristics have one essential peculiarity. The property of recurrence — the property that the closure of the trajectory is minimal — is quite independent of the character of the dependence on time. In fact, every dynamical system having the same trajectories as one whose trajectories form a minimal set, will also consist of recurrent motions. Now in case of almost periodic motion in compact sets we have a group space; however, we cannot assert that every dynamical system which has the same trajectories in the group space thus obtained, will consist of almost periodic motions.

This circumstance led Markov to make the following definition. We will call a dynamical system *harmonizable* if its minimal sets are compact, and if it is possible to change the time in such a way that all the motions become almost periodic.

This definition, in turn, raises the question: Is every compact minimal set harmonizable? Example 8.14 gives, as is easily shown, a negative answer to this question. Moreover, Markov has constructed an example of a locally connected minimal set which is also a manifold and at the same time is not a group space.

9. Asymptotic Trajectories

The following definition of an asymptotic trajectory is due to V. V. Nemickii.

9.01. Definition. A trajectory $f(p, I)$ is called *positively asymptotic* if the set Ω_p is not empty but the intersection $f(p, I^+)\cdot \Omega_p$ is empty.

Negative asymptoticity is defined similarly.

In a compact dynamical system every trajectory $f(p, I)$ is stable Lagrange, so that no set Ω_p is empty. There are then just two possibilities. Either $f(p, I^+) \cdot \Omega_p = 0$, in which case $f(p, I)$ is positively asymptotic; or $f(p, I^+) \cdot \Omega_p \neq 0$, in which case (according to Theorem 4.03) $f(p, I)$ is stable P^+. Thus we have the following alternative.

9.02. Theorem 1. *In a compact dynamical system every trajectory is either positively asymptotic or positively stable in the sense of Poisson.*

9.03. Corollary. *In a compact dynamical system every trajectory approaches its ω-limit set uniformly as $t \to +\infty$.*

By this we mean that for each p and for every $\varepsilon > 0$ one can find a $t_0(p, \varepsilon)$ such that for all $t > t_0$, $\varrho[f(p, t), \Omega_p] < \varepsilon$. For an asymptotic trajectory which is stable L^+ the uniform convergence of $f(p, t)$ to Ω_p was established in Theorem 3.07. For Poisson stable trajectories the corollary is trivially true since then $f(p, I) \subset \Omega_p$.

In certain cases the character of the motion in the ω-limit set of an asymptotic trajectory is determined by supplementary conditions referring to the nature of the approach of the trajectory to its limit set.

9.04. Definition. The half-trajectory $f(p, I^+)$ *uniformly approximates the set* Ω_p if, given any $\varepsilon > 0$, there corresponds a $T(\varepsilon) > 0$ such that if $L \subset f(p, I^+)$ is an arbitrary arc of length greater than or equal to $T(\varepsilon)$ and q is an arbitrary point of Ω_p, then $\varrho(q, L) < \varepsilon$.

9.05. Theorem (Nemickii). *Let $f(p, t)$ be stable L^+. Then in order that the set Ω_p of ω-limit points of $f(p, t)$ be a minimal set, it is necessary and sufficient that $f(p, I^+)$ uniformly approximate Ω_p.*

Sufficiency: Let q be any point in Ω_p. It is sufficient to show that $f(q, I)$ is dense in Ω_p. Assume the contrary; that is, let $r \in \Omega_p$ be such that $\varrho[r, \overline{f(q, I)}] = \alpha > 0$. Next, obtain the number $T(\alpha/2)$ from the hypothesis of uniform approximation. Further, from the continuity of the dynamical system the point q and the numbers $\alpha/2$ and $T(\alpha/2)$ determine an $\eta > 0$ with the property that $\varrho(q, s) < \eta$ implies

$$\varrho[f(q, t), f(s, t)] < \frac{\alpha}{2}$$

for $0 \leqq t \leqq T(\alpha/2)$. Since $q \in \Omega_p$, there is on $f(p, I^+)$ a point $p_1 = f(p, t_1)$, $t_1 > 0$, for which $\varrho(p_1, q) < \eta$. Because the arc $f(p; t_1, t_1 + T)$ approximates Ω_p to within $\alpha/2$, there is a t_2, $0 \leqq t_2 \leqq T$, such that on putting $p_2 = f(p_1, t_2) = f(p, t_1 + t_2)$ we have $\varrho(r, p_2) < \alpha/2$. Let finally $q' = f(q, t_2)$; then

$$\varrho(q', p_2) = \varrho[f(q, t_2), f(p_1, t_2)] < \frac{\alpha}{2}.$$

From these inequalities there follows $\varrho(q', r) < \alpha$, which contradicts the choice of r.

Necessity: Suppose that $f(p, I^+)$ does not uniformly approximate the minimal set Ω_p. Then there exists an $\alpha > 0$, a sequence of intervals (t_1, t_1'), (t_2, t_2'), ..., (t_n, t_n'), ..., $t_n > 0$, $t_n' - t_n \to \infty$, and a sequence of points $\{q_n\} \subset \Omega_p$ such that

$$\varrho[f(p; t_n, t_n'), q_n] > \alpha.$$

Since Ω_p is compact, $\{q_n\}$ has at least one limit point q_0. In order not to complicate the notation, suppose that $q_0 = \lim_{n\to\infty} q_n$ and that $\varrho(q_n, q_0) < \alpha/3$ for all n. Then for arbitrary n

$$\varrho[f(p; t_n, t_n'), q_0] > \frac{2\alpha}{3}.$$

Consider the sequence of points

$$p_n = f\left(p, t_n + \frac{t_n' - t_n}{2}\right) = f(p, \tau_n), \qquad (n = 1, 2, \ldots).$$

Clearly $\tau_n \to \infty$. Since $f(p, t)$ is stable L^+ the sequence $\{p_n\}$ has a limit point $p_0 \in \Omega_p$. For simplicity, we again assume that $p_0 = \lim_{n\to\infty} p_n$.

Now $f(p_0, I)$ is dense in Ω_p, since by hypothesis Ω_p is minimal. Therefore there exists a τ_0 such that $\varrho[f(p_0, \tau_0), q_0] < \alpha/3$. From this we shall obtain a contradiction. Choose $\eta > 0$ from continuity in such a way that $\varrho(p_0, r) < \eta$ implies $\varrho[f(p_0, \tau_0), f(r, \tau_0)] < \alpha/3$. Choose N so large that $\varrho(p_0, p_N) < \eta$ and $\frac{1}{2}(t_N' - t_N) > \tau_0$. From the second condition there follows

$$f(p_N, \tau_0) = f(p, t_N + \tau_0) \in f(p; t_N, t_N'),$$

and hence

$$\varrho[f(p_N, \tau_0), q_0] > \frac{2\alpha}{3}.$$

On the other hand, $\varrho[f(p_N, \tau_0), f(p_0, \tau_0)] < \alpha/3$ and

$$\varrho[f(p_N, \tau_0), q_0] \leqq \varrho[f(p_N, \tau_0), f(p_0, \tau_0)] + \varrho[f(p_0, \tau_0), q_0] < \frac{2\alpha}{3}.$$

This contradiction proves the theorem.

As a special case of minimal sets we have the closures of almost periodic motions. By adding a supplementary condition to Theorem 9.05 we can obtain an asymptotic motion whose ω-limit set is a minimal set of almost periodic motions.

9.06. THEOREM. *Let $f(p, t)$ be stable L^+. In order that Ω_p be a minimal set of almost periodic motions it is sufficient that $f(p, I^+)$ uniformly approximate Ω_p and that $f(p, t)$ be uniformly positively Lyapunov stable relative to $f(p, I^+)$.*

According to Theorem 9.05, Ω_p is minimal. We must prove that the motions in Ω_p are almost periodic.

Given $\varepsilon > 0$, let $\delta = \delta(\varepsilon/3)$ be determined by the uniform Lyapunov stability in such a way that if for some $t_1 > 0$, $t_2 > 0$, $\varrho[f(p, t_1), f(p, t_2)] < \delta$, then

$$\varrho[f(p, t_1 + t), f(p, t_2 + t)] < \frac{\varepsilon}{3}$$

for all $t > 0$.

Let $q_1 \in \Omega_p$, $q_2 \in \Omega_p$ be such that $\varrho(q_1, q_2) < \delta/3$. Let $\bar{t} > 0$ be an arbitrary number. We wish to estimate $\varrho[f(q_1, \bar{t}), f(q_2, \bar{t})]$.

By continuity, we choose an η such that from $\varrho(q_1, r) < \eta$, $\varrho(q_2, s) < \eta$ there follows

$$\varrho[f(q_1, \bar{t}), f(r, \bar{t})] < \frac{\varepsilon}{3}, \quad \varrho[f(q_2, \bar{t}), f(s, \bar{t})] < \frac{\varepsilon}{3}.$$

Let $\sigma = \min[\eta, \delta/3]$. From the definition of Ω_p it follows that there exist a $p_1 = f(p, t_1)$ and a $p_2 = f(p, t_2)$, where $t_1 > 0$, $t_2 > 0$, for which

$$\varrho(p_1, q_1) < \sigma, \quad \varrho(p_2, q_2) < \sigma.$$

Then

$$\varrho(p_1, p_2) < \varrho(p_1, q_1) + \varrho(q_1, q_2) + \varrho(q_2, p_2) < \sigma + \sigma + \frac{\delta}{3} < \delta.$$

By the choice of δ and the Lyapunov stability we have

$$\varrho[f(p_1, \bar{t}), f(p_2, \bar{t})] < \frac{\varepsilon}{3}.$$

From the continuity and the choice of p_1, p_2 we have

$$\varrho[f(q_1, \bar{t}), f(p_1, \bar{t})] < \frac{\varepsilon}{3}, \varrho[f(q_2, \bar{t}), f(p_2, \bar{t})] < \frac{\varepsilon}{3}.$$

The last three inequalities now yield

$$\varrho[f(q_1, \bar{t}), f(q_2, \bar{t})] < \varepsilon$$

for arbitrary q_1, q_2 such that $\varrho(q_1, q_2) < \delta/3$ and for arbitrary $\bar{t} > 0$. This proves that the motions in Ω_p are positively Lyapunov stable.

Now Birkhoff's theorems prove that the motions in Ω_p are all recurrent. From Theorem 8.11 it follows that they must in fact be almost periodic.

What conditions on $f(p, t)$ are necessary in order that the motions in Ω_p be almost periodic remains an open question.

10. Completely Unstable Dynamical Systems

From the preceding sections it is apparent that the general theory of dynamical systems has received its greatest development in the direction of the investigation of systems stable according to Lagrange.

Nemickii has studied the motions of a class of dynamical systems whose properties are just the opposite of those for stable systems; these are the *completely unstable* systems. In Nemickii's work systems of such a type are considered in a space E_n, where they are given by the system of differential equations

$$\frac{dx_i}{dt} = X_i(x_1, x_2, \ldots, x_n), \qquad (i = 1, 2, \ldots, n),$$

whose right-hand sides are defined for all values of the variables and satisfy conditions of uniqueness. Later these results were generalized by M. V. Bebutov to general dynamical systems M defined in a locally compact metric space R. This required the introduction of a new auxiliary apparatus — Bebutov's theory of tubes and sections, which was presented in section 2.

Let a dynamical system M be given in a locally compact metric space R. We recall that a motion is called positively (negatively) stable according to Lagrange if its half-trajectory $f(p;\ 0, +\infty)$ or $f(p;\ 0, -\infty)$ respectively lies in a compact set of the space R. A motion is *unstable according to Lagrange* if it is neither positively nor negatively stable according to Lagrange. If all the motions $f(p, t)$ of the system M are unstable according to Lagrange, we shall then call this system *unstable*. We introduce a new definition.

10.01. DEFINITION. The system M is called *completely unstable* if all its points are wandering.

We recall (section 5) that a point p is called *wandering* if there exist a $\delta > 0$ and a $T > 0$ such that for $|t| \geqq T$ we have

$$S(p, \delta) \cdot f(S(p, \delta), t) = 0.$$

If a system is completely unstable, then it is unstable.

Indeed, if some motion $f(p_0, t)$, for example, were stable L^+ then the set Ω_{p_0} of its ω-limit points would be nonempty. Every point $q \in \Omega_{p_0}$ is nonwandering. In fact, for any $\varepsilon > 0$ consider $S(q, \varepsilon)$ and suppose that $f(p_0, t_0) \in S(q, \varepsilon)$. According to the definition of an ω-limit point, for any $T > 0$ there can be found a $t > T$ such that $f(p_0, t_0 + t) \in S(q, \varepsilon)$, but $f(p_0, t_0 + t) \in f(S(q, \varepsilon), t)$. Consequently, the intersection $S(q, \varepsilon) \cdot f(S(q, \varepsilon), t)$ is nonempty, i.e. q is a nonwandering point. Thus the system M is not completely unstable.

In the space E^2 the converse proposition is also valid.

10.02. THEOREM. *In the plane every unstable system defined by differential equations is completely unstable.*

Let it be given that all motions $f(p, t)$ of a dynamical system, defined in E^2, are unstable and assume that a point p_0 is nonwandering. By hypothesis, p_0 is not a rest point. As in lemma 1.12 of Chapter II, we construct at the point p_0 the normal ap_0b to the trajectory $f(p_0, t)$ of length 2ε with center at p_0, wherein ε is chosen such that in the circle $\overline{S(p_0, 4\varepsilon)}$ the direction of the field (X_1, X_2) differs from the direction of the tangent at the point p_0 by not more than $\pi/4$ and such that in this circle there hold the inequalities

$$\frac{2}{3}\,[X_1^2(p_0) + X_2^2(p_0)] \leqq X_1^2(p) + X_2^2(p) \leqq \frac{3}{2}\,[X_1^2(p_0) + X_2^2(p_0)].$$

Then, in the first place, every trajectory $f(p, t)$ having a point inside $S(p_0, \varepsilon/\sqrt{2})$ will intersect ap_0b in a point p' and the arc $pp' \subset S(p_0, \varepsilon)$. Secondly, there exists a $t_0 > 0$ such that for $t = t_0$

$$f(S(p_0, \varepsilon), t) \cdot S(p_0, \varepsilon) = 0;$$

for example,

$$t_0 = \frac{\sqrt{6}\,\varepsilon}{[X_1^2(p_0) + X_2^2(p_0)]^{1/2}},$$

since in the course of this interval all points lying in $S(p_0, \varepsilon)$ pass out of it, but do not leave $S(p_0, 4\varepsilon)$. By assumption there can be found a $t_1 > t_0$ such that

$$S\left(p_0, \frac{\varepsilon}{\sqrt{2}}\right) \cdot f\left(S\left(p_0, \frac{\varepsilon}{\sqrt{2}}\right), t_1\right) \neq 0,$$

and this means that there can be found a $p \in S(p_0, \varepsilon/\sqrt{2})$ such that $f(p, t_1) \in S(p_0, \varepsilon/\sqrt{2})$. Let the arc $f(p, t)$, while not leaving $S(p_0, \varepsilon)$, intersect the normal ap_0b in the point q and the arc $f(p, t_1 + t)$, $t > 0$, until departing from $S(p_0, \varepsilon)$, intersect ap_0b in the point q_1. Consider the domain D bounded by the arc of the trajectory $f(p, t)$ from q to q_1 and the segment qq_1 of the normal. If points lying on the arc qq_1 enter the domain D for an increase in t, then such a point, as $t \to +\infty$, cannot leave D and therefore the corresponding motion is positively stable according to Lagrange. If these points enter D for a decrease in t, then the motions corresponding to them are stable L^-. The contradiction proves the theorem.

But in E^3 there can exist unstable motions which are not completely unstable. We present an example.

10.03. Example. In the plane xOy we consider a one-parameter family of spirals of Archimedes, filling the whole plane except the origin, whose equations in polar coordinates are: $\varrho = \theta - a$, $\theta > a$ (a is a parameter, $0 \leqq a < 2\pi$).

Next we take that portion of the surface of revolution $z = (x^2 + y^2 - 1)^2/(x^2 + y^2)$ which projects onto the circle $x^2 + y^2 \leqq 1$. This surface, obviously, lies above the plane xOy, is tangent to it along the circle $x^2 + y^2 = 1$, and $z \to \infty$ when $\varrho = [x^2 + y^2]^{1/2} \to 0$. We project parallel to the z-axis onto this piece of the surface the parts of the spirals of the family lying in the circle $x^2 + y^2 \leqq 1$; outside the circle $x^2 + y^2 = 1$ the respective curves are extended as plane spirals. The curves so obtained have everywhere a continuous tangent and along them $z \to \infty$ when

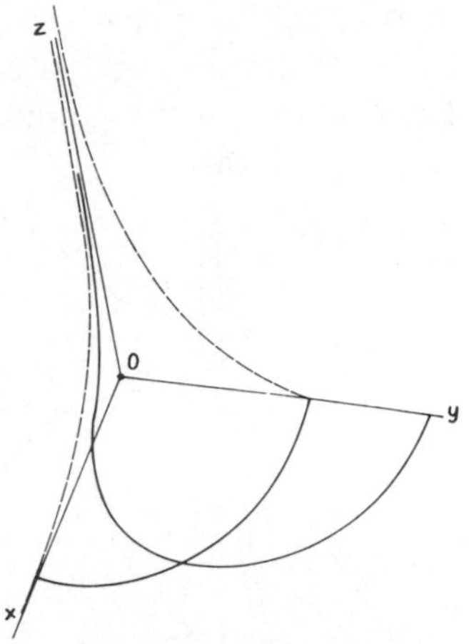

Fig. 31

$\varrho \to 0$ (fig. 31). Finally, we subject this system of curves to parallel translations along the z-axis with parameter b.

The two-parameter family of curves so obtained can be represented by the equations

$$x = (\theta - a) \cos \theta, \quad y = (\theta - a) \sin \theta, \quad z = f(\theta - a) + b,$$

where $f(\alpha) = (\alpha^2 - 1)^2/\alpha^2$ for $0 < \alpha \leqq 1$ and $f(\alpha) = 0$ for $\alpha \geqq 1$; moreover, $0 \leqq a < 2\pi$, $-\infty < b < +\infty$. We add to it the straight line $x = y = 0$. Through each point of space there passes a unique curve of the family and the tangents to the curves form a continuous field of directions.

We subject the space $Oxyz$ to the tranformation

$$x_1 = e^x - 2, \quad y_1 = y, \quad z_1 = z,$$

under which it is mapped in a one-one way on the half-space $x_1 > -2$. At the same time the vector field remains continuous for $x_1 > -2$; in the plane $x_1 = -2$ the transformed field of directions continuously approaches the field given by the straight lines $z_1 = b$. In the half-space $x_1 < -2$ we complete the transformed family by straight lines parallel to the y_1-axis (fig. 32.) Along the curves so obtained, which fill the entire space $Ox_1y_1z_1$, we define a dynamical system, for example, as the system of motions with a constant velocity $ds/dt = 1$; e.g., for $x_1 > -2$, $z_1(t) \to +\infty$ when $t \to -\infty$.

As $t \to -\infty$ the motion taking place along a trajectory transformed from a spiral falls, beginning with a certain t, on the plane $z_1 = b$ and has as α-limit points all the points of the line $x_1 = -2$,

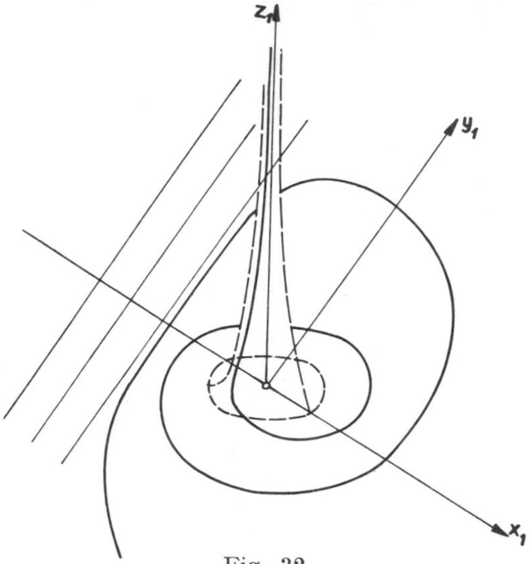

Fig. 32

$z_1 = b$; all these points, therefore, are nonwandering, i.e. the system is not completely unstable. Moreover, all the motions are unstable according to Lagrange in both a positive and negative direction.

The simplest example of a completely unstable system is the system of motions taking place along a family of parallel lines. Our chief purpose will be the establishing of necessary and sufficient conditions that there should exist a topological transformation of the space R into a separable Hilbert space E^∞ (in the case where $R = E^n$, a transformation into E^{n+1}) under which the trajectories of a given system map into a family of parallel lines (infinite in both directions).

10.04. THEOREM. *The complete instability of a dynamical system given in a locally compact metric space R is a necessary condition for the possibility of a one-one bi-continuous mapping of its trajectories onto a family of parallel lines in E^∞ (or in E^m, where $m > n$, if $R = E^n$).*

Let $x = \Phi(p)$, where $p \in R$ and $x = (\xi_0, \xi_1, \ldots, \xi_n, \ldots)$,

$\sum_{n=0}^{\infty} \xi_n^2 < +\infty$, be a homeomorphic mapping of the space R onto a subset $X \subset E^\infty$, under which the trajectories $f(p, t)$ pass into the lines $\xi_1 = c_1, \xi_2 = c_2, \ldots, \xi_n = c_n, \ldots, (\sum_{n=1}^{\infty} c_n^2 < +\infty)$, wherein the variable coordinate ξ_0 varies monotonically for increasing t; we assume for definiteness that ξ_0 increases with t. Then in the set X a dynamical system is defined by the motions $f_1(p, t) = \Phi(f(p, t))$. Suppose that $p \in R$ is an arbitrary point; we take the closure, $\overline{S(p, \varepsilon)}$ of its compact neighborhood. The image of this latter, $\Phi(\overline{S(p, \varepsilon)})$, is a closed, compact set in E^∞. Because of the compactness of the set $\Phi(\overline{S(p, \varepsilon)})$ and the absence of rest points in the system $f_1(p, t) = f_1(\Phi(p), t)$, there can be found a $T > 0$ such that for $t > T$ we have

$$0 = \Phi(\overline{S(p, \varepsilon)}) \cdot f_1(\Phi(\overline{S(p, \varepsilon)}, t) = \Phi(\overline{S(p, \varepsilon)}) \cdot \Phi[f(\overline{S(p, \varepsilon)}, t)].$$

Passing to the inverse transformation Φ^{-1} in the space R, we obtain $\overline{S(p, \varepsilon)} \cdot f(\overline{S(p, \varepsilon)}, t) = 0$ for $t > T$, i.e., p is necessarily a wandering point. The theorem is proved.

However, complete instability, as it will be shown in the sequel, is still insufficient for the possibility of a mapping of a system onto a family of parallel lines. For an explanation of this matter, we introduce a definition.

10.05. Definition. A dynamical system has an *improper saddle point* ("col") if there exist a sequence of points $\{p_n\}$ and sequences of unboundedly increasing numbers $\{\tau_n\}$ and $\{t_n\}$ such that $p_n \to p$, $f(p_n, t_n) \to q$, $0 < \tau_n < t_n$, and $\{f(p_n, \tau_n)\}$ contains no convergent sequence.

In the case where $R = E_n$ this definition reduces to that introduced by Nemickii for a "*saddle at infinity*".

10.06. Lemma. *If a system is unstable and has no improper saddle point and if $p_n \to p$, $q_n = f(p_n, t_n) \to q$, then $\{t_n\}$ is bounded.*

We let $A_n = f(p_n; 0, t_n)$ and $A = \sum_{n=1}^{\infty} A_n$. We shall show that A is compact in R. Assume the contrary; then there exists a sequence $\{q_k\} \subset A$ containing no convergent subsequence. Since each A_n is compact it can contain only a finite number of points of $\{q_k\}$. Consequently, there exist two unbounded, increasing sequences of natural numbers $\{n_k\}$ and $\{l_k\}$ such that

$$q_{n_k} \in A_{l_k}, \quad \text{i.e., } q_{n_k} = f(p_{l_k}, \tau_{l_k}), \quad 0 < \tau_{l_k} < t_{l_k}.$$

But then we have

$$p_{l_k} \to p, \ f(p_{l_k}, t_{l_k}) \to q,$$

and $f(p_{l_k}, \tau_{l_k})$ contains no convergent sequence, i.e., contrary to supposition, the system has an improper saddle point. Thus A is compact in R and therefore $\bar{A}$ is compact in itself.

Assume now that $\{t_n\}$ is unbounded; then without loss of generality it may be assumed that $t_n \to +\infty$. Let $t \geqq 0$ be an arbitrary number; we choose N such that $t_n > t$ for $n \geqq N$. Then for $n \geqq N$ we have $f(p_n, t) \in A$, and because $f(p_n, t) \to f(p, t)$ we obtain

$$f(p, t) \in \bar{A} \quad \text{for any} \quad t \geqq 0,$$

i.e. the motion $f(p, t)$ is positively stable according to Lagrange, which contradicts the instability of the system.

10.07. Corollary. *Under the conditions of Lemma* 10.06 *there hold the relations*

$$(10.08) \qquad t_n \to t_0 \quad \text{and} \quad q = f(p, t_0).$$

Assume that the sequence $\{t_n\}$ does not converge; then, because of its boundedness, there can be found two subsequences $\{t_{n_k}\}$ and $\{t_{n_l}\}$ such that $\lim_{k\to\infty} t_{n_k} = t'$, $\lim_{l\to\infty} t_{n_l} = t''$, $t' \neq t''$. Then we would have

$$\lim_{k\to\infty} f(p_{n_k}, t_{n_k}) = \lim_{k\to\infty} q_{n_k} = f(p, t') = q;$$
$$\lim_{l\to\infty} f(p_{n_l}, t_{n_l}) = \lim_{l\to\infty} q_{n_l} = f(p, t'') = q,$$

i.e.

$$f(p, t') = f(p, t''),$$

which is impossible, since an unstable system does not contain periodic motions. Consequently $t_n \to t_0$ and $q = f(p, t_0)$.

The following theorem establishes the connection between instability, complete instability, and an improper saddle point.

10.09. Theorem. *An unstable system without an improper saddle point is completely unstable.*

Assume the contrary; let an unstable system without an improper saddle point contain some nonwandering point p. We assign two sequences of positive numbers

$$T_1 < T_2 < \ldots < T_n < \ldots, \quad T_n \to +\infty,$$
$$\varepsilon_1 > \varepsilon_2 > \ldots > \varepsilon_n > \ldots, \quad \varepsilon_n \to 0.$$

Then from the definition of a nonwandering point there follows the existence for each n of a p_n and a t_n such that

$$t_n > T_n, \quad \varrho(p_n, p) < \varepsilon_n, \quad \varrho(p, f(p_n, t_n)) < \varepsilon_n.$$

From this we obtain

$$p_n \to p, \quad f(p_n, t_n) \to p, \quad t_n \to +\infty,$$

which contradicts lemma 10.06.

On the other hand, a completely unstable dynamical system can have an improper saddle point as the following example shows.

10.10. Example. The system defined for $-\infty < x < +\infty$, $-\infty < y < +\infty$ by the differential equations

$$\frac{dx}{dt} = \sin y, \qquad \frac{dy}{dt} = \cos^2 y,$$

has as trajectories the curves $x + C = 1/\cos y$ and the lines $y = k\pi + (\pi/2)$, $k = 0, \pm 1, \ldots$ We confine our consideration to the strip R: $-\pi/2 < y < \pi/2$ (fig. **33**); we construct the sphere $\overline{S(0, N)}$ (a compact set). Obviously $R = \sum_{N=1}^{\infty} S(0, N)$.

Whatever the value of N, the arc containing the points $p_n(0, -(\pi/2) + \alpha_n)$ and $q_n(0, (\pi/2) - \alpha_n)$ passes beyond the boundary of $S(0, N)$ if one chooses $\alpha_n < \arcsin 1/(N + 1)$. Thus $p_n \to (0, -\pi/2)$, $q_n \to (0, \pi/2)$, and therefore this system has an improper saddle point (saddle at infinity). Moreover, this is a completely unstable system. For a proof it is sufficient to recall that from regional recurrence there follows (Theorem 5.10) the existence of motions stable according to Poisson, motions which in the present case obviously do not exist.

The system in example 10.10 cannot be mapped onto a family of parallel lines. There holds the general fact:

10.11 Theorem. *If a completely unstable dynamical system has an improper saddle point, then it cannot be mapped onto a family of parallel lines.*

We note at the start that if p_n and q_n lie on the same trajectory and $p_n \to p$ and $q_n \to q$ and if the system can be mapped topologically onto a family of parallel lines in $E^\infty\{\xi_0, \xi_1, \ldots, \xi_k, \ldots\}$ (or in E^m), then p and q lie on the same trajectory.

In fact, let the mapping under discussion be Φ. Then, because of its continuity, $\Phi(p_n) \to \Phi(p)$. Through $\Phi(p_n)$ and $\Phi(p)$ there pass lines of the family. Let these be $\xi_1 = \xi_1^{(n)}$, $\xi_2 = \xi_2^{(n)}, \ldots$ and $\xi_1 = \xi_1^{(0)}$, $\xi_2 = \xi_2^{(0)}, \ldots$ respectively. According to the hypothesis of the theorem, $\Phi(q_n)$ lies on the line passing through $\Phi(p_n)$, $(\xi_i = \xi_i^{(n)})$; consequently the limit point $\Phi(q)$ lies on the limit line passing through the point $\Phi(p)$, i.e. on the line $\xi_i = \xi_i^{(0)}$. But if

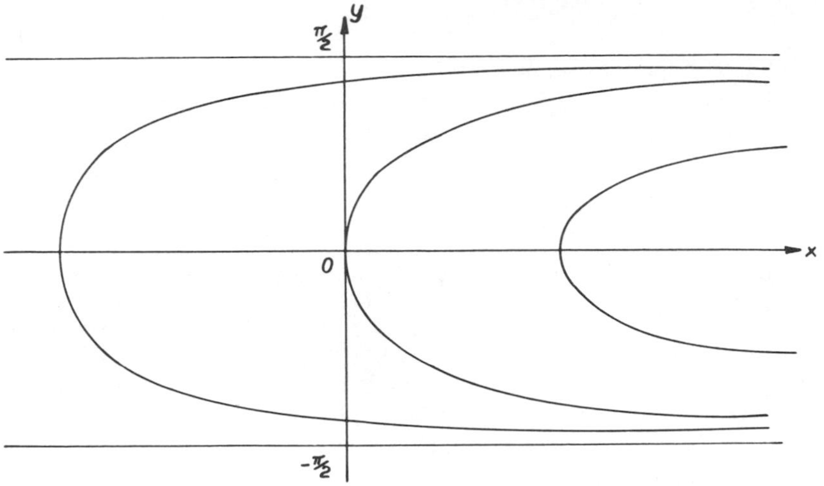

Fig. 33

$\Phi(p)$ and $\Phi(q)$ lie on the same line, then from the one-one property of the mapping Φ the points p and q lie on the same trajectory.

We proceed to the proof of the theorem. The space R is locally compact. As in theorem 1.18 of the present chapter, let

$$R = \sum_{n=1}^{\infty} F_n,$$

where $F_1 \subset F_2 \subset \ldots$ is an increasing sequence of closed, compact sets. From the definition of an improper saddle point there can be found a sequence of pairs of points $\{p_k, q_k\}$ $(k = 1, 2, \ldots;$ $q_k = f(p_k, t_k)$, $t_k > 0)$ and a sequence of numbers $\tau_k (0 < \tau_k < t_k)$ such that $\lim_{k\to\infty} p_k = p'$, $\lim_{k\to\infty} q_k = q'$, $f(p_k, \tau_k) \in R - F_k$.

We shall show that p' and q' do not lie on the same trajectory. We show at the start that $\lim_{k\to\infty} t_k = \infty$. Assume the contrary; let there exist a $T > 0$ such that $0 < t_k \leqq T$, $k = 1, 2, \ldots$. Then the set $P = \{\overline{f(p_k;\ 0,\ T)}\}$ is compact since each arc $f(p_k;\ 0,\ T)$ is compact, and, because of the continuity of the system $\lim_{k\to\infty} f(p_k;\ 0,\ T) = f(p';\ 0,\ T)$. But then, since $0 < \tau_k < t_k \leqq T$, the point $f(p_k, \tau_k) \in P$, which contradicts the choice of these points. Thus $\lim_{k\to\infty} t_k = \infty$.

Suppose that p' and q' lie on the same trajectory $q' = f(p', t')$. Consider the sequence of points $q'_n = f(p_n, t')$; obviously $\lim_{n\to\infty} q'_n = q'$. Thus in any neighborhood $S(q', \delta)$ of the point q' there can be found a pair of points (q_n, q'_n) such that $q_n = f(p_n, t_n) = f(q'_n, t_n - t')$, wherein $t'_n - t'$ is arbitrarily large for sufficiently large n, i. e. $S(q', \delta) \cdot f(S(q', \delta), t)$ is nonempty for arbitrarily large t, i.e. the point q' is nonwandering and this contradicts the hypothesis. The theorem is proved.

Thus, from Theorems **10.04** and **10.11**, it follows that complete instability and the absence of an improper saddle point are necessary conditions for the possibility of a mapping of a dynamical system onto a family of parallel lines. It is the purpose of our further investigations to show that these conditions are also sufficient.

Recall Theorems 2.15 and 2.16, which are valid in any dynamical system defined in a locally compact space R with a countable base. In the case of a completely unstable dynamical system one may introduce the concept of a *section of an invariant set.*

10.12. DEFINITION. Let there be given an invariant set $E \subset R$. We shall call a set F which is closed in E a *section of the set* E if for every point $q \in E$ there exists one and only number t_q such that $f(q, t_q) \in F$.

10.13. THEOREM. *In a completely unstable dynamical system for every point $p \in R$ there can be found a $\delta > 0$ such that the invariant set (infinite tube)*

$$\Phi = f(\overline{S(p, \delta)}; I)$$

has a compact section.

From the complete instability of the system there follows the existence of numbers $\alpha > 0$ and $T > 0$ such that for $|t| > T$.

$$\overline{S(p, \alpha)} \cdot f(\overline{S(p, \alpha)}, t) = 0.$$

We choose $\varepsilon' > 0$ such that $\varepsilon' < \alpha$ and $\overline{S(p, \varepsilon')}$ is compact. Since a completely unstable system contains no periodic motions, then, proceeding from $\varepsilon < \varepsilon'$, we can find, on the basis of Theorems 2.14 and 2.15, a $\delta > 0$ such that the finite tube

$$\Phi_1 = f(\overline{S(p, \delta)}; -T, T)$$

has a local section F. We note that since in the construction of

Theorem 2.14, the set $F \subset \overline{S(p, \varepsilon)}$, and since by definition F is closed, then F is compact.

We shall prove that F is a section of the tube Φ. Suppose that $q \in \Phi$; then there exists a t' such that $f(q, t') = q' \in \overline{S(p, \delta)}$, and since $q' \in \Phi_1$, then, according to the property of a local section, there can be found a t'' such that $f(q', t'') \in F$. Setting $t' + t'' = t$, we have

$$f(q, t) \in F.$$

We shall show that such a t can be found only once. Assume the contrary; let

$$f(q, t_1) \in F, \quad f(q, t_2) \in F, \quad t_2 - t_1 = t > 0.$$

Setting $f(q, t_1) = q_1$, we obtain

$$q_1 \in F, \quad f(q_1, t) \in F.$$

Since $q_1 \in \Phi_1$ and F is a local section of Φ_1, then $|t| > 2T$.

On the other hand, by construction

$$F \subset \overline{S(p, \varepsilon)} \subset \overline{S(p, \varepsilon')} \subset \overline{S(p, \alpha)},$$

and we obtain for $|t| > 2T > T$,

$$\overline{S(p, \alpha)} \cdot f(\overline{S(p, \alpha)}, t) \supset F \cdot f(F, t) \supset f(q_1, t),$$

which contradicts the choice of the numbers α and T. Thus it has been proved that F is a compact section of the tube Φ.

10.14. LEMMA. *If F is a closed compact set in an unstable dynamical system which has no improper saddle point, then the tube $\Phi = f(F; I)$ is a closed set.*

Suppose that $\{p_n\} \subset \Phi$ and $\lim_{n\to\infty} p_n = p$; we shall show that $p \in \Phi$. According to the definition of a tube, for every point p_n there exists a number t_n such that $q_n = f(p_n, t_n) \in F$. Because of the compactness of F, the sequence $\{q_n\}$ has a limit point $q \in F$; we assume, in order not to complicate the notation, that $\lim_{n\to\infty} q_n = q$. Then, according to the Corollary 10.07, we obtain $q = f(p, t)$, i.e. $p = f(q, -t) \in \Phi$. The lemma is proved.

10.15. COROLLARY. *If an infinite tube Φ of an unstable dynamical system without an improper saddle point has a compact section, then Φ is closed.*

10.16. LEMMA. *If a closed, invariant set Φ has a compact section*

F and if $\{p_n\} \subset \Phi$, $\lim_{n\to\infty} p_n = p$, then denoting by $q_n = f(p_n, \tau_n)$ the corresponding points on the section, $q_n \in F$, we have

$$\lim_{n\to\infty} q_n = q = f(p, \tau) \in F, \quad \textit{where} \quad \tau = \lim_{n\to\infty} \tau_n.$$

Because of the compactness of the set F there exists a convergent subsequence $\{q_{n_k}\}$; let $\lim_{k\to\infty} q_{n_k} = q$. From the Corollary 10.07 we have

$$\lim_{q\to\infty} \tau_{n_k} = \tau' \quad \text{and} \quad q = f(p, \tau') \in F.$$

From the definition of a section there follows: $f(p, \tau') = q = f(p, \tau)$, i.e. $\tau' = \tau$. Since this is valid for any convergent subsequence $\{q_{n_k}\}$, there follows:

$$\lim_{k\to\infty} q_n = q, \qquad \lim_{n\to\infty} \tau_n = \tau,$$

which it was required to prove.

Our purpose is the construction of a section of an entire dynamical system which is unstable without an improper saddle point.

10.17. THEOREM. *Let there be given in an unstable system without an improper saddle poini two invariant sets Φ_1 and Φ_2 having compact sections F_1 and F_2; then the invariant set $\Phi_1 + \Phi_2$ has a compact section F, moreover, F can be chosen such that $F \supset F_1$.*

By the Corollary 10.15 the tubes Φ_1 and Φ_2 are closed sets. If $\Phi_1 \cdot \Phi_2 = 0$, then $F_1 + F_2$ will be the section sought. Now let $\Phi_1 \cdot \Phi_2 = \Phi_3 \neq 0$. Put

$$F_1 \cdot \Phi_3 = F_1^*, \qquad F_2 \cdot \Phi_3 = F_2^*.$$

From the compactness of the sections F_1 and F_2 and the fact that the set Φ_3 is closed, the sets F_1^* and F_2^* are compact. We define over F_2^* a real function $\varphi(p)$ in the following manner: suppose that $p \subset F_2^*$; then there exists a unique value t_p such that $q = f(p, t_p) \subset F_1^*$; we set

$$\varphi(p) = t_p.$$

This function is continuous. Indeed, suppose that $\{p_n\} \subset F_2^*$, $p_n \to p$, then $q_n = f(p_n, t_n) \in F_1^*$, where $t_n = t_{p_n} = \varphi(p_n)$ and $f(p, t) \in F_1^*$, $t = \varphi(p)$. From Lemma 10.16 it follows that in such a case

$$\lim_{n\to\infty} t_n = t \quad \text{and} \quad \lim_{n\to\infty} q_n = q = f(p, t) \in F_1^*,$$

i.e.

$$\lim_{n\to\infty} \varphi(p_n) = \varphi(p),$$

which proves the continuity of the function $\varphi(p)$.

We define next over the compact set $F_2 \supset F_2^*$ the continuous function $\psi(p)$, coinciding over F_2^* with $\varphi(p)$ (this continuation is possible; see the theorem of Brouwer-Urysohn in section 1 of this chapter) and we set

$$F_3 = \{f(p, \psi(p)), \ p \in F_2\}.$$

Defined in such a manner, the set $F_3 \supset F_1^*$ and is intersected by every trajectory of Φ_2 in one and only one point.

We shall prove that F_3 is compact. Suppose that $\{q_n\} \subset F_3$; then $q_n = f(p_n, \psi(p_n))$, $p_n \in F_2$. Because of the compactness of F_2 there exists a convergent sequence $\{p_{n_k}\}$, $\lim_{k\to\infty} p_{n_k} = p \in F_2$; then as a consequence of the continuity of the function $\psi(p)$,

$$\lim_{k\to\infty} \psi(p_{n_k}) = \psi(p) \quad \text{and} \quad \lim_{k\to\infty} q_{n_k} = f(p, \psi(p)) \in F_3.$$

Thus F_3 is a section of the tube Φ_2. We set

$$F = F_1 + F_3.$$

Obviously, F is compact; it is a section of the tube $\Phi_1 + \Phi_2$, since every trajectory of this latter tube has one and only one point of intersection with it; for $\Phi_1 - \Phi_3$ and $\Phi_2 - \Phi_3$ this is obvious, for Φ_3 we have

$$F_1 \cdot \Phi_3 = F_3 \cdot \Phi_3 = F_1^*.$$

Finally, $F \supset F_1$. The theorem is proved.

10.18. Theorem. *Every unstable dynamical system without an improper saddle point has a section.*

Since by Theorem 10.09 the system is completely unstable, then by Theorem 10.13 for every point $p \in R$ there can be found a $\delta > 0$ such that the tube $f(\overline{S(p, \delta)}; I)$ has a section. Since R is a space which satisfies the second axiom of countability, there can be found a system of such neighborhoods $\{S(p_n, \delta_n)\}$ that

$$R = \sum_{n=1}^{\infty} S(p_n, \delta_n).$$

We construct the invariant sets

$$\Phi_n = f(\overline{S(p_n, \delta_n)}; I), \qquad (n = 1, 2, \ldots)$$

and denote by F_n the compact section of the tube Φ_n.

We next construct the sequence of compact sets

$$F^{(1)} \subset F^{(2)} \subset \ldots \subset F^{(n)} \subset \ldots$$

in the following manner:

$$F^{(1)} = F_1$$

is the section of the set Φ_1; we assume next that we have constructed $F^{(n)}$, a compact section of the invariant set

$$\Phi^{(n)} = \sum_{h=1}^{n} \Phi_h,$$

and let

$$\Phi^{(n+1)} = \sum_{h=1}^{n+1} \Phi_h = \Phi^{(n)} + \Phi_{n+1}.$$

Then as the set $F^{(n+1)}$ we take, in accordance with Theorem 10.17, a compact section of the set $\Phi^{(n+1)}$, for which $F^{(n+1)} \supset F^{(n)}$.

It is clear that

$$\sum_{n=1}^{\infty} \Phi^{(n)} = R.$$

Next, put

$$F = \sum_{n=1}^{\infty} F^{(n)}.$$

The set F has one and only one point in common with each trajectory of R. In fact, let $p \in R$ be an arbitrary point and let n_1 be the smallest number such that $p \in \overline{S(p_{n_1}, \delta_{n_1})}$. Then $p \in \Phi^{(n_1)}$ and from the definition of the section $F^{(n_1)}$ there can be found a unique value t_1 such that $f(p, t_1) = q_1 \in F^{(n_1)} \in F$. If there should exist another point $q_2 = f(p, t_2) \in F$, then there could be found a natural number $n_2 > n_1$ such that $q_2 \in F^{(n_2)}$, and since $F^{(n_2)} \supset F^{(n_1)}$, then

$$q_1 \in F^{(n_2)},$$

which contradicts the definition of the section $F^{(n_2)}$.

We shall show that F is closed, i. e. it is a section in R. Suppose that $\{q_n\} \subset F$ and $\lim_{n\to\infty} q_n = q$; then there can be found an n_0 such that

$$q \in S(p_{n_0}, \delta_{n_0})$$

and an N such that for $n \geqq N$,

$$q_n \in S(p_{n_0}, \delta_{n_0}).$$

Therefore, for $n \geqq N$, we have

$$q_n \in \Phi^{(n_0)} \cdot F = F^{(n_0)}.$$

Then, since $F^{(n_0)}$ is closed, we find

$$q \in F^{(n_0)} \subset F,$$

which it was required to prove.

10.19. THEOREM. *Every unstable dynamical system without a saddle point at infinity can be mapped topologically onto a family of parallel lines in Hilbert space.*

Let F be a section of the system. We define a mapping Ψ of the space R into E^∞ in the following manner. Let $p \in R$ be any point and let $p^* = f(p, -t)$ be a point of the corresponding trajectory lying in F. There exists a mapping of the section F onto the Hilbert space, $\Psi_1(p^*) = q^* = (\xi_1, \xi_2, \ldots, \xi_n, \ldots)$ (Theorem 1.27). We then set

$$\Psi(p) = (t, \xi_1, \xi_2, \ldots, \xi_n, \ldots) = q \in E^\infty.$$

We set

$$\Psi(R) = R^* \subset E^\infty.$$

Under the mapping Ψ the trajectories go over into the lines $\xi_1 = \text{const.}$, $\xi_2 = \text{const.} \ldots$, $\xi_n = \text{const.}, \ldots$ From the property of a section it follows that to each point p there corresponds a unique point q; further, if there be given two points q_1 and q_2, $q_1 \in R^*$ and $q_2 \in R^*$, then either their coordinates ξ_n do not coincide, and then the corresponding points p lie on different trajectories, or the coordinates $\xi_1, \xi_2, \ldots, \xi_n, \ldots$ coincide, with the t coordinates differing, and then to them there correspond two distinct points of the same trajectory. Thus the correspondence between R and R^* is one-one. We shall show that it is bi-continuous.

1. Suppose that $\{p_n\} \subset R$, $\lim_{n\to\infty} p_n = p$. We denote the corresponding points on the section F by $p_n^* = f(p_n, -t_n)$. Then

there exists an n_0 such that $p \in \Phi^{(n_0)}$ (the notation is as in Theorem 10.18), and, since $\{p_n\}$ converges, there can be found a natural number N such that for $n \geqq N$ also $p_n \in \Phi^{(n_0)}$ and then $p_n^* \in F^{(n_0)}$.

Since the section $F^{(n_0)}$ is compact, it follows from Lemma 10.16 that there exist

$$\lim_{n\to\infty} t_n = t, \qquad \lim_{n\to\infty} p_n^* = f(p, -t).$$

From this, setting

$$\Psi(p) = (t, \xi_1, \xi_2, \ldots, \xi_i, \ldots),$$

and

$$\Psi(p_n) = (t_n, \xi_1^{(n)}, \xi_2^{(n)}, \ldots, \xi_1^{(n)}, \ldots),$$

we have:

$$\lim_{n\to\infty} \varrho_1[\Psi(p_n), \Psi(p)] = \lim_{n\to\infty} [(t_n - t)^2 + \sum_{i=1}^{\infty} (\xi_i^{(n)} - \xi_i)^2]^{1/2} = 0.$$

2. Suppose that $\{q_n\} \subset R^*$ and $\lim_{n\to\infty} q_n = q$, i.e.

$$\lim_{n\to\infty} [(t_n - t)^2 + \sum_{i=1}^{\infty} (\xi_i^{(n)} - \xi_i)^2]^{1/2} = 0.$$

From this

$$\lim_{n\to\infty} t_n = t, \quad \lim_{n\to\infty} \sum_{i=1}^{\infty} (\xi_i^{(n)} - \xi_i)^2 = 0.$$

We set

$$\Psi^{-1}(q_n) = p_n \in R, \quad \Psi^{-1}(q) = p \in R,$$

$$f(p_n, -t_n) = p_n^* \in F, \quad f(p, -t) = p^* \in F.$$

From the fact that $\lim_{n\to\infty} (\xi_i^{(n)} - \xi_i) = 0$ and from the bi-continuity of the mapping Ψ_1 of the set F into E^∞, it follows that $\lim_{n\to\infty} p_n^* = p^*$ and further $\lim_{n\to\infty} f(p_n^*, t_n) = f(p^*, t)$, i.e.

$$\lim_{n\to\infty} p_n = \lim_{n\to\infty} \Psi^{-1}(q_n) = p = \Psi^{-1}(q).$$

The bi-continuity has been proved.

From a comparison of the results of Theorems 10.04, 10.09, 10.11, and 10.19 there follows the

10.20. FUNDAMENTAL THEOREM. *In order that a dynamical system given in a locally compact metric space R which satisfies the*

second axiom of countability should be homeomorphic to a system of parallel lines in a Hilbert space, it is necessary and sufficient that it be unstable and have no improper saddle point.

10.21. REMARK. If the system is given in the space $E^{(n)}$, then it is mapped homeomorphically onto a family of lines in $E^{(n+1)}$ in the following manner: suppose that $p \in E_n$ and $q = f(p, -t) \in F$ (a section of the system). Then, if the coordinates of the point q be $(\xi_1, \xi_2, \ldots, \xi_n)$,

$$\Psi(p) = (t, \xi_1, \xi_2, \ldots, \xi_n).$$

11. Dynamical Systems Stable According to Lyapunov

Let R be a locally compact separable metric space. We also may assume that R is connected, for otherwise the results to be obtained will refer to each component of R. Suppose that a dynamical system $f(p, t)$ is defined on R. Suppose further that every point $p \in R$ is Lyapunov stable relative to R, i.e. to every $p \in R$ and $\varepsilon > 0$, there corresponds a $\delta(p, \varepsilon)$ such that $\varrho(p, q) < \delta$ implies $\varrho[f(p, t), f(q, t)] < \varepsilon$ for $-\infty < t < +\infty$.

Such systems (in an n-dimensional space) have been investigated by M. V. Bebutov.

Before presenting the fundamental result we must prove a series of lemmas concerning Lyapunov stable systems.

11.01. LEMMA. *If the motion $f(p, t)$ is stable L^+ or stable L^- then its closure is a compact minimal set.*

Suppose e. g. that $f(p, t)$ is stable L^+. Then Ω_p is not empty and is compact. Let $q \in \Omega_p$. Since Ω_p is closed and invariant, $\overline{f(q; I)} \subset \Omega_p \subset \overline{f(p; I)}$.

On the other hand, since $q \in \overline{f(p, I)}$ there exists a sequence $p_n = f(p, t_n)$ such that $\lim_{n\to\infty} p_n = q$. Corresponding to any $\varepsilon > 0$ we may determine a number $\delta(q, \varepsilon)$ from the Lyapunov stability of q. Let us choose n_0 such that $\varrho(q, p_n) < \delta$ for $n \geqq n_0$. Then $\varrho[f(q, -t_n), p] < \varepsilon$ for $n \geqq n_0$. That is to say, p is the limit of the points $q_n = f(q, -t_n)$. Thus $p \in \overline{f(q; I)}$; hence $\overline{f(p; I)} \subset \overline{f(q; I)}$ and therefore $\overline{f(p; I)} = \overline{f(q; I)}$. In other words, every trajectory through $q \in \overline{f(p; I)}$ is dense in this closure; hence $\overline{f(p; I)}$ is minimal and, since it is contained in Ω_p, is compact. This proves the lemma.

11.02. COROLLARY. *If $f(p, t)$ is unstable L^+ (or unstable L^-) then it is Lagrange unstable in both directions.*

11.03. LEMMA. *If $f(p, t)$ is Lagrange unstable then it has no α- or ω-limit points.*

Let q_0 be an ω-limit point for $f(p, t)$. Since R is locally compact there is an $\alpha > 0$ such that $\overline{S(q_0, 2\alpha)}$ is compact. Since $q_0 \in \Omega_p$ there is such a t_0 that $p_0 = f(p, t_0) \subset S(q_0, \alpha/4)$. Also there is a sequence $\{t_n\}$, $t_n \to +\infty$, for which $\lim_{n\to\infty} f(p_0, t_n) = q_0$.

On the other hand, it follows from the instability of $f(p_0, t)$ that there is a sequence $\{\tau_n\}$, $\tau_n \to +\infty$, which is such that the sequence $\{f(p_0, \tau_n)\}$ contains no convergent subsequence. Now determine the numbers θ_n by the requirement that $\varrho[p_0, f(p_0, \theta_n)] = \alpha$ while $\varrho[p_0, f(p_0, t)] > \alpha$ for $\theta_n < t \leqq \tau_n$. Since $f(p_0, \theta_n) \in \overline{S(q_0, 2\alpha)}$ we may extract a convergent subsequence from the sequence $\{f(p_0, \theta_n)\}$. For simplicity of notation let us assume that $f(p_0, \theta_n)$ converges to q_1. Now the numbers $\tau_n - \theta_n$ increase indefinitely; otherwise there would exist a convergent subsequence $\{\tau_{n_k} - \theta_{n_k}\}$, with $\lim_{n\to\infty} (\tau_{n_k} - \theta_{n_k}) = T$, and then

$$\lim_{k\to\infty} f(p_0, \tau_{n_k}) = \lim_{k\to\infty} f[f(p_0, \theta_{n_k}), \tau_{n_k} - \theta_{n_k}] = f(q_1, T),$$

which would contradict the choice of the sequence $\{\tau_n\}$.

Now choose $\delta(q_1, \alpha/4)$ with reference to the Lyapunov stability of q_1, and fix n_0 so that for $n \geqq n_0$ we have the inequality

$$\varrho[q_1, f(p_0, \theta_n)] < \delta.$$

Let $t > \theta_{n_0}$; choose $n_1 > n_0$ in such a way that $0 < t - \theta_{n_0} < \tau_{n_1} - \theta_{n_1}$. Then $\theta_{n_1} < \theta_{n_1} + t - \theta_{n_0} < \tau_{n_1}$, so that $\varrho\,[p_0, f(p_0, \theta_{n_1} + t - \theta_{n_0})] > \alpha$. Therefore

$$\begin{aligned} \varrho[f(p_0, t), q_0] &\geqq \varrho[p_0, f(p_0, \theta_{n_1} + t - \theta_{n_0})] \\ &- \varrho(p_0, q_0) - \varrho[f(p_0, \theta_{n_1} + t - \theta_{n_0}), f(q_1, t - \theta_{n_0})] \\ &- \varrho[f(q_1, t - \theta_{n_0}), f(p_0, \theta_{n_0} + t - \theta_{n_0})] \\ &> \alpha - (\alpha/4) - (\alpha/4) - (\alpha/4) = \alpha/4. \end{aligned}$$

This holds for an arbitrarily large $t > \theta_{n_0}$ which contradicts the condition that $\varrho[f(p_0, t_n), q_0]$ approach zero as $t_n \to \infty$. Hence the lemma is proved.

11.04. LEMMA. *The set M_1 of points on motions stable L is open.*

Let $q \in M_1$. Then $\overline{f(q; I)}$ is compact. Owing to local compactness, there is an $\varepsilon > 0$ for which $S(\overline{f(q; I)}, \varepsilon) = \bar{S}$ is compact. For that ε let $\delta(q, \varepsilon)$ be chosen according to the Lyapunov stability. Then

$\varrho(q, r) < \delta$ implies $f(r, I) \subset \bar{S}$, so that $f(r, t)$ is stable L. Thus q is an interior point of M_1, which proves the lemma.

11.05. LEMMA. *The set M_2 of points on the unstable motions is open.*

Assume the contrary; then there is a $p \in M_2$ which is a limit point of M_1. By Lemma 11.01, if $q \in M_1$, then $f(q, t)$ is recurrent. Take an $\varepsilon > 0$ and according to the Lyapunov stability find a corresponding $\delta(p, \varepsilon/3) \leqq \varepsilon/3$. Then let $q_0 \in M_1$, $\varrho(p, q_0) < \delta$. Because of the Poisson stability of the recurrent motion $f(q_0, t)$, there will correspond to any $T > 0$ a $t_1 > T$ for which $\varrho[q_0, f(q_0, t_1)] < \varepsilon/3$. Now Lyapunov stability implies that

$$\varrho[f(p, t_1), f(q_0, t_1)] < \frac{\varepsilon}{3}.$$

Hence

$$\varrho[p, f(p, t_1)] < \varrho(p, q_0) + \varrho[q_0, f(q_0, t_1)] + \varrho[f(q_0, t_1), f(p, t_1)] < \varepsilon;$$

that is, p is Poisson stable, which contradicts Lemma 11.03.

The preceding lemmas yield theorems concerning Lyapunov stable dynamical systems.

11.06. THEOREM. *In a connected locally compact space there are exactly two types of Lyapunov stable systems: either all motions are stable L or else all motions are unstable in both directions.*

Assume that both M_1 and M_2 are nonempty. Then $R = M_1 + M_2$, where (by Lemmas 11.04, 11.05) M_1, M_2 are open and $M_1 \cdot M_2 = 0$. This however contradicts the connectedness of R and proves the theorem.

11.07. THEOREM. *If a connected dynamical system is Lyapunov stable and Lagrange stable then each of its motions is either a rest point, a periodic motion, or an almost periodic motion. The whole system is either a single minimal set of almost periodic motions (or a periodic motion or a rest point), or else is decomposed into a sum of uncountably many such minimal sets.*

According to Lemma 11.01, every motion $f(p, t)$ belongs to a certain compact minimal set Σ and so is recurrent. But by Theorem 8.12, a Lagrange stable recurrent motion is here almost periodic.

There are now two possibilities:

1. In a certain set Σ the point p_0 is an interior point, that is to say there exists some sphere $S(p_0, \alpha) \subset \Sigma$. Then every point of

Σ is interior (cf. the final theorem of Section 7). Thus Σ is an open set; but by definition Σ, as a minimal closed invariant set, is closed. Hence $\Sigma = R$ (R is connected) and the whole system consists of a single minimal set of almost periodic motions.

2. There remains the case when no minimal set Σ is dense anywhere in R. Now the locally compact space R can be represented as a sum $R = \sum_{n=1}^{\infty} F_n$ of a countable number of compact sets F_n. Each F_n is a space and so cannot be decomposed into a finite or countable sum of nowhere dense sets. Hence R is decomposed into an uncountable sum of minimal sets of almost periodic motions (or periodic motions or rest points).

11.08. Example. Let R be a three-dimensional torus with coordinates (x_1, x_2, x_3) taken modulo 1. The differential equations of motion are taken as

$$\frac{dx_1}{dt} = 1, \quad \frac{dx_2}{dt} = \alpha, \quad \frac{dx_3}{dt} = \beta,$$

or, in finite terms

$$x_1 = x_1^0 + t, \quad x_2 = x_2^0 + \alpha t, \quad x_3 = x_3^0 + \beta t.$$

Upon introducing the Euclidean metric (coordinate differences being taken modulo 1 in such a way as to obtain the smallest value), one can easily see that $\varrho(p, q) = \varrho[f(p, t), f(q, t)]$ for all t, so that Lyapunov stability holds.

If α, β and β/α are irrational, then the whole surface of the torus is a minimal set.

If $\alpha = p/q$ is rational and β irrational, then the projection on the torus (x_1, x_2) produces on it a family of closed curves which close after p turns along the x_2 circumference and q turns along the x_1 circumference. The topological product of each of these curves with the x_3 circumference gives rise to a two-dimensional torus on which each motion is everywhere dense and almost periodic. This decomposition into a 1-parameter family of minimal sets corresponds to the fact that the system admits one first integral whose left-hand side is single valued on the whole space:

$$qx_1 - px_2 = qx_1^0 - px_2^0 \equiv c.$$

If α and β are rational and are reduced to a common denominator $\alpha = p_1/q$, $\beta = p_2/q$, then each integral curve is closed, making q

turns along the x_1 circumference, p_1 along the x_2 circumference, and p_2 along the x_3 circumference. We have then a 2-parameter family of minimal sets, each of which consists of a single periodic motion. Analytically this corresponds to the fact that there are two single valued integrals,

$$qx_1 - p_1x_2 = c_1, \qquad qx_2 - p_2x_3 = c_2.$$

11.09. THEOREM. *If the system is Lagrange unstable then it is homeomorphic to a system of parallel lines in Hilbert space.*

By the fundamental theorems of Section 10, it is sufficient to show that the system has no saddle points at infinity. For suppose that the system has saddle points; then there exist a sequence of points $\{p_n\}$ and sequences of numbers $\{t_n\}$, $\{\tau_n\}$, $0 < \tau_n < t_n$, with the property that $p_n \to p$, $f(p_n, t_n) \to q$, while $f(p_n, \tau_n)$ has no convergent subsequence. (The sequence $\{t_n\}$ is unbounded of course, for otherwise $\{\tau_n\}$ would also be bounded and $\{f(p, \tau_n)\}$ would contain a convergent subsequence.) Let us show that q is then an ω-limit point for $f(p, t)$. Given an $\varepsilon > 0$ determine a $\delta(p, \varepsilon/2)$ relative to the Lyapunov stability and an n_0 such that for $n \geqq n_0$

$$\varrho(p, p_n) < \delta, \quad \varrho[f(p_n, t_n), q] < \frac{\varepsilon}{2}.$$

Then for $n \geqq n_0$ we have

$$\varrho[f(p, t_n), q] < \varrho[f(p, t_n), f(p_n, t_n)] + \varrho[f(p_n, t_n), q] < \varepsilon.$$

Then $q = \lim_{n\to\infty} f(p, t_n)$ and so $f(p, t)$ has q as an ω-limit point. This contradicts Lemma 11.03 and so proves the theorem.

To summarize the present section: if a dynamical system is defined on a connected locally compact metric space R and is Lyapunov stable, then there are only two possibilities for it; *either the system is homeomorphic to a family of straight lines, or else all motions are almost periodic and constitute either a single minimal set or an uncountable number of minimal sets.*

CHAPTER VI

Systems with an Integral Invariant

1. Definition of an Integral Invariant

We shall consider the motions of a dynamical system given by the differential equations

$$(1.01)\qquad \frac{dx_i}{dt} = X_i(x_1, x_2, \ldots, x_n) \qquad (i = 1, 2, \ldots, n).$$

The functions X_i are defined in some closed domain R of the "phase space" $(x_1, x_2, \ldots, x_n)$; we shall regard them as being continuously differentiable with respect to all the arguments. Then the initial values $x_1^{(0)}, x_2^{(0)}, \ldots, x_n^{(0)}$ for $t = t_0$ determine a motion of the system (1.01):

$$(1.02)\quad x_i = \varphi_i(t - t_0; x_1^{(0)}, x_2^{(0)}, \ldots, x_n^{(0)}) \qquad (i = 1, 2, \ldots, n),$$

where the φ_i have continuous partial derivatives with respect to the initial values $(x_1^{(0)}, x_2^{(0)}, \ldots, x_n^{(0)})$. We shall denote the motion (1.02) more concisely thus:

$$x = f(x_0, t).$$

An *integral invariant* (of the nth order), according to Poincaré, is an expression of the form

$$(1.03)\qquad \iint_D \cdots \int M(x_1, x_2, \ldots, x_n)\, dx_1\, dx_2 \ldots dx_n,$$

where the integration is extended over any domain D, if this expression possesses the property

$$(1.04)\qquad \iint_D \cdots \int M(x_1, x_2, \ldots, x_n) dx_1\, dx_2 \ldots dx_n$$
$$= \iint_{D_t} \cdots \int M(x_1, x_2, \ldots, x_n)\, dx_1\, dx_2 \ldots dx_n.$$

Here $D_t = f(D, t)$ is the domain occupied at the instant t by the points which for $t = 0$ occupy the domain D.

Poincaré also gave a simple dynamical interpretation of the condition (1.04) characterizing an integral invariant. We shall consider the system in three-dimensional space

$$\frac{dx}{dt} = X(x, y, z), \quad \frac{dy}{dt} = Y(x, y, z), \quad \frac{dz}{dt} = Z(x, y, z), \tag{1.05}$$

and shall interpret it as a system of equations determining the velocity of the steady state motion of a fluid filling the space R. If $\varrho(x, y, z)$ denotes the density of the fluid at the point (x, y, z), then the integral

$$\iiint_D \varrho(x, y, z)\, dx\, dy\, dz \tag{1.06}$$

represents the mass of the fluid filling the domain D; the expression (1.06) is an integral invariant since this mass of fluid remains unaltered when the particles of the fluid, having undergone a displacement along their trajectories for a time interval t, occupy the domain D_t. Thus, for the system (1.05) defining a steady state motion of the fluid, there exists an integral invariant in which the function M of the formula (1.03) is the density of the fluid. If the fluid is incompressible, then $\varrho(x, y, z) =$ const. and we have

$$\iiint_D dx\, dy\, dz = \iiint_{D_t} dx\, dy\, dz. \tag{1.07}$$

Thus, *in the case of an incompressible fluid, the volume is an integral invariant.*

We now introduce a partial differential equation satisfied by the function M of the formula (1.03) — the "density of the integral invariant". We first take a local point of view. We choose a closed domain D lying wholly within R and a time interval $(-T, T)$ so small that $D_t \subset R$ for $-T < t < T$. Next we take within this interval a fixed instant t and a small increment h such that $t + h \in (-T, T)$. We set

$$I(t) = \iint_{D_t} \cdots \int M(x_1, x_2, \ldots, x_n)\, dx_1 dx_2 \ldots dx_n \tag{1.08}$$

Then one can write

$$(1.09) \qquad I(t+h) = \iint_{D_{t+h}} \cdots \int M(x_1', x_2', \ldots, x_n')dx_1' dx_2' \ldots dx_n',$$

where $(x_1', x_2', \ldots, x_n')$ is a point of the domain D_{t+h}. By virtue of the uniqueness of the solution and the continuous dependence on the initial conditions, the formulas which are obtained from (1.02) if one sets $t - t_0 = h$ and denotes the coordinates for the value t by x_i and for the instant $t+h$ by x_i', i.e.

$$(1.10) \qquad x_i' = \varphi_i(h;\, x_1, x_2, \ldots, x_n) \qquad (i = 1, 2, \ldots, n),$$

determine a one-one correspondence between the points of the domains D_t and D_{t+h}. Therefore in the expression (1.09) one can pass from the variables x_i' to the variables x_i, at the same time replacing the domain D_{t+h} by the domain D_t. From the conditions imposed on X_i there follows the existence of the continuous partial derivatives

$$\frac{\partial x_i'}{\partial x_j} = \frac{\partial \varphi_i}{\partial x_j} \qquad (i, j = 1, 2, \ldots, n);$$

therefore, by the formula for a change of variables, the multiple integral (1.09) takes the form

$$(1.11) \qquad \begin{aligned} I(t+h) = \iint_{D_t} \cdots \int M[\varphi_1(h;\, x_1, x_2, \ldots, x_n), \ldots, \varphi_n(h;\, x_1, x_2, \ldots, x_n)] \\ \cdot \frac{D(x_1', x_2', \ldots, x_n')}{D(x_1, x_2, \ldots, x_n)} dx_1\, dx_2 \ldots dx_n. \end{aligned}$$

We shall assume that the function M also admits continuous partial derivatives with respect to all its arguments. Under this assumption we shall compute the integrand in formula (1.11). First of all we note that because of the differentiability with respect to h of the functions φ_i we have

$$\begin{aligned} &M[\varphi_1(h;\, x_1, x_2, \ldots, x_n), \ldots, \varphi_n(h;\, x_1, x_2, \ldots, x_n)] \\ &\qquad = M\left[x_1 + h\left(\frac{\partial \varphi_1}{\partial h}\right)_{h=0} + o(h), \ldots, x_n + h\left(\frac{\partial \varphi_n}{\partial h}\right)_{h=0} + o(h)\right], \end{aligned}$$

where $o(h)$ denotes in general functions whose quotients when divided by h tend to zero uniformly with respect to $x_1, x_2, \ldots, x_n$ in the domain D_t as $h \to 0$. Noting that the expressions (1.02) are solutions of the system (1.01) and, consequently,

$$\left(\frac{\partial \varphi_i}{\partial h}\right)_{h=0} = X_i(x_1, x_2, \ldots, x_n) \qquad (i = 1, 2, \ldots, n),$$

we have, because of the existence of the complete differential of M (and this follows from the existence of the continuous partial derivatives),

$$(1.12) \qquad M(x'_1, x'_2, \ldots, x'_n)$$
$$= M(x_1, x_2, \ldots, x_n) + h \sum_{i=1}^{n} X_i(x_1, x_2, \ldots, x_n) \frac{\partial M(x_1, x_2, \ldots, x_n)}{\partial x_i} + o(h).$$

Furthermore, the functions $\partial x'_i/\partial x_j$ $(i, j, = 1, 2, \ldots, n)$, as already mentioned, are continuous functions of $h, x_1, x_2, \ldots, x_n$; moreover, they admit continuous derivatives with respect to h; and for $h = 0$ the functions are equal to δ_{ij} (*i.e.* 0 for $i \neq j$ and 1 for $i = j$). Their derivatives satisfy variational equations [1]

$$\frac{d}{dh}\frac{\partial x'_i}{\partial x_j} = \sum_{k=1}^{n} \frac{\partial X_i(x'_1, x'_2, \ldots, x'_n)}{\partial x'_k} \frac{\partial x'_k}{\partial x_j} \qquad (i, j = 1, 2, \ldots, n);$$

whence

$$\frac{\partial x'_i}{\partial x_j} = \delta_{ij} + h \sum_{k=1}^{n} \frac{\partial X_i(x_1, x_2, \ldots x_n)}{\partial x_k} \left(\frac{\partial x'_k}{\partial x_j}\right)_{h=0} + o(h)$$
$$= \delta_{ij} + h \frac{\partial X_i(x_1, x_2, \ldots, x_n)}{\partial x_j} + o(h).$$

Substituting these values into the Jacobian of the transformation and selecting in the determinant the term not containing h and the terms of the first order with respect to h, we obtain.

$$(1.13) \qquad \frac{D(x'_1, x'_2, \ldots, x'_n)}{D(x_1, x_2, \ldots, x_n)} = 1 + h \sum_{i=1}^{n} \frac{\partial X_i(x_1, x_2, \ldots, x_n)}{\partial x_i} + o(h).$$

Substituting the values (1.12) and (1.13) into (1.11), we find

$$I(t+h) = \iint_{D_t} \cdots \int \left\{M + h\left[\sum_{i=1}^{n} X_i \frac{\partial M}{\partial x_i} + M \sum_{i=1}^{n} \frac{\partial X_i}{\partial x_i}\right] + o(h)\right\} dx_1\, dx_2 \ldots dx$$

[1]See V. V. Stepanov, *Course in Differential Equations*, Chap. VII, § 3.

We compute the derivative $I'(t)$:

$$I'(t) = \lim_{h\to 0}\frac{I(t+h) - I(t)}{h}$$

$$= \lim_{h\to 0}\frac{1}{h}\iint_{D_t}\cdots\int\Big\{h\Big[\sum_{i=1}^{n} X_i\frac{\partial M}{\partial x_i} + M\sum_{i=1}^{n}\frac{\partial X_i}{\partial x_i}\Big] + o(h)\Big\}\,dx_1\ldots dx_n$$

$$= \iint_{D_t}\cdots\int\Big[\sum_{i=1}^{n}\frac{\partial(MX_i)}{\partial x_i}\Big]\,dx_1\ldots dx_n.$$

Now let $I(t)$ be an integral invariant; then the equality

$$I'(t) = 0$$

holds for any (sufficiently small) domain D. From this we obtain a necessary condition for the density M:

$$\sum_{i=1}^{n}\frac{\partial(MX_i)}{\partial x_i} = 0. \tag{1.14}$$

It is easily seen that if, conversely, the condition (1.14) is fulfilled identically, then the expression (1.03) is an integral invariant.

The condition (1.14) is the partial differential equation for M. From the existence theorem for such an equation one can assert that for the system under investigation a *local invariant integral always exists*. However, this fact does not enable us to draw the desired conclusions of a qualitative character concerning the dynamical system. In fact, we shall consider only positive integral invariants (or at least non-negative). For a sufficiently small domain D we can satisfy this restriction in the following manner. For a unique determination of the solution of the partial differential equation (1.14) one must assign the Cauchy data. If the point $(x_1^{(0)}, x_2^{(0)}, \ldots, x_n^{(0)})$ is not critical for the system (1.01), it be assumed that for at least one i, $x_i \neq 0$ in some neighborhood of it. Let $i = 1$. Then we can assign for $x_1 = x_1^{(0)}$ the initial condition $M = \varphi(x_2, x_3, \ldots, x_n)$, where $\varphi > 0$; hence, because of the continuity of M as a solution of (1.14), M will be positive for values of x_1 in a sufficiently small neighborhood of the point $x_1 = x_1^{(0)}$.

But we are considering a system of the form (1.01) in some domain R of the space $(x_1, x_2, \ldots, x_n)$ (or in some n-dimensional manifold) which is an invariant set of the system; *i.e.*, if the initial point lies in R, so does the entire trajectory. We shall call an expression of the form (1.03) an *integral invariant* only in the case when $M > 0$ in the whole domain R; then the equality (1.04) holds for any domain D and for any value of $t(-\infty < t < +\infty)$. Furthermore, we shall introduce the added restriction

$$\int\int_R \cdots \int M \, dx_1 \, dx_2 \ldots dx_n < +\infty.$$

If such an integral invariant exists, then its density M satisfies the equation (1.14). But it is impossible to prove its existence for any system satisfying only the condition of differentiability. The existence of an integral invariant in this sense is an additional restriction imposed on the system (1.01).

1.15. NOTES: The case when the right-hand sides of the equations depend on t is reduced to the form (1.01) by a preliminary replacement of t by x_{n+1} and the introduction of the supplementary equation $dx_{n+1}/dt = 1$. The condition (1.14), after a return to the variable t, takes the form

$$\frac{\partial M}{\partial t} + \sum_{i=1}^{n} \frac{\partial (MX_i)}{\partial x_i} = 0. \tag{1.16}$$

1.17. A function M, satisfying equation (1.14) or (1.16), is called a *multiplier of Jacobi.*

1.18. The condition that the system (1.01) admit volume as an n-dimensional integral invariant, *i.e.* $M = 1$, is

$$\sum_{i=1}^{n} \frac{\partial X_i}{\partial x_i} = 0.$$

The equations in the Hamiltonian form,

$$\frac{dp_i}{dt} = -\frac{\partial H}{\partial q_i}, \quad \frac{dq_i}{dt} = \frac{\partial H}{\partial p_i} \qquad (i = 1, 2, \ldots, n),$$

where

$$H = H(p_1, p_2, \ldots, p_n, q_1, q_2, \ldots, q_n),$$

obviously belong to this class.

1.19. Any system (1.01) with an integral invariant in which $M > 0$ can be reduced to the case $M = 1$. For this it is necessary to transform the independent variable (time) by means of the formula $d\tau = dt/M$; then the equations take the form

$$\frac{dx_i}{d\tau} = MX_i = X'_i(x_1, x_2, \ldots, x_n);$$

and because of (1.14)

$$\sum_{i=1}^{n} \frac{\partial X'_i}{\partial x_i} = 0.$$

The gist of the transformation of the type indicated lies in the fact that, while not altering the trajectories of the particles, we multiply the velocity at the point $(x_1, x_2, \ldots, x_n)$ by the value of the function M at this point. However, this transformation does not simplify the theory significantly.

1.20. If the right-hand sides of the equations (1.01) are subject only to Lipschitz conditions with respect to $x_1, x_2, \ldots, x_n$, and if an integral invariant exists, then the function M has partial derivatives almost everywhere (and has bounded derived numbers everywhere); the condition (1.14) is fulfilled almost everywhere (V. V. Stepanov, *Compositio Math.* 3).

We consider in particular a dynamical system for $n = 2$:

$$\frac{dx}{dt} = X(x, y), \quad \frac{dy}{dt} = Y(x, y). \tag{1.21}$$

We shall deduce the necessary and sufficient condition under which area will be an integral invariant for the system (1.21). From equation (1.14) the condition $M = 1$ gives

$$\frac{\partial X}{\partial x} + \frac{\partial Y}{\partial y} = 0. \tag{1.22}$$

Obviously, this is the condition that the expression

$$Ydx - Xdy \tag{1.23}$$

be an exact differential.

In the general case $M \neq 1$ equation (1.14) gives

$$\frac{\partial (MX)}{\partial x} + \frac{\partial (MY)}{\partial y} = 0.$$

This is the equation for an integrating factor M of the expression

(1.23); thus the system (1.21) in the presence of an integral invariant must possess an integrating factor which is continuous and positive over the entire invariant set under consideration.

As another example, we consider the system of linear equations

$$\frac{dx}{dt} = ax + by, \qquad \frac{dy}{dt} = cx + dy, \tag{1.24}$$

where a, b, c, d are constants, with the critical point (0, 0). The condition (1.22) for invariance of area gives

$$a + d = 0.$$

Reducing the system (1.24) to normal form [2], we obtain for λ the equation

$$\lambda^2 - (a + d)\lambda + ad - bc = 0,$$

or, in our case,

$$\lambda = \pm\sqrt{-ad + bc}.$$

Thus, invariance of area for the system (1.24) holds only in case the critical point is a center (imaginary roots) or a saddle, where for the latter case there must be $\lambda_2 = -\lambda_1$.

From what has been said it follows that the methods of the local theory of differential equations do not enable one, in the general case, to establish the existence of a non-negative integral invariant; and these will have great significance in this chapter for the investigation of dynamical systems. There exists a series of investigations giving conditions under which a dynamical system will have an integral invariant in the sense mentioned above. These conditions appear as restrictions imposed on the system.

We shall proceed along a more abstract route. As in the preceding chapter, we shall consider here dynamical systems in the metric space R. Frequently we shall assume that this space has a countable base. In certain cases we shall regard R as compact or locally compact. The role of the integral invariant in these abstract dynamical systems is played by invariant measure. But before the introduction of invariant measure it is necessary to establish a theory of measure in general. The following section is devoted to an exposition of one such theory.

[2]See V. V. Stepanov, *Course in Differential Equations,* Chap. II, § 2.

2. Carathéodory Measure

Let there be introduced in the metric space R a Carathéodory measure. This measure μA (*outer measure*) is defined for any set $A \subset R$ by the following axioms:

I. $\mu A \geqq 0$; moreover, *there exist sets of positive finite measure, and the measure of the empty set is zero.*

II. *If* $A \subset B$, *then* $\mu A \leqq \mu B$.

III. *For any countable sequence of sets there holds the inequality*

$$\mu\{\sum_{i=1}^{\infty} A_i\} \leqq \sum_{i=1}^{\infty} \mu A_i.$$

IV. *If* $\varrho(A, B) > 0$, *then* $\mu(A + B) = \mu A + \mu B$.

2.01. THEOREM. *Let* $G \subset R$ *be an open set distinct from the entire space; i.e.,* its complement, the closed set $F = R - G$, is nonempty. *Let* $B \subset G$ *be any set for which* μB *is finite; we consider the sequence of open sets* $G_n = \{p;\ \varrho(p, F) > 1/n\}$ $(n = 1, 2, \ldots)$, *and set* $B_n = B \cdot G_n$. *Then* $\lim_{n\to\infty} \mu B_n = \mu B$.

Obviously we have $B_1 \subset B_2 \subset \ldots \subset B_n \subset \ldots \subset B$, whence, by Axiom II,

$$\mu B_1 \leqq \mu B_2 \leqq \ldots \leqq \mu B_n \leqq \ldots \leqq \mu B.$$

Consequently, there exists a limit

$$\lim_{n\to\infty} \mu B_n = \lambda \leqq \mu B. \tag{2.02}$$

We introduce the notation: $B = B_n + R_n$, $C_n = B_{n+1} - B_n$. Thus

$$C_n = \left\{p \in B;\ \frac{1}{n} \geqq \varrho(p, F) \geqq \frac{1}{n+1}\right\}$$

and

$$R_n = C_n + C_{n+1} + C_{n+2} + \ldots$$

We note from the triangle axiom that if $p_n \in C_n$ and $p_{n+2} \in C_{n+2}$ then

$$\varrho(p_n, p_{n+2}) \geqq \varrho(p_n, F) - \varrho(p_{n+2}, F) \geqq \frac{1}{n+1} - \frac{1}{n+2},$$

whence

$$\varrho(C_n, C_{n+2}) \geqq \frac{1}{(n+1)(n+2)} > 0.$$

Therefore, because of the measure Axioms II and IV, we have for any k:

$$\mu C_1 + \mu C_3 + \ldots + \mu C_{2k-1} = \mu(C_1 + C_3 + \ldots + C_{2k-1}) \leqq \mu B,$$
$$\mu C_2 + \mu C_4 + \ldots + \mu C_{2k} = \mu(C_2 + C_4 + \ldots + C_{2k}) \leqq \mu B.$$

Therefore the series $\sum_{k=1}^{\infty} \mu C_k$ is convergent. Next, from Axiom III we have

$$\mu B \leqq \mu B_n + \mu R_n, \tag{2.03}$$

$$\mu R_n \leqq \sum_{k=n}^{\infty} \mu C_k.$$

Because of the convergence of the series of measure μC_n we have $\lim_{n\to\infty} \mu R_n = 0$ and, passing to the limit as $n \to \infty$ in the inequality (2.03), we obtain

$$\mu B \leqq \lambda. \tag{2.04}$$

A comparison of equations (2.02) and (2.04) gives

$$\mu B = \lim_{n\to\infty} \mu B_n.$$

2.05. Corollary. Let W be any set for which $\mu W < +\infty$ and let G, F and G_n have the same meaning as in Theorem 2.01. We have the identical representation

$$W = WG + (W - WG).$$

Furthermore, we have

$$W - WG \subset F, \quad WG_n \subset G_n, \quad WG_n \subset WG;$$

therefore

$$\varrho(W - WG,\ WG_n) \geqq \frac{1}{n} > 0.$$

By Axiom III,

$$\mu W \leqq \mu WG + \mu(W - WG). \tag{2.06}$$

On the other hand, by Axioms II and IV,

$$\mu W \geqq \mu\{WG_n + (W - WG)\} = \mu WG_n + \mu(W - WG).$$

Passing to the limit in this equation as $n \to \infty$, we have, from what has been proved in Theorem 2.01,

$$\mu W \geqq \mu WG + \mu(W - WG).$$

Comparing with equation (2.06), we have

$$\mu W = \mu WG + \mu(W - WG)$$

for any open set G and any $W \subset R$ for which $\mu W < +\infty$.

2.07. Definition. A set $A \subset R$ is called *measurable* if for any set W such that $\mu W < +\infty$ we have

$$\mu W = \mu WA + \mu(W - WA).$$

Thus we have shown that *for any measure* μ, *open sets are measurable.*

2.08. A consequence of the definition of measurability. *If a set A is measurable, $A \cdot B = 0$, and the set $A + B$ has a finite measure, then $\mu(A + B) = \mu A + \mu B$.*

For the proof it is sufficient to set $A + B = W$ and note that $W - WA = B$.

Since in the applications we shall be interested in measurable sets almost exclusively, we shall present a series of theorems concerning measurable sets.

2.09. Theorem. *If A is measurable, then $R - A = A'$ is also measurable.*

Let W be an arbitrary set; then $A'W = W - AW$, $W - A'W = AW$ and therefore

$$\mu A'W + \mu(W - A'W) = \mu(W - AW) + \mu AW = \mu W,$$

whence the theorem follows directly.

2.10. Corollary. *All closed sets are measurable.*

2.11. Theorem 3. *The intersection $D = AB$ of two measurable sets A and B is measurable.*

By hypothesis, for any set W of finite measure we have, because A is measurable,

$$\mu W = \mu AW + \mu(W - AW). \tag{2.12}$$

Taking the set AW in the role of the set W, we have by virtue of the measurability of B,

$$\mu AW = \mu ABW + \mu(AW - ABW). \tag{2.13}$$

Next we take instead of W the set $W - ABW$ and write the condition that A be measurable. As a preliminary we note that

$$A(W - WAB) = AW - ABW$$

and

$$W - WAB - A(W - WAB) = W - ABW - (AW - ABW)$$
$$= W - AW.$$

We then have

$$\mu(W - ABW) = \mu(AW - ABW) + \mu(W - AW). \tag{2.14}$$

Comparing (2.12), (2.13) and (2.14), we obtain the condition that AB be measurable:

$$\mu W = \mu ABW + \mu(W - ABW).$$

2.15. COROLLARY. *The union of two measurable sets is measurable.*

Let A and B be measurable; then $A + B = R - (R - A)(R - B)$. By Theorems 2.09 and 2.11 we find that $A + B$ is measurable.

By the method of complete induction it is easy to prove that the intersection and union of any finite number of measurable sets are measurable.

2.16. COROLLARY. *If A_1 and A_2 are measurable sets without common points, then $\mu(A_1 + A_2) = \mu A_1 + \mu A_2$.*

If one of the sets A_1 and A_2 has an infinite measure, then both sides of the equation are infinite (the left because of Axiom II or III). If both sets have a finite measure, then by Axiom III we have

$$\mu(A_1 + A_2) \leqq \mu A_1 + \mu A_2 < \infty,$$

and from Corollary 2.15 and the measurability of the set A_1, taking $W = A_1 + A_2$, we find

$$\mu(A_1 + A_2) = \mu A_1 + \mu A_2.$$

This conclusion can be extended by complete induction to the union of any finite number of measurable sets without common points.

2.17. THEOREM. *The intersection of a countable number of measurable sets is measurable.*

Let there be given the measurable sets $A_1, A_2, \ldots, A_n, \ldots$. It is required to prove the measurability of the set

$$\Omega = \prod_{n=1}^{\infty} A_n.$$

First we replace the system $\{A_n\}$ by a new system $\{B_n\}$, where $B_n = A_1 \cdot A_2 \cdot \ldots \cdot A_n$. We have $B_1 \supset B_2 \supset \ldots \supset B_n \supset \ldots$. According to Theorem 2.11, all the B_n are measurable; obviously, $\Omega = \prod_{n=1}^{\infty} B_n$. Let W be any set of finite measure; we set $W_n = B_n W$ $(n = 1, 2, \ldots)$ and $W_0 = \Omega W$. Obviously, we have $W_1 \supset W_2 \supset \ldots \supset W_n \supset \ldots \supset W_0$, whence, by Axiom II, we obtain

$$\mu W_1 \geqq \mu W_2 \geqq \ldots \geqq \mu W_n \geqq \ldots \geqq \mu W_0.$$

Consequently, there exists

$$\lim_{n\to\infty} \mu W_n = \lambda \geqq \mu W_0.$$

Next we note that the set W can be represented thus:

$$W = W_0 + (W - W_1) + (W_1 - W_2) + \ldots + (W_n - W_{n+1}) + \ldots,$$

wherein the terms have no common points. From Axiom III we deduce

$$\mu W \leqq \mu W_0 + \mu(W - W_1) + \mu(W_1 - W_2) + \ldots + \mu(W_n - W_{n+1}) + \ldots.$$

From the measurability of the sets B_n there follows

$$\mu(W - W_1) = \mu(W - B_1 W) = \mu W - \mu B_1 W = \mu W - \mu W_1;$$
$$\mu(W_n - W_{n+1}) = \mu(W_n - B_{n+1} W_n) = \mu W_n - \mu B_{n+1} W_n = \mu W_n - \mu W_{n+1} \qquad (n = 1, 2, \ldots).$$

From this we obtain an estimate for μW:

$$\mu W \leqq \mu W_0 + \mu W - \lim_{n\to\infty} \mu W_n,$$

or

$$\mu W_0 \geqq \lambda.$$

Thus we have proved that

$$\mu W_0 = \lambda = \lim_{n\to\infty} \mu W_n. \tag{2.18}$$

Next, from the representation

$$W - W_0 = (W - W_1) + (W_1 - W_2) + \ldots + (W_n - W_{n+1}) + \ldots$$

there follows, by virtue of Axiom III:

$$\mu(W - W_0) \leqq \mu(W - W_1) + \mu(W_1 - W_2) + \ldots + \mu(W_n - W_{n+1}) + \ldots = \mu W - \lambda = \mu W - \mu W_0,$$

whence

$$\mu W \geqq \mu W_0 + \mu(W - W_0).$$

On the other hand, by the same Axiom III,

$$\mu W \leqq \mu W_0 + \mu(W - W_0),$$

and therefore

$$\mu W = \mu \Omega W + \mu(W - \Omega W),$$

which proves the measurability of the set Ω.

2.19. Corollary. Assuming that $\mu B_k < \infty$ for some $k \geqq 1$, we have from the relation (2.18), taking $W = B_k$ and noting that in this case $W_n = B_n$ $(n > k)$ and $W_0 = \Omega$,

$$\mu \Omega = \lim_{n \to \infty} \mu B_n.$$

2.20. Theorem. *If $A_1, A_2, \ldots, A_n, \ldots$ are measurable sets, then $\sum_{n=1}^{\infty} A_n$ is measurable.*

The proof follows directly from Theorems 2.09 and 2.17 and the general relation

$$\sum_{n=1}^{\infty} A_n = R - \prod_{n=1}^{\infty} (R - A_n).$$

2.21. Corollary. *If the measurable sets $A_1, A_2, \ldots, A_n, \ldots$ satisfy the condition $A_1 \subset A_2 \subset \ldots \subset A_n \subset \ldots$* and $\sum_{n=1}^{\infty} A_n = A$, *then $\mu A = \lim_{n \to \infty} \mu A_n$.*

In fact, by Theorem 2.20, the set A is measurable, and by Axiom II, $\mu A \geqq \lim_{n \to \infty} \mu A_n$. If $\lim_{n \to \infty} \mu A_n = \infty$ the assertion is obvious. If $\lim_{n \to \infty} \mu A_n = \lambda < \infty$, then on setting $A - A_n = R_n$ we have $R_n \supset R_{n+1}$. From this, because the sets A, A_n, and R_n are measurable, we find $\mu A = \mu A_n + \mu R_n$, and since R_n converges to the null set, from the corollary of Theorem 4 we have $\lim_{n \to \infty} \mu R_n = 0$. Therefore, $\mu A = \lim_{n \to \infty} \mu A_n$.

2.22. Corollary. *If the measurable sets $A_1, A_2, \ldots, A_n, \ldots$ have no points in common and $A = \sum_{n=1}^{\infty} A_n$, then $\mu A = \sum_{n=1}^{\infty} \mu A_n$.*

Setting $B_n = \sum_{k=1}^{n} A_k$, we have $B_{n+1} \supset B_n$, from which, by Corollary 2.21, we obtain $\mu A = \lim_{n \to \infty} \mu B_n$. But from Corollary

2.16 we have $\mu B_n = \sum_{k=1}^{n} \mu A_k$, whence the assertion follows.

Sets obtained from open sets by the application of a countable number of operations of addition (forming unions) and multiplication (forming intersections) constitute a class of sets *measurable in the sense of Borel* (*measurable* B) in the space R. From Theorems 2.17 and 2.20, by the method of transfinite induction we obtain the theorem:

2.23. THEOREM. *For any Carathéodory measure* μ *all sets measurable* B *are* μ*-measurable.*

Finally, in order to relate the measure of any set $A \subset R$ to the measure of Borel sets, Carathéodory introduces an axiom (of regularity).

V. *The* (outer) *measure of any set* $A \subset R$ *is equal to the greatest lower bound of the measures of Borel sets containing* A.

In certain cases of particular importance for applications the Carathéodory measure for any set A can be defined as a generalized outer measure of Lebesgue, *i.e.*, as the greatest lower bound of the measures of open sets containing A. The most important of these cases is that when the measure of the entire space R is finite.

2.24. THEOREM. *If* $\mu R < \infty$, *then the* (regular) *measure of any set* $A \subset R$ *is equal to the greatest lower bound of the measures of open sets containing* A.

First, by the method of complete induction, we shall prove the theorem for two particular cases.

1. Let there be given the μ-measurable sets $A_1 \supset A_2 \supset \dots \supset A_n \supset \dots$, $\prod_{n=1}^{\infty} A_n = A$ (A is measurable according to Theorem 2.17). We assume that the theorem is true for A_n and shall prove its validity for A. According to the Corollary of Theorem 2.17 (because of the finiteness of the measure), for any $\varepsilon > 0$ there can be found an n such that $\mu A_n < \mu A + \varepsilon/2$; by assumption there exists an open set $G \supset A_n$ such that $\mu G < \mu A_n + \varepsilon/2$. From the two inequalities there follows $\mu G < \mu A + \varepsilon$, wherein $G \supset A$. The assertion has been proved.

2. Let there be given the measurable sets $A_1 \subset A_2 \subset \dots \subset A_n \subset \dots$, $\sum_{n=1}^{\infty} A_n = A$. Then the set A is measurable and $\lim_{n\to\infty} \mu A_n = \mu A$. We assume that the theorem is true for every A_n and shall show its validity for A. Let there be given $\varepsilon > 0$. For each A_n ($n = 1, 2, \dots$) we select an open set $G_n \supset A_n$ such that $\mu G_n < \mu A_n + \varepsilon/2^{n+1}$. Because of the measurability of the sets G_n and A_n we have:

$\mu(G_n - A_n) = \mu G_n - \mu A_n < \varepsilon/2^{n+1}$. We set $\sum_{n=1}^{\infty} G_n = G$. Obviously, G is an open set and $G \supset A$.

We compute

$$\mu(G_1 + G_2 + \ldots + G_n) = \mu G_n + \mu(G_{n-1} - G_{n-1} \cdot G_n)$$
$$+ \mu(G_{n-2} - G_{n-2} \cdot \sum_{k=n-2}^{n} G_k) + \ldots + \mu(G_1 - G_1 \sum_{k=2}^{n} G_n)$$
$$\leqq \mu A_n + \mu(G_n - A_n) + \mu(G_{n-1} - A_{n-1})$$
$$+ \mu(G_{n-2} - A_{n-2}) + \ldots + \mu(G_1 - A_1) \leqq \mu A_n + \varepsilon.$$

Passing to the limit as $n \to \infty$, we obtain

$$\mu G \leqq \mu A + \varepsilon.$$

Thus μA is the greatest lower bound of the measures of open sets $G \supset A$. The proposition is proved.

We proceed now to the general case.

Since the proposition, obviously, is valid for open sets and retains its validity for sets obtained from open sets by the operations of addition and multiplication, it is valid for all sets measurable B. Finally, from Axiom V, it follows directly that for a regular measure the outer measure of any set $A \subset R$ is equal to the greatest lower bound of the measures of the open sets containing A.

Theorem 2.24 is easily generalized to the case when the space has an infinite measure, but can be represented as a countable union of sets each of which has a finite measure.

2.25. Note. From the proof of Theorem 2.24 and of preceding theorems, we notice that it is sufficient that a regular measure μ satisfying Axioms I—V be assigned for all open sets of the space R (which is assumed to have a finite measure or to consist of a countable number of spaces of finite measure); in such a case this measure can be extended over all sets $A \subset R$.

Finally, if the space R has a countable base $\{U_n\}$ possessing the property that the intersection of any two sets of the base lies in the base, then it is sufficient to assign a measure over the (open) sets U_n. In fact, there will then be defined $\mu(U_n + U_m) = \mu U_n + \mu U_m - \mu U_n U_m$, and analogously for the sum of any number of sets of the base. Next, any open set G can be represented as a countable sum of sets of the base $G = \sum_{k=1}^{\infty} U_{n_k}$, and we define $\mu G = \lim_{m\to\infty} \mu \sum_{k=1}^{m} U_{n_k}$. It is easily verified that if the measure over $\{U_n\}$ satisfies the conditions I—IV, then also the extended measure satisfies them.

We proceed with the presentation of certain concepts related to the *metric theory of functions*.

A function $\varphi(p)$, where $p \in R$ and the value of the function φ is a real number, is called *μ-measurable* if for any $\alpha, -\infty < \alpha < +\infty$, the set $\{p;\ \varphi(p) > \alpha\}$ is μ-measurable.

It is easily seen that if a function $\varphi(p)$ is μ-measurable then the set $\{p;\ \alpha < \varphi(p) < \beta\}$, where α and β are any real numbers, and also the set $\{p;\ \varphi(p) = \alpha\}$, are μ-measurable.

Next, for a bounded, measurable function $\varphi(p)$ there is defined the *Lebesgue integral*: if $m \leqq \varphi(p) \leqq M$, we divide the interval (m, M) by intermediate points $m = l_0 < l_1 < \ldots < l_n = M$ and form the sum

$$\sum_{i=0}^{n-1} l_i \cdot \mu E_i + \sum_{i=0}^{n} l_i \cdot \mu E'_i,$$

where

$$E_i = \{p;\ l_i < \varphi(p) < l_{i+1}\}, \qquad E'_i = \{p;\ \varphi(p) = l_i\}.$$

It is easily proved that these sums have a unique limit when $n \to \infty$ and the greatest of the differences $l_{i+1} - l_i$ tends to zero. We shall call this limit the integral of Lebesgue (of Radon) and shall denote it by

$$\int_R \varphi(p) d\mu \text{ or } \int_R \varphi(p) \mu(dp).$$

The definition of the Lebesgue integral generalizes to unbounded functions. If $\varphi(p) \geqq 0$ and is μ-measurable, then we introduce the functions

$$\varphi_n(p) = \begin{cases} \varphi(p), & \varphi(p) \leqq n, \\ n, & \varphi(p) > n. \end{cases}$$

Then we define

$$\int_R \varphi(p) d\mu = \lim_{n\to\infty} \int_R \varphi_n(p) d\mu;$$

moreover, there exists either a finite or an infinite limit since the right-hand side does not decrease for increasing n; if the limit is finite the function $\varphi(p)$ is *μ-summable*.

Analogously, there is defined the integral of a nonpositive measurable function.

Finally, if there be given any μ-measurable function $\varphi(p)$, then, as is usual, we represent it in the form of the sum of a non-negative and a nonpositive function

$$\varphi(p) = \varphi_1(p) + \varphi_2(p),$$

where

$$\varphi_1(p) = \begin{cases} \varphi(p) & \text{if } \varphi(p) > 0, \\ 0 & \text{if } \varphi(p) \leqq 0; \end{cases}$$

$$\varphi_2(p) = \begin{cases} 0 & \text{if } \varphi(p) \geqq 0, \\ \varphi(p) & \text{if } \varphi(p) < 0; \end{cases}$$

then, by definition,

$$\int_R \varphi(p) d\mu = \int_R \varphi_1(p) d\mu + \int_R \varphi_2(p) d\mu$$

if both integrals on the right-hand side are finite. In this case $\varphi(p)$ is called a μ-summable function.

The Lebesgue integral defined in this manner possesses a series of proporties of the ordinary Lebesgue integral; we shall not enumerate them.

In the sequel we shall frequently be concerned with functions of the form $F(p, t)$, where p is a point of space and t is a real number. We shall need a theorem concerning the possibility of an interchange of the order of integration with respect to p and t: Fubini's Theorem. We shall present its proof for integrals of the form which we shall encounter.

We shall consider the space $R \times I$ — the topological product of the metric space R with a countable base and the space I of the real variable t $(-\infty < t < +\infty)$. The points of this space are represented by an ordered combination of the point p and the number t: $(p, t) \in R \times I$. The space $R \times I$ can be regarded as metric if, for example, there be defined in it the distance

$$\varrho[(p_1, t_1), (p_2, t_2)] = [\varrho^2(p_1, p_2) + (t_1 - t_2)^2]^{1/2}.$$

Furthermore, if the base for R is $\{U_n\}$, then the space $R \times I$ also has a countable base; one may take for this base the totality of open sets $\{U_n \times \Delta_i\}$, where $\{\Delta_i\}$ is the set of all open intervals of the space I with rational extremities. In addition it is clear that if the base $\{U_n\}$ possesses the property that the intersection $U_n \cdot U_m$

also belongs to the base, then the base $\{U_n \times \Delta_i\}$ of the space $R \times I$ possesses the same property. Over the sets of the base $\{U_n \times \Delta_i\}$ we define a measure ν:

$$\nu(U_n \times \Delta_i) = \mu U_n \cdot \text{mes}\, \Delta_i,$$

where μ denotes the Carathéodory measure in the space R and "mes" denotes the usual Lebesgue linear measure over I (mes Δ_i is the length of the interval Δ_i). From the note to Theorem 2.24, a measure defined in this way extends as a Carathéodory measure over all sets of the space $R \times I$ wherein, if the space R is a the union of not more than a countable number of sets of finite measure μ, the space $R \times I$ is the union of a countable number of sets of finite measure. After these preliminary remarks we proceed to the formulation and proof of Fubini's theorem.

2.26. THEOREM. (Fubini) *If $F(p, t)$ is a non-negative ν-measurable function, then there holds the equality*:

$$\int_{R\times I} F(p, t)d\nu = \int_I dt \int_R F(p, t)d\mu = \int_R d\mu \int_I F(p, t)dt.$$

Moreover, the inner integrals may have no sense for a set of values of t and, correspondingly, of p of measure zero.

The fundamental difficulty which must be overcome consists in proving this theorem when the integrand is the characteristic function of some measurable set, *i. e.* a function equal to 1 over the given set and zero outside it. Indeed, from this case it is easy to pass to the case of a function taking on only a finite number of values — this can be written as a linear combination of a finite number of characteristic functions. Furthermore, every measurable bounded function can be approximated with arbitrary precision by a function assuming only a finite number of values. Finally, for the passage to an unbounded non-negative function we employ truncated functions F_n analogous to those already introduced here in the definition of the Lebesgue integral [3].

1. If $\varphi(p, t)$ is the characteristic function of the set $U_n \times \Delta_i$, then according to the definition of the measure ν we have

$$\nu(U_n \times \Delta_i) = \int_{R\times I} \varphi(p, t)d\nu = \mu U \cdot \text{mes}\, \Delta_i = \int_R d\mu \int_I \varphi(p, t)dt$$
$$= \int_I dt \int_R \varphi(p, t)d\mu.$$

[3]For the proof of Fubini's theorem for double integrals, see de la Vallée Poussin, *Cours d'Analyse Infinitésimale*, 2-ème éd., t. II, § 106-110.

In precisely the same way, if $\varphi(p, t)$ is the characteristic function for the set $(U_n \times \Delta_i) + (U_m \times \Delta_j)$, then from the formula defining measure,

$$\nu(U_n \times \Delta_i + U_m \times \Delta_f)$$
$$= \nu(U_n \times \Delta_i) + \nu(U_m \times \Delta_j) - \nu[(U_n \times \Delta_i) \cdot (U_m \times \Delta_j)]$$

and in view of the fact that $(U_n \times \Delta_i) \cdot (U_m \times \Delta_j) = (U_n \cdot U_m) \times (\Delta_i \cdot \Delta_j)$ we again obtain

$$\text{(2.27)} \qquad \int_{R\times I} \varphi(p, t) d\nu = \int_R d\mu \int_I \varphi(p, t) dt = \int_I dt \int_R \varphi(p, t) d\mu.$$

This formula is easily generalized to open sets consisting of the union of a finite number of sets of the base $R \times I$.

2. Now let G be any open set of the space $R \times I$. Denoting for simplicity the base $\{U_n \times \Delta_j\}$ of the space $R \times I$ by $\{V_n\}$, we have

$$G = \sum_{k=1}^{\infty} V_{n_k},$$

whence

$$\nu G = \lim_{m\to\infty} \nu(\sum_{k=1}^{m} V_{n_k}).$$

We set $\sum_{k=1}^{m} V_{n_k} = G_m$ and let $\varphi_m(p, t)$ be the characteristic function for G_m and $\varphi(p, t)$ the characteristic function for G. According to what has been proved, we have

$$\nu G_m = \int_{R\times I} \varphi_m(p, t) d\nu = \int_R d\mu \int_I \varphi_m(p, t) dt = \int_I dt \int_R \varphi_m(p, t) d\mu.$$

Noting that $\lim_{m\to\infty} \varphi_m(p, t) = \varphi(p, t)$ and that in the Lebesgue integral of a bounded function one can pass to the limit under the integral sign, we obtain

$$G = \int_{R\times I} \varphi(p, t) d\nu = \int_R d\mu \int_I \varphi(p, t) dt = \int_I dt \int_R \varphi(p, t) d\mu,$$

i.e., formula (2.27) is also valid in this case.

The same limiting transition will prove the validity of the formula for any set measurable B wherein both inner integrals exist for all values of the parameter.

3. Finally, let $A \subset R \times I$ be any ν-measurable set and $\varphi(p, t)$ its characteristic function. For simplicity we assume that $A \subset R_1 \times I_1$, where $R_1 \times I_1$ is one of the countable number of

sets of finite measure forming the space $R \times I$; the passage to the general case is obtained by a countable summation.

For definiteness let $\nu(R_1 \times I_1) = 1$. According to the definition of ν-measure there exists a sequence of open sets (here, as everywhere in the sequel, relative to $R_1 \times I_1$ regarded as the space),

$$G_1 \supset G_2 \supset \ldots \supset G_n \supset \ldots \supset A,$$

such that

$$\lim_{n\to\infty} \nu G_n = \nu A.$$

Denoting the characteristic function of A by $\varphi(p, t)$ and that of G_n by $\varphi_n(p, t)$, according to what has preceded we have

$$\nu G_n = \int_{R_1\times I_1} \varphi_n(p, t)d\nu = \int_{I_1} dt \int_{R_1} \varphi_n(p, t)d\mu. \tag{2.28}$$

Noting that $\varphi_n \geqq \varphi$ and that the nonincreasing sequence of bounded functions φ_n converges to a limit function $\varphi'(p, t)$, wherein $\varphi'(p, t) \geqq \varphi(p, t)$, on passing to the limit under the integral sign we obtain

$$\nu A = \int_{I_1} dt \int_{R_1} \varphi'(p, t)d\mu. \tag{2.29}$$

Completely analogously, on including the measurable set $R_1 \times I_1 - A$ in a system of open sets

$$\Gamma_1 \supset \Gamma_2 \supset \ldots \supset \Gamma_n \supset \ldots \supset R_1 \times I_1 - A,$$

denoting the characteristic function for Γ_n by $1 - \psi_n(p, t)$ and setting $\lim_{n\to\infty} \psi_n = \varphi''(p, t)$, wherein $\psi_n(p, t) \leqq \varphi''(p, t) \leqq \varphi'(p, t)$ we have

$$\nu(R_1 \times I_1 - A) = \int_{I_1} dt \int_{R_1} (1 - \varphi''(p, t))d\mu,$$

whence

$$\nu A = \int_{I_1} dt \int_{R_1} \varphi''(p, t)d\mu. \tag{2.30}$$

Comparing the equalities (2.29) and (2.30), we find

$$\int_{I_1} dt \int_{R_1} [\varphi'(p, t) - \varphi''(p, t)]d\mu = 0. \tag{2.31}$$

In this last expression $\int_{R_1}[\varphi'(p, t) - \varphi''(p, t)]d\mu$ is a non-negative function of t; because of equation (2.31) it is equal to zero for almost all values of $t \in I_1$; *i. e.*, for almost all values of $t \in I_1$

$$\int_{R_1} \varphi'(p, t)d\mu = \int_{R_1} \varphi''(p, t)d\mu,$$

and since we have $\varphi' \geqq \varphi \geqq \varphi''$, then for almost all values of t,

$$\int_{R_1} \varphi'(p, t)d\mu = \int_{R_1} \varphi(p, t)d\mu,$$

Substituting the expression thus found in the inner integral (2.29) and noting that its value over a set of values of t of measure zero does not influence the magnitude of the integral, we find

$$\nu A = \int_{R_1 \times I_1} \varphi(p, t)d\nu = \int_{I_1} dt \int_{R_1} \varphi(p, t)d\mu. \tag{2.32}$$

Rewriting formula (2.28) in the form

$$\nu G_n = \int_{R_1 \times I_1} \varphi_n d\nu = \int_{R_1} d\mu \int_{I_1} \varphi_n(p, t)dt$$

and repeating the same arguments, we obtain

$$\int_{R_1 \times I_1} \varphi(p, t)d\nu = \int_{R_1} d\mu \int_{I_1} \varphi(p, t)dt. \tag{2.33}$$

The formulas (2.32) and (2.33) completely prove the theorem of Fubini for the characteristic functions of sets belonging to a topological product space of finite measure.

The extension to characteristic functions of any ν-measurable sets contained in $R \times I$ and also to any measurable non-negative functions is accomplished as shown above.

The measure ν bears for us an auxiliary character, and in applications we shall employ a corollary of Fubini's theorem on the possibility of interchanging the order of integration in a double integral.

2.34. COROLLARY. *If a non-negative function $F(p, t)$ is ν-measurable in the space $R \times I$, then there holds the equality*

$$\int_I dt \int_R F(p, t)d\mu = \int_R d\mu \int_I F(p, t)dt.$$

In particular, if one of the two integrals is finite, then the other is likewise finite.

2.35. NOTE. If $F(p, t)$ is any function ν-summable in the space $R \times I$, then, as usual, we represent it in the form of a sum of a non-negative function and a non-positive function, $F = F_1 + F_2$, and the theorem remains valid.

2.36. NOTE. We have carried out the proof of Fubini's theorem for the topological product of a metric space with a countable base and a one-dimensional Euclidean space because this case will be encountered in the sequel. The proof would not be altered if we were to have two metric spaces with countable bases, R_1 and R_2, with Carathéodory measures μ_1 and μ_2, each R being the union of not more than a countable number of sets of finite measure.

1. Recurrence Theorems.

Let a dynamical system $f(p, t)$ be given in a metric space R. A measure μ defined in the space R is called *invariant* (with respect to the system $f(p,t)$) if for any μ-measurable set A there holds the equality

$$(3.01) \qquad \mu f(A, t) = \mu A \qquad (-\infty < t < +\infty).$$

From the property (3.01) it follows that images of a measurable set are measurable. This invariant measure is the natural generalization of the integral invariant considered in section 1 for systems of differential equations.

Systems with an invariant measure possess a series of properties distinguishing them from general dynamical systems. In this section we shall consider the theorem of Poincaré-Carathéodory on recurrence.

Let there exist in a space R of a dynamical system an invariant measure μ. Suppose that the measure of the entire space is finite; for simplicity we set $\mu R = 1$. The recurrence theorem separates naturally into two parts.

3.02. THEOREM (RECURRENCE OF SETS). *Let $A \subset R$ be a measurable set, $\mu A = m > 0$. Then there can be found positive and negative values of t ($|t| \geqq 1$) such that $\mu[A \cdot f(A, t)] > 0$.*

For the proof we consider the positions of the set $f(A, t)$ for integral values of $t (t = 0, \pm 1, \pm 2, \ldots)$ and introduce the notation

$$A_n = f(A, n) \qquad (n = 0, \pm 1, \pm 2, \ldots).$$

Because of invariance we have

$$\mu A_n = \mu A = m > 0.$$

If it be assumed, for example, that the sets $A_0, A_1, \ldots, A_k$ intersect in pairs only in sets of measure zero, we then obtain

$$\mu(A_0 + A_1 + \ldots + A_k) = km,$$

which, when $k > 1/m$, leads to a contradiction of the assumption that $R = 1$.

Thus there exist two sets A_i, A_j $(i \neq j)$ such that

$$\mu(A_i \cdot A_j) > 0. \tag{3.03}$$

Suppose that $i < j$; then $0 \leqq i < j \leqq k$. On applying to the set $A_i \cdot A_j$ the transformation $f(p, -i)$, we obtain from (3.03) that

$$\mu(A_0 \cdot f(A_0, j - i)) > 0,$$

which proves the assertion, since $j - i \geqq 1$; moreover, we can choose

$$j - i \leqq \left[\frac{1}{m}\right] + 1.$$

If one applies the transformation $f(p, -j)$, to (3.03), one obtains

$$\mu(A_0 \cdot f(A_0, i - j)) > 0, \qquad i - j \leqq 1.$$

The theorem is proved.

3.04. NOTE. By the same method it is easy to prove that the values of t for which $\mu(A \cdot f(A, t)) > 0$ can be arbitrarily large in absolute value. In fact, let T, $T > 0$ be any preassigned number; we choose an integer $N > T$ and consider the sequence of sets

$$A_0, A_N, A_{2N}, \ldots, A_{kN}, \ldots.$$

The preceding argument leads to the relation

$$\mu(A_0 \cdot f(A_0, N(j - i)) > 0, \qquad N(j - i) \geqq N > T,$$

and analogously for values $t < -T$.

3.05. THEOREM (RECURRENCE OF POINTS). *If in a space R with a countable base we have $\mu R = 1$ for an invariant measure μ, then almost all points $p \in R$ (in the sense of the measure μ) are stable according to Poisson, i. e.* denoting the set of points unstable according to Poisson by $\mathscr{E}$, we have $\mu\mathscr{E} = 0$.

We first take any measurable set A such that $\mu A = m > 0$. As in the preceding theorem, we set

$$A_n = f(A, n) \qquad (n = 0, \pm 1, \pm 2, \ldots).$$

Next we construct the sets

$$A_0 \cdot A_1 = A_{01},\ A_0 \cdot A_2 = A_{02}, \ldots, A_0 \cdot A_n = A_{0n}, \ldots$$

$$(3.06) \qquad \ldots, A_0 - \sum_{i=1}^{\infty} A_{0i} = A_{0\infty},$$

$$A_1 \cdot A_2 = A_{12},\ A_1 \cdot A_n = A_{1n}, \ldots, A_1 - \sum_{i=2}^{\infty} A_{1i} = A_{1\infty},$$

. . . .

We shall prove that $\mu A_{0\infty} = 0$. Suppose that $\mu A_{0\infty} = l > 0$. Since $f(A_i, 1) = A_{i+1}$, then $f(A_{0i}, 1) = A_{1,\, i+1}$ and furthermore, $f(A_{0\infty}, 1) = A_{1\infty}$. Repeating the same reasoning for the sets $A_{1\infty}, A_{2\infty}, \ldots$ and taking into consideration the invariance of the measure μ, we obtain

$$f(A_{0\infty}, n) = A_{n\infty} \quad (n = 1, 2, \ldots); \qquad \mu A_{0\infty} = \mu A_{1\infty} = \ldots = \mu A_{n\infty} = \ldots = l.$$

Moreover, by construction we have

$$A_{0\infty} \cdot A_i = 0 \qquad (i = 1, 2, \ldots),$$

whence, since $A_i \supset A_{i\infty}$,

$$A_{0\infty} \cdot A_{i\infty} = 0 \qquad (i = 1, 2, \ldots).$$

Analogously,

$$A_{i\infty} \cdot A_{j\infty} = 0 \qquad (j = i + 1,\ i + 2, \ldots).$$

Thus the sets $A_{i\infty}$ $(i = 0, 1, 2, \ldots)$, taken in pairs, have no common points. Therefore, the assumption $\mu A_{0\infty} = l > 0$ contradicts the finiteness of the measure of the space R. The assertion is proved.

We now take a countable defining system of neighborhoods $\{U^{(n)}\}$ of the space R and construct for each $U^{(n)}$ the set $U^{(n)}_{0\infty}$ according to the scheme (3.06); according to what has been proved, $\mu U^{(n)}_{0\infty} = 0$ $(n = 1, 2, \ldots)$. We define the set

$$\mathscr{E} = \sum_{n=1}^{\infty} U^{(n)}_{0\infty}.$$

Obviously, $\mu\mathscr{E} = 0$. We assert that all points $p \in R - \mathscr{E}$ are stable P^-.

Indeed, suppose that $p \in R - \mathscr{E}$. Then from the definition of the set $\mathscr{E}$, for any neighborhood $U^{(k)}$ containing the point p there can be found images of this neighborhood containing p, *i.e.*, there can be found a natural number m such that

$$p \in f(U^{(k)}, m).$$

Applying to both parts of this inclusion the operation $f(p, -m)$ we obtain

$$f(p, -m) \in U^{(k)}.$$

Since $U^{(k)}$ is any neighborhood containing p and $-m \leqq -1$, then from this it follows that p is stable P^-.

Analogously, repeating the same arguments, beginning with the scheme (3.06) for the sets $A_0, A_{-1}, \ldots, A_{-n}, \ldots$ and defining the set $A_{0, -\infty}$, we construct first the sets $U^{(n)}_{0, -\infty}$ $(n = 1, 2, \ldots)$ and then the set

$$\mathscr{E}_1 = \sum_{n=1}^{\infty} U^{(n)}_{0, -\infty} \qquad (\mu\mathscr{E}_1 = 0),$$

such that any point $p \in R - \mathscr{E}_1$ is stable P^+. In this way, every point belonging to the set $R - (\mathscr{E} + \mathscr{E}_1)$ is stable P, while $\mu(\mathscr{E} + \mathscr{E}_1) = 0$. Theorem 3.05 is proved.

3.07. Note. Every point p of the set $\mathscr{E}$, as it is easy to see from its construction, is unstable P^- with respect to the discrete sequence of values $t = -1, -2, \ldots$, since it belongs to one of the sets $U^{(n)}_{0\infty}$ and therefore none of the points $f(p, -n)$ belongs to the neighborhood $U^{(n)}$. From a remark in section 4 of Chapter IV all these points are therefore unstable P^- for a continuous variation of t as $t \to -\infty$. Thus the set $\mathscr{E} + \mathscr{E}_1$ is the set of all points $p \in R$ unstable P in the usual sense.

3.08. Note. The closure of the set of points lying on the stable trajectories defines a set of central motions (Chapter V, section 5). In many cases important for applications every open nonempty set has a positive measure. Such, for example, are the systems of differential equations in Euclidean space considered in section 1 and in general, systems in a Euclidean space with an integral invariant of the form

$$\iint_D \cdots \int M dx_1 dx_2 \ldots dx_n,$$

where M is a positive measurable function. From what has been proved it follows that in systems with an invariant measure of such a type the set of points unstable P is nowhere dense in R and, consequently, *the set C of central motions fills the whole space.*

Under the general definition of an invariant measure this situation does not hold and we can only assert that the open set $R - C$ appearing as the complement of the set of central motions has a μ-measure equal to zero.

A. Khintchine has sharpened Poincaré's Theorem 3.02. As we have seen, Theorem 3.02 asserts only that for any set $E\,(\mu E > 0)$ there exist values of t arbitrarily large in absolute value such that $\mu(E \cdot f(E, t)) > 0$. But Theorem 3.02 does not provide an estimate of the upper limit for the measure of the intersection, and also gives no information as to how dense are those values of t for which this measure exceeds a certain positive quantity. These facts are established by the theorem of Khintchine. We shall present a proof of this theorem due to Visser.

3.09 LEMMA (Visser). *If in the space R, in which there is defined a measure μ for which $\mu R = 1$, there is given a system of measurable sets $E_1, E_2, \ldots, E_i, \ldots$,*

$$E_i \subset R, \quad \mu E_i \geqq m > 0 \qquad (i = 1, 2, \ldots),$$

then there can be found at least two sets E_i, E_j $(i \neq j)$ such that $\mu(E_i \cdot E_j) > \lambda m^2$, where λ is any number less than 1.

We shall explain the significance of the quantity m^2 in the statement of the theorem. Let $m = 1/k$, where k is an integer; let $\mu E_i = m$ $(i = 1, 2, \ldots, k)$ and let $E_1, E_2, \ldots, E_k$ intersect in pairs in sets of measure zero. Then $\mu(E_1 + E_2 + \ldots + E_k) = 1$ and, if $\mu E_{k+1} = m$, E_{k+1} intersects at least one E_i $(i = 1, 2, \ldots, k)$ in a set of measure $\geqq (1/k)\, m = m^2$.

The number λ in the hypothesis of the theorem cannot be taken greater than 1 as the following example shows: on the segment $[0, 1]$ let

$$E_1 = \left[0, \frac{1}{2}\right], \quad E_2 = \left[0, \frac{1}{4}\right] + \left[\frac{1}{2}, \frac{3}{4}\right], \ldots,$$

$$E_n = \left[0, \frac{1}{2^n}\right] + \left[\frac{2}{2^n}, \frac{3}{2^n}\right] + \ldots + \left[\frac{2^n - 2}{2^n}, \frac{2^n - 1}{2^n}\right], \ldots$$

Here mes $E_i = \frac{1}{2}$; mes $(E_i \cdot E_j) = \frac{1}{4} (i \neq j)$.

PROOF OF THE LEMMA. Given $\mu E_i \geqq m$, suppose that $\mu(E_i \cdot E_j) \leqq l$ for any pair of sets if $i \neq j$. We shall estimate the quantity l from below. We take a finite number of sets of the system

$E_1, E_2, \ldots, E_k$ $(k > 1)$. From the hypothesis of the theorem, for $k = 2, 3, \ldots$ we have

$$1 = \mu R \geqq \mu \sum_{i=1}^{k} E_i \geqq \sum_{i=1}^{k} \mu E_i - \sum_{\substack{i,j=1 \\ i \neq j}}^{k} \mu(E_i E_j) \geqq km - \frac{k(k-1)}{2} l.$$

From this

$$l \geqq \frac{2(km-1)}{k(k-1)}.$$

This estimate is crude for large k since the denominator increases faster than the numerator. We shall find the most favorable value of k (*i.e.* that giving the largest value of the right-hand side):

$$\frac{km-1}{k(k-1)} \geqq \frac{(k-1)m-1}{(k-1)(k-2)}; \quad \frac{km-1}{k(k-1)} > \frac{(k+1)m-1}{k(k+1)}.$$

From these equations we have

$$mk \leqq 2, \qquad mk > 2 - m,$$

or $2/m - 1 < k \leqq 2/m$. Thus the most favorable value is $k = [2/m] = 2/m - \theta$, where $0 \leqq \theta < 1$. Then $km = 2 - \theta m$.

We shall estimate l from below, making use of the value found for k. Substituting this value in the inequality for l, we find

$$l \geqq \frac{2m^2(1-\theta m)}{(2-\theta m)(2-m-\theta m)}.$$

We have

$$\frac{1-\theta m}{2-m-\theta m} > \frac{1}{2}, \qquad \frac{2}{2-\theta m} > 1.$$

Strengthening the inequality, we find

$$l \geqq \tfrac{1}{2} m^2.$$

Thus if $l < \frac{1}{2}m^2$, then there can be found sets E_i, E_j $(i \neq j)$ such that $\mu(E_i \cdot E_j) > l$.

In order to replace the factor $\frac{1}{2}$ by the factor λ, where λ is subjected to the single condition $\lambda < 1$, we have recourse to the following procedure. We construct a set R^N as the topological product of N copies of the set R, every point $p \in R^N$ is a combination $(p_1, p_2, \ldots, p_N)$, where *i.e.* $p_i \in R$ $(i = 1, 2, \ldots, N)$; to the set

$E^N = \{(p_1, p_2, \ldots, p_N)\}$, $p_i \in E \subset R$, we assign the measure $\mu^N E = (\mu E)^N$. We replace the sets E_i by E_i^N and apply to them the part of the lemma which has been proved. There can be found sets E_i^N and E_j^N $(i \neq j)$ such that $\mu^N(E_i^N \cdot E_j^N) > \frac{1}{2}m^{2N}$. But according to the definition of μ^N we have

$$\mu^N(E_i^N \cdot E_j^N) = [\mu(E_i \cdot E_j)]^N.$$

Consequently,

$$\mu(E_i \cdot E_j) > \sqrt[N]{\tfrac{1}{2}} \,.\, m^2.$$

For a given $\lambda < 1$ there can always be found a natural number N such that

$$\sqrt[N]{\tfrac{1}{2}} > \lambda.$$

Thus,

$$\mu(E_i \cdot E_j) > \lambda m^2,$$

which it was required to prove.

3.10. Theorem of Khintchine. *Under the hypotheses of Theorem* (3.02) *for any measurable set* E, $\mu E = m > 0$, *the inequality*

$$\mu(t) = \mu(E \cdot f(E, t)) > \lambda m^2$$

is fulfilled for a relatively dense set of values of t *on the axis* $-\infty < t < +\infty$ (*for any* $\lambda < 1$).

Assume that the proposition is not true; then there exist a measurable set E, $\mu E = m > 0$, a number $\lambda_0 < 1$, and arbitrarily large intervals of the t-axis where the inequality

$$\mu(t) = \mu(E \cdot f(E, t)) \leqq \lambda_0 m^2 \tag{3.11}$$

is fulfilled. Let Δ_1 be an interval of length L_1 such that for $t \in \Delta_1$ there holds the inequality (3.11); let its midpoint be l_1. There exists a Δ_2 of length $L_2 \geqq 2\,|l_1|$ $(\Delta_1 \cdot \Delta_2) = 0)$, where again (3.11) is fulfilled. Let its midpoint be l_2; since $t = 0$ does not lie in Δ_2, $|l_2| > |l_1|$. We denote, in general, by Δ_n $(\Delta_i \cdot \Delta_n = 0$ if $i < n)$ the interval of length $L_n \geqq 2\,|l_{n-1}|$ in which the inequality (3.11) holds and we denote its midpoint by l_n; $|l_n| > |l_{n-1}|$. Since the number $l_j - l_i$ lies in the interval Δ_j $(j > i)$, then by assumption

$$\mu(E \cdot f(E, l_j - l_i)) \leqq \lambda_0 m^2.$$

From this, because of the invariance of the measure,

$$\mu(f(E, l_i) \cdot f(E, l_j)) \leqq \lambda_0 m^2 \quad (i < j),$$

i.e., the sets E, $f(E, l_1), \ldots, f(E, l_n), \ldots$ satisfy an inequality of the form (3.11), which contradicts the lemma. The theorem is proved.

3.12. NOTE. By employing the theory of spectral decompositions, Khintchine proved more. Namely, he proved the representation $\mu(t) = m\int_{-\infty}^{\infty} e^{itx}\, d\varphi(x)$, where $\varphi(x)$ is a distribution function, *i.e.* a nondecreasing function of x for which $\varphi(-\infty) = 0, \varphi(+\infty) = 1$; from this it follows that $\mu(t)$ is the sum of an almost periodic function and a function whose mean square on the interval $(-\infty, +\infty)$ is equal to zero.

4. Theorems of E. Hopf

The theorems of E. Hopf constitute a generalization of the second part of Poincaré's theorem on recurrence to the case where the measure of the whole space R is infinite. Obviously, in this case it is impossible to assert that in the presence of an invariant measure almost all the motions are stable according to Poisson. To see this, it is sufficient to consider in the n-dimensional, unbounded Euclidean space the system of differential equations admitting an invariant volume:

$$\frac{dx_1}{dt} = 1, \ \frac{dx_2}{dt} = 0, \ldots, \frac{dx_n}{dt} = 0.$$

All its solutions $x_1 = x_1^{(0)} + t$, $x_2 = x_2^{(0)}, \ldots, x_n = x_n^{(0)}$ go to infinity as $t \to \infty$ and $t \to -\infty$ and the trajectories have no limit point at a finite distance, *i.e.*, they are unstable P.

For the formulation of Hopf's theorems it is necessary to introduce the concept of a *departing point*. We shall consider a *locally compact space R with a countable base* in which motions have been defined. We shall say that a point p is departing as $t \to +\infty$ if the trajectory $f(p, t)$ has no ω-limit points; analogously, p is a departing point as $t \to -\infty$ if $f(p, t)$ has no α-limit points. It is obvious that if a point is departing as $t \to +\infty$ or as $t \to -\infty$, then the same is true of all other points on the same trajectory.

4.01. THEOREM I OF HOPF. *Let there be given a locally compact metric space of motions with a countable base. Let there be defined in it an invariant measure μ having the following properties:* $\mu R = +\infty$, *but for any compact set $F \subset R$ the measure μF is finite. Then almost all points $p \in R$ as $t \to +\infty$ are either stable according to Poisson or are departing.*

We shall show first that in the proof one may restrict his attention to integral values of t only. It is obvious that if $f(p, t)$ has no ω- or α-limit points then the sequence $\{f(p, n)\}$ $(n = 1, 2, \ldots)$ also has no corresponding limit points. We shall prove the converse proposition.

4.02. LEMMA. *If the sequence $\{f(p, n)\}$ has no limit points as $n \to \infty$,* then *$f(p, t)$ has no ω-limit points.*

Assume the contrary: let there exist a sequence of values $t_1 < t_2 < \ldots t_n < \ldots$, $\lim_{n\to\infty} t_n = +\infty$, such that

$$\lim_{n\to\infty} f(p, t_n) = q.$$

Denoting by k_n the largest integer not exceeding t_n, we have $t_n = k_n + \sigma_n$, $0 \leqq \sigma_n < 1$. Since the set of numbers $\{\sigma_n\}$ is bounded, it has a limit point σ, $0 \leqq \sigma \leqq 1$, and there exists a subsequence converging to σ. In order not to complicate the notation we shall assume that the choice of this subsequence has already been made and that $\lim_{n\to\infty} \sigma_n = \sigma$.

Thus we have

$$\lim_{n\to\infty} f(p, k_n + \sigma_n) = q.$$

From this, because of the continuous dependence of f on the initial conditions, for trajectories near to $f(p, t)$ we find

$$\lim_{n\to\infty} f(p, k_n + \sigma_n - \sigma) = f(q, -\sigma),$$

i.e., for any $\varepsilon > 0$ and $n > N_1(\varepsilon)$ we have

$$\varrho[f(p, k_n + \sigma_n - \sigma), f(q, -\sigma)] < \frac{\varepsilon}{2}.$$

Since the points $f(p, k_n + \sigma_n - \sigma)$ fall, beginning with a certain n, in a compact neighborhood of the point $f(q, -\sigma)$, the continuity being uniform in this neighborhood, and since $\sigma_n - \sigma \to 0$, then for $n > N_2(\varepsilon)$ we have

$$\varrho[f(p, k_n + \sigma_n - \sigma), f(p, k_n)] < \frac{\varepsilon}{2}.$$

From the two inequalities there follows for $n \geqq \max [N_1, N_2]$,

$$\varrho[f(p, k_n), f(p, -\sigma)] < \varepsilon,$$

i.e. the sequence $\{f(p, k_n)\}$ with integral arguments k_n has a limit point $f(q, -\sigma)$, contrary to hypothesis. The lemma is proved.

Obviously, an analogous proposition is valid for α-limit points.

Thus, in order to determine the set of all points p departing as $t \to +\infty$ it is sufficient to consider the set of points departing for the sequence $t = 1, 2, \ldots, n, \ldots$. We saw in Section 4 of Chapter V that in just such a way the set of points stable according to Poisson as $t \to +\infty$ is identical with the set of points stable according to Poisson with respect to the sequence $t = 1, 2, \ldots, n, \ldots$. Thus, in the sequel we can restrict our consideration to the sequence

$$\{f(p, n)\} \qquad (p \in R;\ n = 1, 2, \ldots). \tag{4.03}$$

PROOF OF THE THEOREM. Let a defining, countable system of neighborhoods for R be $U_1, U_2, \ldots, U_n, \ldots$. The set of points $p \in R$ unstable P^+, as we know (section 3), is

$$V^+ = \sum_{n=1}^{\infty} U_n^*,$$

where

$$U_n^* = U_n - U_n \cdot \sum_{m=1}^{\infty} f(U_n, -m).$$

We now determine among the sets of points unstable P^+ the set of points nondeparting as $n \to +\infty$. We have seen that the set U_n^* possesses the property

$$U_n^* \cdot f(U_n^*, -k) = 0 \qquad (k = 1, 2, \ldots),$$

or, applying the operation $f(p, k)$,

$$U_n^* \cdot f(U_n^*, k) = 0 \qquad (k = 1, 2, \ldots).$$

We introduce temporarily the concept of a *point departing (nondeparting) as $t \to +\infty$ from a compact set $F \subset R$*; such we shall call a point for which the sequence (4.03) has no limit points in F.

Let A be a set such that $A \cdot f(A, k) = 0$ $(k = 1, 2, \ldots)$. We construct the set of points $p \in A$, nondeparting from F as $t \to +\infty$. We denote

$$F \cdot f(A, k) = D_k \text{ and } D_k^* = f(D_k, -k) = f(F, -k) \cdot A \subset A.$$

Every point $p \in D_k^*$ is a point belonging for $t = 0$ to the set A and for $t = k$ to the set F; the set of those points $p \in A$ which do not depart from F is the set of points for which, for an infinite set of values of $k > 0$,

$$f(p, k) \in F,$$

i.e. it is the set

$$W^+(A, F) = \limsup_{k \to \infty} D_k^* = \prod_{l=1}^{\infty} \sum_{k=l}^{\infty} D_k^*.$$

Assigning a sequence of compact sets

$$F_1 \subset F_2 \subset \ldots \subset F_m \subset \ldots, \quad \text{where } \lim_{m \to \infty} F_m = R,$$

we obtain the set of nondeparting points contained in A as the sum

$$W^+(A) = \sum_{m-1}^{\infty} W^+(A, F_m).$$

In order to determine the set of points nondeparting as $t \to +\infty$ and unstable P^+ in the entire space R, it is sufficient for the defining system of neighborhoods $\{U_n\}$ to construct

$$W^+ = \sum_{n=1}^{\infty} W^+(U_n^*).$$

We proceed to the computation of the measure of this set. By hypothesis the measure of any compact set F is finite: $\mu F < +\infty$. Furthermore, all the sets D_k have no points in common and are contained in F. Therefore $\sum_{k=1}^{\infty} \mu D_k \leqq m < +\infty$, *i.e.* the series

$$\sum_{k=1}^{\infty} \mu D_k$$

converges. Because of the invariance of the measure we have

$$\mu D_k = \mu D_k^*,$$

i.e. the series $\sum_{k=1}^{\infty} \mu D_k^*$ also converges.

But we have

$$W^+(A, F) = \prod_{l=1}^{\infty} \sum_{k=l}^{\infty} D_k^* \subset \sum_{k=l}^{\infty} D_k^*$$

for any l and, because of the convergence of the series, one can take l so large that

$$\mu \sum_{k=l}^{\infty} D_k^* \leqq \sum_{k=l}^{\infty} \mu D_k^* < \varepsilon,$$

where $\varepsilon > 0$ is an arbitrary number; thus

$$\mu W^+(A, F) < \varepsilon,$$

i.e.,

$$\mu W^+(A, F) = 0.$$

Thus, if $A \cdot f(A, k) = 0$ $(k = 1, 2, \ldots)$, the measure of the set of points contained in A and nondeparting as $t \to +\infty$ from any compact set F is equal to zero. Furthermore,

$$\mu W^+(A) = \mu \sum_{m=1}^{\infty} W^+(A, F_m) \leqq \sum_{m=1}^{\infty} \mu W^+(A, F_m) = 0.$$

Therefore, setting consecutively $A = U_1^*, U_2^*, \ldots$ and summing, we find for the measure of the set of points nondeparting as $t \to +\infty$ and unstable P^+,

$$\mu W^+ = 0.$$

The theorem is proved.

4.04. THEOREM II OF HOPF. *Under the hypotheses of Theorem I. (4.01), almost all motions departing as $t \to +\infty$ ($t \to -\infty$) depart also as $t \to -\infty$ ($t \to +\infty$); almost all motions stable P^+ (P^-) are also stable $P^-(P^+)$.*

We have seen that all motions defined by the points

$$p \in V^+ = \sum_{n=1}^{\infty} U_n^*,$$

with the possible exception of a set of measure 0, are departing as $t \to +\infty$. For the proof we employed only the fact that $U_n^* \cdot f(U_n^*, k) = 0$ for $k = 1, 2, \ldots$. But from this last relation, as it was shown above, there follows

$$U_n^* \cdot f(U_n^*, -k) = 0 \qquad (k = 1, 2, \ldots).$$

From this, by a completely analogous argument, we obtain that almost all points of the set V^+ are departing as $t \to -\infty$.

Consequently, and all the more so, almost all points of the set of points departing as $t \to +\infty$, $V^+ - W^+ \subset V^+$, are departing also as $t \to -\infty$.

Applying the same arguments to V^- we find that almost all motions departing as $t \to -\infty$ depart also as $t \to +\infty$. The first part of the theorem is proved.

For the proof of the second part we note that we have two decompositions of the space,

$$(4.05) \qquad \begin{aligned} R &= S^+ + (V^+ - W^+) + W^+ = S^+ + V^+, \\ R &= S^- + (V^- - W^-) + W^- = S^- + V^-, \end{aligned}$$

into sets (S) of points stable according to Poisson, sets ($V - W$) of departing points and, finally, sets (W) of points simultaneously unstable and nondeparting as $t \to +\infty$ and $t \to -\infty$. Moreover, according to what has been proved, $\mu W^+ = \mu W^- = 0$.

The set $S^+ \cdot V^-$ has measure 0 since almost all motions leaving V^- are departing as $t \to +\infty$ and, therefore, are unstable P^+. Analogously, $\mu(S^- \cdot V^+) = 0$.

Multiplying the respective members of the equalities (4.05) we obtain the decomposition of R:

$$R = S^+ \cdot S^- + (V^+ - W^+) \cdot (V^- - W^-) + E, \quad \text{where } E = 0,$$

i.e. except for a set of measure zero all points of R are either stable P or depart both as $t \to +\infty$ and as $t \to -\infty$. This completely proves Hopf's Theorem II.

5. G. D. Birkhoff's Ergodic Theorem.

5.01. First part of the ergodic theorem. In problems of statistical mechanics the probability of finding a point in some given domain of the phase space plays an important role. *This probability is defined as the limit as $T \to +\infty$ of the ratio of the time spent by the moving point $f(p, t)$ in the neigborhood under consideration to the duration of the whole time interval under consideration,* i.e. *to T.*

In order to express this quantity analytically we consider a phase space R with an invariant measure μ, $\mu R = 1$, and in it a measurable set E. We introduce the characteristic function $\varphi_E(p)$ of the set E, *i.e.*, the function defined by the conditions

$$\varphi_E(p) = \begin{cases} 1, & p \in E, \\ 0, & p \in R - E. \end{cases}$$

Then the set of instants of time of the interval $(0, T)$ for which the points $f(p, t) \in E$ is obviously expressed by the integral $\int_0^T \varphi_E(f(p, t))dt$; the limiting value under discussion is

$$\lim_{T\to\infty} \frac{1}{T} \int_0^T \varphi_E(f(p, t))dt.$$

Birkhoff's theorem first of all asserts the existence of this limit for almost all initial positions of the point p. Without complicating the proof, we follow Khintchine and in place of the characteristic function consider any function $\varphi(p)$, μ-measurable and absolutely summable (with respect to the measure μ), *i.e.* for which there exists the integral

$$\int_R |\varphi(p)| d\mu.$$

In this manner the first part of the ergodic theorem of Birkhoff-Khintchine may be formulated thus:

5.02. *If in a phase space R there is defined an invariant measure μA, $\mu A = \mu f(A, t)$ and $\mu R = 1$, then for any absolutely summable function $\varphi(p)$ the time mean*

$$\lim_{T\to\infty} \frac{1}{T} \int_0^T \varphi(f(p, t))dt$$

exists for all p, with the possible exception of points $p \in \mathscr{E}$, where $\mu\mathscr{E} = 0$.

We shall investigate first the question of the existence of $\int_0^t \varphi(f(p, t))dt = \Phi(p, t)$. If the function $\varphi(p)$ is bounded and measurable in the sense of Borel, then $\varphi(f(p, t))$ for *any* p is also bounded and measurable B, *i. e.* it is a summable function of t.

We shall assume only that $\varphi(p)$ is measurable and summable. In our case of an invariant measure μ it is easy to verify that $\varphi(f(p, t))$ is measurable in the space $R \times I$ with respect to the measure ν (see section 2).

In fact,

$$A_1 = \{(p, t);\ \varphi(f(p, t)) > \alpha\}$$

is obtained as a union of sets $f(A, -t)$, $-\infty < t < +\infty$, where

$$A = \{p;\ \varphi(p) > \alpha\};$$

for, if we set $f(p, t) = q$, from the inequality $\varphi(f(p, t)) > \alpha$ there follows $q \in A$, $p \in f(A, -t)$. Since by hypothesis A is measurable μ, there exist Borel sets B_1 and B_2, $B_1 \supset A \supset B_2$, $\mu B_1 - \varepsilon < \mu A < \mu B_2 + \varepsilon$. If we now consider, for example, the sets

$$A^* = \{(p, t);\ p \in f(A, -t),\ t_1 \leqq t \leqq t_2\},$$
$$B_1^* = \{(p, t);\ p \in f(B_1, -t),\ t_1 \leqq t \leqq t_2\},$$
$$B_2^* = \{(p, t);\ p \in f(B_2, -t),\ t_1 \leqq t \leqq t_2\},$$

then, obviously, $B_1^* \supset A^* \supset B_2^*$. Furthermore, B_1^* and B_2^*, as sets measurable B, are measurable ν, and from Fubini's Theorem we have, because of the invariance of the measure μ,

$$\nu B_1^* = \int_{t_1}^{t_2} dt \int_{p \in f(B_1, -t)} \mu(dp) = (t_2 - t_1)\mu B_1,$$

and analogously $\nu B_2^* = (t_2 - t_1)\mu B_2$.

Thus the set A^* is measurable B since it can be included between two sets whose measures differ by an arbitrarily small quantity. But then A_1 is also measurable as the sum of a countable number of sets of the form A.

We note further that because of the invariance of the measure μ we have

$$(5.03)\qquad \int_R |\varphi(f(p,t))|\,\mu(dp) = \int_R |\varphi(p)|\mu(dp) = \int_R |\varphi(p)|\,d\mu,$$

since, for example,

$$E_i(t) = \{p;\ l_{i-1} < |\varphi(f(p,t))| < l_i\}$$
$$= \{p = f(q, -t);\ l_{i-1} < |\varphi(q)| < l_i\} = f(E_i(0), -t),$$

and since $\mu E_i = \mu f(E_i, t)$, from the definition of the Lebesgue integral there follows the equality (5.03). This argument we write concisely as

$$\int_R |\varphi(f(p,t))|\,\mu(dp) = \int_R |\varphi(q)|\mu(df(q, -t)) = \int_R |\varphi(q)|\mu(dq).$$

Consider now the integral

$$\int_{R\times(0,T)} |\varphi(f(p,t))|\,d\nu = \int_0^T dt \int_R |\varphi(f(p,t))|\,\mu(dp) = \int_0^T dt \int_R |\varphi(p)|\mu(dp)$$
$$= T\int_R |\varphi(p)|\,d\mu.$$

From the condition of the summability of φ it follows that it has a finite value.

Applying Fubini's theorem, we obtain

$$T\int_R |\varphi(p)|d\mu = \int_R d\mu \int_0^T |\varphi(f(p,t))|dt,$$

wherein, because of the same theorem, the inner integral exists (and is finite) for almost all $p \in R$. Noting that if $\int_0^T |\varphi(f(p, t))|dt$ exists for a given T it exists also for every $T' < T$, on assigning T the sequence of values $T_1 < T_2 < \ldots < T_n < \ldots$, $T_n \to +\infty$, we obtain that $\int_0^t |\varphi(f(p, t))|d\mu$ exists for all positive values of t for any point $p \in R$ except, perhaps, for points of a set of μ-measure zero, and, consequently, for these same points $\int_0^t \varphi(f(p, t))d\mu$ also exists.

After these remarks we proceed to the proof of the ergodic theorem. The proof will be conducted with the aid of consecutive reductions to simpler propositions.

First reduction. It is sufficient to prove that the limit under discussion in the theorem exists for almost all p when t passes through integral values. Indeed, we shall estimate the difference (here $[t]$ is the largest integer $\leqq t$)

$$\left| \frac{1}{t}\int_0^t \varphi(f(p,t))\,dt - \frac{1}{[t]}\int_0^{[t]} \varphi(f(p,t))\,dt \right|$$

$$\leqq \left| \frac{1}{t}\int_0^t \varphi(f(p, t))dt - \frac{1}{t}\int_0^{[t]} \varphi(f(p,t))\,dt \right| + \left| \left(\frac{1}{t} - \frac{1}{[t]}\right) \int_0^{[t]} \varphi(f(p,t))dt \right|$$

$$\leqq \left| \frac{1}{t}\int_{[t]}^t \varphi(f(p,t))\,dt \right| + \frac{1}{t} \left| \frac{1}{[t]}\int_0^{[t]} \varphi(f(p,t))dt \right|.$$

If $\lim_{n\to\infty} (1/n) \int_0^n \varphi(f(p, t))dt$ exists and is finite, then the second term in the last member tends to 0 as $t \to +\infty$; we shall estimate the first term:

$$\left| \frac{1}{t}\int_{[t]}^t \varphi(f(p,t))\,dt \right| \leqq \frac{1}{t}\int_{[t]}^{[t+1]} |\varphi(f(p,t))|dt$$

$$= \frac{1}{t}\left\{ \int_0^{[t]+1} |\varphi(f(p,t))|\,dt - \int_0^{[t]} |\varphi(f(p,t))|dt \right\}$$

$$= \frac{[t]+1}{t} \cdot \frac{1}{[t]+1}\int_0^{[t+1]} |\varphi(f(p,t))|\,dt - \frac{[t]}{t} \cdot \frac{1}{[t]}\int_0^{[t]} |\varphi(f(p,t))|\,dt.$$

If it is proved that the mean value with respect to integral values of the argument exists for almost all values p for any integrable function, that is, in particular, for $|\varphi(p)|$, then the last expression tends to zero wherever this limit exists, since as $t \to +\infty$ the limit of the minuend is equal to the limit of the subtrahend.

Thus it is sufficient to investigate the question concerning the existence of

$$\lim_{n\to\infty} \frac{1}{n} \int_0^n \varphi(f(p, t))dt, \tag{5.04}$$

where n is a natural number.

Second reduction. We consider the totality $\{(\alpha_n, \beta_n)\}$ of all nondegenerate intervals on the real line with rational end points. Setting

$$\limsup_{n\to\infty} \frac{1}{n} \int_0^n \varphi(f(p, t))dt = \psi^*(p),$$

$$\liminf_{n\to\infty} \frac{1}{n} \int_0^n \varphi(f(p, t))dt = \psi_*(p),$$

we consider the sets $V_n = \{p;\ \psi^*(p) > \beta_n,\ \psi_*(p) < \alpha\}$. If the equalities $\mu V_n = 0$ $(n = 1, 2, \ldots)$ hold, then for $V' = \sum_{n=1}^{\infty} V_n$ also $\mu V' = 0$. If $p \in R - V'$, then between $\psi_*(p)$ and $\psi^*(p)$ it is impossible to insert an interval with rational end points, *i.e.* in general there is no interval and consequently $\psi_*(p) = \psi^*(p)$.

Thus, in order to prove the theorem, it is sufficient to show a contradiction to the assumption that there exists a set V_n, $\mu V_n > 0$, such that for $p \in V_n$ we have $\psi_*(p) < \alpha_n < \beta_n < \psi^*(p)$.

Therefore, proving the theorem from the contrary, we must assume the existence of two numbers $\alpha < \beta$ and of a set S, $\mu S > 0$, such that for $p \in S$ we have

$$\psi^*(p) > \beta, \qquad \psi^*(p) < \alpha,$$

and must show that this assumption leads to a contradiction.

We note that the set S is an invariant set. In fact, setting

$$\int_0^t \varphi(f(p, t))dt = F(p, t),$$

we have for any point $f(p, r)$

$$\frac{F(f(p, r), k)}{k} = \frac{F(p, r+k) - F(p, r)}{k} = \frac{F(p, r+k)}{r+k}\left(1 + \frac{r}{k}\right) - \frac{F(p, r)}{k},$$

and since by hypothesis

$$\limsup_{k\to\infty} \frac{F(p, r+k)}{r+k} > \beta, \quad \lim_{k\to\infty}\left(1+\frac{r}{k}\right) = 1, \quad \lim_{k\to\infty} \frac{F(p, r)}{k} = 0,$$

then

$$\limsup_{k\to\infty} \frac{F(f(p, r), k)}{k} > \beta.$$

Analogously we obtain

$$\liminf_{k\to\infty} \frac{F(f(p, r), k)}{k} < \alpha.$$

Thus, after the second reduction and in view of the last remark, we conclude that if the theorem is false there exist two numbers $\alpha < \beta$ and an invariant set S, where $\mu S > 0$, such that for every $p \in S$

$$\limsup_{n\to\infty} \frac{F(p, n)}{n} > \beta, \qquad \liminf_{n\to\infty} \frac{F(p, n)}{n} < \alpha.$$

We shall show that this assumption leads to a contradiction.

If $p \in S$ there exist values of n such that $F(p, n)/n > \beta$; the smallest of these values of n we denote by l. The set of those points $p \in S$ for which l does not exceed a given number k we shall call S_k. Obviously, $S_{k+1} \supset S_k$ and

$$\lim_{k\to\infty} S_k = S.$$

Therefore, if $\mu S > 0$, there can be found a k such that $\mu S_k > 0$. We fix this number k.

The remaining reasoning follows Kolmogoroff.

We shall call a segment $[a, b]$ of the number axis, where a and b are integers and $b > a$, a *singular segment* for a given point p and the number β if

$$\frac{F(p,b)-F(p,a)}{b-a} > \beta, \quad \text{but} \; \frac{F(p,b')-F(p,a)}{b'-a} \leqq \beta \text{ for } a < b' < b.$$

Singular segments for a given point p cannot partially overlap. Indeed, suppose that the segments $[a, b]$, $[a', b']$ are singular and $a < a' < b < b'$. Consider the relation

$$\frac{F(p,b)-F(p,a)}{b-a} = \frac{\dfrac{F(p,b)-F(p,a')}{b-a'}(b-a') + \dfrac{F(p,a')-F(p,a)}{a'-a}(a'-a)}{b-a}$$

Of the two ratios in the numerator of this last expression, at least one is greater than β, since otherwise the left-hand side would be less than or equal to β. But if $[F(p, a') - F(p, a)]/(a' - a) < \beta$, then the segment $[a, b]$ would not be singular and if $[F(p, b) - F(p, a')]/(b - a') > \beta$, then the segment $[a', b']$ would not be singular. The assertion is proved.

For the number k which we have fixed, we shall say that a singular segment of length not exceeding k is *k-singular*, if it is not properly contained in any singular segment of length $\leqq k$. Every singular segment of length $\leqq k$ is contained in one and only one k-singular segment; this will be the greatest segment of length not exceeding k containing the given one. It is determined in a unique way since singular segments do not partially overlap.

One may define the set S_k in these terms as the set of points $p \in S$ to which there correspond singular segments of length not exceeding k with the left ends at the origin. This system of singular segments can be replaced by another defining the same set S_k. Namely, each singular segment of the form $[0, h]$, where $h \leqq k$, lies inside a unique k-singular segment $[a, b]$, where $a \leqq 0 < b$. We shall show that, conversely, inside each k-singular segment $[a, b]$ such that $a \leqq 0 < b$ there lies a k-singular segment of the form $[0, h]$, where obviously $h \leqq k$. In fact, if $a = 0$, then $[a, b]$ is the required segment; if $a < 0$, then we have

$$\beta < \frac{F(p,b) - F(p,a)}{b-a} = \frac{\dfrac{F(p,b) - F(p,0)}{b-0} \cdot b + \dfrac{F(p,0) - F(p,a)}{0-a} \cdot (-a)}{b-a}$$

$$= \frac{\dfrac{F(p, b)}{b} \cdot b + \dfrac{F(p, 0) - F(p, a)}{0-a} \cdot (-a)}{b-a}.$$

Since the segment $[a, b]$ is singular, and since $0 < b$, we have $[F(p, 0) - F(p - a)]/(0 - a) \leqq \beta$. Consequently, $F(p, b)/b > \beta$. If now for all b' $(0 < b' < b)$ there is fulfilled the inequality $F(p, b')/b' \leqq \beta$, then the required segment is $[0, b]$; if for some $b' > 0$ there holds the opposite inequality, then the smallest of them, b'', gives a singular segment $[0, b'']$ of the required form. Thus we have the desired system of k-singular segments defining S_k.

Changing the notation, we write $-a = r$, $b - a = l$, where

$0 \leqq r < l \leqq k$; we determine S_{rl} as the set of points $p \in S_k$ to which there corresponds the k-singular segment $[-r, l-r]$, *i. e.*, $[F(p, l-r) - F(p, -r)]/l > \beta$, and for $0 < l' < l$ the opposite inequality holds. Since every point $p \in S_k$ corresponds to one and only one segment $[-r, l-r]$, then we have a decomposition of S_k into disjoint sets:

$$S_k = \sum_{l=1}^{k} \sum_{r=0}^{l-1} S_{rl}.$$

We shall now explain what the set $f(S_{rl}, m)$ represents. This is the totality of points p such that $f(p, -m) \in S_{rl}$, *i.e.* the condition

$$\frac{F(f(p, -m), l-r) - F(f(p, -m), -r)}{l} > \beta$$

is fulfilled, and for l', where $0 < l' < l$, there is fulfilled the opposite inequality. We compute

$$F(f(p, m), n) = \int_0^n \varphi[f(f(p, m), t)]dt = \int_0^n \varphi[f(p, m+t)]dt$$

$$= \int_m^{m+n} \varphi(f(p, t))dt = F(p, m+n) - F(p, m).$$

Thus $f(S_{rl}, m)$ is the set of points for which

$$\frac{F(p, l-r-m) - F(p, -r-m)}{l} > \beta,$$

while for values $l' < l$ the opposite inequality is valid.

Thus we have

$$f(S_{rl}, m) = S_{r+m,\, l},$$

where $S_{r+m,\, l} \subset S_k$ provided that $0 \leqq r+m < l$. Obviously ,the segment $[-r-m, -r-m+l]$ is singular for $p \in S_{r+m,\, l}$.

We proceed to the basic point of the proof. The idea consists in passing from integrals with respect to the time to an integral with respect to the set $S_k \subset R$ and in introducing the measure of this set into the inequality. Namely, we integrate $F(p, 1) = \int_0^1 \varphi(f(p, t))dt$ over the set S_k:

$$\int_{S_k} F(p, 1)d\mu = \sum_{l=1}^{k} \sum_{r=0}^{l-1} \int_{S_{rl}} F(p, 1)d\mu.$$

Noting that $S_{rl} = f(S_{0l}, r)$, we have:

$$\int_{S_{rl}} F(p,\ 1)d\mu = \int_{p\epsilon f(S_{0l},\ r)} F(p,\ 1)d\mu = \int_{p'\epsilon S_{0l}} F(f(p',\ r),\ 1)d\mu$$
$$= \int_{S_{0l}} [F(p,\ r+1) - F(p,\ r)]d\mu.$$

Thus

$$\int_{S_k} F(p,\ 1)d\mu = \sum_{l=1}^{k} \sum_{r=0}^{l-1} \int_{S_{0l}} [F(p,\ r+1) - F(p,\ r)]d\mu$$
$$= \sum_{l=1}^{k} \int_{S_{0l}} F(p,\ l)d\mu.$$

Since the segments $[0,\ l]$ are singular for points $p \in S_{0l}$, then for these $F(p,\ l)/l > \beta$, and we obtain

$$\int_{S_k} F(p,\ 1)d\mu > \beta \sum_{l=1}^{k} l \cdot \mu S_{0l}.$$

But since $S_{rl} = f(S_{0l},\ r)$ $(r = 1,\ 2, \ldots,\ l-1)$, then $\mu S_{0l} = \mu S_{rl}$ and we can write

$$\int_{S_k} F(p,\ 1)d\mu > \beta \cdot \sum_{l=1}^{k} \sum_{r=0}^{l-1} \mu S_{rl} = \beta \cdot \mu S_k.$$

Since

$$S = \lim_{k\to\infty} S_k,$$

then

$$\int_S F(p,\ 1)d\mu \geqq \beta \cdot \mu S.$$

By analogous reasoning, but proceeding from the inequality

$$\liminf_{n\to\infty} \frac{F(p,\ n)}{n} < \alpha \quad \text{for } p \in S,$$

we obtain

$$\int_S F(p,\ 1)d\mu \leqq \alpha \cdot \mu S, \text{ where } \alpha < \beta.$$

Since we have assumed $\mu S > 0$, this is impossible; and the contradiction thus obtained proves the theorem.

Passing to the second part of the ergodic theorem of Birkhoff, we introduce as a preliminary the concept of an *irreducible* (or *transitive*) dynamical system.

5.05. DEFINITION. A system $f(p,\ t)$, $p \in R$, is called *irreducible* with respect to the measure μ if it is impossible to represent R as the sum of two measurable invariant sets of positive measure without common points; in other words, if A is invariant and measurable and $\mu A > 0$, then $\mu(R - A) = 0$.

Example of an irreducible set. Such an example is afforded by uniform motions on a torus $\mathfrak{T}(0 \leqq \varphi < 1,\ 0 \leqq \theta < 1)$ with the ratio of velocity components irrational (see Example 4.04, of Chapter V) and with the invariant measure $\iint_A d\theta d\varphi = \mu A$, so that $\mu\mathfrak{T} = 1$.

Indeed, let there be an invariant set $A \subset \mathfrak{T}$, $\mu A > 0$; then the intersection of the set A with the meridian $\varphi = 0$ gives an invariant set E and its linear measure mes $E > 0$, which is easily verified on noting that $\mu A = \int_0^1 d\varphi \int_E d\theta$. Consequently, the set E has a point of denseness θ_0, *i.e.* for any $\varepsilon > 0$ there can be found a $\delta > 0$ such that $(1/2\delta)$ mes $E \cdot (\theta_0 - \delta,\ \theta_0 + \delta) > 1 - \varepsilon$. Since the set of points $\{\theta_n\} = \{\theta_0 + n\alpha\}$ is everywhere dense on $\varphi = 0$, there exists a natural number N such that for every point θ there can be found a θ_i $(0 \leqq i < N)$ satisfying the inequality $|\theta - \theta_i| < \delta$, *i.e.* the intervals $(\theta_i - \delta, \theta_i + \delta)$ $(i = 0, 2, \ldots, N)$ cover the entire circle $\varphi = 0$. Noting that mes E is an invariant measure, we find $(1/2\delta)$ mes $E \cdot (\theta_i - \delta,\ \theta_i + \delta) > 1 - \varepsilon$. From this it follows that mes $E > 1 - \varepsilon$, or, because of the arbitrariness of the number ε, mes $E = 1$. Consequently, $\mu A = 1$. Thus, assuming that $\mu A > 0$, we have obtained $\mu A = 1$, *i.e.*, the system is irreducible.

SECOND PART OF THE ERGODIC THEOREM. *If in the space R with an invariant measure μ, $\mu R = 1$, the system $f(p, t)$ is irreducible (transitive), then the mean time*

$$\lim_{T\to\infty} \frac{1}{T}\int_0^T \varphi(f(p,t))\,dt = \psi(p) \tag{5.06}$$

has one and the same value for almost all points $p \in R$.

We note that the function $\psi(p)$ is defined almost everywhere in R and is measurable (as a limit of measurable functions). Furthermore, this function is invariant, *i.e.* it assumes a constant value along every trajectory (on which it is defined):

$$\psi(f(p,t)) = \psi(p).$$

Indeed, if the limit (5.06) exists for a point p, then we have for any fixed t_0:

$$\psi(f(p, t_0)) - \psi(p) = \lim_{T\to\infty} \frac{1}{T}\int_{t_0}^{t_0+T} \varphi(f(p, t))dt - \lim_{T\to\infty}\frac{1}{T}\int_0^T \varphi(f(p,t))dt$$

$$= \lim_{T\to\infty}\Big\{\frac{T+t_0}{T}\cdot\frac{1}{T+t_0}\int_0^{t_0+T} \varphi(f(p, t))dt$$

$$-\frac{1}{T}\int_0^{t_0}\varphi(f(p, t))dt - \frac{1}{T}\int_0^T \varphi(f(p, t))dt\Big\} = \psi(p) - \psi(p) = 0.$$

We shall prove that under the condition of transitivity the function $\psi(p)$ is a constant almost everywhere. Assuming the contrary, we denote by M the least upper bound of the function $\psi(p)$ over R computed on neglecting a set of measure zero [4] and analogously we denote by m the greatest lower bound of the function $\psi(p)$ on neglecting a set of measure zero. From the assumption there follows $M > m$.

Let α satisfy the inequalities $m < \alpha < M$. We obtain

$$\mu\{p;\ \psi(p) < \alpha\} = \mu E_\alpha > 0$$

and

$$\mu(R - E_\alpha) = \mu\{p;\ \psi(p) \geqq \alpha\} > 0.$$

Because of the invariance of the function $\psi(p)$, the set E_α and its complement are invariant and we have a decomposition of R into two invariant sets of positive measure which contradicts the condition of irreducibility. The theorem is proved.

6. Supplements to the Ergodic Theorem

6.01. Properties of the invariant function $\psi(p)$. In order to compute the mean value which is constant almost everywhere in the case of an irreducible dynamical system, and also for the deduction of other consequences of the ergodic theorem, it is necessary to study certain properties of the function $\psi(p)$ defined with respect to a given summable function $\varphi(p)$ for all $p \in R$ except, perhaps, for a set of μ-measure zero:

$$\psi(p) = \lim_{T\to\infty}\frac{1}{T}\int_0^T \varphi(f(p, t))dt. \tag{5.06}$$

[4]This means that $\mu\{p;\ \psi(p) > M\} = 0$, while for any $\varepsilon > 0$ we have $\mu\{p;\ \psi(p) > M - \varepsilon\} > 0$.

We have already seen that this function is invariant.

6.02. LEMMA. *The family of functions*

$$\varphi_T(p) = \frac{1}{T}\int_0^T \varphi(f(p, t))dt$$

is summable in R uniformly with respect to the parameter T; i.e., for any $\varepsilon > 0$ there exists a $\delta > 0$ such that, if $\mu A < \delta$,

$$\int_A |\varphi_T(p)| d\mu < \varepsilon.$$

Indeed, applying Fubini's Theorem, we have

$$\int_A |\varphi_T(p)| d\mu = \int_A \left| \frac{1}{T}\int_0^T \varphi(f(p, t))dt \right| d\mu \leqq \int_A d\mu \int_0^T \frac{1}{T} |\varphi(f(p,t))| dt$$

$$(6.03) \qquad = \frac{1}{T}\int_0^T dt \int_A |\varphi(f(p, t))| d\mu = \frac{1}{T}\int_0^T dt \int_{p\epsilon f(A,\, t)} |\varphi(p)| d\mu.$$

But because of the summability of $\varphi(p)$ in R, for a given ε and for $\varphi(p)$ there exists a $\delta > 0$ such that

$$\int_A |\varphi(p)| d\mu < \varepsilon$$

when $\mu A < \delta$. Choosing this δ and noting that if $\mu A < \delta$ then, from the invariance of the measure μ, $\mu f(A, t) < \delta$, we find

$$\int_A |\varphi_T(p)| d\mu < \frac{1}{T}\int_0^T \varepsilon\, dt = \varepsilon$$

for any T which it was required to prove.

6.04. COROLLARY. Since for non-negative functions the integral of the limit is less than or equal to the limit of the integral (Fatou's lemma), from the inequality (6.03) we obtain

$$\int_A |\psi(p)| d\mu \leqq \sup_{0 \leqq t < \infty} \int_{f(A,\ ;} |\varphi(p)| d\mu.$$

From the lemma just proved we can obtain an important result.

6.05. THEOREM. *The value of $\int_R \psi(p) d\mu$ (and the value of the function $\psi(p)$ in the case of an irreducible system) is equal to $\int_R \varphi(p) d\mu$.*

We have

$$\psi(p) = \lim_{T\to\infty} \varphi_T(p).$$

We assign $\varepsilon > 0$ arbitrarily and for the number $\varepsilon/3$ and the function $\varphi(p)$ we choose a number $\delta > 0$ according to the lemma. On the basis of a theorem of Lebesgue, for the numbers $\varepsilon/3$ and δ there exists a $T_0(\varepsilon/3, \delta)$ such that for $T > T_0$ we have

$$\mu E = \mu\left\{p;\ |\psi(p) - \varphi_T(p)| \geqq \frac{\varepsilon}{3}\right\} < \delta.$$

Now we estimate for $T > T_0$ the difference

$$\int_R \psi(p)d\mu - \int_R \varphi_T(p)d\mu.$$

We have

$$\left|\int_R \psi(p)d\mu - \int_R \varphi_T(p)d\mu\right| \leqq \int_R |\psi(p) - \varphi_T(p)|d\mu$$
$$\leqq \int_{R-E} |\psi(p) - \varphi_T(p)|d\mu + \int_E |\psi(p)|d\mu + \int_E |\varphi_T(p)|d\mu.$$

Because of the choice of T, the first integral $< \int_R (\varepsilon/3)d\mu < \varepsilon/3$, and from the choice of δ, on the basis of the lemma and its corollary, each of the succeeding integrals is less than $\varepsilon/3$, whence

$$\left|\int_R \psi(p)d\mu - \int_R \varphi_T(p)d\right| < \varepsilon, \quad \text{for} \quad T > T_0,$$

i.e.,

$$\int_R \psi(p)d\mu = \lim_{T\to\infty} \int_R \varphi_T(p)d\mu.$$

But on the other hand, according to Fubini's theorem

$$\int_R \varphi_T(p)d\mu = \int_R d\mu \frac{1}{T}\int_0^T \varphi(f(p, t))dt = \frac{1}{T}\int_0^T dt \int_R \varphi(f(p, t))\mu(dp)$$
$$= \frac{1}{T}\int_0^T dt \int_{p'\epsilon R} \varphi(p')\mu(df(p', -t)),$$

or, in view of the invariance of the measure μ,

$$\int_R \varphi_T(p)d\mu = \frac{1}{T}\int_0^T dt \int_R \varphi(p)\mu(dp) = \int_R \varphi(p)d\mu.$$

Thus, finally,

$$\int_R \psi(p)d\mu = \int_R \varphi(p)d\mu. \tag{6.06}$$

If the system is irreducible, then almost everywhere in R we have $\psi(p) = c$ (a constant), and remembering that $\mu R = 1$, we obtain

$$c = \int_R \varphi(p) d\mu.$$

7. Statistical Ergodic Theorems

E. Hopf used this name for theorems which assert the existence of limits of the form (5.06) of the preceding section, but in the mean with respect to the space R.

7.01. THEOREM A. *The limiting relation*

$$\lim_{T\to\infty} \int_R \left| \frac{1}{T} \int_\alpha^{\alpha+T} \varphi(f(p, t)) dt - \psi(p) \right| \mu(dp) = 0 \tag{7.02}$$

holds uniformly with respect to $\alpha(-\infty < \alpha < +\infty)$.

Introducing for the proof the variable point $q = f(p, \alpha)$ instead of p, noting that because of the invariance of the function $\psi(p)$ we have $\psi(p) = \psi(q)$, and taking into consideration the invariance of the measure μ, we obtain the equality

$$\int_R \left| \frac{1}{T} \int_\alpha^{\alpha+T} \varphi(f(p, t)) dt - \psi(p) \right| \mu(dp) = \int_R \left| \frac{1}{T} \int_0^T \varphi(f(q, t)) dt - \psi(q) \right| \mu(dq)$$

Consequently, to prove the uniformity of the limit (7.02) with respect to α, it is sufficient to prove the existence of the ordinary limit in the case when $\alpha = 0$, *i.e.* to prove that

$$\lim_{T\to\infty} \int_R |\varphi_T(p) - \psi(p)| d\mu = 0.$$

But in the proof of the theorem of the preceding section we have already estimated this integral and have seen that it can be made arbitrarily small for sufficiently large T, and this proves Theorem A.

7.03. THEOREM B (J. von Neumann). *If $\varphi(p)$ is a measurable function which is square integrable in R with respect to an invariant measure μ, then $\psi(p)$ is also a function with an integrable square and the limiting relation*

$$\lim_{T\to\infty} \int_A \left[\frac{1}{T} \int_\alpha^{\alpha+T} \varphi(f(p, t)) dt - \psi(p)) \right]^2 d\mu = 0.$$

holds uniformly.

We shall begin with the same remark as in Theorem A, namely, for the proof of Theorem B it is sufficient to consider the limit of the integral

$$\int_R |\varphi_T(p) - \psi(p)|^2 \, d\mu.$$

We consider the integral

$$\int_A \varphi_T^2(p) d\mu,$$

where $A \subset R$ is any μ-measurable set. Applying Schwarz's inequality to the inner integral, then applying Fubini's Theorem, and employing the invariance of the measure μ, we have

$$\begin{aligned}
\int_A \varphi_T^2(p) d\mu &= \int_A \left\{ \frac{1}{T} \int_0^T \varphi(f(p, t)) dt \right\}^2 \mu(dp) \\
&\leqq \int_A \left\{ \frac{1}{T} \int_0^T 1 \cdot dt \cdot \frac{1}{T} \int_0^T \varphi^2(f(p, t)) dt \right\} \mu(dp) \\
&= \frac{1}{T} \int_0^T dt \int_A \varphi^2(f(p, t)) \mu(dp) \\
&= \frac{1}{T} \int_0^T dt \int_{p \in f(A, t)} \varphi^2(p) \mu(df(p, -t)) \\
&= \frac{1}{T} \int_0^T dt \int_{p \in f(A, t)} \varphi^2(p) \mu(dp).
\end{aligned}$$

From this, because of the summability of the function $\varphi^2(p)$, it follows that for any $\varepsilon > 0$ there exists a $\delta > 0$ such that if $\mu A = \mu(f(A, t)) < \delta$ then

$$\int_A \varphi_T^2(p) d\mu \leqq \frac{1}{T} \int_0^T dt \int_{f(A, t)} \varphi^2(p) d\mu < \varepsilon,$$

i.e. the uniform summability of the family of functions $\{\varphi_T^2(p)\}$ is established.

In particular, noting that almost everywhere in R

$$\lim_{T \to \infty} \varphi_T(p) = \psi(p),$$

according to Fatou's Lemma, for $\mu A < \delta$ we have

$$\int_A \psi^2(p) d\mu < \varepsilon,$$

whence follows the summability of the function $\psi^2(p)$.

Thus, the uniformly summable family of functions $\varphi_T^2(p)$ converges almost everywhere to the summable function $\psi^2(p)$. Applying arguments to this family analogous to those employed in the lemma, we obtain for any $\varepsilon > 0$:

$$\int_R \{\varphi_T(p) - \psi(p)\}^2 d\mu = \int_E (\varphi_T - \psi)^2 \, d\mu + \int_{R-E} (\varphi_T - \psi)^2 \, d\mu,$$

where

$$E = \left\{ p;\ |\psi(p) - \varphi_T(p)| < \frac{\varepsilon}{\sqrt{5}} \right\}.$$

For the first integral we have

$$\int_E (\varphi_T - \psi)^2 d\mu < \int_R \frac{\varepsilon^2}{5} \, d\mu = \frac{\varepsilon^2}{5}.$$

We estimate next the second integral:

$$\int_{R-E} (\varphi_T - \psi)^2 \, d\mu \leqq 2 \int_{R-E} \varphi_T^2 d\mu + 2 \int_{R-E} \psi^2 \, d\mu.$$

Choosing T sufficiently large, we can, according to a theorem of Lebesgue, make $\mu(R - E)$ arbitrarily small, so that from the uniform summability of the functions φ_T^2 and the summability of the function ψ^2 we shall have

$$\int_{R-E} \varphi_T^2 d\mu < \frac{\varepsilon^2}{5}, \qquad \int_{R-E} \psi^2 d\mu < \frac{\varepsilon^2}{5},$$

i.e.

$$\int_{R-E} (\varphi_T - \psi)^2 d\mu < \frac{4\varepsilon^2}{5}.$$

From this follows

$$\int_R (\varphi_T - \psi)^2 \, d\mu < \varepsilon^2,$$

and this, from the remark made at the beginning, proves von Neumann's theorem.

8. Generalizations of the Ergodic Theorem

Birkhoff's ergodic theorem actually presupposes that the measure of the whole space is finite. By a normalization of the measure (on multiplication by a positive number) this case reduces to that when $\mu R = 1$. However, even if $\mu R = +\infty$ for an invariant measure μ, one can obtain results concerning the mean time of a summable function $\varphi(p)$.

We shall impose on R the same restrictions as in the theorems of Hopf (section 4 of the present chapter). Namely, R is a locally compact metric space with a countable base possessing an invariant measure μ; furthermore, it may be that $\mu R = +\infty$, but for any compact set $F \subset R$ we have $\mu F < +\infty$.

As in section 4, we shall confine our consideration to the passage to the limit as $t \to +\infty$. Then, by Hopf's Theorem I, almost all points are stable P^+ or are departing. From the definition of a departing point $p \in R$, for any compact set $F \subset R$ there can be found a $t_0 > 0$ such that $f(p, t) \cdot F = 0$ for $t > t_0$. Therefore, if $\varphi_F(p)$ is the characteristic function of the set F, then $\int_0^T \varphi_F(f(p, t))dt \leqq t_0$ for $T > 0$, whence, in particular, there follows

$$\lim_{T\to\infty} \frac{1}{T}\int_0^T \varphi_F(f(p,t))dt = 0.$$

Thus, the generalization of Birkhoff's theorem to departing points presents no interest. Therefore, *in the sequel we shall restrict our consideration to the set of points* $R_1 \subset R$ *stable* P^+. We know that R_1 is an invariant set.

8.01. Lemma. *Let* $g(p) > 0$ *be a continuous function and suppose that* $p \in R_1$. *Then*

$$\lim_{T\to\infty} \int_0^T g(f(p,t))dt = \infty.$$

Let p be some fixed point of R_1. According to the hypothesis $g(p) = \alpha > 0$. If p is a rest point, the assertion is obvious since the integral under consideration is equal to αT.

If p is not a rest point there exists a neighborhood $S(p, k)$ $(k > 0)$ such that for arbitrarily large values of t we shall have $f(p, t) \cdot S(p, k) = 0$ (in the contrary case the point p would be the unique ω-limit point for the motion $f(p, t)$, *i.e.* a rest point, which is absurd). On the other hand, because of the Poisson stability, for any $\varepsilon > 0$ there can be found arbitrarily large values of t for which $f(p, t) \in S(p, \varepsilon)$.

We choose $\varepsilon < k/2$ and so small that $\overline{S(p, 2\varepsilon)}$ is compact and that for $q \in S(p, 2\varepsilon)$ we shall have $g(q) > \alpha/2$ (because of the continuity of the function g). From the remark just made it follows that there can be found two sequences of numbers $\{t_n\}$ and $\{t^{(n)}\}$,

$$0 < t_1 < t^{(1)} < t_2 < t^{(2)} < \ldots < t_n < t^{(n)} < \ldots,$$

$$\lim_{n\to\infty} t_n = \lim_{n\to\infty} t^{(n)} = +\infty,$$

such that

$$f(p, t_n) \cdot S(p, 2\varepsilon) = 0 \quad \text{and} \quad f(p, t^{(n)}) \in S(p, \varepsilon) \qquad (n = 1, 2, \ldots).$$

Thus in the time between t_{n-1} and t_n the moving point $f(p, t)$ traverses at least twice a route between the surface of the sphere $S(p, \varepsilon)$ and the surface of the sphere $S(p, 2\varepsilon)$. The time length of each such path is greater than some positive τ_0; this follows from the uniform continuity of the function $f(p, t)$ over the compact set $\overline{S(p, 2\varepsilon)}$ for $|t| \leqq M$ [5]. Thus we have

$$\int_0^{t_n} g(f(p, t))dt > 2(n-1)\tau_0(\alpha/2),$$

i.e.

$$\lim_{T\to\infty} \int_0^T g(f(p, t))dt = +\infty,$$

which it was required to prove.

For the deduction of generalizations of Birkhoff's theorem we shall consider positive continuous functions $g(p)$ such that

$$\int_R g(p)d\mu < +\infty.$$

We shall show the existence in the space R of such a continuous function $0 < g(p) \leqq 1$ which has the value $g(p) = 1$ over some given compact set F_0.

Indeed, because R has a countable base and is locally compact, one can construct a sequence of compact sets

$$F_0 \subset F_1 \subset \ldots \subset F_n \subset \ldots, \lim_{n\to\infty} F_n = R;$$

besides, we shall assume that for each n there exists an $\varepsilon_n > 0$

[5] See the footnote to Theorem 7.10, Chap. V.

such that $S(F_n, \varepsilon_n) \subset F_{n+1}$.[6] Let $\mu F_0 = m_0$, $\mu(F_1 - F_0) = m_1, \ldots$, $\mu(F_n - F_{n-1}) = m_n$. All these numbers are finite. Next we select constants $\alpha_0 = 1 > \alpha_1 > \ldots > \alpha_n > \ldots$, $\lim_{n\to\infty} \alpha_n = 0$, such that the series

$$\sum_{n=1}^{\infty} \alpha_{n-1}\, m_n$$

converges. If $p \in F_n - F_{n-1}$, then let $\varrho(p, F_{n-1}) = d_n$, $\varrho(p, R - F_n) = \delta_n$, wherein $d_n + \delta_n \geqq \varepsilon_{n-1} > 0$.

Then we define: $g(p) = 1$ for $p \in F_0$ and $g(p) = (\delta_n\alpha_{n-1} + d_n\alpha_n)/(\delta_n + d_n)$ for $p \in F_n - F_{n-1}$. It is easy to verify that the function constructed in this manner possesses all the required properties; in particular, for $p \in F_n - F_{n-1}$ we have $g(p) \leqq \alpha_{n-1}$ and therefore

$$\int_R g(p)d\mu = \int_{F_0} d\mu + \sum_{n=1}^{\infty} \int_{F_n - F_{n-1}} g(p)d\mu \leqq \alpha_0 m_0 + \sum_{n=1}^{\infty} \alpha_{n-1} m_n < +\infty.$$

8.02. The generalized theorem of Birkhoff. *If in a locally compact space R with a countable base there exists an invariant measure μ such that* (perhaps) $\mu R = \infty$ *but $\mu F < \infty$ for any compact set $F \subset R$, and if $g = g(p)$ is a continuous, bounded positive function $(0 < g(p) \leqq 1)$ for which $\int_R g d\mu < +\infty$, then for almost all points*

[6] We shall prove the lemma: *If F is a compact set of a locally compact space R, there can be found an $\varepsilon > 0$ such that $\overline{S(F, \varepsilon)}$ is compact.*

Indeed, for every point $p \in F$ there can be found a neighborhood whose closure is compact; from these neighborhoods (according to the Corollary 1.13, Chap. V); one can select a finite number covering F: $\sum_{k=1}^{m} U_k = U \supset F$. The set U contains F and its closure is compact. Every point $p \in F$ together with its neighborhood lies in U. We take for each point p the largest number ε_p such that $S(p, \varepsilon_p) \subset U$. The lemma will be proved if one shows that $\inf_{p \in F} \varepsilon_k = \varepsilon > 0$. Assuming the contrary, we have a sequence $\{p_n\} \subset F$, $\varepsilon_{p_n} \to 0$. Because of the compactness of F the sequence $\{p_n\}$ has a limit point $p_0 \in F$; for simplicity we shall suppose that $p_n \to p_0$. To the point p_0 there corresponds $\varepsilon_{p_0} > 0$. Let N be chosen such that $\varrho(p_n, p_0) < \varepsilon_{p_0}/2$ for $n > N$. Then $S(p_n, \varepsilon_{p_0}/2) \subset S(p_0, \varepsilon_{p_0}) \subset U$, *i.e.*, $\varepsilon_{p_n} > \varepsilon_{p_0}/2$ contrary to assumption, which proves the lemma.

Now the construction of a sequence $\{F_n\}$ with the required property is easily carried out. If the initial sequence does not satisfy the additional condition, then we take $F_0^* = F_0$; letting $F_1^*, F_2^*, \ldots, F_n^*$ be already defined, we find ε_n such that $S(F_n^*, \varepsilon_n)$ is compact and set

$$F_{n+1}^* = F_n^* + \overline{S(F_n^*, \varepsilon_n)}.$$

$p \in R$ stable P^+, and for any measurable function $\varphi(p)$ summable over any compact set, the limit (finite or infinite)

$$\lim_{T\to\infty} \frac{\int_0^T \varphi(f(p,t))\,dt}{\int_0^T g(f(p,t))\,dt} = \psi(p) \tag{8.03}$$

exists.

The proof, with slight changes, follows along the same lines as the proof of the fundamental theorem of Birkhoff-Khintchine. The change consists chiefly in the fact that in the denominator of the ratio there stands in place of the time T the integral $\int_0^T g(f(p,t))dt = \tau(p, \tau)$, wherein, from the lemma of this section, $\tau(p, T) < T$, $\lim_{T\to\infty} \tau(p, T) = \infty$ ("modified time", see the remark at the end of this section).

We shall consider from this point of view the course of the proof of section 5. The first reduction was based on the fact that $\lim_{t\to\infty} [t]/t = \lim_{t\to\infty} ([t]+1)/t = 1$. In our case there corresponds to this fact the limiting equality

$$\lim_{t\to\infty} \frac{\tau[p, [t])}{\tau(p, t)} = \lim_{t\to\infty} \frac{\tau(p, [t]+1)}{\tau(p, t)} = 1$$

which follows from the boundedness of the function $g(p)$ and from the circumstance that $\lim_{t\to\infty} \tau(t) = \infty$. Thus it is sufficient for us to prove the relation (8.03) for the case when T is equal to a natural number n.

The second reduction goes through without change and the proof reduces to showing the impossibility of the existence of an invariant set $S \subset R_1$ of positive measure for whose points, in the notation of section 5: $\int_0^t \varphi(f(p, t))dt = F(p, t)$, we have

$$\limsup_{n\to\infty} \frac{F(p, n)}{\tau(p, n)} > \beta, \quad \liminf \frac{F(p, n)}{\tau(p, n)} < \alpha; \quad \alpha < \beta.$$

Analogously with section 5, we define the segment $[a, b]$, where a, b are integers, to be singular if

$$\frac{F(p,b) - F(p,a)}{\tau(p,b) - \tau(p,a)} > \beta \quad \text{and} \quad \frac{F(p,b') - F(p,a)}{\tau(p,b') - \tau(p,a)} \leqq \beta \quad \text{for } a < b' < b.$$

The properties of singular segments are preserved intact, as are the definitions and properties of the sets S_k and S_{kl}.

Minor changes occur only in the latter stage of the proof, to which we now pass. The formula

$$\int_{S_k} F(p,\ 1)d\mu = \sum_{l=1}^{k} \int_{S_{0l}} F(p,\ l)d\mu \tag{8.04}$$

obviously remains valid as one depending on the properties of the sets S_r under the application of a displacement along integral intervals. Along with this we shall obtain an analogous formula if we replace the function $\varphi(p)$ by $g(p)$:

$$\int_{S_k} \tau(p,\ 1)d\mu = \sum_{l=1}^{k} \int_{S_{0l}} \tau(p,\ l)d\mu. \tag{8.05}$$

But since the segments $[0,\ l]$ are singular for $p \in S_{0l}$, there holds the inequality

$$F(p,\ l) > \beta \cdot \tau(p,\ l), \text{ whence } \int_{S_{0l}} F(p,\ l)d\mu > \beta \int_{S_{0l}} \tau(p,\ l)d\mu.$$

Comparing this inequality with the relations (8.04) and (8.05), we obtain

$$\begin{aligned} \int_{S_k} F(p,\ 1)d\mu > \beta \int_{S_k} \tau(p,\ 1)d\mu &= \beta \int_{S_k} d\mu \int_0^1 g(f(p,t))dt \\ &= \beta \int_0^1 dt \int_{S_k} g(f(p,t))\mu\,(dp). \end{aligned} \tag{8.06}$$

In equation (8.06) we pass to the limit as $k \to \infty$. We note that $\lim_{k\to\infty} S_k = S$ and that the set S is invariant; therefore, in view of the invariance of the measure μ, we have

$$\begin{aligned} &\int_S F(p,\ 1)d\mu = \int_S \mu(dp) \int_0^1 \varphi(f(p,t))dt \\ &= \int_0^1 dt \int_S \varphi(f(p,t))\mu\,(df(p,t)) = \int_0^1 dt \int_S \varphi(p)\mu(dp) = \int_S \varphi(p)d\mu \end{aligned}$$

and analogously

$$\lim_{k\to\infty} \int_{S_k} \tau(p,\ 1)\ \mu(dp) = \int_S g(p)d\mu.$$

Thus, from equation (8.06) we obtain

$$\int_S \varphi(p)d\mu \geqq \beta \int_S g(p)d\mu. \tag{8.07}$$

We remark, finally, that in view of the fact that $\mu S > 0$ and $g(p) > 0$ the last integral is different from zero (it also differs from $+\infty$, since $g(p)$ has been chosen such that $\int_R g(p)d\mu < +\infty$.

Going through the same reasoning, but starting from a consideration of those segments $[a, b]$ where for $p \in S$ there is fulfilled the inequality

$$\frac{\int_0^T \varphi(f(p, t)dt}{\int_0^T g(f(p, t))dt} < \alpha$$

we arrive at the inequality

$$\int_S \varphi(p)d\mu \leqq \alpha \int_S g(p)d\mu. \tag{8.08}$$

The contradiction of the inequalities (8.07) and (8.08) proves the generalized theorem of Birkhoff.

8.09. COROLLARY. If, for $p \in S$, where S is an invariant set, we have

$$\lim_{T\to\infty} \frac{\int_0^T \varphi(f(p, t))dt}{\int_0^T g(f(p, t))\,dt} < \alpha, \quad \text{then} \int_S \varphi(p)d\mu \leqq \alpha \int_S g(p)d\mu$$

and analogously for the sign $>$.

8.10. COROLLARY. We have proved the existence of the limit function $\psi(p)$ for almost all $p \in R_1$. It is easy to show that this function is invariant, *i.e.*,

$$\psi(f(p, t)) = \psi(p).$$

We shall make a series of essential remarks.

In the sequel we shall assume that

$$\int_{R_1} |\varphi(p)|d\mu < +\infty.$$

We define a decomposition of the space R_1 into invariant sets E_i,

$$E_i = \{p_i;\ l_i \leqq \psi(p) + l_{i+1}\},$$

corresponding to a decomposition of the number axis by the numbers

$$\ldots l_{-n} < l_{-n+1} < \ldots < l_0 < l_1 < \ldots < l_n < \ldots,$$

where

$$l_{i+1} - l_i = d.$$

We now obtain two series of inequalities (the first series by virtue of Corollary 8.09, the second from the theorem of the mean):

$$\left.\begin{array}{ll}(8.11) & l_i\int_{E_i} g(p)d\mu \leqq \int_{E_i} \varphi(p)d\mu \leqq l_{i+1}\int_{E_i} g(p)d\mu \\ (8.12) & l_i\int_{E_i} g(p)d\mu \leqq \int_{E_i} \psi(p)g(p)d\mu < l_{i+1}\int_{E_i} g(p)d\mu\end{array}\right\} i=0,\pm 1,\pm 2,\ldots$$

We sum the first series of inequalities; because of the summability of $|\varphi(p)|$ the series of the center terms of the inequalities (8.11) converges absolutely. Therefore, the series of the first and third parts of the inequalities (8.11) or (8.12) also converges, since the differences between the corresponding sums do not exceed $d\cdot\int_{R_1} g(p)d\mu$. From this follows the absolute convergence of the series

$$\sum_{i=1}^{+\infty}\int_{E_i}\psi(p)\,g(p)\,d\mu = \int_{R_1}\psi(p)\,g(p)\,d\mu.$$

Finally, letting d tend to zero, we obtain the desired equality

$$\int_{R_1}\varphi(p)d\mu = \int_{R_1}\psi(p)g(p)d\mu. \tag{8.13}$$

This formula is a generalization of formula (6.06), of the present chapter (cf. *Cor.* 8.09) into which it passes if, under the assumption $\mu R_1 < +\infty$, one sets $g(p) = 1$.

We shall point out one application of formula (8.13); *we assume that $\mu R_1 = \infty$ and that R_1 contains no invariant sets of finite positive measure.* This case presents the chief interest since if R_1 contains an invariant set E, $\mu E < +\infty$, then considering the motions $f(p, t)$, $p \in E$, we find those conditions fulfilled under which Birkhoff's fundamental theorem can be applied.

We shall prove the following proposition: *if $\int_{R_1}|\varphi(p)|d\mu < +\infty$, then for almost all points $p \in R_1$ the equality*

$$\lim_{T\to\infty}\frac{1}{T}\int_0^T\varphi(f(p,t))dt = 0. \tag{8.14}$$

holds.

We introduce the special function $g(p)$ defined as above: $0 < g(p) \leqq 1$ and $g(p) = 1$ when $p \in F_0$, where F_0 is an arbitrarily assigned compact set; $\int_{R_1} g(p)d\mu < +\infty$. It is obvious that $\int_0^T g(f(p, t))dt \leqq T$, whence

$$\left|\frac{1}{T}\int_0^T \varphi(f(p,t))dt\right| \leqq \frac{1}{T}\int_0^T |\varphi(f(p,t))|dt \leqq \frac{\int_0^T |\varphi(f(p,t))|dt}{\int_0^T g(f(p,t))dt}.$$

Therefore

$$(8.15)\qquad \limsup_{T\to\infty}\left|\frac{1}{T}\int_0^T \varphi(f(p,t))dt\right| \leqq \lim_{T\to\infty}\frac{\int_0^T |\varphi(f(p,t))|dt}{\int_0^T g(f(p,t))dt} = \psi_g^*(p)$$

where $\psi_g^*(p)$ is defined almost everywhere in R_1 and the inequality remains valid for any function $g(p)$ satisfying the conditions indicated. Returning to formula (8.13), we find

$$(8.16)\qquad \int_{R_1} \psi_g^*(p)g(p)d\mu \leqq \int_{R_1} |\varphi(p)|d\mu = M < +\infty,$$

where M is a constant independent of the choice of the function $g(p)$.

We assume that over the set E, $\mu E > 0$, we have

$$\limsup_{T\to\infty}\left|\frac{1}{T}\int_0^T \varphi(f(p,t))dt\right| \geqq \alpha > 0.$$

The set E is invariant; consequently, according to the hypothesis, $\mu E = \infty$. It follows from the inequality (8.15) that the inequality $\psi_g^*(p) \geqq \alpha$ holds over E, where α, obviously, is independent of the choice of the function $g(p)$.

On the other hand, since $\mu E = +\infty$, we can choose the compact set F_0 entering in the definition of $g(p)$ in such a way that $\mu(E \cdot F_0) > M/\alpha$. Then we obtain

$$\int_{R_1} \psi_g^*(p)g(p)d\mu \geqq \int_E \psi_g^*(p)g(p)d\mu \geqq \int_{F_0 \cdot E} \psi_g^*(p)g(p)d\mu$$
$$> \alpha \int_{F_0 \cdot E} g(p)d\mu > \alpha \cdot \frac{M}{\alpha} = M.$$

This inequality contradicts the inequality (8.16). The equality (8.14) is thus proved.

From equation (8.14), setting $\varphi(p)$ equal to the characteristic function of any set A contained in a compact part of the space R, we see that the probability of the stay of the point p in the set A, under our supplementary hypothesis, is equal to zero for almost all points. In fact, as we saw in our introductory remark, this situation holds also for motions unstable according to Poisson, which were excluded in our later arguments.

8.17. The case of an irreducible system. Consider the case of an irreducible (transitive) system in which $\mu R = +\infty$. The unique invariant set of positive measure is in this case the entire space (or the whole space with the exception of a set of measure zero). The preceding argument is also applicable to every function $\varphi(p)$ with a finite integral $\int_R |\varphi(p)| d\mu$ (in particular, for every characteristic function of a set compact in this space. For almost all $p \in R$ we have

$$\lim_{T\to\infty} \frac{1}{T} \int_0^T \varphi(f(p, t))dt = 0. \tag{8.14}$$

Next, introducing a function $g(p)$, $\int_R g(p)d\mu < +\infty$, we have, according to the general formula,

$$\lim_{T\to\infty} \frac{\int_0^T \varphi(f(p, t))\, dt}{\int_0^T g(f(p, t))\, dt} = \psi_g(p) \tag{8.18}$$

for almost all points $p \in R$. But it is easy to see in the given case that the invariant function $\psi_g(p)$ has a constant value almost everywhere, since, assuming the contrary, we would have a decomposition of R into two invariant sets of positive measure.

Thus, in the case of irreducibility, $\psi_g(p) = C$ for almost all points, where the constant C depends on the choice of the function $g(p)$. Namely, there follows from formula (8.13)

$$C = \frac{\int_R \varphi(p)d\mu}{\int_R g(p)d\mu}. \tag{8.19}$$

Finally, although by formula (8.14) the probability of the stay of almost every point p for $0 \leqq t < +\infty$ in a compact set (of finite measure) is equal to zero, still we can estimate the *ratio* of the mean times of the stay of the moving point in two sets E_1 and E_2 compact in the space R and, consequently, having a finite measure.

Suppose that $F_0 \supset E_1 + E_2$ is a compact (in itself) set. We construct the special function $g(p)$ indicated above, equal to 1 for $p \in F_0$ and such that $\int_R g(p)d\mu < +\infty$. According to the generalized theorem of Birkhoff and by virtue of (8.19), we have for every integrable function $\varphi(p)$ and almost all $p \in R$,

$$\lim_{T\to\infty} \frac{\int_0^T \varphi(f(p,t))\,dt}{\int_0^T g(f(p,t))\,dt} = C$$

where

$$C = \frac{\int_R \varphi(p)\,d\mu}{\int_R g(p)\,d\mu}.$$

We introduce φ_1 and φ_2, the characteristic functions of the sets E_1 and E_2. Our purpose is to compute

$$\lim_{T\to\infty} \frac{\frac{1}{T}\int_0^T \varphi_1(f(p,t))\,dt}{\frac{1}{T}\int_0^T \varphi_2(f(p,t))\,dt}.$$

This limit can be found on the basis of the preceding relationships:

$$\lim_{T\to\infty} \frac{\frac{1}{T}\int_0^T \varphi_1(f(p,t))\,dt}{\frac{1}{T}\int_0^T \varphi_2(f(p,t))\,dt} = \lim_{T\to\infty} \left\{ \frac{\int_0^T \varphi_1(f(p,t))\,dt}{\int_0^T g(f(p,t))\,dt} : \frac{\int_0^T \varphi_2(f(p,t))\,dt}{\int_0^T g(f(p,t))\,dt} \right\}$$

$$= \lim_{T\to\infty} \frac{\int_0^T \varphi_1(f(p,t))\,dt}{\int_0^T g(f(p,t))\,dt} : \lim_{T\to\infty} \frac{\int_0^T \varphi_2(f(p,t))\,dt}{\int_0^T g(f(p,t))\,dt} = \frac{\int_R \varphi_1(p)\,d\mu}{\int_R g(p)\,d\mu} : \frac{\int_R \varphi_2(p)\,d\mu}{\int_R g(p)\,d\mu} = \frac{\mu E_1}{\mu E_2}.$$

Thus, *in the case of a system irreducible in R, the mean times of the stay of the point $f(p, t)$ in the two sets E_1 and E_2 are in the same ratio as the measures of these sets.*

8.20. Remark. The function $g(p)$ introduced in this section serves as a "transformation of the time"; in fact, if one makes a change of variables, introducing the new "time"

$$\tau(p, t) = \int_0^t g(f(p, t))\,dt,$$

then formula (1) of the present section reduces to the classical formula of Birkhoff. Furthermore, it turns out that the new dynamical system $f_1(p, t)$ obtained after the transformation of the time possesses the invariant set measure

$$\mu_1 E = \int_E g(p)\mu(dp).$$

It is clear that under the condition $\int_A g(p)\,d\mu < +\infty$ the new measure of the entire space is finite; $\mu_1 R < +\infty$. Thus, with the help of a transformation of the time, the generalized theorem of Birkhoff reduces to the classical, and, in particular, formula (8.13), by virtue of formulas (8.11) and (8.12), takes the form

$$\int_R \varphi(p) d\mu_1 = \int_R \psi(p) d\mu_1.$$

The theory of the time transformation and of invariant measure is presented in the memoir of M. V. Bebutov and V. V. Stepanov, *Sur la mesure invariante dans les systèmes dynamiques qui ne diffèrent que par le temps*, Mat. Sbornik (N.S.) **7** (49), 143-164 (1940).

8.21. Example. We shall show the existence of an irreducible (transitive) dynamical system with an invariant measure which is infinite for the whole space and finite for any compact set.

Consider on the torus $\mathfrak{T}(\varphi, \theta)$: $0 \leqq \varphi < 1$, $0 \leqq \theta < 1$, $\varphi + k \equiv \varphi$, $\theta + k' \equiv \theta$ (k and k' integers), motions determined by the differential equations

$$\frac{d\varphi}{dt} = \Phi(\varphi, \theta), \qquad \frac{d\theta}{dt} = \alpha\Phi(\varphi, \theta),$$

where α is an irrational number and Φ is a continuously differentiable function periodic with respect to both arguments (a continuous point function on the torus), wherein $\Phi(0, 0) = 0$ and for the remaining points $\Phi(\varphi, \theta) > 0$. Trajectories not passing through the point (0, 0) are given by the equations

$$\theta = \theta_0 + \alpha\varphi; \qquad (-\infty < \varphi < +\infty;\ \theta_0 \neq 0 \pmod 1);$$

they are stable according to Poisson. We shall regard as the space R the surface of the torus without the point (0, 0). From section 1 it follows that this system admits the integral invariant

$$\iint_E \frac{d\varphi d\theta}{\Phi(\varphi, \theta)},$$

i.e., there exists an invariant measure

$$\mu E = \iint_E \frac{d\varphi d\theta}{\Phi(\varphi, \theta)}.$$

We choose the function $\Phi(\varphi, \theta)$ such that

$$\int_0^1 \int_0^1 \frac{d\varphi \, d\theta}{\Phi(\varphi, \theta)} = +\infty,$$

for example, $\Phi = \sin^2 \pi\varphi + \sin^2 \pi\theta$. Then $\mu R = +\infty$, and since any compact (closed) set F does not have $(0, 0)$ as its limit point, $\mu F < +\infty$.

Finally, this system is irreducible. In fact, in section 5 we proved the irreducibility of the system

$$\frac{d\varphi}{dt} = 1, \ \frac{d\theta}{dt} = \alpha,$$

with the invariant measure $\mu_1 E = \iint\limits_E d\varphi d\theta$. But, under the transition from the measure μ_1 to the measure μ and inversely, sets of measure zero go into sets of measure zero and invariant sets of both systems coincide (with a precision to within a trajectory in the second system passing through $(0, 0)$, but this trajectory, obviously, has μ and μ_1-measure zero). Thus the irreducibility of our first system has been proved.

We remark, finally, that the transition from the first system to the second can be regarded as a change in the time according to the formula

$$dt' = \Phi(\varphi, \ \theta)dt.$$

Thus the function $\Phi(\varphi, \theta)$ plays the role of the function $g(p)$ for the transition from the generalized theorem of Birkhoff to the classical.

9. Invariant Measures of an Arbitrary Dynamical System

The investigations presented in the preceding sections of this chapter assumed the *a priori* knowledge of a measure which was invariant for the given dynamical system.

N. M. Kryloff and N. N. Bogoliuboff, in a memoir *La théorie générale de la mesure et son application à l'étude des systèmes dynamiques de la mécanique non linéaire* (Ann. of Math (2) **38**, 65–113 (1937)), have given a construction, for a very broad class of dynamical systems, of a measure invariant with respect to a given dynamical system. In this section we shall present the most significant of the very important results obtained by these authors.

We shall consider dynamical systems in a compact metric space R. In this space we consider the set of all possible measures μ satisfying the conditions presented in section 2 of this chapter. In the sequel we shall restrict ourselves to those measures for which μR is finite. By means of multiplication by a corresponding positive number we reduce this case to that for which

$$\mu R = 1.$$

The authors call measures satisfying this last condition *normalized measures.*

9.01. DEFINITION. The sequence of measures $\{\mu_n\}$ *converges (weakly) to the measure* μ if for any continuous function $\varphi(p)$ of the point $p \in R$ we have

$$\lim_{n\to\infty} \int_R \varphi(p) d\mu_n = \int_R \varphi(p) d\mu.$$

A fundamental fact for the following theory is that a set of normalized measures in a compact space R is compact. Before proving this theorem we shall prove certain auxiliary propositions.

Consider the set of continuous functions $\{\varphi(p)\}$, $p \in R$. This set forms a metric space if one defines in it the distance of two functions $\varphi_1(p)$ and $\varphi_2(p)$:

$$\varrho(\varphi_1, \varphi_2) = \max_{p \in R} |\varphi_1(p) - \varphi_2(p)|.$$

9.02. THEOREM. *In the space of continuous functions* $\{\varphi(p)\}$, $p \in R$, *there exists a countable, everywhere dense set (a fundamental system of functions).*

For the triplet of natural numbers s, r, n we denote by $\Phi(s, r, n)$ the subset of the set $\{\varphi\}$ satisfying the conditions

$$|\varphi(p)| \leqq s; \; |\varphi(p) - \varphi(q)| \leqq \frac{1}{n}, \text{ if } \varrho(p, q) \leqq \frac{1}{r}.$$

Because of the compactness of the space R there exists in it a $(1/r)$-net; let this consist of the points $p_1, p_2, \ldots, p_{N_r}$. Since the set of values assumed by the functions $\varphi(p) \in \Phi(s, r, n)$ fills, at any point p_i, an interval of length $2s$, there can be found a system $\Phi^*(s, r, n)$ of not more than $(sn)^{N_r} = l$ functions $\varphi_1^*, \varphi_2^*, \ldots, \varphi_l^*$ of $\Phi(s, r, n)$ such that for any $\varphi \in \Phi(s, r, n)$ there exists a φ_i^* of the system $\Phi^*(s, r, n)$ such that

$$|\varphi(p_k) - \varphi_i^*(p_k)| \leqq \frac{1}{n} \qquad (k = 1, 2, \ldots, N_r).$$

Then the countable system of functions

$$\Phi^* = \sum_{s, r, n=1}^{\infty} \Phi^*(s, r, n)$$

will be a fundamental system. In fact, let there be given any continuous function $\varphi(p)$ and an arbitrary $\varepsilon > 0$. We take integers $s_0 \geqq \max_{p \in R} |\varphi(p)|$, $n_0 > 3/\varepsilon$. Finally, for the number $1/n_0$ we find a $\delta > 0$ such that from $\varrho(p, q) < \delta$ there follows $|\varphi(p) - \varphi(q)| < 1/n$, and we take the integer $r_0 < 1/\delta$. It is obvious that for such a choice $\varphi \in \Phi(s_0, r_0, n_0)$ and therefore there can be found in the system $\Phi^*(s, r, n)$ a function $\varphi_\nu^*(p)$ such that in the points of the corresponding $(1/r_0)$-net we shall have

$$|\varphi(p_k) - \varphi_\nu^*(p_k)| \leqq \frac{1}{n_0} \qquad (k = 1, 2, \ldots, N).$$

If now one takes any point $p \in R$ there can be found a point p_k such that $\varrho(p, p_k) \leqq 1/r_0$ and we have

$$|\varphi(p) - \varphi_\nu^*(p)| \leqq |\varphi(p) - \varphi(p_k)| + |\varphi(p_k) - \varphi_\nu^*(p)|$$

$$+ |\varphi_\nu^*(p_k) - \varphi_\nu^*(p)| < \frac{\varepsilon}{3} + \frac{\varepsilon}{3} + \frac{\varepsilon}{3} = \varepsilon.$$

The theorem is proved.

The integral

$$\int_R \varphi(p) d\mu = A\varphi$$

entering in the definition of convergence of measures, where $\varphi(p)$ is a continuous function, is obviously a *linear functional* of φ, *i.e.*, it is *distributive*, $A(\varphi_1 + \varphi_2) = A\varphi_1 + A\varphi_2$, and *bounded*, $|A\varphi| \leqq \max |\varphi| \cdot \mu R$. Besides, this functional is *positive, i. e.*, $A\varphi \geqq 0$ if $\varphi(p) \geqq 0$; finally, if the measure is normalized, then for $\varphi \equiv 1$ we have $A\varphi = 1$.

We shall prove the converse of this fact.

9.03. Theorem (Riesz-Radon). *Every positive linear functional $A\varphi$ defined for continuous functions $\{\varphi(p)\}$, $p \in R$, is expressed by an integral*

$$\int_R \varphi(p)d\mu,$$

where μ *is a certain measure; moreover, if* $A1 = 1$, *then* $\mu R = 1$, i.e., *the measure is normalized.*

According to the hypothesis, $A\varphi$ is defined for continuous functions φ; we shall extend its definition to characteristic functions of open sets $G \subset R$, *i. e.* to functions $\varphi_G(p)$ defined thus: $\varphi_G(p) = 1$ if $p \in G$; $\varphi_G(p) = 0$ if $p \in R - G$. With this object we represent $\varphi_G(p)$ as the limit of a nondecreasing sequence of continuous functions $\{\varphi_n(p)\}$, defining, for example, $\varphi_n(p)$ in the following way: $\varphi_n(p) = 1$, if $\varrho\ (p, R - G) \geqq 1/n$, $\varphi_n(p) = 0$ for $p \in R - G$; finally, if $0 \leqq \varrho(p, R - G) \leqq 1/n$, then $\varphi_n(p) = n \cdot \varrho(p, R - G)$.

We shall have

$$A\varphi_1 \leqq A\varphi_2 \leqq \ldots \leqq A\varphi_n \leqq \ldots \leqq 1.$$

Therefore the sequence of numbers $A\varphi_n$ has a limit; we shall put, by definition,

$$A\varphi_G = \lim_{n\to\infty} A\varphi_n.$$

It can be shown that this same limit is obtained for whatever nondecreasing sequence of continuous functions φ_n we may take provided only $\lim_{n\to\infty} \varphi_n = \varphi_G$.[7]

[7]In fact, along with $\{\varphi_n\}$ let there be a nondecreasing sequence of continuous functions $\{\psi_n(p)\}$, $\lim_{n\to\infty} \psi_n(p) = \varphi_G(p)$. We shall show that for any φ_n and any $\varepsilon > 0$ there can be found an m such that $\psi_m(p) > \varphi_n(p) - \varepsilon$, $p \in R$.

Indeed, it follows from the convergence of the sequence $\{\psi_n\}$ that for any fixed point p_0 there can be found an m_0 such that $\psi_{m_0}(p_0) > \varphi_G(p_0) - \varepsilon/3 \geqq \varphi_n(p_0) - \varepsilon/3$. Because of the continuity of the functions φ_n and ψ_{m_0} at the point p_0, there can be found a $\delta_0 > 0$ such that for $\varrho(p_0, p) < \delta_0$ simultaneously $|\varphi_n(p) - \varphi_n(p_0)| < \varepsilon/3$ and $|\psi_{m_0}(p) - \psi_{m_0}(p_0)| < \varepsilon/3$. From this $\psi_{m_0}(p) > \varphi_n(p) - \varepsilon$ if $\varrho(p, p_0) < \delta$. Since the function ψ_m does not decrease with an increase in m, for every $m > m_0$ we have $\psi_m(p) > \varphi_n(p) - \varepsilon$.

Thus each point $p_0 \in R$ is the center of some sphere of radius S_0 with the indicated property. Because of the compactness of R one can choose a finite number of spheres $S(p_1, \delta_1), \ldots, S(p_N, \delta_N)$ covering the whole of R. Let the corresponding values of m be $m_1, m_2, \ldots, m_N$. Taking $m = \max\ [m_1, m_2, \ldots, m_N]$ we shall have

$$\psi_m(p) > \varphi_n(p) - \varepsilon \qquad \text{for any } p \in R.$$

Then

$$A\psi_m > A\varphi_n - \varepsilon,$$

and since under our conditions the same argument can be given with the sequences $\{\varphi_n\}$ and $\{\psi_n\}$ interchanged we obtain

$$\lim_{m\to\infty} A\psi_m = \lim_{n\to\infty} A\varphi_n.$$

Next we define $A\varphi$ for characteristic functions φ_F of closed sets $F \subset R$; namely, since the set $R - F$ is open, we set

$$A\varphi_F = 1 - A\varphi_{R-F}.$$

The G-functional defined in such a way for the characteristic functions of open sets we shall call the *measure* of these open sets:

$$A\varphi_G = \mu G.$$

It follows from the boundedness of the functional A that $\mu R < +\infty$, and, in particular, if $A1 = 1$, then $\mu R = 1$. It is easy to verify that the function μ defined in this way for open sets satisfies the Axioms I—IV for a measure. From a theorem of section 2 of this chapter, this measure can be extended over all Borel sets and then, with the aid of Axiom V, over any sets, as a regular Carathéodory-Lebesgue measure.

We shall show, finally, that $A\varphi$ is expressed by a Lebesgue-Radon integral for a continuous $\varphi(p)$. We divide the interval of variation of φ into N equal parts by the points of division: $\min \varphi = l_0, l_1, l_2, \ldots, l_N = \max \varphi$. We construct an auxiliary function φ_N, setting $\varphi_N(p) = \varphi(p) = l_i$ $(i = 1, 2, \ldots, N)$ over the closed sets F_i for which $\varphi(p) = l_i$ and setting $\varphi_N(p) = l_i$ $(i = 0, 1, 2, \ldots, N - 1)$ over open sets G_i for which $l_i < \varphi(p) < l_{i+1}$.

Because of the linearity and additive property of the functional A extended over closed and open sets, we have

$$|A\varphi - A\varphi_N| \leqq \frac{\max \varphi - \min \varphi}{N},$$

$$A\varphi_N = \sum_{i=1}^{N} l_i \cdot A\varphi_{F_i} + \sum_{i=0}^{N-1} l_i \cdot A\varphi_{G_i} = \sum_{i=1}^{N} l_i \cdot \mu F_i + \sum_{i=0}^{N-1} l_i \cdot \mu G_i.$$

Noting that the last expression represents a Lebesgue sum, we obtain for $N \to \infty$:

$$A\varphi = \int_R \varphi(p)\, d\mu.$$

9.04. THEOREM. *The set of normalized measures $\{\mu\}$ in a compact space R is weakly compact.*

Let there be given a countable sequence of normalized measures

$$\mu_1, \mu_2, \ldots, \mu_n, \ldots$$

We enumerate the functions of the fundamental system Φ^* satisfying the condition $|\varphi^*| \leqq 1$:

$$\varphi_1^*,\ \varphi_2^*, \ldots, \varphi_n^*, \ldots$$

The set of numbers

$$\int_R \varphi_1^*(p) d\mu_n \qquad (n = 1, 2, \ldots)$$

is contained between -1 and $+1$; from it there can be chosen a convergent subsequence. Let the corresponding measures, the former order being conserved, be $\mu_1^{(1)}, \mu_2^{(1)}, \ldots, \mu_n^{(1)}, \ldots$

We consider next the bounded sequence of numbers

$$\int_R \varphi_2^* \, d\mu_n^{(1)} \qquad (n = 1, 2, \ldots).$$

From it we again choose a convergent subsequence. Let the corresponding measures, the original order being preserved, be $\mu_1^{(2)}, \mu_2^{(2)}, \ldots, \mu_n^{(2)}, \ldots$ Continuing this same process, we obtain sequences of measures $\mu_1^{(k)}, \mu_2^{(k)}, \ldots, \mu_n^{(k)}, \ldots$ possessing the property that there exists the limit

$$\lim_{n\to\infty} \int_R \varphi_i^* \, d\mu_n^{(k)} \qquad (i = 1, 2, \ldots, k).$$

Finally, taking the diagonal sequence of the table $\{\mu_n^{(k)}\}$, we obtain the sequence of measures

$$\mu_1^{(1)} = \mu^{(1)},\ \mu_2^{(2)} = \mu^{(2)}, \ldots, \mu_k^{(k)} = \mu^{(k)}, \ldots,$$

satisfying the property that for any φ_n^* $(n = 1, 2, \ldots)$ there exists the limit

$$\lim_{n\to\infty} \int_R \varphi_n^* \, d\mu^{(k)} = \lim_{k\to\infty} A^{(k)} \varphi_n^*.$$

It is easy to convince oneself that this limit exists for every continuous function φ such that $|\varphi| \leqq 1$. In fact, let there be given any number $\varepsilon > 0$; there can be found a function φ_n^* such that $|\varphi - \varphi_n^*| < \varepsilon/3$. Then for any k we have

$$|A^{(k)}\varphi - A^{(k)}\varphi_n^*| < \frac{\varepsilon}{3}.$$

On the other hand, because of the convergence of the sequence $A^{(k)}\varphi_n^*$, there exists an N such that for $k \geqq N$ and $m \geqq 0$

$$|A^{(k)}\varphi_n^* - A^{(k+m)}\varphi_n^*| < \frac{\varepsilon}{3}.$$

Comparing these inequalities, we find

$$|A^{(k)}\varphi - A^{(k+m)}\varphi| \leqq |A^{(k)}\varphi - A^{(k)}\varphi_n^*| + |A^{(k)}\varphi_n^* - A^{(k+m)}\varphi_n^*| + |A^{(k+m)}\varphi_n^* - A^{(k+m)}\varphi| < \varepsilon$$

for $k \geqq N$ and $m \geqq 0$, *i. e.*, $\lim_{k\to\infty} A^{(k)}\varphi$ exists for any φ such that $|\varphi| \leqq 1$.

We denote this limit by $A\varphi$ and determine it for any continuous function φ by the condition

$$A\varphi = \max |\varphi| \cdot A \frac{\varphi}{\max |\varphi|}.$$

It is easy to see that $A\varphi$ is a linear functional, and $A1 = 1$; consequently, because of Theorem 9.03, there exists a normalized measure μ such that

$$A\varphi = \int_R \varphi(p)\, d\mu.$$

Thus we have obtained $\mu^{(k)} \to \mu$ as $k \to \infty$ in the sense of weak convergence. The theorem is proved.

Every measure to which some subsequence of a given set of measures converges (weakly) we shall call a *limiting* measure for this set; if a countable sequence $\{\mu_n\}$ has a unique limiting measure μ, then we shall say that there exists a *limit* of the measures, and in this circumstance we shall write

$$\lim_{n\to\infty} \mu_n = \mu.$$

We proceed to the formulation and proof of the fundamental theorem of Kryloff and Bogoliuboff. We remark first of all that a space always possesses a normalized measure; it is sufficient to single out any point $p \in R$ and set

$$m_p A = 1 \quad \text{if} \quad p \in A; \; m_p A = 0 \quad \text{if} \quad p \in R - A.^{8}$$

[8]The measures $\{m_p\}$ possess the important property: every normalized measure m is a limiting measure (in the sense of weak convergence) for the linear combinations

$$\sum_{i=1}^{n} \alpha_i m_{p_i}, \quad \text{where } \alpha_i > 0 \text{ and } \sum_{i=1}^{n} \alpha_i = 1.$$

In fact, let there be given a normalized measure $m(E)$ in a compact metric space R. We construct in R a countable, everywhere dense set of points $\{p_n\}$ such that finite systems of them, $\{p_1, p_2, \ldots, p_{N_k}\}$, form $(1/k)$-nets ($k = 1, 2, 3, \ldots$) and such that for $N_{k-1} < i < N_k$ the point

9.05 Theorem (Kryloff and Bogoliuboff). *In a compact phase space R of a dynamical system f(p, t) there exists an invariant* (normalized) *measure.*

$$p_i \in R - \sum_{l=1}^{i-1} S\left(p_l, \frac{1}{k}\right).$$

We shall form the measure m_k as a linear combination of the measures m_{p_i} $(i = 1, 2, \ldots, N_k)$ in the following manner. We construct the sets

$$S\left(p_1, \frac{1}{k}\right) = E_1^{(k)}, \quad S\left(p_i, \frac{1}{k}\right) - S\left(p_i, \frac{1}{k}\right) \cdot \sum_{l=1}^{i-1} S\left(p_l, \frac{1}{k}\right) = E_i^{(k)} \quad (i = 2, 3, \ldots, n)$$

Obviously, $E_i^{(k)} E_j^{(k)} = 0$ if $j \neq i$. From the choice of the point p_i, for $N_{k-1} < i < N_k$ we have $p_i \in E_i^{(k)}$. We write

$$m\, E_i^{(k)} = \alpha_i^{(k)}.$$

From the property of a $1/k$ — net we have $\sum_{i=1}^{N_k} \alpha_i^{(k)} = mR = 1$. Now we define

$$m_k = \sum_{i=1}^{N_k} \alpha_i^{(k)} m_{p_i}.$$

We shall prove now that m_k converges (weakly) as $k \to \infty$ to the measure m. Indeed, let $\varphi(p)$ be any continuous function and $\varepsilon > 0$ be arbitrary. We determine an integer k such that from the inequality $\varrho(p, q) < 1/k$ there should follow $|\varphi(p) - \varphi(q)| < \varepsilon$.

We estimate the difference

$$\left| \int_R \varphi(p) m(dp) - \int_R \varphi(p) m_k(dp) \right|$$

$$\leqq \sum_{i=1}^{N_k} \left| \int_{E_i^k} \varphi(p) m(dp) - \int_{E_i^k} \varphi(p) m_k(dp) \right| = \sum_{i=1}^{N_k} \left| \int_{E_i^k} \varphi(p) m(dp) - \alpha_i^{(k)} \varphi(p_i) \right|$$

$$\leqq \sum_{i=1}^{N_k} |\varphi(\bar{p}_i) - \varphi(p_i)| m\, E_i^{(k)},$$

where $\bar{p}_i \in E_i^{(k)} \subset S(p_i, 1/k)$, whence $|\varphi(\bar{p}_i) - \varphi(p_i)| < \varepsilon$ and, consequently,

$$\left| \int_R \varphi(p) m(dp) - \int_R \varphi(p) m_k(dp) \right| < \varepsilon \sum_{i=1}^{N_k} m E_i^{(k)} = \varepsilon.$$

The assertion is proved.

Considering the linear combinations with rational coefficients r_i,

$$\sum_{i=1}^{n} r_i m_i, \quad r_i > 0, \quad \sum_{i=1}^{n} r_i = 1,$$

we conclude that the space of normalized measures possesses a countable, everywhere dense set. This space, as we have seen, is compact. It is metrizable; one may for example introduce a metric as follows: Let m_1 and m_2 be two normalized measures; let $\{\varphi_n^*(p)\}$ be a fundamental system of bounded functions, $|\varphi^*(p)| \leqq 1$. We define the distance

$$(m_1, m_2) = \sum_{n=1}^{\infty} \frac{1}{2^n} \left| \int_R \varphi_n^*(p) m_1(dp) - \int_R \varphi_n^*(p) m_2(dp) \right|.$$

It is easy to see that all the axioms of a metric space are fulfilled here.

Thus the space of normalized measures in a compact space satisfying the second countability axiom is itself a separable, compact metric space.

Let m be any normalized measure in the compact space R. For a given fixed τ and any continuous function $\varphi(p)$ we define the positive linear functional

$$A_\tau \varphi = \frac{1}{\tau}\int_0^\tau dt \int_R \varphi(f(q,t))m(dq).$$

According to Theorem 9.03 it defines a normalized measure m_τ and we have

$$\frac{1}{\tau}\int_0^\tau dt \int_R \varphi(f(q,t))m(dq) = \int_R \varphi(p) m_\tau(dp).$$

From Theorem 9.04 the set of measures $\{m_\tau\}$ is compact and therefore there can be chosen a convergent subsequence from any sequence with τ increasing without bound (for example, from the sequence m_n); let this convergent subsequence be $\{m_{\tau_n}\}$, $\lim_{n\to\infty} \tau_n = \infty$. We denote the limit measure of this subsequence by

$$\lim_{n\to\infty} m_{\tau_n} = \mu^*;$$

for any continuous function $\varphi(p)$ we have

$$\begin{aligned} \lim_{n\to\infty} \frac{1}{\tau_n}\int_0^{\tau_n} dt \int_R \varphi(f(q,\ t))m(dq) &= \lim_{n\to\infty}\int_R \varphi(p)m_{\tau_n}(dp) \\ &= \int_R \varphi(p)\mu^*(dp). \end{aligned} \tag{9.06}$$

The measure μ^* is invariant. In fact, first of all, in order that μ^* be an invariant measure it is necessary and sufficient that for any t_0 and any continuous function $\varphi(p)$ there should hold the relation

$$\int_R \varphi(p)\mu^*(dp) = \int_R \varphi(f(p,\ t_0))\mu^*(dp). \tag{9.07}$$

Indeed, in the proof of Theorem 9.03 we saw that the functional (9.06) retains its meaning in the case when φ is the characteristic function φ_G of an open set, whence, assuming the relation (9.07), we obtain for any open set G

$$\int_R \varphi_G(p)\mu^*(dp) = \mu^* G = \int_R \varphi_G(f(p,t_0))\mu^*(dp) = \mu^* f(G, -t_0)$$

i.e.,

$$\mu^* G = \mu^* f(G, -t_0).$$

Furthermore, this last relation extends over all measurable sets.

Conversely, if the last relation holds for all measurable sets, then the integrals on both sides of equation (9.07), regarded as Lebesgue integrals, are equal.

Thus it is sufficient for us to prove the relation (9.07), or, because of formula (9.06), the relation

$$\lim_{n\to\infty} \frac{1}{\tau_n}\int_0^{\tau_n} dt \int_R \varphi(f(q,t))m(dq) = \lim_{n\to\infty} \frac{1}{\tau_n}\int_0^{\tau_n} dt \int_R \varphi(f(q, t_0+t))m(dq) \tag{9.08}$$

By Fubini's Theorem the functions following the limit operator in this last equation can be rewritten in the form

$$\int_R m(dq) \frac{1}{\tau_n}\int_0^{\tau_n} \varphi(f(q,t))dt \quad \text{and} \quad \int_R m(dq) \frac{1}{\tau_n}\int_0^{\tau_n} \varphi(f(q,t_0+t))dt.$$

We estimate the difference

$$\left|\frac{1}{\tau_n}\int_0^{\tau_n} \varphi(f(p,t))dt - \frac{1}{\tau_n}\int_0^{\tau_n} \varphi(f(q,t_0+t))dt\right|$$
$$\leqq \frac{1}{\tau_n}\left\{\left|\int_0^{t_0} \varphi(f(q,t))dt\right| + \left|\int_{\tau_n}^{\tau_n+t_0} \varphi(f(q,t))dt\right|\right\} \leqq \frac{2|t_0|M}{\tau_n}$$

where M is the maximum of $|\varphi(q)|$ over R; M is finite since $\varphi(p)$ is continuous and the space R is compact. For a sufficiently large τ_n this difference is arbitrarily small in absolute value; whence, by virtue of the existence of the limit (9.06), there follows the relation (9.08), *i. e.*, the relation (9.07). The assertion is proved.

A further problem investigated by Kryloff and Bogoliuboff consists in studying the totality of all invariant measures admitting a given dynamical system and also in singling out from the dynamical system that part of it which has the measure 1 for any normalized invariant measure.

9.09. Definition. A set $E \subset R$ has the *probability zero* if for every invariant normalized measure μ we have $\mu E = 0$. If, for some measure, $\mu E > 0$, then E has a *positive probability*; in particular, if for every invariant normalized measure $\mu E = 1$, then the set E has the *maximum probability*.

9.10. Definition. A point $p \in R$ is called *quasi-regular* if, for an arbitrary continuous function $\varphi(p)$, there exists

$$\lim_{\tau\to\infty}\frac{1}{\tau}\int_0^\tau \varphi(f(p, t))dt. \tag{9.11}$$

9.12. THEOREM. *The set U of quasi-regular points is invariant and has maximum probability.*

The invariance of the set U follows from the estimate of the difference

$$\left|\frac{1}{\tau}\int_0^\tau \varphi(f(p, t))dt - \frac{1}{\tau}\int_0^\tau \varphi(f(p, t_0 + t))dt\right| \leqq \frac{2|t_0|M}{\tau}$$

introduced in the proof of the preceding theorem, whence it follows that

$$\lim_{\tau\to\infty}\frac{1}{\tau}\int_0^\tau \varphi(f(p, t))\,dt = \lim_{\tau\to\infty}\frac{1}{\tau}\int_0^\tau \varphi(f(p, t_0 + t))\,dt;$$

i. e. the point $f(p, t_0)$ is quasi-regular whenever the point p is.

We shall prove that U has maximum probability. We take the fundamental system of functions $\{\varphi_n^*\}$. Let E_n be the set of points such that for $p \in E_n$ there exists no limit (9.11) for $\varphi = \varphi_n^*$.

By Birkhoff's Theorem the set E_n has the probability zero. We set $E = \sum_{n=1}^\infty E_n$; E also has the probability zero.

We shall show that every point $p \in R - E$ is quasi-regular, *i. e.*, that $R - E = U$. Let $\varphi(p)$ be an arbitrary continuous function and let $\varepsilon > 0$ be an arbitrary number; suppose that $p_0 \in R - E$. Then we can find a $\varphi_n^*(p)$ such that $|\varphi_n^*(p) - \varphi(p)| < \varepsilon/3$. Next, because of the fact that $p_0 \in R - E_n$, there can be found a T such that for $\tau_1 > T$ and $\tau_2 > T$ we have

$$\left|\frac{1}{\tau_2}\int_0^{\tau_2} \varphi_n^*(f(p_0, t))\,dt - \frac{1}{\tau_1}\int_0^{\tau_1} \varphi_n^*(f(p_0, t))\,dt\right| < \frac{\varepsilon}{3}.$$

But then

$$\left|\frac{1}{\tau_2}\int_0^{\tau_2} \varphi(f(p_0, t))\,dt - \frac{1}{\tau_1}\int_0^{\tau_1} \varphi(f(p_0, t))\,dt\right|$$
$$\leqq \left|\frac{1}{\tau_2}\int_0^{\tau_2}\varphi dt - \frac{1}{\tau_2}\int_0^{\tau_2}\varphi_n^* dt\right| + \left|\frac{1}{\tau_1}\int_0^{\tau_1}\varphi dt - \frac{1}{\tau_1}\int_0^{\tau_1}\varphi_n^* dt\right|$$
$$+ \left|\frac{1}{\tau_2}\int_0^{\tau_2}\varphi_n^* dt - \frac{1}{\tau_1}\int_0^{\tau_1}\varphi_n^* dt\right| < \varepsilon$$

i. e. the limit (9.11) exists, which we were to prove.

In the sequel the original measure, on the basis of which invariant measures will be constructed, will be the previously mentioned measure relative to some point $p \in R$:

$$m_p(A) = \begin{cases} 1, & \text{if } p \in A, \\ 0, & \text{if } p \in R - A. \end{cases}$$

Starting with this normalized measure we shall construct, as in Theorem (9.05), the measure $m_{p,\,\tau}$ defined by the equation (for any continuous $\varphi(q)$):

$$\int_R \varphi(q) m_{p,\,\tau}(dq) = \frac{1}{\tau}\int_0^\tau dt \int_R \varphi(f(q, t))\, m_p(dq).$$

But from the definition of the measure m_p, we have

$$\int_R \varphi(f(q, t))\, m_p(dq) = \varphi(f(p, t)),$$

whence we obtain for a definition of the measure $m_{p,\,\tau}$:

$$\int_R \varphi(q) m_{p,\,\tau}(dq) = \frac{1}{\tau}\int_0^\tau \varphi(f(p, t))dt.$$

In the following we shall consider only quasi-regular points p, $p \in U$; in this case the limit as $\tau \to \infty$ in the first part of the last equation exists; consequently, there exists an invariant normalized measure μ_p,

$$\mu_p = \lim_{\tau \to \infty} m_{p,\,\tau},$$

and this measure is defined by the equation ($\varphi(p)$ is any continuous function):

$$\int_R \varphi(q)\mu_p(dq) = \lim_{\tau \to \infty} \frac{1}{\tau}\int_0^\tau \varphi(f(p, t))dt. \tag{9.13}$$

The measure $\mu_p(A)$ is called the *individual measure* corresponding to the quasi-regular point $p \in U$.

We remark that the set U, as a set of points over which a countable set of continuous functions of p and τ has a limit as $\tau \to \infty$, is measurable B.

For any given set A measurable B, the measure $\mu_p(A)$, regarded as a function of the point $p \in U$, is also measurable B. In fact, for a continuous function $\varphi(p)$ the right-hand side of equation (9.13) and, consequently, also the left, is a measurable function of p as the limit for $\tau \to \infty$ of continuous functions of p over a measurable set U. Furthermore, the transition on the left-hand side of equation (9.13) to the measure of an open set is connected with the passage to the limit with respect to a sequence of continuous functions $\varphi(p)$ converging to a characteristic function of an open set; so that $\varphi_p(G)$ is a function of p measurable B. Finally, the definition of the measure of any set measurable B is associated with not more than a countable number of passages to the limit under which measurability according to Borel is preserved. The assertion is proved.

We introduce now the relation between the individual measure μ_p and any invariant (normalized) measure μ. Starting with any continuous function $\varphi(p)$, we obtain, because of the invariance of the measure μ and the fact that U has maximum probability,

$$\int_R \varphi(q)\mu(dq) = \int_R \varphi(f(q,t))\mu(dq) = \frac{1}{\tau}\int_0^\tau dt \int_R \varphi(f(q,t))\mu(dq)$$

$$= \int_U \mu(dq)\frac{1}{\tau}\int_0^\tau \varphi(f(q,t))\,dt = \int_U \mu(dq)\int_R \varphi(r) m_{q,\tau}(dr);$$

whence, in view of the existence of a limit on the right-hand side as $\tau \to \infty$ and on the basis of formula (9.13), we deduce

$$\int_R \varphi(q)\mu(dq) = \int_U \mu(dq)\int_R \varphi(r)\mu_q(dr). \tag{9.14}$$

Passing from the continuous function $\varphi(p)$ to the characteristic function of an open set G, we find from the relation (5) that

$$\mu G = \int_U \mu_q(G)\mu(dq).$$

For any set $A \subset R$ measurable B the integral

$$\int_U \mu_q(A)\mu(dq)$$

exists by virtue of the measurability of the function $\mu_q(A)$. This integral, obviously, defines a measure which, according to what has been proved, for open sets, coincides with the measure μ; therefore we obtain the desired relation

$$\mu A = \int_U \mu_q(A)\mu(dq). \tag{9.15}$$

From the relation (9.15) follows the validity of the equation (9.14) for any function $\varphi(p)$ measurable B.

9.16. LEMMA. *The set U_T of points $p \in U$ for which the equation*

$$(9.17) \qquad \int_U \left\{ \int_R \varphi(r)\mu_q(dr) - \int_R \varphi(r)\mu_p(dr) \right\}^2 \mu_p(dq) = 0$$

is fulfilled for any continuous function $\varphi(p)$ *has maximum probability.*

Let $\varphi(p)$ be some continuous function. Because of the fact that the left-hand side of the inequality (9.17) is non-negative, it is sufficient for us to prove that for any invariant measure μ we shall have

$$\int_U \left[\int_U \left\{ \int_R \varphi(r)\mu_q(dr) - \int_R \varphi(r)\mu_p(dr) \right\}^2 \mu_p(dq) \right] \mu(dp) = 0.$$

We shall prove this last relation.

According to formula (9.13), we have for a given $\varphi(p)$, if $q \in U$:

$$\int_R \varphi(r)\mu_q(dr) = \lim_{\tau\to\infty} \frac{1}{\tau} \int_0^\tau \varphi(f(q,t))\,dt = \psi(q)$$

and analogously for $p \in U$:

$$\int_R \varphi(r)\mu_p(dr) = \psi(p).$$

Therefore we must establish the equality

$$\int_U \mu(dp) \int_U (\psi(q) - \psi(p))^2 \mu_p(dq) = 0.$$

Removing the parentheses in the inner integral, we obtain

$$(9.18) \qquad \begin{aligned} &\int_U \{\psi^2(q) - 2\psi(p)\psi(q) + \psi^2(p)\}\mu_p(dq) \\ &\quad = \int_U \psi^2(q)\mu_p(dq) - 2\psi(p) \int_U \psi(q)\mu_p(dq) + \psi^2(p). \end{aligned}$$

We compute next

$$\begin{aligned} \int_U \psi(q)\mu_p(dq) &= \lim_{\tau\to\infty} \int_U \left(\frac{1}{\tau} \int_0^\tau \varphi(f(q,t)dt) \mu_p(dq) \right. \\ &= \lim_{\tau\to\infty} \frac{1}{\tau} \int_0^\tau dt \int_U \varphi(f(q,t))\mu_p(dq). \end{aligned}$$

Since μ_p is an invariant measure, then for any t

$$\int_U \varphi(f(q,t))\,d\mu_p(q) = \int_U \varphi(q)\,d\mu_p(q);$$

consequently,

$$\int_U \psi(q)\,d\mu_p(q) = \int_U \varphi(q)\,d\mu_p(q) = \lim_{\tau\to\infty} \frac{1}{\tau} \int_0^\tau \varphi(f(p,t))dt = \psi(p).$$

The expression (9.18) is thus equal to

$$\int_U \psi^2(q)\mu_p(dq) - \psi^2(p),$$

and we must prove the vanishing of the integral

$$\int_U \Big[\int_U \psi^2(q)\mu_p(dq) - \psi^2(p) \Big] \mu(dp)$$
$$= \int_U \int_U \psi^2(q)\mu_p(dq)\mu(dp) - \int_U \psi^2(p)\mu(dp).$$

But because of the relation (9.14) for the measurable function $\psi^2(p)$, the first integral is equal to $\int_U \psi^2(p)\mu(dp)$ and so the expression vanishes.

Thus for each continuous function $\varphi(p)$ the set of points where the relation (9.17) is fulfilled has the maximum probability.

Taking functions of a fundamental system $\varphi_1^*, \varphi_2^*, \ldots, \varphi_n^*, \ldots$ successively as φ, and denoting by E_n the set of points p for which

$$\int_U \Big\{ \int_R \varphi_n^*(r)\mu_q(dr) - \int_R \varphi_n^*(r)\mu_q(dr) \Big\}^2 \mu_p(dq) > 0,$$

we have $\mu E_n = 0$ for any invariant measure μ.

We define the set

$$U_T = U - \sum_{n=1}^{\infty} E_n.$$

It has maximum probability. If $p \in U_T$, then for every continuous function $\varphi(p)$ the equality (9.17) holds. In fact, for the fundamental system this follows from the definition, and for any continuous function, from the possibility of approximating it uniformly by the functions $\{\varphi_n^*\}$. The lemma is proved.

We remark in conclusion that because of the invariance of the measures μ_p the set U_T is invariant.

The geometrical meaning of the lemma is as follows: *for almost all points p (in the sense of any invariant measure) the set of those points q for which the individual measures μ_q are different from the individual measure μ_p form a set of μ_p-measure zero.*

9.19. DEFINITION. An invariant measure μ is *transitive* if for any decomposition of R into a sum of two disjoint measurable invariant sets A and $R - A$ it follows from $\mu A > 0$ that $\mu(R - A) = 0$.

9.20. THEOREM. *If $p \in U_T$* (see the Lemma 9.16), *then the measure μ_p is transitive.*

Suppose that $p \in A \subset U_T$, where A is an invariant, measurable set, and let $\varphi(r)$ be a continuous function. By (9.17),

$$\int_R \varphi(r)\mu_q(dr) = \int_R \varphi(r)\mu_p(dr),$$

with the possible exception of points of a set $\{q\}$ of μ_p-measure zero; thus on multiplying both sides of this equality by $\varphi_A(q)\mu_p(dq)$, where $\varphi_A(q)$ is the characteristic function of the set A, and integrating both sides with respect to the set U, we find

$$\int_U \left\{ \int_R \varphi(r)\mu_q(dr) \right\} \varphi_A(q)\mu_p(dq) = \int_U \left\{ \int_R \varphi(r)\mu_p(dr) \right\} \varphi_A(q)\mu_p(dq).$$

The right-hand side of this equation, obviously, is equal to

$$\int_R \varphi(r)\mu_p(dr) \cdot \int_U \varphi_A(q)\mu_p(dq) = \mu_p(A) \cdot \int_R \varphi(r)\mu_p(dr).$$

We now transform the left-hand side. In view of the fact that A is an invariant set and therefore $\varphi_A(f(q, t)) = \varphi_A(q)$, we obtain on applying formula (9.13)

$$\int_U \left\{ \int_R \varphi(r)\mu_q(dr) \right\} \varphi_A(q)\mu_p(dq)$$

$$= \int_U \left\{ \lim_{\tau\to\infty} \frac{1}{\tau} \int_0^\tau \varphi(f(q,t))dt \right\} \varphi_A(q)\mu_p(dq)$$

$$= \lim_{\tau\to\infty} \frac{1}{\tau} \int_0^\tau \left\{ \int_U \varphi(f(q,t))\varphi_A(f(q,t))\mu_p(dq) \right\} dt$$

Since μ_p is an invariant measure, the inner integral does not depend on t. It is equal to

$$\int_U \varphi(q)\varphi_A(q)\mu_p(dq),$$

and we shall have

$$\int_U \left\{ \int_R \varphi(r)\mu_q(dr) \right\} \varphi_A(q)\mu_p(dq) = \int_U \varphi(q)\varphi_A(q)\mu_p(dq).$$

Therefore,

$$\mu_p A \cdot \int_R \varphi(r)\mu_p(dr) = \int_A \varphi(q)\mu_p(dq).$$

This last equality, proved for a continuous function $\varphi(r)$, is valid also for any bounded measurable function. Setting, in particular $\varphi(r) = \varphi_A(r)$, we find

$$\mu_p A \cdot \mu_p A = \mu_p A \ \text{ or } \ \mu_p(A)[\mu_p(A) - 1] = 0,$$

whence either $\mu_p A = 0$ or $\mu_p A = 1$.

The theorem is proved.

It may be formulated thus: *the set of points for which the correspond-*

ing individual measures are transitive[9] *has maximum probability.*

9.21. Definition. A point $p \in U$ is a *density point* if for any $\varepsilon > 0$ we have $\mu_p(S(p, \varepsilon)) > 0$.

9.22. Theorem. *The set U_D of all density points is invariant and has maximum probability.*

For the construction of the set $U_D \subset U$ we construct in R an ε-net for $\varepsilon = 1/m$ $(m = 1, 2, 3, \ldots)$; let this be $\{p_1^{(m)}, p_2^{(m)}, \ldots, p_{N_m}^{(m)}\}$. Then

$$R = \sum_{n=1}^{N_m} S\left(p_n^{(m)}, \frac{1}{m}\right) \equiv \sum_{n=1}^{N_m} S_{nm}.$$

For each point $p_n^{(m)}$ we construct the continuous function

$$\varphi_{nm}(p) = \begin{cases} 1, \ p \in S_{nm}; \\ 2 - m\varrho(p, p_n^{(m)}), \ \dfrac{1}{m} \leqq \varrho(p, p_n^{(m)}) \leqq \dfrac{2}{m}; \\ 0, \ \varrho(p, p_n^{(m)}) \geqq \dfrac{2}{m}. \end{cases}$$

For a point $p \in U$ we set

$$\Phi_{nm}(p) = \lim_{\tau\to\infty} \frac{1}{\tau} \int_0^\tau \varphi_{nm}(f(p, t))\, dt$$

(according to the second part of Birkhoff's Theorem this is an invariant function) and we define the (invariant) set $E_{nm} = \{p;\ \Phi_{nm}(p) = 0\}$.

Let μ be any invariant measure; we have (making use of the invariance of the measure μ and of the sets E_{nm}):

$$\begin{aligned} 0 &= \int_{E_{nm}} \Phi_{nm}(p)\mu(dp) \\ &= \int_{E_{nm}} \left(\lim_{\tau\to\infty} \frac{1}{\tau}\int_0^\tau \varphi_{nm}(f(p, t))\, dt\right) \mu(dp) = \lim_{\tau\to\infty} \frac{1}{\tau}\int_0^\tau dt \int_{E_{nm}} \varphi_{nm}(f(p, t))\mu(dp) \\ &= \int_{E_{nm}} \varphi_{nm}(p)\mu(dp) \geqq \int_{E_{nm}\cdot S_{nm}} \varphi_{nm}(p)\mu(dp). \end{aligned}$$

Since for $p \in S_{nm}$ we have $\varphi_{nm} = 1$, we obtain

$$\mu(E_{nm} \cdot S_{nm}) = 0.$$

We define

[9] We shall call these points *transitive*.

$$U_D = U - \sum_{m=1}^{\infty} \sum_{n=1}^{N_m} E_{nm} \cdot S_{nm}.$$

Since, for any invariant measure μ, $\mu(U - U_D) = 0$, U_D has maximum probability.

We shall show that every density point $p \in U$ lies in the set U_D. Let S_{nm} be any of the spheres containing p; there can be found an $\varepsilon > 0$ such that $S(p, \varepsilon) \subset S_{nm}$. According to the definition of a density point, we have [10]:

$$0 < \mu_p(S(p, \varepsilon)) = \int_R \chi_{S(p,\varepsilon)}(r)\mu_p(dr) < \int_R \varphi_{nm}(r)\mu_p(dr)$$

$$= \lim_{\tau\to\infty} \frac{1}{\tau}\int_0^\tau \varphi_{nm}(f(p, t))dt = \Phi_{nm}(p),$$

i. e., if the density point $p \in S_{nm}$, then $p \notin E_{nm}$, and therefore

$$p \in U - \sum_{m=1}^{\infty} \sum_{n=1}^{N_m} E_{nm} \cdot S_{nm} = U_D.$$

We shall show next that if p is not a density point then $p \notin U_D$. According to assumption there exists an $\varepsilon > 0$ such that $\mu_p(S(p, \varepsilon)) = 0$. For this ε we determine an m satisfying the condition $1/m < \varepsilon/4$. By construction there can be found a point $p_n^{(m)}$ such that $\varrho(p, p_n^{(m)}) < 1/m$. Then

$$p \in S\left(p_n^{(m)}, \frac{1}{m}\right) \subset S\left(p_n^{(m)}, \frac{2}{m}\right) \subset S(p, \varepsilon).$$

We have:

$$0 = \mu_p(S(p, \varepsilon)) = \int_R \chi_{S(p,\varepsilon)}(r)\mu_p(dr)$$

$$\geqq \int_R \varphi_{nm}(r)\mu_p(dr) = \lim_{\tau\to\infty} \frac{1}{\tau}\int_0^\tau \varphi_{nm}(f(p, t))\, dt = \Phi_{nm}(p).$$

Thus $\Phi_{nm}(p) = 0$; *i. e.*, $p \in E_{nm}$. Since $p \in S_{nm}$,

$$p \in E_{nm} \cdot S_{nm}; \quad i.\ e.\ p \notin U_D$$

which it was required to prove.

Finally, the set U_D is invariant since, if $p \in U_D$, for a given t and $\varepsilon > 0$ there can be found an $\varepsilon_1 > 0$ such that

$$f(S(p, \varepsilon_1), t) \subset S(f(p, t), \varepsilon);$$

hence, because of the invariance of the measure μ_p,

[10]Here and in the sequel $\chi_E(p)$ denotes the characteristic function of the set E.

$$\mu_p[S(f(p,t),\varepsilon)] \geqq \mu_p[f(S(p,\varepsilon_1),t)] = \mu_p[S(p,\varepsilon_1)] > 0,$$

i. e. $f(p,t)$ is also a density point for any t.

9.23. DEFINITION. The points $p \in U_T \cdot U_D = U_R$ are called *regular*; they are those density points which are at the same time transitive.

From Theorems 9.20 and 9.22 the set U_R of regular points is invariant and has maximum probability.

9.24. THEOREM. *The set $\overline{U}_R$ (the closure of the set of regular points) is a minimal center of attraction*[11] *for the system $f(p, t)$.*

We denote the characteristic function of the set $S(U_R, \varepsilon)$ by $\varphi_S(p) = \chi_{S(U_R,\varepsilon)}(p)$. According to the definition of a center of attraction, for any $\varepsilon > 0$ and any $p \in R$ we must have

$$\mathbf{P}[f(p,t) \in S(U_R,\varepsilon)] \equiv \lim_{\tau\to\infty} \frac{1}{\tau}\int_0^\tau \varphi_S(f(p,t))\,dt = 1. \tag{9.25}$$

For regular points $p \in U_R$ equation (9.25) is obvious. Assume now that there can be found a nonregular point p_0 and a number $\gamma (0 < \gamma \leqq 1)$ such that the equation (9.25) is not fulfilled, *i. e.*

$$\mathbf{P}[f(p_0,t) \in S(U_R,\varepsilon)] = 1 - \gamma < 1.$$

That is, there can be found a sequence of numbers $\{\tau_n\}$, $\lim_{n\to\infty} \tau_n = \infty$, such that

$$\lim_{n\to\infty} \frac{1}{\tau_n}\int_0^{\tau_n} \varphi_S(f(p,t))\,dt = 1 - \gamma.$$

Denoting again by $m_{p_0}(A)$ the measure which is equal to 1 if $p_0 \in A$ and equal to 0 if $p_0 \in R - A$, we form the sequence of measures m_{p_0,τ_n} as in the construction of individual measures. Now this sequence is not convergent in general since the point p_0 may also fail to be quasi-regular, but because of Theorem 9.04 (on compactness) there exists a sequence $\{\tau_n'\} \subset \{\tau_n\}$ such that the $m_{p_0,\tau_n'}$ converge (weakly) to $\mu^*_{p_0}$, where $\mu^*_{p_0}$ is an invariant measure. Furthermore, from Theorems 9.20 and 9.22 we have

$$\mu^*_{p_0} U_R = 1, \qquad \mu^*_{p_0}(R - U_R) = 0.$$

We construct the continuous function $\varphi(p)$:

[11] We employ the terminology of section 6, Chap. V. Kryloff and Bogoliuboff use the expression: "the motions $f(p, t)$ are statistically asymptotic to the set U_R".

$$\varphi(p) = \begin{cases} 1, & p \in \overline{U}_R \\ 1 - \dfrac{1}{\varepsilon}\varrho(p, \overline{U}_R), & 0 < \varrho(p, \overline{U}_R) \leqq \varepsilon \\ 0, & \varrho(p, \overline{U}_R) \geqq \varepsilon. \end{cases}$$

For the measure μ_p^* we obtain, applying the reasoning leading to formula (**9.13**),

$$\lim_{n\to\infty} \frac{1}{\tau_n'} \int_0^{\tau_n'} \varphi(f(p, t))dt = \int_R \varphi(q)\mu_{p_0}^*(dq).$$

In view of the choice of the function $\varphi(p)$ we have, for the right-hand side of the last equation,

$$\int_R \varphi(q)\mu_{p_0}^*(dq) \geqq \int_R \chi_{\overline{U}_R}(q)\mu_{p_0}^*(dq) = \mu_{p_0}^*(\overline{U}_R) = 1,$$

and for the left-hand side,

$$\lim_{n\to\infty} \frac{1}{\tau_n'} \int_0^{\tau_n'} \varphi(f(p, t))dt \leqq \lim_{n\to\infty} \frac{1}{\tau_n'} \int_0^{\tau_n'} \varphi_S(f(p, t))\,dt.$$

Combining these results, we obtain

$$\lim_{n\to\infty} \frac{1}{\tau_n'} \int_0^{\tau_n'} \varphi_S(f(p, t))\,dt = 1,$$

which contradicts the choice of the sequence $\{\tau_n\} \supset \{\tau_n'\}$. Thus we have proved that $\overline{U}_R$ is a center of attraction.

We shall show that it is a minimal center of attraction. Assume that there exists a center of attraction M forming a proper part of the set $\overline{U}_R$. Since M, according to assumption, is a closed set, there can be found a point $p \in U_R$, $\varrho(p, M) = \alpha > 0$; then

$$S\left(p, \frac{\alpha}{2}\right) \cdot S\left(M, \frac{\alpha}{2}\right) = 0.$$

Since p is a density point, $\mu_p S(p, \alpha/2) > 0$; therefore $\mu_p S(M, \alpha/2) < \mu_p(R) = 1$, whence

$$\mathbf{P}\left[f(p, t) \in S\left(M, \frac{\alpha}{2}\right)\right] < 1;$$

i. e., M is not a center of attraction.

The theorem is proved.

9.26. Example. Consider, in a domain R:

$$x^2 + y^2 \leqq 1$$

of the plane E^2, the motions defined by the differential equations

$$\frac{dx}{dt} = -y + x(1 - x^2 - y^2), \qquad \frac{dy}{dt} = x + y(1 - x^2 - y^2);$$

or, in polar coordinates,

$$\frac{d\theta}{dt} = 1, \qquad \frac{dr}{dt} = r(1 - r^2).$$

We have a rest point $r = 0$ and a limit cycle $r = 1$; all the motions for which the initial value of r_0 satisfies the condition $0 < r_0 < 1$ approach the limit cycle as $t \to +\infty$. Here all points are quasi-regular. For the rest point this is obvious; next, taking at first $\tau = 2n\pi$, where n is a natural number, we have for any initial point (r_0, θ_0), $r_0 \neq 0$, and any function $\varphi(r, \theta)$ continuous in the circle $r \leqq 1$, *i, e.* periodic with respect to θ with the period 2π:

$$\frac{1}{\tau}\int_0^\tau \varphi(r(t),\ \theta(t))dt = \frac{1}{2n\pi}\sum_{k=0}^{n-1}\int_{2k\pi}^{2(k+1)\pi} \varphi(r(t),\ \theta(t))dt$$

$$= \frac{1}{2n\pi}\sum_{k=0}^{n-1}\int_0^{2\pi} \varphi(r(t + 2k\pi),\ \theta(t + 2k\pi))\, dt.$$

But from the law of the motion $\lim_{k\to\infty} r(t + 2k\pi) = 1$, $\theta(t + 2k\pi) = \theta(t) = \theta_0 + t$; therefore

$$\lim_{\tau\to\infty} \frac{1}{\tau}\int_0^\tau \varphi(r(t), \theta(t))dt = \frac{1}{2\pi}\int_0^{2\pi} \varphi(1, \theta_0 + t)dt = \frac{1}{2\pi}\int_0^{2\pi} \varphi(1, \theta)d\theta.$$

The last equation holds in view of the periodicity of the function φ with respect to the argument θ.

Since the function $\varphi(r, \theta)$, being continuous, is bounded in the domain R, then obviously the limit we are considering exists and is equal to one and the same value for any particular approach of τ toward ∞. Consequently, every point $p \in R$ is quasi-regular.

Thus we have two invariant measures, first a μ_0-measure corresponding to the point $O(r = 0)$, which is defined thus: $\mu_0 A = 1$, $O \in A$; $\mu_0(A) = 0$, $O \in R - A$; secondly, each of the points $p(r_0, \theta_0)$, for $0 < r_0 \leqq 1$, determines an invariant measure μ_p, one and the same for all these points, since, from the last equation and formula (9.13):

$$\int_R \varphi(r,\ \theta) d\mu_p = \frac{1}{2\pi}\int_0^{2\pi} \varphi(1,\ \theta) d\theta,$$

where the right-hand side does not depend on $(r_0,\ \theta_0)$.

From the last equation there can be found an evident expression for $\mu_p A$. In fact, we take the equation of the functionals defined for continuous functions. Let A be any set measurable B. From the remark to the Theorem of Riesz-Radon these functionals are extended in a single-valued way to measurable, bounded functions — in particular, to the characteristic function $\varphi_A(r,\ \theta)$ of the set A, and we obtain

$$\mu_p A = \int_R \varphi_A(r,\ \theta) d\mu_p = \frac{1}{2\pi}\int_0^{2\pi} \varphi_A(1, \theta) d\theta = \frac{1}{2\pi} \text{ mes } \{A \cdot (r = 1)\}.$$

In particular, $\mu_p\{r = 1\} = 1$, and $\mu_p\{A\} = 0$ if $A \cdot \{r = 1\} = 0$.

The points $(r_0,\ \theta_0)$ for $0 < r_0 < 1$ are not density points with respect to the measure μ_p since, choosing $\varepsilon < (1 - r_0)/2$ and setting

$$\varphi(r, \theta) = \begin{cases} 1, & \varrho[(r_0, \theta_0),\ (r, \theta)] \leqq \varepsilon; \\ 2 - \dfrac{1}{\varepsilon}\, \varrho[(r_0, \theta_0),\ (r, \theta)], & \varepsilon \leqq \varrho[(r_0, \theta_0),\ (r, \theta)] \leqq 2\varepsilon; \\ 0, & \varrho[(r_0, \theta_0),\ (r, \theta)] \geqq 2\varepsilon; \end{cases}$$

we find

$$0 = \int_R \varphi d\mu_p = \lim_{\tau\to\infty} \frac{1}{\tau}\int_0^{\tau} \varphi(r(t), \theta(t)) dt \geqq \lim_{\tau\to\infty} \frac{1}{\tau}\int_0^{\tau} \varphi_S(r(t), \theta(t)) dt$$
$$= \mathbf{P}[(r(t), \theta(t)) \in S((r_0, \theta_0), \varepsilon)],$$

where φ_S is the characteristic function of the set $S((r_0,\ \theta_0), \varepsilon)$. Thus $\mathbf{P} = 0$.

The full significance of the individual measures μ_p is made clear by the following theorem.

9.27. THEOREM. *Every normalized, invariant, transitive measure μ coincides with an individual measure μ_p, where p is any point of some invariant set $\mathscr{E}_p$.*

We single out from R the set F of points such that for $p \in F$ we have $\mu(S(p,\ \varepsilon)) > 0$ for any $\varepsilon > 0$. It is easy to prove that F is a closed, invariant set and that $\mu(R - F) = 0$; *i. e.* $\mu F = 1$.

In view of the transitivity of the measure μ there exists a set $\mathscr{E}_p \subset F$, $\mu\mathscr{E}_p = 1$, such that for any point $p \in \mathscr{E}_p$ the time mean for any continuous function $\varphi(p)$ has a constant value, *i.e.* (from the second part of Birkhoff's Theorem)

$$\lim_{\tau\to\infty} \frac{1}{\tau}\int_0^\tau \varphi(f(p, t))\,dt = \int_R \varphi(q)\mu(dq) = \int_{\mathscr{E}_p} \varphi(q)\mu(dq).$$

Comparing this result with the definition (9.14) of the invariant measure μ_p, we see that $\mu = \mu_p$, where $p \in \mathscr{E}_p$. The theorem is proved.

9.28. Corollary. *The set $\mathscr{E}_p$ consists of regular points, i.e., $\mathscr{E}_p \subset U_R$.*

In fact, in the construction of the set $\mathscr{E}_p$ we chose all the points where there existed a time mean, *i.e.* $\mathscr{E}_p \subset U$; next we selected the density points with respect to μ, *i. e.* with respect to μ_p; thus $\mathscr{E}_p \subset U_D$. Finally, $\mathscr{E}_p \subset U_T$ since the condition (9.17) is fulfilled. Indeed, if $q \in \mathscr{E}_p$, then $\int_R \varphi(r)\mu_q(dr) = \int_R \varphi(r)\mu_p(dr)$; the set of points $q \in U - \mathscr{E}_p$ has μ_p-measure zero. Therefore the integral (9.17) is equal to zero and, consequently,

$$\mathscr{E}_p \subset U_D \cdot U_T = U_R.$$

9.29. Corollary. The set of regular points U_R decomposes into a system of invariant sets $\{\mathscr{E}\}$ without common points each of which unites points with identical individual measures. We denote by $\mu_{\mathscr{E}}$ this common individual measure for $p \in \mathscr{E}$ and we shall call each of the sets $\mathscr{E}$ *ergodic*. The set of all measures corresponding to the ergodic sets is called a *fundamental system of invariant measures* and is denoted by Σ_μ. In Example 9.26 there were two such measures. There may be an infinity or even a continuum of them.

9.30. Example. A system is given in E_2 by the differential equations

$$\frac{dx}{dt} = -y, \qquad \frac{dy}{dt} = x,$$

or, in polar coordinates

$$\frac{dr}{dt} = 0, \qquad \frac{d\theta}{dt} = 1.$$

By carrying out computations analogous to those in Example 9.26 it is easy to verify that all the points lying on the circle $r = a$, where a is a constant, have the common individual measure

$$\mu_p(A) = \frac{1}{2\pi} \text{ mes } (A \cdot \{r = a\}),$$

and are density points; therefore they form an ergodic set. We obtain a continuum of ergodic sets and the fundamental system contains a continuum of distinct measures.

We have already obtained in formula (9.15) the connection between any invariant measure μ and a fundamental system of invariant measures:

$$\mu A = \int_U \mu_q(A)\mu(dq),$$

where A is any set measurable B.

We note that every linear combination of fundamental measures of the form

$$\mu = \sum_{i=1}^{n} \alpha_i \mu_{p_i} \qquad (\alpha_i > 0, \sum_{i=1}^{n} \alpha_i = 1) \tag{9.31}$$

is an invariant, normalized measure. Furthermore, from the characteristic property of an invariant measure μ (φ is any continuous function),

$$\int_R \varphi(f(p, t))\mu(dp) = \int_R \varphi(p)\mu(dp),$$

it follows that every limit measure for the sequence (9.31) is also an invariant measure.

We shall show the general form of all invariant measures. Let $m(E)$ be any (in general not invariant) measure normalized over the set U of quasi-regular points: $m(U) = m(R) = 1$. Then any invariant normalized measure can be expressed in the form

$$\mu(E) = \int_U \mu_p(E)m(dp). \tag{9.32}$$

In fact, obviously, $\mu(R) = 1$. Furthermore, μ is invariant; indeed,

$$\mu(f(E, t)) = \int_U \mu_p(f(E, t))m(dp) = \int_U \mu_p(E)m(dp) = \mu(E)$$

in view of the invariance of the measure $\mu_p(E)$. Finally, according to formula (9.15), any invariant measure $\mu(E)$ is expressed by the integral

$$\mu(E) = \int_U \mu_p(E)\mu(dp),$$

i. e. by an expression of the form (9.32).

By virtue of a remark at the beginning of this section (see the footnote, preceding Th. 9.05) the measure m is limiting (in the sense of weak convergence) for the sequence of measures

$$m_n = \sum_{i=1}^{n} \alpha_i m_{p_i}, \quad p_i \subset U, \quad \text{where } \alpha_i > 0, \ \sum_{i=1}^{n} \alpha_i = 1.$$

From this it follows that the measure μ is (weakly) limiting for the measures

$$\mu_n(E) = \int_U \mu_p(E) m_n(dp) = \sum_{i=1}^{n} \alpha_i \int_U \mu_p(E) m_{p_i}(dp) = \sum_{i=1}^{n} \alpha_i \mu_{p_i}(E),$$

i. e., any invariant, normalized measure is a limit of measures of the type (9.31).

9.33. DEFINITION. A system is called *strictly ergodic* if it consists of a unique ergodic set or, what is the same, if there exists in the system a unique invariant transitive measure and all points of the system are density points with respect to this measure.

9.34. THEOREM. *Every minimal set consisting of almost periodic motions is strictly ergodic.*

For an almost periodic motion, the function $\varphi(f(p, t))$, $\varphi(p)$ being any continuous function, is almost periodic in the sense of Bohr with respect to the variable t; *i. e.* we assert that there exists for a given $\varepsilon > 0$ a relatively dense set of displacements $\{\tau\}$ such that

$$|\varphi(f(p, t + \tau)) - \varphi(f(p, t))| < \varepsilon.$$

This follows easily from the fact that $\varphi(p)$ is uniformly continuous over a compact set $\overline{f(p; I)}$; therefore, for a given $\varepsilon > 0$ there exists a δ such that from $\varrho(r, q) < \delta$ and $r \in f(p; I)$, $q \in f(p; I)$ there follows $|\varphi(r) - \varphi(q)| < \varepsilon$. Then each displacement $\tau(\delta)$ such that $\varrho(f(p, t + \tau), f(p, t)) < \delta$ will be the desired ε-displacement for $\varphi(f(p, t))$.

According to a theorem of Bohr, the mean value

$$\lim_{T\to\infty} \frac{1}{T} \int_0^T \varphi(f(p, t))\,dt$$

exists; *i. e.* all points $q \in f(p; I)$ are quasi-regular.

Next, for any point $q \in \overline{f(p; I)}$ there can be found a sequence $\{p_n\}$ converging to it, $p_n = f(p, t_n)$, and since furthermore $f(p, t_n + t)$ converges uniformly in t to $f(q, t)$ (see Chapter V, Theorem 8.06), then the same mean value exists for q as well, *i. e.*

$$\lim_{T\to\infty}\frac{1}{T}\int_0^T \varphi(f(q,t))\,dt = \lim_{T\to\infty}\frac{1}{T}\int_0^T \varphi(f(p,t))\,dt;$$

in other words, there exists over $\overline{f(p; I)}$ a unique invariant measure $\mu(A)$ equal to the probability of the stay of the point p in the set A as $t \to \infty$.

Finally, every point of the almost periodic motion is a density point with respect to this invariant measure. In fact, we describe around the point p the two spheres, $S(p, \varepsilon)$ and $S(p, 2\varepsilon)$, where $\varepsilon > 0$ is arbitrary. If $f(p, t_0) \in S(p, \varepsilon)$ there can be found an arc $f(p, [t_1, t_2]) \subset S(p, 2\varepsilon)$, wherein $t_1 < t_0 < t_2$ and $t_2 - t_1 \geqq \alpha(\varepsilon) > 0$.[12] Since the points of return of the motion $f(p, t)$ into $S(p, \varepsilon)$ are relatively dense with the interval $L(\varepsilon)$, then

$$\mathbf{P}[f(p, t) \in S(p, 2\varepsilon)] \geqq \frac{\alpha(\varepsilon)}{L} > 0,$$

and this shows that p is a density point. The theorem is proved.

We note that in this case $U_R = \overline{f(p; I)}$.

The example considered in section 7, Chap. V, of a minimal set of not almost periodic motions gives us the strictly ergodic case; in fact, if after mapping the meridian $\varphi = 0$ onto the circle Γ we map all the motions onto the torus $\mathfrak{T}(\varphi, \theta)$ with the meridian Γ, then from the motions on the complement of the perfect set are obtained almost periodic motions, everywhere dense, filling the torus. (The fact that two initial motions corresponding to initial points at the two ends of an adjacent interval of the set are identified as one has no effect on the measure.) The unique invariant measure of this almost periodic motion is transformed together with its property of being strictly ergodic into initial motions over the set A.

In the general case it cannot be asserted that every minimal set is strictly ergodic. A. A. Markoff has constructed an example of a

[12] The greatest lower bound of the intervals of time beginning with an entry of $f(p, t)$ into $S(p, \varepsilon)$ and ending with a departure from $S(p, 2\varepsilon)$ is greater than zero. Assuming the contrary, we could find a sequence of pairs of points $\{p'_n, p''_n\}$ such that

$$\varrho(p'_n, p) = \varepsilon, \quad \varrho(p''_n, p) = 2\varepsilon; \quad p''_n = f(p'_n, t_n), \quad \lim_{n\to\infty} t_n = 0.$$

Because of the compactness of the space $\overline{f(p, t)}$, the set $\{p'_n\}$ has a limit point p'. Without loss of generality one can assume that $\lim_{n\to\infty} p'_n = p'$.

We have $p'' = \lim_{n\to\infty} p''_n = \lim_{n\to\infty} f(p'_n, t_n) = p'$. But $\varrho(p', p) = \varepsilon$ and $\varrho(p'', p) = 2\varepsilon$. The contradiction proves our assertion.

minimal set in which a certain motion is not quasi-regular, *i. e.* there exist at least two invariant measures.

9.35. Example of a minimal set which is not strictly ergodic (Markoff). For the construction of this example we introduce a metric space R_U which presents great interest for a whole series of questions related to dynamical systems. The points of the space are continuous functions $\varphi(x)$ defined over the entire infinite number axis $-\infty < x < +\infty$. As the distance of the two points $\varphi(x)$ and $\psi(x)$ we take

$$(9.36) \qquad \varrho(\varphi, \psi) = \sup_{-\infty < x < +\infty} \min \left[|\varphi(x) - \psi(x)|, \frac{1}{|x|} \right].$$

It is a simple matter to interpret this definition geometrically. We construct in one figure the graphs of the functions $y = |\varphi(x) - \psi(x)|$

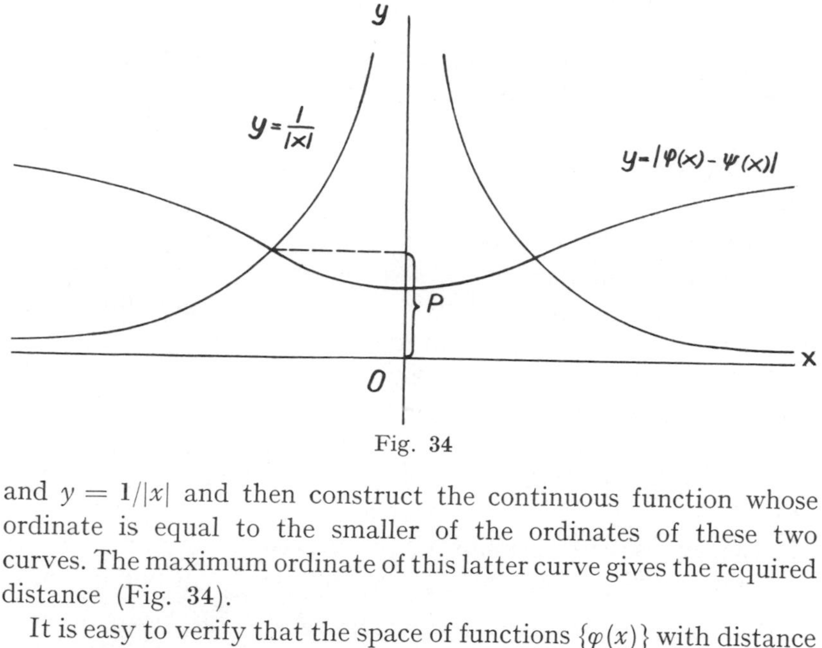

Fig. 34

and $y = 1/|x|$ and then construct the continuous function whose ordinate is equal to the smaller of the ordinates of these two curves. The maximum ordinate of this latter curve gives the required distance (Fig. 34).

It is easy to verify that the space of functions $\{\varphi(x)\}$ with distance defined in such a way is metric.

We note, finally, that in this metric the inequality $\varrho(\varphi, \psi) < \varepsilon$ is equivalent to the inequality $|\varphi(x) - \psi(x)| < \varepsilon$ for $|x| \leqq 1/\varepsilon$.

Consequently, the limiting equality $\lim_{n\to\infty} \varrho(\varphi_n(x), \varphi(x)) = 0$ means that the sequence $\{\varphi_n(x)\}$ converges to $\varphi(x)$ for $-\infty < x < +\infty$ and, moreover, converges uniformly in every bounded interval.

It follows from this remark that the metric space R_U has a countable base, *i. e.* a countable, everywhere dense set of function-points. Namely, it suffices to take the set of all polynomials in x with rational coefficients. In fact, for any $\varepsilon > 0$ and any continuous function $\varphi(x)$ there can be found a polynomial of this set approximating this function over the interval $(-1/\varepsilon, 1/\varepsilon)$ with a precision of at most $\mathscr{E}$.

We define a dynamical system in the space R_U in the following manner. If the point $p \equiv \varphi(x)$, then $f(p, t) \equiv \varphi(x + t)$, or $f(\varphi(x), t) \equiv \varphi(x + t)$; *i. e.* to a displacement of the point along the trajectory over an interval of time t there corresponds a translation of the function along the x-axis (a change of the argument x) of length t. This transformation $f(p, t)$ satisfies all the conditions for a dynamical system. This fact is obvious for the group properties, leaving only the property of continuity to be verified.

Suppose that $\varrho[\varphi_n(x), \varphi(x)] \to 0$ and $t_n \to t$; thus we have a sequence of functions $\varphi_n(x + t_n)$ and a function $\varphi(x + t)$. It is required to prove that

$$\lim_{n\to\infty} \varrho[\varphi_n(x + t_n), \varphi(x + t)] = 0.$$

Let there be given an arbitrary $\varepsilon > 0$. By virtue of the first limiting equality there can be found an N_1 such that for $n > N_1$, with the given t and x satisfying the inequality $|x| \leqq 2/\varepsilon$, we shall have $|\varphi_n(x + t) - \varphi(x + t)| < \varepsilon/2$. Next, because of the uniform continuity of the function $\varphi(x)$ over the interval $-t - 2/\varepsilon \leqq x \leqq -t + 2/\varepsilon$, there can be found a δ greater than 0 and less than $1/\varepsilon$ such that

$$|\varphi(x') - \varphi(x'')| < \frac{\varepsilon}{2}$$

if

$$-t - \frac{2}{\varepsilon} \leqq x' \leqq -t + \frac{2}{\varepsilon}, \quad -t - \frac{2}{\varepsilon} \leqq x'' \leqq -t + \frac{2}{\varepsilon} \quad \text{and} \quad |x' - x''| < \delta.$$

We choose an N_2 such that $|t_n - t| < \delta$ for $n \geqq N_2$. Then, if $N = \max\,[N_1, N_2]$, for $-1/\varepsilon \leqq x \leqq 1/\varepsilon$ and $n \geqq N$ we have

$$|\varphi_n(x + t_n) - \varphi(x + t)| \leqq |\varphi_n(x + t_n) - \varphi(x + t_n)| + |\varphi(x + t_n) - \varphi(x + t)| < \varepsilon,$$

(because $x + t_n = \xi + t$, $|\xi| \leqq 2/\varepsilon$), i.e.

$$\varrho[\varphi_n(x + t_n), \varphi(x + t)] = \varrho[f(p_n, t_n), f(p, t)] < \varepsilon.$$

The continuity is proved.

We remark that in order for the motion defined in the space R_U by the function $\varphi(x + t)$ to be stable according to Lagrange it is necessary and sufficient that the function $\varphi(x)$ be bounded and uniformly continuous over the interval $(-\infty, \infty)$. This assertion is an immediate consequence of the condition of Arzelà for the compactness of a family of functions over a finite interval.

We proceed to Markoff's example. Consider a motion in the space R_U with an initial point defined by the following function $\varphi(x)$:

We take a nondecreasing sequence of natural numbers $\{\alpha_n\}$ subject to the condition

$$\sum_{n=1}^{\infty} \frac{1}{\alpha_n} < +\infty.$$

Then we construct the sequence of numbers $\{\beta_n\}$:

$$\beta_1 = 2\alpha_1 + 1, \quad \beta_n = \prod_{k=1}^{n} (2\alpha_k + 1).$$

Every integer N can be represented in a unique way in the form

$$N = c_0 + c_1\beta_1 + c_2\beta_2 + \ldots + c_m\beta_m, \tag{9.37}$$

where the integral coefficients c_i are subjected to the condition $|c_i| \leqq \alpha_{i+1}$. In fact, we shall find $c_0, c_1, c_2, \ldots$ consecutively. First dividing N by $2\alpha_1 + 1$, we choose the quotient so that the remainder c_0 is that of lowest absolute value, *i.e.* $|c_0| \leqq \alpha_1$; next, dividing the quotient so obtained by $2\alpha_2 + 1$, we again choose the new quotient so that the remainder c_1 is that of lowest absolute value, so that $|c_1| \leqq \alpha_2$, etc.

We can regard the decomposition (9.37) as infinite, wherein $c_{m+1} = c_{m+2} = \ldots = 0$. Next we define an integer function: $\chi(N) = k$ if, in the decomposition (9.37) for N, the coefficient $c_k = 0$ while none of the preceding coefficients are equal to zero. Now we can define $\varphi(N)$ for integral values of the argument

$$\varphi(N) = (-1)^{\chi(N)}.$$

Finally, for nonintegral values of x we define $\varphi(x)$ by a linear interpolation between adjacent integral values of the argument. The function $\varphi(x)$ so defined satisfies the conditions

$$|\varphi(x)| \leqq 1, \; |\varphi(x') - \varphi(x'')| \leqq 2|x' - x''|;$$

i.e. it is bounded and uniformly continuous over the interval $(-\infty, \infty)$. From the remark made above the motion

$$f(p, t) \equiv \varphi(x + t) \tag{9.38}$$

with the initial point $p \equiv \varphi(x)$ is stable according to Lagrange.

We shall show that the motion (13) is recurrent. For this, from Theorem 7.09, Chap. V, it suffices to prove that for any $\varepsilon > 0$ the set of values τ for which

$$\varrho[\varphi(x), \varphi(x + \tau)] < \varepsilon$$

is relatively dense.

An easy arithmetical computation shows that the totality of numbers N for which $c_n = c_{n+1} = c_{n+2} = \ldots = 0$ is the set of all integral points of the segment $[-(\beta_n - 1)/2, (\beta_n - 1)/2]$. For a given $\varepsilon > 0$ we can find an n such that $(\beta_n - 1)/2 > 1/\varepsilon$. We shall show that one can take for τ any number which is a multiple of β_{n+1}; $\tau = m\beta_{n+1}$, m an integer; i. e. that

$$\varrho[\varphi(x), \varphi(x + m\beta_{n+1})] < \varepsilon.$$

In fact, let $\xi = c_0 + c_1\beta_1 + \ldots + c_{n-1}\beta_{n-1}$ be a whole number; then

$$\xi \in \left[-\frac{\beta_n - 1}{2}, \frac{\beta_n - 1}{2}\right];$$

because of the choice of τ we have the decomposition

$$\tau = c'_{n+1}\beta_{n+1} + c'_{n+2}\beta_{n+2} + \ldots,$$

where $|c'_{n+k}| \leqq \alpha_{n+k+1}$, for $k \geqq 1$, whence

$$\xi + \tau = c_0 + c_1\beta_1 + \ldots + c_{n-1}\beta_{n-1} + c'_{n+1}\beta_{n+1} + c'_{n+2}\beta_{n+2} + \ldots$$

On comparing the decomposition of ξ and $\xi + \tau$, we see that both these numbers have c_k as the first coefficient equal to zero if $k \leqq n$, since in both numbers the coefficients $c_0, c_1, \ldots, c_n$ coincide. Therefore, for any integer $\xi \in [-(\beta_n - 1)/2, (\beta_n - 1)/2]$ we have $\varphi(\xi) = \varphi(\xi + \tau)$. From the definition of the function $\varphi(x)$ for nonintegral x we also have

$$\varphi(x) = \varphi(x + \tau), \quad \text{if } x \in \left[-\frac{\beta_n - 1}{2}, \frac{\beta_n - 1}{2}\right], \quad \text{and } \tau = m\beta_{n+1}.$$

Thus, by virtue of the choice of the number β_n and of the definition of distance, there follows

$$\varrho[\varphi(x), \varphi(x+\tau)] < \varepsilon.$$

The set of numbers $\{\tau\}$, forming an arithmetical progression, is relatively dense. Thus the recurrence of the motion (9.38) is proved.

We shall show that the minimal set containing the recurrent motion (9.38) is *not strictly ergodic*. For this purpose we define in the space R_U a continuous function $\Phi(q)$ in the following manner: if $q = \psi(x)$, then $\Phi(q) = \Phi(\psi(x)) = \psi(0)$. We note that

$$\Phi(f(q, t)) = \Phi(\psi(x+t)) = \psi(t)$$

We shall show that for this function the expression

$$\frac{1}{\tau}\int_0^\tau \Phi(f(p, t))dt \equiv \frac{1}{\tau}\int_0^\tau \varphi(t)dt. \tag{9.39}$$

has no limit as $\tau \to \infty$.

In fact, first of all, because of the evenness of the function $\varphi(t)$, the expression (9.39) can be replaced by the expression

$$\frac{1}{2\tau}\int_{-\tau}^{+\tau} \varphi(t)dt. \tag{9.40}$$

Next, if m_1 and m_2 are any integers and $m_1 < m_2$, the expressions

$$\int_{m_2-\frac{1}{2}}^{m_2+\frac{1}{2}} \varphi(t)dt \quad \text{and} \quad \sum_{n=m_1}^{m_2} \varphi(n)$$

are either equal or differ by not more than 1/2. Indeed, the last sum is equal to

$$\int_{m_1-\frac{1}{2}}^{m_2+\frac{1}{2}} \psi(t)dt,$$

where $\psi(t) = \operatorname{sgn} \varphi(t)$, since in each interval $(k - \frac{1}{2}, k + \frac{1}{2})$ (k an integer) $\psi(t)$ is equal to the value of $\varphi(k)$. If $\varphi(k) = \varphi(k+1)$, then $\varphi(t) = \psi(t)$ for $k \leqq t \leqq k+1$ and therefore $\int_k^{k+1} \varphi(t)dt = \int_k^{k+1} \psi(t)dt$; if $\varphi(k) = -\varphi(k+1)$, then

$$\int_k^{k+1} \varphi(t)dt = 0 = \int_k^{k+1} \psi(t)dt.$$

Consequently, the integrals of φ and ψ can differ only over the intervals $(m_1 - \frac{1}{2}, m_1)$ and $(m_2, m_2 + \frac{1}{2})$ in case $\varphi(k)$ changes sign in passing from $m_1 - 1$ to m_1, or from m_2 to $m_2 + 1$; but each of these intervals gives in these cases a value for $\int |\varphi(t) - \psi(t)| dt$ not exceeding $\frac{1}{4}$, and this proves our assertion.

Therefore, for an estimate of the expression

$$I(2N+1) = \frac{1}{2N+1}\int_{-N-\frac{1}{2}}^{N+\frac{1}{2}} \varphi(t)dt,$$

(N is a natural number) we shall compute the sum

$$S(2N+1) = \frac{1}{2N+1}\sum_{k=-N}^{N} \varphi(k).$$

We set $2N + 1 = \beta_n$; *i.e.* $-(\beta_n - 1)/2 \leqq k \leqq (\beta_n - 1)/2$. It is easy to see that all the numbers k of this interval have decompositions

$$k = c_0 + c_1\beta_1 + \ldots + c_{n-1}\beta_{n-1},$$

where the coefficients, independently of one another, pass through the values

$$|c_0| \leqq \alpha_1, \ |c_1| \leqq \alpha_2, \ldots, |c_{n-1}| \leqq \alpha_n,$$

$(2\alpha_1 + 1)(2\alpha_2 + 1) \ldots (2\alpha_n + 1) = \beta_n$ numbers in all. Those for which $c_0 = 0$ and, consequently, $\varphi = 1$, correspond to a fixed value $c_0 = 0$ while $c_1, c_2, \ldots, c_{n-1}$ are arbitrary, therefore there are $\beta_n/(2\alpha_1 + 1)$ of them. Next, those numbers k for which $c_0 \neq 0$, $c_1 = 0$, *i. e.* $\varphi = -1$, correspond to the values $c_0 \neq 0$, $c_1 = 0$. The remaining coefficients are arbitrary; the number of such k's is $\beta_n/(2\alpha_2 + 1) \cdot 2\alpha_1/(2\alpha_1 + 1)$. In general, in the interval considered, the number of quantities k for which, for $l \leqq n - 1$, there hold $c_l = 0$, $c_0 \neq 0$, $c_1 \neq 0, \ldots, c_{l-1} \neq 0$, is equal to

$$\frac{\beta_n}{2\alpha_l + 1} \cdot \frac{2\alpha_1}{2\alpha_1 + 1} \cdot \frac{2\alpha_2}{2\alpha_2 + 1} \cdot \frac{2\alpha_{l-1}}{2\alpha_{l-1} + 1}$$

where for such a k we have $\varphi(k) = (-1)^l$. Finally, the number of those quantities k ($|k| \leqq (\beta_n - 1))/2$ for which

$$c_0 \cdot c_1 \cdot \ldots \cdot c_{n-1} \neq 0$$

is

$$\beta_n \cdot \frac{2\alpha_1}{2\alpha_1 + 1} \cdot \frac{2\alpha_2}{2\alpha_2 + 1} \cdot \frac{2\alpha_n}{2\alpha_n + 1}.$$

We thus have

$$S(\beta_n) = \frac{1}{\beta_n}\left\{\frac{\beta_n}{2\alpha_1 + 1} - \frac{\beta_n}{2\alpha_2 + 1} \cdot \frac{2\alpha_1}{2\alpha_1 + 1} + \ldots \right.$$

$$+ (-1)^{n-1}\frac{\beta_n}{2\alpha_n + 1} \cdot \frac{2\alpha_1}{2\alpha_1 + 1} \cdot \frac{2\alpha_2}{2\alpha_2 + 1} \cdots \frac{2\alpha_{n-1}}{2\alpha_{n-1} + 1}$$

$$\left. + (-1)^n\beta_n \cdot \frac{2\alpha_1}{2\alpha_1 + 1} \cdot \frac{2\alpha_2}{2\alpha_2 + 1} \cdots \frac{2\alpha_n}{2\alpha_n + 1}\right\};$$

or

$$S(\beta_n) = U_0 - U_1 + U_2 - \ldots + (-1)^{n-1} U_{n-1} + (-1)^n \prod\nolimits_n ,$$

where

$$U_0 = \frac{1}{2\alpha_1 + 1}, \quad U_m = \frac{2\alpha_1 \,.\, 2\alpha_2 \ldots 2\alpha_m}{(2\alpha_1 + 1)(2\alpha_2 + 1) \ldots (2\alpha_m + 1)(2\alpha_{m+1} + 1)}$$

$(m = 1, 2, \ldots, n - 1)$ and

$$\prod\nolimits_n = \prod_{l=1}^{n} \frac{2\alpha_l}{2\alpha_l + 1}.$$

We note that the infinite product

$$\prod = \lim_{n\to\infty} \prod\nolimits_n = \prod_{l+1}^{\infty} \frac{2\alpha_l}{2\alpha_l + 1}$$

converges (*i. e.* $\prod > 0$) since it can be represented in the form

$$\prod_{l=1}^{\infty} \left(1 - \frac{1}{2\alpha_l + 1}\right),$$

and the series

$$\sum_{l=1}^{\infty} \frac{1}{2\alpha_l + 1}$$

converges simultaneously with the series

$$\sum_{l=1}^{\infty} \frac{1}{\alpha_l}.$$

The infinite series $u_0 - u_1 + u_2 - \ldots$ converges, since

$$u_{m+1} = u_m \cdot \frac{2\alpha_{m+1}}{2\alpha_{m+2} + 1} < u_m \quad \text{and} \quad \lim_{m\to\infty} U_m = \lim_{m\to\infty} \prod\nolimits_m \cdot \frac{1}{2\alpha_{m+1} + 1} = 0.$$

We set $u_0 - u_1 + u_2 - \ldots = \sigma$.

We consider two cases: n even and n odd.

I. $n = 2m$. We have

$$S(\beta_{2m}) = u_0 - u_1 + u_2 - \ldots - u_{2m-1} + \prod\nolimits_{2m},$$

$$\lim_{m\to\infty} S(\beta_{2m}) = \sigma + \prod = S'.$$

II. $n = 2m + 1$. In this case

$$S(\beta_{2m+1}) = u_0 - u_1 + \dots + u_{2m} - \textstyle\prod_{2m+1},$$
$$\lim_{m\to\infty} S(\beta_{2m+1}) = \sigma - \textstyle\prod = S''.$$

Since $\prod \neq 0$, then $S' \neq S''$.

Passing from sums to integrals, we obtain the result

$$\lim_{m\to\infty} \frac{1}{\tau_m}\int_0^{\tau_m} \Phi(f(p, t))\,dt = S', \text{ if } \tau_m = \frac{\beta_{2m} - 1}{2};$$

$$\lim_{m\to\infty} \frac{1}{\tau'_m}\int_0^{\tau'_m} \Phi(f(p, t))\,dt = S'' \text{ if } \tau'_m = \frac{\beta_{2m+1} - 1}{2}.$$

Thus the point $p \equiv \varphi(x)$ is not quasi-regular; it defines *more than one* individual measure and, consequently, the minimal set $\overline{f(p; I)}$ is not strictly ergodic.

BIBLIOGRAPHY TO PART TWO

CHAPTER V

1. BIRKHOFF, G. D. Quelques théorèmes sur le mouvement des systèmes dynamiques. *Bull. Soc. Math. France 40*, 1–19, 1912. Collected Mathematical Papers I, 654–672.
2. BIRKHOFF, G. D. Über gewisse Zentralbewegungen dynamischer Systeme. *Ges. Wiss. Göttingen. Nachr., Math.-Phys. Klasse 1926*, 81–92. *1927. Collected Mathematical Papers II*, 283–294.
3. BIRKHOFF, G. D. *Dynamical Systems*, Chap. 7. New York, American Math. Soc., 1927.
4. BIRKHOFF, G. D. Some unsolved problems of theoretical dynamics. *Science 94*, 1–3, 1941. *Collected Mathematical Papers II*, 710–712.
5. MARKOFF, A. A. Sur une propriété générale des ensembles minimaux de M. Birkhoff. *C. R. Acad. Sci. Paris 193*, 823–825, 1931.
6. MARKOFF, A. A. On a general property of minimal sets. *Rusk. Astron. Zhurnal*, 1932.
7. MARKOFF, A. A. Stabilität im Liapounoffschen Sinne und Fastperiodizität. *Math. Z. 36*, 708–738, 1933.
8. FRANKLIN, P. Almost periodic recurrent motions. *Math. Z. 30*, 325–331, 1929.
9. STEPANOFF, V. V., and TYCHONOFF, A. Sur les espaces des fonctions presque périodiques. *C. R. Acad. Sci. Paris 196*, 1199–1201, 1933.
10. STEPANOFF, V. V., and TYCHONOFF, A. Über die Räume der fastperiodischen Funktionen. *Mat. Sbornik 41*, 166–178, 1934.
11. MOORE, R. L. On the generation of a simple surface by means of a set of equicontinuous curves. *Fund. Math. 4*, 106–117, 1923.
12. CHETAEV, N. On stability in the sense of Poisson. *Zap. Kazansk. Matem. Obsch.*, 1929.
13. WAZEWSKI, T., and ZAREMBA, S. Sur les ensembles de condensation des caractéristiques d'un système d'équations différentielles ordinaires. *Ann. Soc. Polon. Math. 15*, 24–33, 1936.
14. WAZEWSKI, T. Sur les intégrales stables non périodiques des systèmes d'équations différentielles. *Ann. Soc. Polon. Math. 13*, 50–52, 1934.
15. URBANSKI, W. Sur la structure de l'ensemble des solutions cycliques d'un système d'équations différentielles. *Ann. Soc. Polon. Math. 13*, 44–49, 1934.
16. URBANSKI, W. Note sur les systèmes quasi-ergodiques. *Ann. Soc. Polon. Math. 13*, 20–23, 1934.
17. CHERRY, T. M. Topological properties of the solutions of ordinary differential equations. *Amer. J. Math. 59*, 957–982, 1937.
18. HILMY, H. Sur les ensembles quasi-minimaux dans les systèmes dynamiques. *Ann. of Math. (2) 37*, 899–907, 1936.
19. HILMY, H. Sur la structure d'ensembles des mouvements stables au sens de Poisson. *Ann. of Math. (2) 37*, 43–45, 1936.
20. HILMY, H. Sur les centres d'attraction minimaux des systèmes dynamiques. *Compositio Math. 3*, 227–238, 1936.

21. Hilmy, H. Sur les théorèmes de récurrence dans la dynamique générale. *Amer. J. Math. 61*, 149–160, 1939.
22. Hilmy, H. Sur une propriété des ensembles minimaux. *C. R. (Doklady) Acad. Sci. URSS (N. S) 14*, 261–262, 1937.
23. Whitney, H. Regular families of curves. *Proc. Nat. Acad. Sci. USA 18*, 275–278; 340–342, 1932.
24. Nemickii, V. V. Sur les systèmes dynamiques instables. *C. R. Acad. Sci. Paris 199*, 19–20, 1934.
25. Nemickii, V. V. Über vollständig unstabile dynamische Systeme. *Ann. Mat. Pura Appl. (4) 14*, 275–286, 1935–1936.
26. Nemickii, V. V. Sur les systèmes de courbes remplissant un espace métrique. *C. R. (Doklady) Acad. Sci. URSS (N. S.) 21*, 99–102, 1938.
27. Nemickii, V. V. Sur les systèmes de courbes remplissant un espace métrique. *Mat. Sbornik (N. S) 6*, 283–292, 1939.
28. Nemickii, V. V. Systèmes dynamiques sur une multiplicité intégrale limite. *C. R. (Doklady) Acad. Sci. URSS (N. S) 47*, 535–538, 1945.
29. Bebutoff, M. Sur la représentation des trajectoires d'un système dynamique sur un système de droites parallèles. *Bull. Math. Univ. Moscou 2, no. 3*, 1939.
30. Bebutoff, M. Sur les systèmes dynamiques stables au sens de Liapounoff. *C. R. (Doklady) Acad. Sci. URSS (N. S) 18*, 155–158, 1938.
31. Bebutoff, M. Sur les systèmes dynamiques dans l'espace des fonctions continues. *C. R. (Doklady) Acad. Sci. URSS (N. S) 27*, 904–906, 1940.
32. Bebutoff, M. Sur les systèmes dynamiques dans l'espace des fonctions continues. *Bull. Math. Univ. Moscou 2, no. 5*, 1939.
33. Bebutoff, M., and Stepanoff, V. V. Sur la mesure invariante dans les systèmes dynamiques qui ne diffèrent que par le temps. *Mat. Sbornik (N. S) 7*, 143–164, 1940.
34. Troickii, S. On dynamical systems defined by an everywhere dense set of recurrent motions. *Ucenye Zapiski Moskov. Gos. Univ. Matematika 15*, 1939 (Russian).
35. Wiener, N., and Wintner, A. On the ergodic dynamics of almost periodic systems. *Amer. J. Math. 63*, 794–824, 1941.
36. Barbachine, E. Sur certaines singularités qui surviennent dans un système dynamique quand l'unicité est en défaut. *C. R. (Doklady) Acad. Sci. URSS (N. S.) 41*, 139–141, 1943.
37. Gottschalk, W. H., and Hedlund, G. A. *Topological Dynamics*. New York, American Math. Soc., 1955.
38. Nemickii, V. V. Topological problems of the theory of dynamical systems. (Translated from *Usp. Mat. Nauk (N. S) 4, no. 6 (34)*, 91–153, 1949). *Translation no. 103, Amer. Math. Soc.* This reference contains many complements to Chapter V, and references to more recent literature.

CHAPTER VI

1. Poincaré, H. *Méthodes nouvelles de la mécanique céleste*, t. III. Paris, 1899.
2. Carathéodory, C. Über den Wiederkehrsatz von Poincaré. *S.-B. Pruess. Akad. Wiss. no. XXXIV*, 580–584, 1919.
3. Birkhoff, G. D. *Dynamical Systems*, Chap. 7. New York, American Math. Soc., 1927.
4. Birkhoff, G. D., and Smith, P. A. Structure analysis of surface transformations. *J. Math. Pures Appl. (9) 7*, 345–379, 1928. *Collected Mathematical Papers II*, 360–395.

5. Birkhoff, G. D. Proof of a recurrence theorem for strongly transitive systems; Proof of the ergodic theorem. *Proc. Nat. Acad. Sci. USA 17*, 650–660, 1931. *Collected Mathematical Papers II*, 398–408.
6. Birkhoff, G. D. What is the ergodic theorem? *Amer. Math. Monthly 49*, 222–226, 1942. *Collected Mathematical Papers II*, 713–717.
7. Khintchine, A. The method of spectral reduction in classical dynamics. *Proc. Nat. Acad. Sci. USA 19*, 567–573, 1933.
8. Khintchine, A. Eine Verschärfung des Poincáreschen „Wiederkehrsatzes." *Compositio Math. 1*, 177–179, 1934.
9. Khintchine, A. Zu Birkhoff's Lösung des Ergodenproblems. *Math. Ann. 107*, 485–488, 1932.
10. Kolmogoroff, A. Ein vereinfachter Beweis des Birkhoff-Khintchineschen Ergodensatzes. *Mat. Sbornik* (*N. S*) *2* (*44*), 367–368, 1937.
11. Kolmogoroff, A. A simplified proof of the ergodic theorem of Birkhoff-Khintchine. *Uspehi Mat. Nauk 5*, 1938 (Russian).
12. Kryloff, N., and Bogoliuboff, N. Les mesures invariantes et transitives dans la mécanique non linéaire. *Mat. Sbornik* (*N. S.*) *1* (*43*), 707–710, 1936.
13. Kryloff, N., and Bogoliuboff, N. La théorie générale de la mesure dans son application à l'étude des systèmes dynamique de la mécanique non linéaire. *Ann. of Math.* (*2*) *38*, 65–113, 1937.
14. Hopf, E. *Ergodentheorie*, Berlin, 1937.
15. Hopf, E. Zwei Sätze über den wahrscheinlichen Verlauf der Bewegungen dynamischer Systeme. *Math. Ann. 103*, 710–719, 1930.
16. Hopf, E. Theory of measure and invariant integrals. *Trans. Amer. Math. Soc. 34*, 373–393, 1932.
17. Stepanoff, V. V. Sur une extension du théorème ergodique. *Compositio Math. 3*, 239–253, 1936.
18. Bebutoff, M. V., and Stepanoff, V. V. Sur le changement du temps dans les systèmes dynamique possédant une mesure invariant. *C. R.* (*Doklady*) *Acad. Sci. URSS* (*N. S.*) *24*, 217–219, 1939.
19. Markoff, A. Quelques théorèmes sur les ensembles abéliens. *C. R.* (*Doklady*) *Acad. Sci. URSS* (*N. S.*) *1*, 311–313, 1936.
20. Markoff, A. Sur l'existence d'un invariant intégral. *C. R.* (*Doklady*) *Acad. Sci. URSS* (*N. S.*) *17*, 459–462, 1937.
21. Markoff, A. On mean values and exterior densities. *Mat. Sbornik* (*N. S*) *4* (*46*), 165–190, 1938.
22. Demidovitch, B. Sur l'existence d'un invariant intégral dans l'ensemble des points périodiques. *C. R.* (*Doklady*) *Acad. Sci. URSS* (*N. S*) *2*, 11–13, 1936.
23. Demidovitch, B. On certain sufficient conditions for the existence of an integral invariant. *Mat. Sbornik* (*N. S*) *3* (*45*), 291–310, 1938 (Russian with French summary).
24. Oxtoby, J. C., and Ulam, S. M. On the existence of a measure invariant under a transformation. *Ann. of Math.* (*2*) *40*, 560–566, 1939.
25. Hilmy, H. Sur les théorèmes de récurrence dans la dynamique générale. *Amer. J. Math. 61*, 149–160, 1939.
26. Fomin, S. On finite invariant measures in dynamical systems. *Mat. Sbornik* (*N. S.*) *12* (*54*), 98–108, 1943 (Russian with English summary).
27. Hurewicz, W. Ergodic theorem without invariant measure. *Ann. of Math.* (*2*) *45*, 192–206, 1944.
28. Visser, C. On Poincaré's recurrence theorem. *Bull. Amer. Math. Soc. 42*, 397–400, 1936.
29. von Neumann, J. Zur Operatorenmethode in der klassischen Mechanik. *Ann. of Math. 33*, 1932.

Index to Part Two

(Index to Part One is on pp. 301-303)